A&A LIBRARY

ASTRONOMY AND ASTROPHYSICS LIBRARY

Springer
Berlin
Heidelberg
New York
Barcelona
Budapest
Hong Kong
London
Milan
Paris
Santa Clara
Singapore
Tokyo

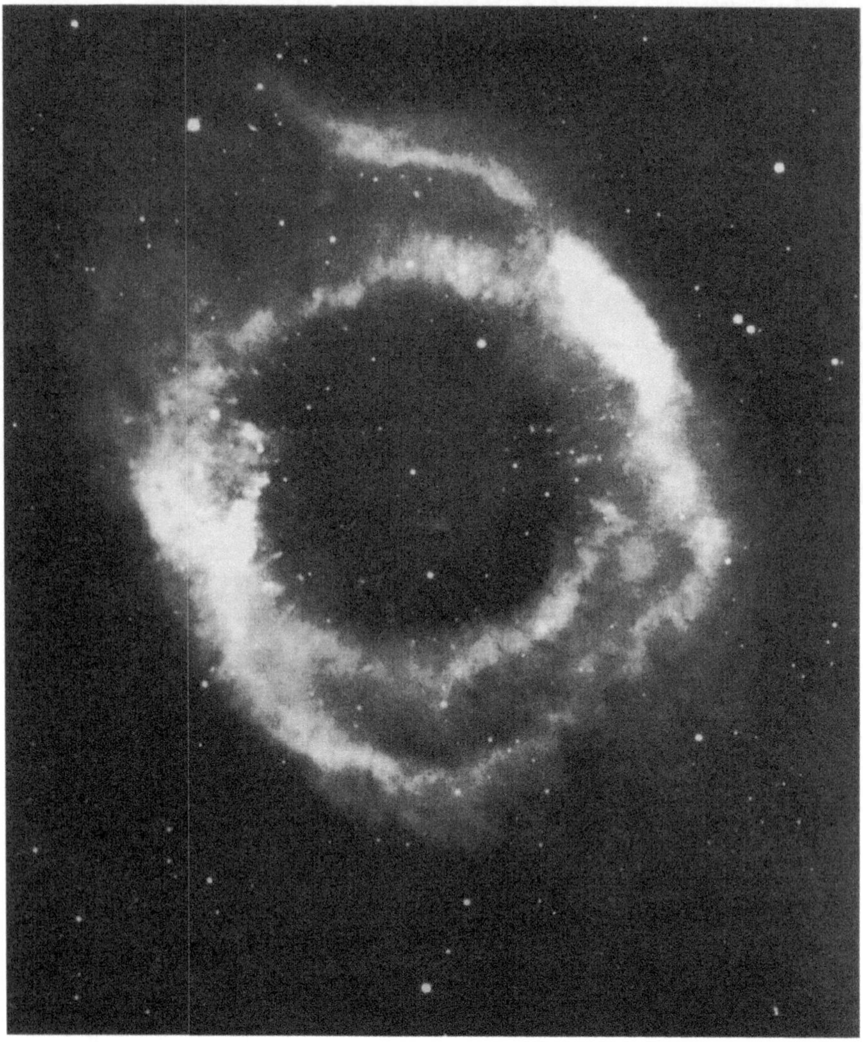

NGC 7293   Giant bipolar planetary nebula, Helix, in Aquarius

Grigor A. Gurzadyan

The Physics and Dynamics of Planetary Nebulae

With 125 Figures, 14 Plates and 93 Tables

Springer

Grigor A. Gurzadyan
Garny Space Astronomy Institute, P.O. Box 370/15
Yerevan 2, Armenia

Series Editors

Immo Appenzeller
Landessternwarte, Königstuhl
D-69117 Heidelberg, Germany

Gerhard Börner
MPI für Physik und Astrophysik
Institut für Astrophysik
Karl-Schwarzschild-Str. 1
D-85748 Garching, Germany

Martin Harwit
511 H Street SW
Washington, DC 20024, USA

Rudolf Kippenhahn
Rautenbreite 2
D-37077 Göttingen, Germany

Peter A. Strittmatter
Steward Observatory
The University of Arizona
Tucson, AZ 85721, USA

Virginia Trimble
Astronomy Program
University of Maryland
College Park, MD 20742, USA
and Department of Physics
University of California
Irvine, CA 92717, USA

Cover picture: The Planetary Nebula NGC 6543, nicknamed the "Cat's Eye Nebula". The image was obtained by the Hubble Space Telescope on September 18, 1994. Courtesy of J. P. Harrington, K. J. Borkowski (University of Maryland), and NASA.

Library of Congress Cataloging-in-Publication Data

Gurzadi͡an, G. A. (Grigor Aramovich) [Planetarnye tumannosti. English] The physics and dynamics of planetary nebulae / Grigor A. Gurzadyan. p. cm. – (Astronomy and astrophysics library, ISSN 0941-7834) Includes bibliographical references and index. ISBN 3-540-60965-2 (alk. paper) 1. Planetary nebulae. I. Title. II. Series. QB855.5.G913 1997 523.1'135–DC20 96-35762

ISSN 0941-7834
ISBN 3-540-60965-2 Springer-Verlag Berlin Heidelberg New York

Typesetting: Camera-ready copy from the author using a Springer $\TeX$ macro package
SPIN: 10500890 55/3144 - 5 4 3 2 1 0 - Printed on acid-free paper

Victor Ambartsumian (left) and the author, 1975

Preface

Planetary nebulae are the classic subject of astrophysics. The physical processes occurring in this highly ionized gaseous medium, the formation of emission lines in clearly specified conditions, the continuous emission extending from the far ultraviolet up to infrared and radio frequencies, the generation of exotic forms of radiation predicted by atomic physics, along with methods for deciphering the observed spectra and detecting physical and kinematic parameters of the radiating medium, etc. – all these problems form the solid foundations of the physical theory of gaseous nebulae. They are an essential part of the arsenal of powerful tools and concepts without which one cannot imagine understanding and interpreting the enormous diversity of processes taking place in the Universe – in gaseous envelopes surrounding the stars of various classes, from cool dwarfs and flare stars up to hot supergiants, as well as in stellar chromospheres and coronae, in atmospheres of unstable and anomalous stars, in circumstellar clouds and gaseous shells born in nova and supernova explosions, in diffuse nebulae and the interstellar medium, in interacting binary systems, in galaxies with emission lines, in quasars, etc.

The last thirty years have seen a turning-point in our knowledge concerning the very nature of planetary nebulae (PNs). The radio emission of PNs was discovered after it was predicted theoretically. On the other hand, the powerful infrared emission discovered both in the continuum and in emission lines was never expected. Shortly after this discovery, the source of this infrared emission was revealed – dust particles. This result marked one of the fundamental discoveries of recent decades – dust particles are a permanent component of any PN. The primordial nature of these particles also seems to be understood – they exist even at the protonebular phase, within the outer layers of cool giants, as embryos of future PNs.

The epoch of space observations, which started for PNs with the first ultraviolet spectrogram obtained by the space observatory *Orion-2* in 1973, allowed access to the most important source of information: resonance lines, located – by the caprice of Nature – for all atoms and ions in the ultraviolet region of spectra, inaccessible to ground based observations. And suddenly we were engulfed by an avalanche of unexpected results.

Among these unpredicted events, the problem of magnetic fields in PNs should be considered of special importance. In the light of modern investigations, both theoretical and observational, it now seems hard to doubt the reality of dipole-like magnetic fields in PNs and their crucial role in the formation of the observed diversity of forms and spatial structures of PNs.

The wide use of CCD techniques has resulted in a marked increase in the discovery of PNs with two envelopes; now almost every second PN has a double envelope. The nature of these configurations is becoming more and more obvious as a result of the dynamic splitting of the single envelope into two, and as an inevitable stage in the evolution of any PN.

The inner kinematics and dynamics of PNs has also become better understood. The crucial role of the joint or differential actions of three effects – the light pressure, the magnetic field and hydrodynamic effects – on the gaseous pressure of an outflow of gaseous matter from the nebula's nucleus has been revealed. Gravity thus seems to play almost no role in the dynamics of PNs.

Astrophysicists have come close to understanding the cosmogonical essence of PNs. According to modern opinion PNs are just one of the stages of the natural evolution of a definite class of stars, the evolution of which proceeds via non-explosive loss of the mass excess. The previous stage of PNs has with high probability been identified with the cool giants, or more precisely, with the Mirids – OH/IR maser sources.

These are only a few of a number of intriguing problems associated with PNs and discussed in the present volume. This is in fact a new monograph compared with the author's book published in Russian under the title *Planetary Nebulae: Physics, Dynamics* (Nauka Academic Publishers, Moscow, 1993), so that the Russian edition preserves its own interest. This volume differs essentially from the author's book *Planetary Nebulae* (Gordon and Breach, NY, 1969) not only in the scope of the problems covered, but also in the general spirit, obviously as a result of the epoch of space astrophysics we are now living in. Specific attention is paid to the theoretical achievements in the interpretation of observations of planetary nebulae, objects which now seem to be less mysterious than before.

It is a great pleasure for the author to express his gratitude to V. A. Ambartsumian for numerous discussions during work on the present monograph. The author should like to acknowledge the valuable discussions he had with Y. Terzian of Cornell University and M. Perinotto of the University of Florence. Special thanks are due to G. Jacoby, Y. Chu, B. Balick and K. Kwitter for supplying the excellent images of planetary nebulae.

Garni, Armenia
July 1996

G. A. Gurzadyan

Contents

1. Global Concepts

1.1 The Shapes of Planetary Nebulae

Planetary nebulae are expanding gaseous envelopes of enormous size in the centre of which the so-called nucleus of the nebula – a high temperature star, the source of the excitation of the nebula's light emission – is located. The expansion takes place when the gravitational force of the central star is completely absent. Planetary nebulae (PNs) belong to the category of galactic objects whose form and spatial structure are not governed by any gravitational forces, including the gravity of the nebula itself. The absolute magnitude of the potential energy of a nebula – an ionized envelope – is much smaller – ten thousand times – compared with the kinetic energy of its expansion or the thermal energy of heated gas, i.e. plasma.

Planetary nebulae are therefore an exception among those objects whose dynamics cannot be explained without consideration of the action of gravitational forces. The form and structure of PNs, as well as their evolutionary variations occurring during a very short period on the cosmogonical scale – a few tens or hundreds of thousands of years – are determined first of all by simultaneously acting internal forces: gas pressure, radiation pressure and the forces of electromagnetic origin.

The brightest PNs are of a disk-like form, and through a telescope they look like planets, with a rather greenish appearance (which explains their name). PNs reveal a great diversity of shapes.

The majority of PNs appear on photographic plates, say as bright rings. However, at long exposures the ring disappears and turns into a continuous disk. Therefore the actual form of a nebula has nothing in common with a ring or torus, but resembles a space shell of finite thickness and depressed gas concentration in its central zone. The observed ring-like form is due to the projection effect of the PN in the sky.

The ideal round-shaped PNs are rather rare. The most common shapes are slightly flattened or elongated. On the basis of early spectrographic observations Campbell and Moore (1918) assumed that the flattening is the result of the rotation of the PN around its minor axis. However, later this viewpoint was not confirmed; more precise and reliable observations failed to detect any trace of nebular rotation.

The flattening or elongation of a PN, as we will see later (Chap. 17), can apparently be explained by the presence of magnetic fields in nebulae.

An example of a ring-like PN is NGC 6720, in the constellation of Lyra, one of the brightest known PNs (the integral stellar magnitude is $9^m.3$), and the gigantic, at least by apparent size, NGC 7293 in Aquarius ($12' \times 12'$).

It is very difficult, if not impossible, to draw conclusions about the real shapes of PNs based only on their apparent forms, i.e. without taking into account the spatial orientation of the nebula relative to the line of sight. For example, PNs whose photographic images appear as true disks or rings may actually be elongated along the line of sight. Therefore, the conclusion that the elongated shapes of PNs are the most common seems to be inevitable. The absolutely spherical shape is an extremely rare exception. In view of all the factors determining the shapes of PNs, the preservation of sphericity during the life of a PN seems to be much more difficult than its distortion, in fact practically impossible.

1.2 The Structure of Nebulae

Among the variety of shapes and structures of PNs, some common regularities can, however, be established. Indeed, the brightness at the edge of the minor axis of a PN is, as a rule, greater than at the edges of the major axis. In some cases this difference is pronounced even on plates, without the need for further measurements. In these cases one speaks of *bipolar* PNs.

There are certain variations of bipolar PNs. As a basic criterion of bipolarity one can take the brightness difference or the brightness ratio at the edges of the minor and major axes of the PN. In some objects this difference is relatively small, i.e. the bipolar structure is represented only by an increase in brightness at the edges of the short axis. This is the case, for example, for the ring PNs NGC 6720 and NGC 7293 (Plates I, V), when the brightness at the edges of the minor axis is one and a half times greater than at the edges of the major axis. Therefore these ring PNs are seen also to be bipolar ones.

At the same time, in the second group of PNs the difference in brightness is so great that it gives the impression that the PN is divided into two bright parts of condensations, so-called "beads", with moon-like shapes located at the edges of the short axis, symmetrically with respect to the centre of the PN: as a result we have a *typical bipolar* PN. Occasionally, these "beads" have sharp edges on the inside. Among the interesting examples of such objects are NGC 3587 (the "Owl"), NGC 2474-5, NGC 7354, NGC 7662, NGC 3195, A 70; some of these are shown on Plates VI, VIII.

Finally, in the third group of bipolar PNs the "beads" are strongly developed and almost completely isolated – NGC 6058, A 80, Mz 2, A 66 and A 19 have such shapes (Plate VII).

Extreme manifestations of bipolarity are the so-called "rectangular" or "hour-glass" objects. These PNs differ from the common bipolar ones since they are strongly compressed along the minor axis and elongated along the major axis; as a result, their contour is close to the rectangular. A characteristic example of this type of nebula is IC 4 406 in the southern sky (Plate XII). The PNs He 2-114 and NGC 6 537, and to some extent Sh 2-71, NGC 7 026 and NGC 650-1, are also "rectangular", at least in their main structure. The impressive examples of "hour-glass" type are NGC 2 346, NGC 6 302 and M 2-9 (Plate V).

Almost half of the PNs are bipolar. On the other hand, as will be shown in Chap. 17, the "beads" can be observed only when the major axis of the PN is almost perpendicular to the line of sight. It follows that among the remaining PNs there are also bipolar ones, whose major axes are oriented almost parallel to the line of sight. This fact once more indicates the decisive role of the relative orientation of the PNs on their apparent shape.

Of special interest are those PNs that consist of two concentric shells around the nucleus. In this case we speak of *two-envelope* PNs. Moreover PNs with *three shells* around the central nucleus have also been discovered. However, the physical essence and the mechanism of the origin of the second and third envelopes are the same (Chap. 17). The ratio between the radii of the outer and inner envelopes covers a wide range; its smallest values differ slightly from unity, and can reach up to several units; there are also known cases when it exceeds even ten.

Typical examples of two-envelope PNs are NGC 2 392, NGC 1 535, NGC 3 242, NGC 6 826 and NGC 7 662 (Plates VIII, IX). In some cases the surface brightness of the second envelope is so small that it is revealed with difficulty (NGC 1 535 and NGC 6 891, Plate VIII) even on long-exposure CCD images. The presence of a rather weak and extended outer – third – envelope was established for one of the interesting PNs, NGC 6 543 (Plate XII), as well as for NGC 2 022 and NGC 6 826 (Plate IX). However, in the latter cases one can also deal with so-called *haloes*, the nature of which is quite different.

Thus more than half of the known PNs are double-envelope ones.

In Chap. 17, which is devoted to the origin of double-envelope PNs, it will be argued that all PNs can be born as single-envelope ones and become double enveloped at the end of their lives. In other words, the appearance of the second envelope around the first, at a definite stage of the life of the PN, is an evolutionary outcome, a phase through which all PNs should pass.

A sparse group includes so-called *spiral-like* PNs. In the majority of cases they are common ring-like or two-envelope objects with two spiral-like arms located symmetrically at both sides of the PN. Typical representatives of such PNs are NGC 4 361, NGC 6 210, A 79 and NGC 6 309 (Plate XII). However, unlike the spiral structure of galaxies, in PNs the spiral structure has no dynamical character. In particular, the spiral structure is not a result of the combination of two types of motion – the radial outflow of matter from the central star-nucleus and rotation of the nebula. The origin of the spiral arms and similar properties can be simply explained within a framework that accepts the essential role of magnetic fields in PNs.

Occasionally the mentioned main shapes can be present within a single nebula in various combinations. For example, NGC 4 361 (Plate XIII) is spiral and at the same time has two envelopes; Mz 2 is bipolar and also has two envelopes, and furthermore there are traces of spiral structure on the second envelope. NGC 650-1 is bipolar and spiral and has two envelopes. One can mention plenty of such examples. The majority of double-envelope PNs are also bipolar.

We should emphasize that, independent of the peculiarities in the shape and structure of PNs, all of them have three-dimensional structure. In other words, flat, two-dimensional forms are not typical for PNs. For example, ring-like PNs are not tori; they are hollow spheres with a definite envelope thickness and with the same mean density of matter in all directions.

However, the true spherical shape and in particular the concentric distribution of matter density, are not peculiar properties of PNs. Therefore, the results obtained for spherical models of PNs cannot, in general, produce a true picture of the distribution of the matter in real, i.e. non-spherical, PNs.

Here it is necessary to make a reservation. Unlike the case of stellar atmospheres, where the construction of a single model is quite possible and completely justified owing to their determination by only a few free parameters – often by just one (temperature), in some cases by two (temperature and gravitation), or in extreme cases by three (temperature, gravitation and chemical composition) – the number of parameters determining a model of PNs is rather large. Therefore to build a single model for all PNs is impossible in principle.

The characteristic feature of PNs that are *small* in absolute dimensions is their clearly outlined shape, with sharp boundaries. This indicates the dominant role of the forces controlling the dynamics of the PNs at a given period of their lives and preventing the disruption or destruction of the latter.

The most numerous group is formed by so-called *stellar* PNs. These objects have rather small *visible* sizes. A fraction of them are of small linear dimensions. They represent early phases in the development of PNs. Others can well be PNs of larger dimensions, of well-known types, whose star-like appearance is just due to the fact that they are located far away from us.

More than half the PNs discovered and registered in catalogues are of stellar type. As a rule they are discovered either with the help of objective prisms, through the characteristic nebular spectrum, or by photometry combinations of different filters.

The angular dimensions of the majority of PNs are smaller than one arc minute. The dimensions of the most interesting objects fall in the range $20''$–$40''$. A very small fraction of the PNs have dimensions larger than $1'$. A few have a huge extent – greater than 10′, and in one case (A 72, Plate XI) the diameter of the PN is 40′!

The problem of the linear dimensions of PN is connected with that of the determination of their distances, as discussed in Chap. 12.

1.3 Luminosity and the Nuclei

The luminosity of a PN is caused by the direct radiation of its central star – the *nucleus* of the PN. In spite of this, the total brightness of the PN exceeds many times the brightness of this star. This is due to the fact that the PN absorbs the invisible ultraviolet part of the energy radiated by the nucleus and re-emits it in the visible band. Since the nuclei of PNs are very hot stars, the largest fraction of their energy is emitted in UV and as a result, the PN appears brighter than the illuminating central star.

The observations give, as a rule, the possibility of determination of the integrated photographic stellar magnitude, m_n, of the PN, i.e. the sum of the brightness of the envelope and the nucleus.

In addition one can also determine the apparent photographic stellar magnitude, m_*, of the central star-nucleus. Since the PNs are completely transparent at frequencies corresponding to visual light, the magnitude m_n in fact represents the total energy emitted by the PN in all directions and per unit time at the frequencies of the lines in the photographic region of the spectra. If so, then the difference $\delta = m_* - m_n$ must be always positive and, as an important parameter, does not depend on the distance to the PN: this parameter determines the temperature and size of the nucleus, as well as the mass of the PN itself.

The integrated apparent magnitude of PN $\mathrm{m_n}$ varies within wide limits – from $6^{\mathrm{m}}.5$ (NGC 7293) up to 18–20^{m} (A 47, A 48, A 58, A 69). The quantity δ varies from $0^{\mathrm{m}}.1$ (NGC 1514, Hb 12) to 7^{m} (NGC 7048, NGC 7293, BD+30°3639). On average, for the majority of PNs $\delta \approx 3$–4^{m}.

An important parameter is the quantity H – the mean surface brightness of the PN: for its determination it is necessary to know m_n and the mean angular diameter of nebula D'' (in arc seconds). The mean value of H varies from 4^{m} (NGC 6153) to $14^{\mathrm{m}}.5$ (NGC 6772, NGC 3587) in a circle of $1''$ diameter. In one case $H = 19^{\mathrm{m}}.4$ (NGC 2474-5). The envelopes of PNs with small apparent dimensions have the largest surface brightnesses.

With the development and introduction of electro-photometric measurement techniques, the centre of attention has moved to measurements of the fluxes $F(\mathrm{H}_\beta)$ in the H_β emission line of hydrogen for each nebula in absolute energy units. Having the observed magnitude of $F(\mathrm{H}_\beta)$ in ergs $\mathrm{cm^{-2}\,s^{-1}}$ we may determine in particular the distances, sizes and masses of PNs, electron concentrations within the nebulae, as well as the absolute bolometric luminosity of the central star-nucleus.

The nuclei of PNs are faint stars. Only few PNs have nuclei brighter than tenth magnitude: NGC 1514 ($m_* = 8^{\rm m}.0$), NGC 6826 ($9^{\rm m}.5$), NGC 2392 ($9^{\rm m}.5$) and A 33 with the variable nucleus within the limits of 7.7–$8^{\rm m}.5$. The photographic stellar magnitudes of only a few nuclei attain $20^{\rm m}$ (NGC 1501). However, there are PNs whose nuclei are mainly invisible because of the high brightness of the nebula itself; hence, they must be fainter than $21^{\rm m}$. For the majority of PNs $m \approx 15$–16.

Considerable progress has been reached in the problem of the determination of the distances of PNs (Chap. 12). As a result, we now have data that are quite reliable for the absolute magnitude of PNs $M_{\rm n}$ and their nuclei M_*. The magnitudes $M_{\rm n}$ vary from $-1^{\rm m}.7$ (NGC 1535), or $2^{\rm m}.9$ (He 2-108), up to $+5^{\rm m}.5$ (Hb 12). The absolute magnitudes M are scattered between the limits $-0^{\rm m}.2$ or $-0^{\rm m}.3$ (IC 4593) up to $+8^{\rm m}$ (NGC 7293) and $+9^{\rm m}.3$ (A 71); the dispersion in the first case ($M_{\rm n}$) is of the order of 8–$9^{\rm m}$, in the second (M_*) of the order of $10^{\rm m}$. The average magnitudes of $M_{\rm n}$ and M_* for one selection of 50 PNs are: $M_{\rm n} \approx 0^{\rm m}.2$ and $M_* \approx 2^{\rm m}.7$, i.e. are positive in both cases. No matter how precise or uncertain the data are, one can draw the following conclusion: the nuclei of PNs are not giants but most probably are dwarfs of medium brightness. This conclusion had already been drawn in the early days of PN investigations and was confirmed by later investigations.

As for the colour index, in the case of PNs it does not have the same physical significance as for stars. For stars it depends basically on one parameter – the colour temperature of the star. In PNs it represents the difference between the stellar magnitudes summed over the intensity of all emission lines and part of the continuous spectrum that lies in the visual region of the spectrum and those summed over the intensity of the emission lines, again from the visual region. But this difference depends on many factors and can vary from one PN to another; but on average it should be of the order of $+0^{\rm m}.7$.

The maximum magnitudes of the radii of the nuclei of PNs are of the order of the Sun's radius, the minimum are of the radius of the white dwarfs, $R_* \approx 0.1 R_\odot$, i.e. the dispersion in the magnitudes of R_* is about two orders. However, the limiting magnitudes of the nuclei temperature T_* may differ tenfold. As a result, the influence of the nucleus's radius and its temperature on its bolometric luminosity will be the same, so long as $L_* \sim R_*^2 T_*^4$.

1.4 The Optical Spectrum

The most characteristic feature of PNs is their spectra in general and in the optical region in particular. These differ strongly from the spectra of stars and other celestial objects. At the same time, the general structure of the spectra is identical for all PNs. Because of this, it is always possible to distinguish unmistakably the spectrum of a PN from, say, that of a peculiar star, even when the apparent size of a PN is too small to enable its identification with photometry.

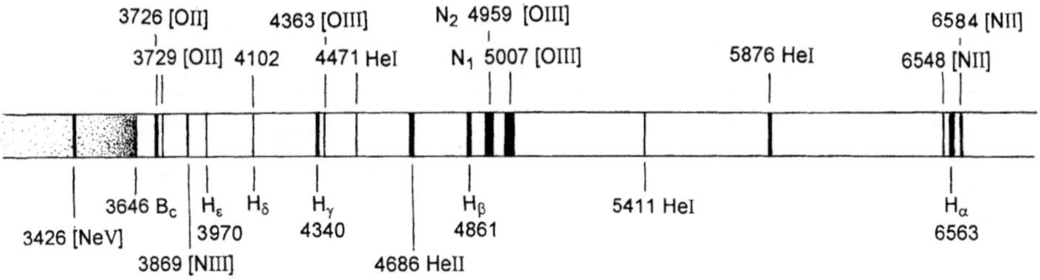

Fig. 1.1. General structure of the spectrum of a planetary nebula in the optical region, 3 300–7 000 Å. Only the most important emission lines, both permitted and forbidden, are shown. The shaded part from the left, beginning from $\lambda = 3\,646$ Å, is the Balmer continuum of hydrogen

The spectrum of a PN in the optical region consists of a series of intense and weak emission lines, superimposed on a weak continuous background. The general structure of the spectrum of a PN in the optical region is shown schematically in Fig. 1.1. The brightest emission lines in all PNs are the so-called "nebulium" lines: N_1, with a wavelength of 5 007 Å, and N_2, with 4 959 Å. Bowen in 1928 was the first to show that these and several other formerly mysterious lines originate from *forbidden* transitions downwards from metastable states in several atoms and ions. As a rule, these transitions are not observable in our laboratories, since it is almost impossible to produce the conditions – an extremely low concentration of matter in a volume – for their excitation. Bowen's discovery in fact served as the basis of one of the most important areas of study in the physics of gaseous nebulae – the theory of their luminescence in forbidden lines.

The lines N_1 and N_2 belong to doubly ionized oxygen O^{++}. Sometimes they are referred to as the OIII doublet and written 5 007 [OIII] and 4 959 [OIII]; the square brackets indicate as a rule that the given line corresponds to a forbidden transition. These lines also appear in the spectra of diffuse PNs, in the spectra of envelopes ejected by novae and supernovae and also in the spectra of non-stationary stars surrounded by gaseous envelopes. However, in PNs the lines N_1 and N_2 are considerably brighter: the intensity ratio of the lines N_1, N_2 and H_β is 10 : 3 : 1, while in diffuse nebulae it varies from 2 : 0.6 : 1 to 0.3 : 0.1 : 1.

The ratio of the intensities of these lines, I_{N_1}/I_{N_2}, is roughly equal to 3. It is constant for all PNs and it does not depend on their physical conditions. However, the ratio $I_{N_1+N_2}/I_{H_\beta}$ varies within very wide limits – from ~ 0.1 up to 40 and higher. In Table 1.1 the magnitudes of this ratio are given for a number of PNs, including the those with the smallest (BD+30°3639) and the largest (HaTr 3) fixed magnitudes of this ratio. It should be noted that the ratio $I_{N_1+N_2}/I_{H_\beta}$ is larger than 20 only in a few cases, and although the large magnitude of the ratio does not always indicate high excitation for the given object, such exceptionally high values, larger than 20 or 30, represent

Table 1.1. The ratio of intensities of emission lines ($N_1 + N_2$) [OIII] to H_β of hydrogen in the spectrum of some planetary nebulae

Nebula	$\frac{N_1+N_2}{H_\beta}$	Nebula	$\frac{N_1+N_2}{H_\beta}$	Nebula	$\frac{N_1+N_2}{H_\beta}$
BD+30°3639	0.06	IC 4997	7.58	NGC 3211	20.8
He 2-162	0.16	NGC 6543	8.50	NGC 3918	21.5
Cn 3-1	0.23	NGC 6826	10.20	NGC 6644	23.5
Mz 3	0.31	NGC 6565	12.7	M 2-21	24.6
He 2-64	0.57	NGC 1535	13.8	Mz 2	26.3
Cn 1-2	1.00	NGC 7009	14.0	He 2-76	28.2
IC 418	1.85	NGC 7662	16.8	A 23	32.0
M 2-24	2.55	NGC 2440	17.9	He 2-84	33.0
M 3-20	3.35	NGC 6818	18.9	Wray 16-122	38.7
Hb 12	4.66	NGC 7027	19.2	HaTr 3	44.5

special interest for the theory. As we shall see later (Chap. 5), in the high excitation nebulae a very small magnitude for this ratio is quite possible.

Doubly ionized oxygen (O^{++}) also produces a considerably weaker forbidden line 4363 [OIII], located just nearby the line H_γ – the third line in the Balmer series of hydrogen (λ = 4340 Å). This line, 4363 [OIII], plays an important role for the diagnosis of gaseous nebulae – it is an indicator of electron temperature of the emitting medium.

In the short wavelength edge of the optical spectrum of PNs there are two lines, both forbidden, belonging to singly ionized oxygen O^+: they are composed of the doublet 3726 [OII] and 2729 [OIII]. These lines are so close to each other that they are not resolved in the usual spectrographs, and blend into one line with equivalent wavelength 3727 Å as a line: 3727 [OII]. Unlike the ratio $I(N_1)/I(N_2)$, the ratio $I(3729)/I(3726)$ depends on the physical conditions in the PN, and in particular on its electron concentration, and may be used as a reliable tool for the determination of electron concentration in PNs.

The ratio of the intensities ($N_1 + N_2$ [OIII])/3727 [OII] is an indicator for the determination of the temperatures of the nuclei of gaseous nebulae (Chap. 5).

The constant feature always present in the spectra of gaseous nebulae is the red doublet of singly ionized nitrogen – forbidden lines 6548 [NII] and 6584 [NII]. These lines fall to the left and right of H_α (6563 Å). In spectrograms obtained with ordinary spectrographs all these three lines merge into one. They can be resolved only by high dispersion spectrographs (as a rule with diffraction gratings). The ratio $I(6584)/I(6548)$ is completely analogous to the ratio $I(N_1)/I(N_2)$, i.e. it is also equal to three and does not depend on the physical conditions in the PN. However, the ratio I(6584 + 6548 [NII])/$I(H_\alpha)$ is different in various PNs and is of the order of unity: i.e. for NGC 7027, it equals 3.2. The line 5755 [NII] is an analog of the line 4363

[OIII], hence, the ratio $I(6\,548 + 6\,584\ \text{[NII]})/I(5\,755\ \text{[NII]})$ in this case will also be an indicator for the determination of the electron temperature of the emitting medium.

The lines of neutral helium 5 876 HeI and 4 471 HeI are observed in the so-called low excitation nebulae, and also in diffuse nebulae.

The strongest line in the optical region of ionized helium 4 686 HeII (analogous to the line 18 751 P_α, the first line of the Paschen series of hydrogen) is a constant content of the spectra of all high excitation nebulae. The largest magnitude of the ratio 4 686 HeII/H_β is of the order of 1.5 (A 30, A 78, NGC 246), the smallest is of the order of 0.02 (NGC 6 543, NGC 6 210, NGC 6 891). In very low excitation class nebulae the line 4 686 HeII in emission is completely absent.

Except for the line 4 686 HeII, ionized helium shows itself in the optical region also by a large number of lines of subordinate series, more than ten, all extremely faint. The strongest line of ionized helium, 1 640 HeII (the analog of 6 563 H_α, the first line of the Balmer series of hydrogen), lies in the ultraviolet region and may be measured only by space observations. Theoretically, the strongest line, of course, is 303 HeII, analogous to the line 1 216 L_α, the first line of the Lyman series of hydrogen. However, the line 303 HeII cannot be registered in principle even in space conditions, outside Earth's atmosphere, because of the completely opaque interstellar medium at $\lambda < 912$ Å.

The constant content of the spectra of high excitation PNs is the forbidden lines of doubly ionized and four times ionized neon – 3 869 [NeIII] and 3 426 [NeV]; the latter is in the region of $\lambda < 3\,646$ Å, i.e. on the background of the Balmer continuum. The maximum magnitude of the ratio 3 486 [NeV]/H_β is 1.5, i.e. of the same order as the maximal magnitude of the ratio 4 686 HeII/H_β. The line 3 426 [NeV] has been registered only in the spectra of extremely high excitation PNs – NGC 6 302, NGC 2 440, NGC 4 361 etc. For NGC 7 027 the ratio 3 426 [NeV]/$H_\beta = 1.25$.

The first four lines of the Balmer series, 6 563 H_α, 4 861 H_β, 4 340 H_γ and 4 120 H_δ are relatively strong and can be easily measured; they are present in the spectrum of all excitation classes of PNs. But the rest of the lines starting from 3 970 H_ε merge with lines of other elements, and their measurement with the usual observational techniques is practically impossible. Hydrogen also produces a continuum, weak in the optical region of the spectrum, considerably stronger beyond the Balmer limit at 3 646 Å. The first lines of the remaining series of hydrogen – Paschen, Brackett, etc. are scattered in the far infrared region, longer than 10 000 Å. The strongest line among them, the resonance line 1 216 L_α, the first line of the Lyman series of hydrogen, lies in the far ultraviolet region and in principle may be registered only in space conditions. However, this line may easily be lost on the background of the powerful Lyman-alpha emission line of the geocorona.

The emission lines enumerated above are typical for all PNs. However, in addition to them, the spectra of a number of PNs reveal several other weak lines, the majority due to forbidden transitions, which belong to various ions of iron, chlorine, sulphur, silicon, argon, calcium, potassium, magnesium, etc.

A detailed list of emission lines and their relative intensities including 200–300 or more lines, have been compiled for many PNs. The essential part of these results are summarized in the catalogue by Kaler (1976) and for the most important emission lines in the catalogue by Acker et al. (1992).

1.5 The Spectrum of Nuclei

The nuclei of PNs typically are high temperature stars. In contrast to the spectra of PNs, the spectra of their nuclei show great diversity. To a certain extent this is natural: it could be partially explained by a large dispersion in absolute brightness of the nuclei. On the other hand, the nuclei of the majority of PNs are still unknown – the faintness of the nuclei requires more detailed spectrographic studies. One can assume, however, that the real dispersion in spectral types of the nuclei must be quite large.

Spectra of nuclei of more than 50 PNs have already been studied, including the ultraviolet region. The variety of the nuclei spectra is too large; it seems hard to find two nuclei with absolutely identical spectra. Nevertheless, according to their spectra, the nuclei of PNs can be classified within the following types.

(a) Wolf–Rayet type, i.e. stars with very broad emission lines indicating a powerful outflow of matter (NGC 7026, NGC 6803).
(b) Of type; emission and absorption lines appear in the spectra simultaneously (NGC 2392, IC 4593).
(c) O type; absorption lines appear in the spectra, but emission lines are absent (NGC 246, IC 2149).
(d) Continuum type (Cont.); the spectrum consists entirely of continuous radiation without any traces of absorption or emission lines (NGC 1535, NGC 7009). In this case one is most likely dealing with white dwarfs. However, according to their absolute magnitude these nuclear white dwarfs are not classical white dwarfs; the latter are much less bright than the white dwarf nuclei of PNs.

It is well known that the classical Wolf-Rayet stars are divided into two subclasses. In one of them, WC, the lines (bands) of carbon predominate, while the nitrogen lines are almost absent. In the stars of the second subclass, WN, on the contrary, the nitrogen lines predominate and the carbon lines are almost absent.

In PNs the nuclei of type WN are almost completely absent; if the nucleus is of Wolf–Rayet type, it is as a rule of WC type.

Sometimes we observe nuclei with the spectrum WC + Of, i.e. a combination of two types of spectral WR star and Of star. Since such a combination is not observed among stars of the main sequence, one can assume the possibility of the existence of a binary system at the centre of PNs. At least four PNs are known whose nuclei belong to this combined type – NGC 6 543, NGC 6 572, NGC 6 826 and IC 4 997 – all are interesting and well known objects.

The nuclei that belong to the stellar type Of are of special interest. These stars have been well studied, and time variations in intensity and shape of the emission lines have been discovered in their spectra.

The nuclei of type O, i.e. with absorption lines, are often observed in extended PNs, i.e., in those which are in the later phase of their lives.

In several cases, the possession by some nuclei of late class stars, A, G, and even K, has been established. These PNs and their nuclei are as follows (Pottasch, 1987):

NGC 1 514	B9–A0 III	NGC 2 346	A5 V
NGC 3 132	A0 V	A 35	G8 III
He 2-36	A2 III	A 63	K V

Most likely there are binary systems in the centres of these nebulae with a visible cool giant without any role and an invisible white dwarf as a source of excitation of the PN.

While speaking about the nuclei belonging to WR, Of or O types, one must keep in mind that this implies the existence only of a similarity between the *spectral* characteristics of these stars and the nuclei of PNs. Actually, there exists a huge difference, about $3–4^{m}$, in absolute magnitude between those two categories of objects. It is important to point out that the nuclei of PNs, although they may resemble dwarfs, are according to the physical processes that occur in their atmospheres, and to some degree in their interiors, in general identical with normal hot stars.

Since we are discussing here the physical nature of the nuclei of PNs we cannot avoid mentioning their non-stationarity. The nuclei can appear in quite a variety of forms – there are differences in the variations of the brightness, in the strength and shapes of the emission lines, and hence in the variations of their outflow power along with short periods of relative calmness, etc.

Variability and various degrees of non-stationarity have been established or suggested for a number of PN nuclei: NGC 1 514, 2 346, 2 392, 6 572, 6 891 IC 2 149, 3 568, 4 593, 4 997, and also A 14, 20, 30, 36, 39, 41, 46, 51, 61, 63, 74, 78, etc. In most cases these is still a question about the variability of the nuclei.

As to the distribution of the nuclei according to spectral type, the situation is roughly as given in Table 1.2 for the nuclei of 70 PNs. The greatest number, 26%, of nuclei have continuous spectra (Cont.). Presumably here we have white dwarfs with accretion disks. Within this group might also be included the six nuclei mentioned above, if it is assumed that the white dwarf with accretion disk may form a binary system with late type stars. Nearly 20% of nuclei have a combined spectrum, WR + Of, and 14% are of WR type.

Table 1.2. Distribution of nuclei of planetary nebulae by spectral type

Nucleus type	Number of nuclei	Distrib. [%]	Nucleus type	Number of nuclei	Distrib. [%]
Cont.	18	26	O	8	12
WR + Of	15	20	sdO	5	7
WC	10	14	Others	4	6
Of	10	14			

1.6 Stratification of Radiation: Observations

The *monochromatic* images of PNs in different emission lines obtained with the help of objective or slitless spectrographs reveal one of their interesting properties. The photometric study of these images allows one to reach certain conclusions about the distribution of radiation of the PN in the emission lines of different atoms and ions. By means of such measurements it has been established that the sizes of the monochromatic images of the PN are *different* for various lines and in average they decrease with increasing ionization potential of the atom that produce a given emission line. This phenomenon is called the *stratification* (lamination) of radiation, first discovered by Wright (1917) at the dawn of observational study of PNs – at the beginning of the century.

Let us turn to Table 1.3, where the most interesting results are shown for several PNs; it is clear that the sizes of monochromatic images decrease with the increase of the ionization potential. The largest sizes are in hydrogen lines H_β with an ionization potential of 13.5 eV, and the smallest sizes are in the line 3 426 [NeV] with an ionization potential of 95 eV; in the last case, the sizes of the monochromatic images are 1.5–2 times smaller compared with the sizes in the [OIII] images.

Table 1.3. Sizes, in arc seconds, of monochromatic images of a number of planetary nebulae in the lines of ions with various ionization potentials (eV)

Nebula	H_β 13.5	N_1+N_2 35.0	[NeIII] 40.9	HeII 54.2	[NeV] 96
IC 2165	5.″3	5.″7	5.″9	4.″4	3.″4
NGC 2392	13	19.7	19.2	14.4	10.8
NGC 2440	5.7	6.4	6.5	5.3	3.1
NGC 6818	17.6	19.7	19.2	14.4	10.8
NGC 6886	3.9	5.0	–	3.4	–
NGC 7026	6.0	6.3	6.1	3.7	–
NGC 7662	13.9	14.1	14.7	12.1	9.1

It is interesting to note that the sizes of monochromatic images in the lines N_1+N_2 are almost the same as the sizes in the hydrogen lines (H_β) although the ionization potential both for doubly ionized oxygen and hydrogen differ by nearly two. This fact has an explanation (Chap. 2).

Besides this, variations in shape while one moves from one image to another have been also discovered. For example, NGC 2 165 seems to be typically bipolar in hydrogen lines as well as in the forbidden lines 6 548 [NII] and 6 584 [NII]. However, the bipolarity is expressed rather weakly in the line 3 869 [NeIII] and disappears completely in 3 426 [NeV]. A similar situation occurs for NGC 7 662; it is bipolar and two-envelope in the lines of hydrogen and doubly ionized oxygen, ring-like in the lines 4 686 HeII and 3 896 [NeIII] and disk-like in the line 3 426 [NeV]. In the case of the ring-like NGC 6 720 (Lyra) the images in the lines N_1 and N_2 are indeed ring-like but the image in 4 686 HeII is considerably smaller and its shape is that of a uniformly illuminated disk, filling the inner region of the ring.

These facts indicate the entire conditionality in the determinations of the shapes and structures of PNs. Actually one cannot judge the shape and, especially, the structure of a PN without taking into account in which lines it is observed. Some lines give us information on the outer regions of a nebula, while others give information on its central parts. The data in Table 1.3 also reveal another remarkable conclusion: in high potential lines, a considerable part of a PN disappears completely.

Thus, one cannot judge the true form of a PN and construct its spatial model based on its images obtained only with one or two lines.

As for the interpretation of this stratification phenomenon two approaches are possible: either various chemical elements have different distributions inside the PN, or the conditions controlling their emission depend on their distance from the exciting star. If the first explanation is true, the sizes of all images that belong to different ions of the same element should be identical. This, as follows from Table 1.3, does not occur. Therefore, one has to accept the second possibility. As will be shown in the next chapter, the stratification of the radiation in the PN is entirely explained by the different conditions under which the atoms and ions radiate.

1.7 Excitation Class

Though the structure of PN spectra is common for all objects, the strengths of the emission lines may vary strongly when one passes from one object to another. This concerns first of all the high potential lines, particularly 4 686 HeII and 3 426 [NeV]. These lines, although completely absent in the spectra of some PNs, are clearly visible in some other PNs and reach the highest strength in yet other PNs. Accordingly we shall speak about PNs of low, middle and high excitation classes: in the last case the ratio, for example, 3 426 [NeV]/H_β may be even larger than unity and reach values of 2–3, and the ratio 4 686 HeII/H_β may be of the order of 1–1.5. The determining factor is first of all the temperature of the central star: in the case of low excitation PNs the nuclei should be of low temperature; on the other hand, in the case of high excitation PNs we must have hot and very hot nuclei. In this way the concept of the excitation class of PN and the problem of the classification of their spectra appears.

Attempts to classify the spectra of gaseous nebulae were undertaken long ago and more than once; nevertheless, it is difficult to assert the role of the excitation class in the physics of gaseous nebulae. In all events it is impossible to relate the excitation class completely to a single parameters of PNs – to the temperature or luminosity of nuclei, to the size or the mass, to the shape or the structure, to the electron temperature or the electron concentration, to the inner kinematics or chemical composition, etc. With such an abundance of parameters, the spectral class of a PN or the class of its excitation itself acquires the status of an independent parameter.

In the case of stars, the spectral class is not an independent parameter, so long as in each case for the given spectral class we understand a definite luminosity of the star or its effective temperature, or a combination of both parameters. The peculiarity of the situation in the case of PNs is this: the spectral or excitation class has a collective character and depends essentially on all the mentioned parameters of the nebula and its nucleus.

Here we shall describe the quantitative way to classify the spectra of gaseous nebulae (Gurzadyan, 1988, 1991). This system is based on the use of the ratio of intensities of emission lines $I(\mathrm{N}_1 + \mathrm{N}_2)/I(4\,686\ \mathrm{HeII})$. In PN this ratio varies within wide limits – from a few units up to 500 or more, and at the same time it is a sensitive indicator for excitation classes. Dividing this interval into parts with various magnitudes of the parameter $\log I(\mathrm{N}_1{+}\mathrm{N}_2)/I(4\,686)$, we end up with a very simple system for the classification of spectra according to the degree or class of excitation.

The excitation class p itself is indicated by integer numbers from 1 up to 12^+, and the whole sequence of PNs is divided into three groups according to the degree of excitation:

Low	– excitation classes	$p = 1$–4,
Middle	– excitation classes	$p = 4$–8,
High	– excitation classes	$p = 9$–12^+.

In the middle and high excitation classes are included PNs in whose spectra the emission line 4 686 HeII is presented definitely. In this case the classification is carried out according to the observed magnitude of the ratio $I(\mathrm{N}_1 + \mathrm{N}_2)/I(4\,686)$ as is shown in Table 1.4: here for each excitation class p a definite magnitude of the logarithm of this ratio, $\log I(\mathrm{N}_1 + \mathrm{N}_2)/I(4\,686)$, corresponds to the limits ± 0.1. For strongly excited PNs one more class, 12^+, is added with the mean value of the mentioned logarithm being equal to 0.6. This relates to objects with extremely high nucleus temperatures, 300 000 K and higher.

Table 1.4. Excitation class p of planetary nebula spectra of low ($p = 1$–4), middle ($p = 4$–8) and high ($p = 9$–12^+) classes of excitation estimated by observed magnitude of the ratio $(N_1 + N_2)/4686$ HeII and $(N_1+N_2)/H_\beta$

p	$\frac{N_1+N_2}{H_\beta}$	p	$\log\frac{N_1+N_2}{4686}$	p	$\log\frac{N_1+N_2}{4686}$
	Low		Middle		High
1	0– 5	4	2.6	9	1.7
2	5–10	5	2.5	10	1.5
3	10–15	6	2.3	11	1.2
4	>15	7	2.1	12	0.9
		8	1.9	12^+	0.6

It must be noted that the transition from middle excitation classes ($p =$ 4–8) to higher ones ($p = 9$–12^+) is accompanied by intensification of high ionization emission lines, particularly of the line 3426 [NeV].

In the case of low excitation PNs ($p = 1$–4) the emission line 4686 HeII is completely absent or hardly noticeable. Therefore we realize the classification in this case according to the observed magnitude of the ratio $I(N_1 + N_2)/I(H_\beta)$; the numerical dependence of this ratio on the excitation class p is also presented in Table 1.4. For the sake of clarity and also for convenience of application, this system is presented in a graphic form in Fig. 1.2 with

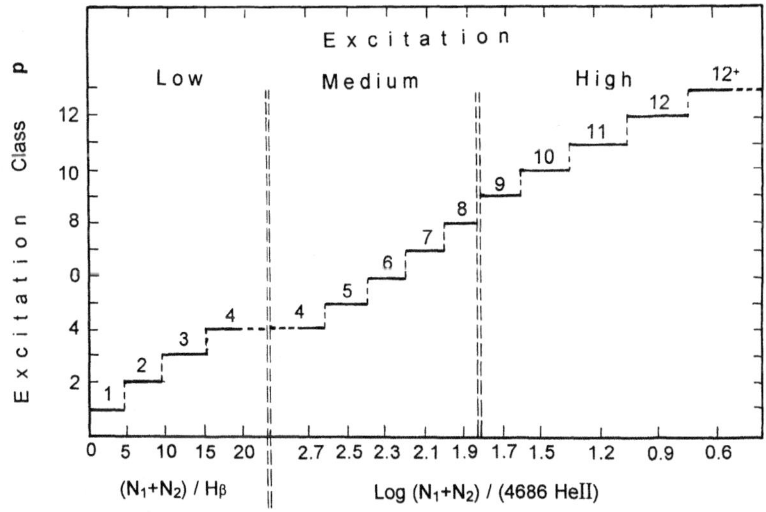

Fig. 1.2. Graphic representation of excitation class, p, of gaseous nebulae showing their dependence on observational parameters $(N_1 + N_2)/H_\beta$ in the case of low excitation nebulae ($p = 1$–4) and $\log(N_1+N_2)/4686$ in the cases of medium ($p =$ 4–8) and high excitation nebulae ($p = 9$–12^+)

an indication of the limits of the parameters $\log I(\mathrm{N}_1+\mathrm{N}_2)/I(4686)$ and $(\mathrm{N}_1+\mathrm{N}_2)/\mathrm{H}_\beta$ separately for each excitation class.

The estimation of the class $p = 4$ is connected with some difficulties; in fact this is a transient class when the line 4686 HeII is at the limit of its disappearance or the ratio $(\mathrm{N}_1+\mathrm{N}_2)/\mathrm{H}_\beta$ is too large at the complete absence of the line 4686 HeII. Therefore he class $p = 4$ should be determined by two means, i.e. by the very large ratio of $(\mathrm{N}_1+\mathrm{N}_2)/\mathrm{H}_\beta(> 15)$ as is shown in Fig. 1.2.

In the case of $p = 4$ and $p = 12^+$ the limiting magnitudes of the corresponding parameters are open: they may be $(\mathrm{N}_1+\mathrm{N}_2)/\mathrm{H}_\beta(> 15)$ or $\log(\mathrm{N}_1+\mathrm{N}_2)/4686$ HeII > 2.6 in the first case, and $\log(\mathrm{N}_1+\mathrm{N}_2)/4686$ HeII < 0.6 in the second.

The classification system proposed here is convenient also because it is based on the use of the strongest and most easily registered lines $\mathrm{N}_1+\mathrm{N}_2$ [OIII], H_β and 4686 HeII. In the case of the latter, 4686 HeII, its very presence or absence is enough to reveal the degree of excitation class. Moreover, the wavelengths of all these three lines differ only slightly from each other, which makes possible the application of the classification even without the correction of line intensities for interstellar extinction.

At the end of Chap. 18 of this book, the Catalogue of Excitation Classes for 750 Galactic Planetary Nebulae, based on this system, is given, where the intensities of necessary emission lines were taken from the catalogue by Acker et al. (1992). For seven PNs in the Magellanic Clouds the excitation classes are as follows:

	LMC N2	LMC N7	LMC N33	LMC N40	SMC P2	SMC P5	SMC P67
p	10	11	9	10	10	10	12^+

if one uses the line intensity data by Aller et al. (1987).

1.8 Classification of the Shapes

The classification of the forms or shapes of PNs is a considerably more difficult problem. In order to work out a classification system of nebular shapes, and to "sort" the individual nebulae among various classes or types, we strongly depend upon observational material, i.e. photographic images obtained with existing observational techniques. On the other hand, as experience shows, the appearance of any powerful telescope or observational facilities provides new and typically more important traits in the structure or shape of a given PN. As a result our views on the class or type of the shape for the same object have to be modified. An obvious example is the history of observing two-envelope PNs, when with the appearance of CCD techniques the outer much fainter envelopes were discovered around those PNs, earlier classified as single envelope ones.

The classification of the shapes of PNs should be based mainly upon the most characteristic traits of their shape and structure. Since PNs from the dynamical point of view are non-stationary objects, it is desirable for this classification to have at the same time an *evolutionary* content, i.e. that it should reflect the stages of development.

As we shall see later, the differences in the nature of the effective forces produce completely different nebular shapes and structure. By considering the influence of these forces as criteria for a morphological classification, we subscribe it by *physical* content, showing the dominating role of one or the other force at a given stage of development of PNs.

Starting from these conditions one can complete and modify the *dynamical* classification system in the following way (Gurzadyan, 1969).

First of all we should retain the class of *planetoid* nebulae, i.e. those which consist of only *one envelope* with a brightness distribution more or less uniform over the disk. We shall denote this class by the symbol I.

By class II we shall denote the *double-envelope* nebulae, regardless of their relative sizes or other peculiarities. These nebulae seem to be the evolutionary relation of the planetoid nebulae (Chap. 13).

Class III denotes the *ring-like* nebula with no evidence of a second envelope, but with an outlined ring shape.

Symbols I, II and III fully characterize certain nebular shapes and structures, and together they acquire an evolutionary meaning: the development of PNs in the direction I $\rightarrow$ II $\rightarrow$ III, i.e. planetoid $\rightarrow$ double-envelope $\rightarrow$ ring-like.

The symbols I, II and III also characterize the sizes of PNs, as visible and absolute: class I nebulae are the smallest, class III the largest.

Another element that should be included in the classification of forms of PN is the *bipolarity* of their structure. Bipolarity can be present in all three types of PN mentioned above. On the other hand, there exist different degrees of development of bipolarity, owing to differences presumably in the influence of the magnetic and inertial forces. Therefore, the class of bipolar nebulae are denoted by the letter B, to which we add the index a, b, c, d, indicating the degree of development of bipolarity. Thus, in PNs of type Ba, bipolarity shows only as an increase in the brightness at the edges of their minor axes (NGC 6720); in type Bb, two "caps" are well developed (NGC 3587); in type Bc, the "caps" are quite clearly separated (A 66). The extreme case of bipolarity, and also of flatness, is the "rectangular" or "hour-glass" PN denoted as Bd (IC 4406, M 2-9, NGC 2346).

By Sp we shall denote the rather exotic class of spiral PNs where we observe two symmetrical spiral-like arms, the result of the joint action of the intrinsic magnetic field of PN and the regular magnetic field of the Galaxy.

Finally, D stands for transition objects, i.e. from planetary to diffuse nebulae, the latter corresponding to the later state in the development of a PN.

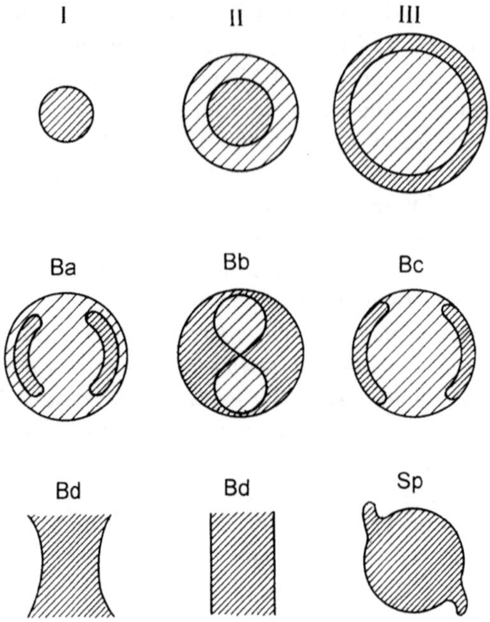

Fig. 1.3. Schematic representation of the shapes of planetary nebulae of various classes

Concerning the flatness, i.e. a deviation from the sphericity in the shape of PNs, no special symbol is introduced to designate it, since *if a nebula is bipolar, then it is also flattened.*

We shall maintain the class of stellar (S) PNs. Here we merge together all PNs with *apparent* dimensions small enough that they cannot be distinguished from stars by their appearance only.

In Fig. 1.3 the different types of PN are depicted schematically, with their designations.

Thus, according to the proposed classification all PNs can be distributed among the following basic types (a typical example of each type is given in parentheses):

I – Planetoid (NGC 6 803)
II – Double envelope (NGC 1 535, NGC 7 662)
III – Ring–like (A 39, A 72)
Ba – Bipolar of first type (NGC 6 720, NGC 7 293)
Bb – Bipolar of second type (NGC 3 587, A 70)
Bc – Bipolar of third type (A 19, A 66)
Bd – "Rectangular" (IC 4 406, NGC 2 346)
Sp – Spiral (NGC 4 361, NGC 6 210)
D – Diffuse (A 16, A 50)
S – Stellar (IC 4 997)

It is important to notice that in the majority of cases one and the same object can belong to several types at the same time. In such cases one must use a combined system of classification. For example, NGC 4361 is spiral (Sp), has a second envelope (II) and also shows an increase in brightness at the edges of the minor axis (Ba). Therefore, this PN belongs to the class Sp II Ba. The symbols are ordered according to the degree in which each aspect shows itself – the most important and strongest feature is in the first place, etc.

More than 150 PNs have been classified according to the above classification system on the basis of their images obtained mainly with the application of CCD techniques (Jewitt et al., 1986; Chu et al., 1987). The results are presented at the end of Chap. 18 in the Catalogue of Shapes of Planetary Nebulae.

1.9 Expansion of Nebulae

The very existence of PNs with shapes strongly differing from one another – from point or stellar-like up to the ring-like of giant sizes – prompts the fact of their expansion. However, direct confirmation of this phenomenon obviously might be the increase of the angular sizes of some PNs established by direct measurements of positions of the edges of the PNs obtained in different epochs. The well known PN in Lyra, NGC 6720, was the first object for which this effect – the secular increasing of angular sizes – in the amount of $0''.9 \pm 0''.1$ in 100 years seems to have been established. The effect of the increase of angular sizes has been established for a dozen PNs (NGC 246, 3242, 6572, 6853, 7009, 7293 etc.) (e.g. Weinberger, 1989).

The first investigations of the internal motions and expansions of PNs were carried out in the early period of their study by the methods of astrospectroscopy. However, the appearance of the most reliable results comes from the period of the first powerful telescopes equipped with slit spectrographs of high spectral resolution at their coudé focuses.

The basic results on the internal kinematics of PNs are as follows. PNs are expanding gaseous envelopes whose expansion velocities are relatively small, of the order of 30 km s^{-1}. This is shown by the shape of their emission lines; they appear split into two components of approximately the same intensity, located symmetrically with respect to the central frequency (Fig. 1.4). The splitting, i.e. the distance between the components, is largest in the centre of the image of the PN and decreases with increasing distance from the centre. At the edges of the PN the two components merge into one. The general outward shape of the spectral line resembles an ellipse whose major axis corresponds to the radius of the PN ($D''/2$) and the minor axis to the expansion velocity: $V = (\Delta\lambda/\lambda)c$. The violet component of the spectral line, marked f in Fig. 1.4(b), obviously corresponds to the radiation coming from the region of the nebula closer to us (front), and the red component r to that from the farther side of the PN. It must be noticed that this method for the determi-

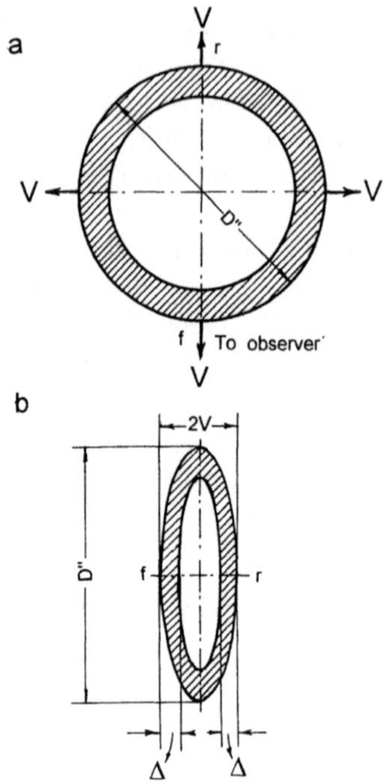

Fig. 1.4. (**a**) Planetary nebula expanding with a velocity V. (**b**) The shape of its emission line

nation of the expansion velocity of the PN does not require knowledge of its radial velocity of motion with respect to the observer.

One should add that the spectral lines would have exactly the same structure and the same behavior if the PNs were contracting. However, this possibility can be rejected without any difficulty.

The fact that both components of the spectral lines have almost the same brightness shows that the radiation from the farther side of the PN can propagate through the whole nebula without any absorption. From this fact one can conclude that PNs are transparent at the frequencies of visual radiation; as we shall see later (Chaps. 2 and 3), this conclusion is in complete accordance with the theory of generation of emission lines in PNs.

With the exception of the hydrogen and, to some extent, helium lines, the widths of the components of the spectral lines Δ (Fig. 1.4) which belong to other elements are almost identical and correspond to a velocity of 10 km s^{-1}. Since the expansion velocity is two or three times greater and almost identical for all PNs, it can be assumed that the *turbulent* velocities in PNs are small compared to the expansion velocities (we shall return to this question in Chaps. 16–17).

The last conclusion is of special interest. It happens that in several PNs the brightness distribution over the disk is far from being uniform. In different cases the fluctuations in brightness distribution reach very large values. To the intensity fluctuations there should correspond, although to a smaller degree, fluctuations in electron (ion) concentration. If we assume that the fluctuations in the distribution of electron concentration are tied to turbulent motions – an assumption which is, generally speaking, quite attractive – then we would expect the presence of fluctuations in the velocities of the internal motions. However, as was pointed out before, this is not observed. Therefore one has to look for another reason to explain the origin of these fluctuations in electron concentration (Chap. 17).

For the determination of the expansion velocities the emission lines HI, HeII, [SII], [OII], [ArIV], [NeIII], [NeV] and more often [OIII] are used. The number of PNs with known magnitudes of expansion velocities is of the order of 300 (up to 1989, Weinberger, 1989). Among them there are objects with almost zero expansion velocity (IC 418, A 35) as well as with a sufficiently high velocity – of the order of $2V_0 = 70$–$80\ \mathrm{km\ s^{-1}}$. An expansion velocity higher than $2V_0 = 100\ \mathrm{km\ s^{-1}}$ has been established for NGC 2 392.

The analysis of the observational data results in a number of interesting and unpredicted discoveries. One of them is related to the gradient of the expansion velocity within the PN, the essence of which is as follows: the expansion velocity in average decreases with the increase of the ionization potential of the ion or atom exciting the given emission line. For three objects this dependence of $2V_0$ ($\mathrm{km\ s^{-1}}$) on the ionization potential (eV) is as follows:

		NGC 6 818	NGC 7 027	NGC 7 662
[OII],	13.6 eV	60.2	47.2	58.0
[OIII],	35.0 eV	56.2	40.9	52.7
[NeV],	96.0 eV	32.6	38.2	38.6

On the basis of the correlation between the size of the monochromatic image of the PN and the ionization potential of an emission line one reaches the conclusion that there exists a gradient in the expansion velocity within the PN, i.e. the outer layers expand faster than the inner ones. Obviously, some force should be responsible for such a difference in expansion velocities.

1.10 The Apparent Distribution of Planetary Nebulae

The apparent distribution of nearly 400 PNs of all types and all sizes with Galactic coordinates is shown in Fig. 1.5. Their concentration towards the Galactic plane is obvious, although not as marked as in the case of the hot giants of classes O–B. The Galactic latitude for isolated PNs reaches as high as 60° or even higher. The strong concentration of PNs in the direction of centre of the Galaxy ($l \sim 325°$) is also quite visible. More than 50 PNs, i.e. nearly one seventh of those PNs, are concentrated in the region of the sky between $l = 320$–$340°$ and $b = 0$ and $b = -10°$.

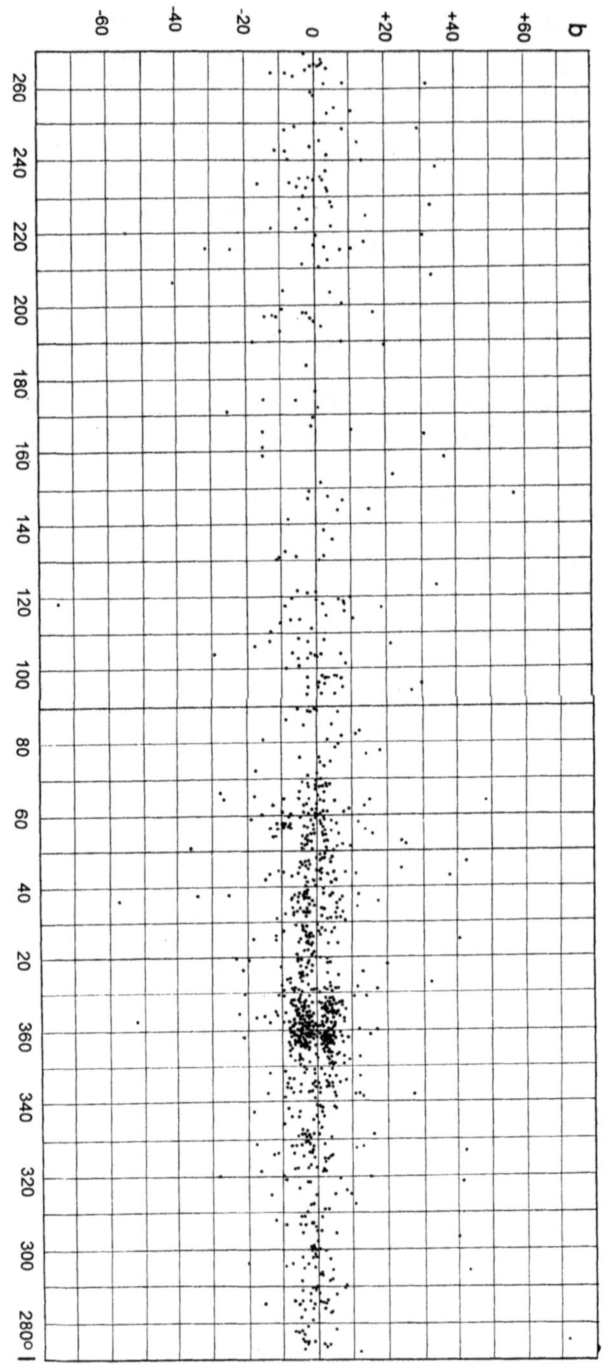

Fig. 1.5. Visible distribution of Galactic planetary nebulae in the Galactic coordinate system (l, b)

PNs with large apparent dimensions are distributed more or less uniformly over the sky, while small and stellar-like ones are concentrated near the Galactic plane. The true Galactic concentration of PNs must obviously be greater that observed since, because of the interstellar absorption of light, weak PNs cannot be discovered near the plane of the Galaxy, while they can be discovered easily at high Galactic latitudes.

The apparent distribution and distances of PNs show that they form a system in the Galaxy which is intermediate between the flat and the spherical subsystems (see Sect. 18.4; e.g. Stasinska et al., 1991).

The sharpness and narrowness of the emission lines in the spectra of PNs permit the measurements of their radial velocities with high accuracy. For 120 PNs such measurements had already been realized in the early days of the spectroscopic investigations of PNs. The radial velocities were quite high. Among these PN eight had a velocity greater than 100 km s^{-1}. The maximum positive velocity has been fixed for NGC 6 644 (+193 km s^{-1}) and the maximum negative velocity for IC 4 732 (−145 km s^{-1}).

1.11 Planetary Nebulae in Other Galaxies

The search and observations of PNs in other galaxies can provide valuable information first of all from the cosmogonical point of view, e.g. regarding the correlation of PNs and the type of galaxy. In addition we can also obtain some idea of the upper limit on absolute brightness of PNs from apparent brightness and the distance to the given galaxy. We mention the upper limit of brightness, since the weak PNs will be out of the limit of their discovery, and hence observational selection effects should be inevitable. No less important are the problems of the chemical composition of PNs in other galaxies.

PNs have been discovered in over fifty galaxies of various types, including both the Large and Small Magellanic Clouds (LMC and SMC), a number of galaxies of the Local Group, such as M 31, M 32, M 81, NGC 6 822, in the Fornax Cluster, the Leo I Group, the Virgo Cluster, in the NGC 5 128 and NGC 1 023 Groups, etc. The number of discovered PNs in each of these galaxies is of the order of a few tens or more. The number of PNs in some galaxies is as follows:

LMC	137	Ford et al., 1989; Barlow, 1989; Morgan, 1994
SMC	74	Meyssonnier et al., 1993; Barlow, 1989
M 32	64	Ford et al., 1989
NGC 3 031	185	Ford et al., 1989
NGC 5 128	> 400	Hui et al., 1993
NGC 4 594	> 300	Hui et al., 1993
NGC 224	104	Jacoby, Ciardullo, 1993
NGC 3 379	30	Ciardullo et al., 1993

The largest number of PNs, 1200 objects, has been discovered in Andromeda (M 31), which shows emission in $N_1 + N_2$ plus H_β lines most of which have no continuum and are believed to be PNs (Meyssonnier et al., 1993). Recently 34 PNs have been discovered in the central part of a late-type spiral galaxy, NGC 300, using CCD and the on-band/off-band filter technique (Soffner et al., 1996).

The total number of PNs discovered up to 1993 in various types of galaxies amounts over two thousand.

The estimated number of PN based on stellar evolution models is different in various galaxies (Peimbert, 1993):

NGC 185	70	NGC 3 377	5 600
SMC	120	M 31	8 000
M 32	140	M 81	9 400
NGC 205	260	NGC 4 486	11 500
LMC	640	NGC 4 374	14 000

In the same manner, the total Galactic PN population is expected to be 9 000 ± 3 000.

The flux, 1.34×10^{-8} ergs $\mathrm{cm}^{-2}\ \mathrm{s}^{-1}$ in the line 5 007 [OIII], for a maximally bright PN at a distance 1 kpc, can be used as a distance indicator when sufficient numbers of PNs will be identified in any given external galaxy at distances exceeding 10 Mpc (Jacoby et al., 1989). A luminosity function $N(M) \sim 10^{1.6M}$ has been proposed to determine the probability of *observed* objects to be brighter than the cutoff magnitude M_* (Bottinelli et al., 1991; Ciardullo et al.,1993). Jacoby et al. (1990) have argued that M_* is independent of the Hubble type. The application of this method leads, for example, to the Virgo Cluster distance ~ 15 Mpc. The galactic distances derived by means of the PN method can reproduce the high accuracy distances from Cepheids, within ~ 10% (Tammann, 1993). The use of 5 007 [OIII] and the PN luminosity function as standard candles to derive distances to galaxies seems to be efficient. These distances, obviously, have important consequences for the estimation of the Hubble Constant and the age of the Universe (Jacoby, Ciardullo, 1993; McMillan et al., 1993).

It is interesting that the distances to remote galaxies can be determined much more accurately, within ~ 5%, using PNs, while distances to closer Galactic PNs usually cannot be measured better than by a factor 2–3, as we shall see in Chap. 7.

The images of a number of PNs in the Small Magellanic Cloud have been obtained with the help of the *Hubble Space Telescope* in the line 5 007 [OIII] (Dopita et al., 1993). As a result, the accurate diameters of these barely resolved PNs have been determined; they were within the limits 2.4×10^{16}–7.0×10^{17} cm. Using the Faint Object Camera on-board the *Hubble Space Telescope*, the highly resolved structural images of three PNs in the SMC, P2, P5 and P85, and of two in the LMC, N66 and N201, have been obtained as well (Blades et al., 1992; Dopita et al., 1984).

The upper limit of the absolute photographic magnitude was established for the PN in SMC: $M_{\mathrm{pg}} = -5^{\mathrm{m}}.5$. In our Galaxy no PN with $M_{\mathrm{pg}} \sim -5^{\mathrm{m}}$

has been discovered, though that does not mean that such powerful objects are absent in general, merely that our knowledge is far from being reliable concerning the distances to Galactic PNs.

The durations of the lives of PNs of extreme luminosities, the largest and the smallest, differ from one another approximately 10 times. Therefore, the expected number of PN must be larger than the above data, namely, nearly 1 400 in LMC and ~ 500 in SMC.

The PNs in our Galaxy are not associated with open stellar clusters and stellar associations; in all cases the PNs in these stellar systems have not yet been discovered. Can one therefore assume that the PNs are only a stage of stellar evolution *far* from the early stage of the birth and formation of stars? In Chap. 18 a positive answer will be given to this question.

As for globular clusters, only in one of them, M 15, has a relatively bright ($13^{m}.8$) PN been discovered (as early as 1928) the edge of the cluster. A special search was carried out later in relation to 43 globular clusters with the help of a slitless spectrograph in combination with an H_α light filter, although this resulted in the discovery of PNs in only one cluster, again in M 15 (Peterson, 1976).

Two PNs in one globular cluster! Of course, this is a very important fact, especially from the cosmogonical point of view, since the globular clusters being the oldest population of the spherical subsystem of the Galaxy are closely related to its evolution.

The globular cluster M 22 is the second in which a PN has been discovered (Borkowski et al., 1993a; Harrington, Paltoglou, 1993).

1.12 Designations, Catalogues, Surveys

PN are designated by their number in Dreyer's New General Catalogue of Clusters and Nebulae – abbreviated NGC – first appearing in 1887. Later on, it was completed by two supplements, in 1894 and 1908: the First Index Catalogue and Second Index Catalogue. They are designated IC I and IC II. If a PN appears in the second supplement, it is denoted, for example, NGC II 2 022, or II 2 022; if in the first, IC I 351 or I 351.

In the period of the NGC catalogues it was impossible to distinguish PNs from galaxies or clusters, because the NGC numbers included all objects – galaxies, clusters of all types as well as diffuse and planetary nebulae.

Beginning in the 1930s and 40s activity in the discovery of PNs essentially increased. Thus, each author presented his discoveries under his own numeration scheme. Especially successful were Minkowski, Abell, Haro, Henize, Kohoutek, and also Hubble, Blanco, Menzel, Perek, Mayall, Merill, Sharpless, The, and others. Some of these authors published more than one list. A spontaneous system of notation based on a combination of the first letter of the author's name with the number of the object according to his list, e.g. M 1-60 or M 3-28 were PNs No. 60 and 28 from the first and third of Minkowski's lists, and these were H 1-4, He 2-186, K 3-45 etc.

Moreover, one and the same PN was often "discovered" independently by two or even three authors. The appearance of confusion seems to be real. The appearance in 1967 of the *Catalogue of Galactic Planetary Nebulae*, by Perek and Kohoutek (1967), put an end to this story. In this fundamental, and in fact first, catalogue of PNs (only !) a new system of designation was introduced. Though not quite appropriate for practical use, but nevertheless quite logical in essence, i.e. according to the Galactic longitude and latitude: e.g. $5{-}6^\circ.1$ (NGC 6 620), $319 + 6^\circ.1$ (He 2-112) etc. This system permits the possibility of including in the catalogue each PN to be newly discovered without disturbing the integrity of the catalogue itself. This catalogue contains an enormous amount of information for all 1 034 known PNs – coordinates and angular sizes, stellar magnitudes of the nebula and its nucleus, spectral data, data for radial velocities, proper motions, references to the source, as well as photographic maps with identification of the nebula.

There are also many other catalogues, surveys, extensive investigations etc., concerning certain aspects of PNs: "Enumeration of Nebulae and Description of their Main Properties" (Acker et al., 1981), "Catalogue of the Expansion Velocities of Planetary Nebulae" (Weinberger, 1989), "Catalogue of Distances of Planetary Nebulae" (O'Dell, 1962; Cahn, Kaler, 1971; Acker, 1978; Daub, 1982; Milne, Aller, 1982; Maciel, 1984), "Analysis of Radial Velocities" (Schneider et al., 1983), "Internal Motions in Planetary Nebulae" (Sabbadin et al., 1986), "Catalogue of Central Stars of Planetary Nebulae with the Information of their Main Properties" (Acker et al., 1982), "Catalogue of Radio Observations of Planetary Nebulae" (Higgs, 1971), etc.

The most complete (1 820 objects) fundamental work entitled the *Strasburg–ESO Catalogue of Galactic Planetary Nebulae* was composed by a group of authors headed by Acker (Acker et al., 1992). This two-volume catalogue is especially valuable since besides identification maps and detailed data on the PNs and their nuclei, it contains a complete bibliography on the papers, reports and communications for each object.

2. The Origin of Emission Lines

2.1 The Nature of Nebular Emission

The spectra of the purely gaseous nebulae are completely different from the spectra of their central stars. The discrepancy is expressed in the structure of the spectra itself as well as in the intensities in certain of its parts. In the visual band the spectrum of the nebula consists of a number of very intense emission lines with a hardly noticeable continuum. The spectrum of the central star-nucleus, on the other hand, consists of a powerful continuous background with hardly noticeable absorption lines or no absorption lines at all. The energy emitted by a nebula in different emission lines considerably exceeds the energy of the corresponding regions in the spectrum of the star-nucleus. Nevertheless, the luminosity of a nebula is entirely due to the radiation of its central star-nucleus.

However, the integrated brightness of a nebula, without its nucleus, is from ten to many hundreds of times greater than the brightness of the central star. Therefore, we have to accept that the nebulae are kept luminous as a result of the energy radiated by the nuclei in invisible regions of the spectrum. Since the nuclei of PNs are high temperature stars, and since most of their energy is emitted in the far ultraviolet region of the spectrum, one can assume that the nebulae derive their energy from this ultraviolet radiation of the central star. The problem is to elucidate how the nebula "transforms" this energy into visible light.

Before answering this question let us see what properties the radiation will possess after travelling a distance many times greater than the radius of the star.

Assume that the nucleus emits as a blackbody of a temperature T_*. The radiation density ρ_ν^* per unit frequency interval in thermodynamic equilibrium temperature T_* is given by Planck's law:

$$\rho_\nu^* = \frac{8\pi h\nu^3}{c^3} \frac{1}{e^{h\nu/kT_*} - 1} . \tag{1}$$

At the surface of the star the radiation density should be half of the above value, since radiation occurs from only one half of the sphere. As the distance to the star increases, this quantity decreases, and at a distance r the radiation density ρ_ν will be given by the equation

$$\rho_\nu = W\rho_\nu^* , \tag{2}$$

where W is the so-called *dilution coefficient.* The physical meaning of W is obvious from (2): it represents the ratio of the radiation density at the point of observation to the radiation density at the surface layers of the star.

Strictly speaking, W depends also on the opacity of the medium, and hence on the frequency ν, and has the following form:

$$W_\nu = \frac{1}{2}\left[1 - \sqrt{1 - \left(\frac{r_*}{r}\right)^2}\,\right] e^{-\tau_\nu} , \tag{3}$$

where r_* is the radius of the star and τ_ν is the optical depth at the frequency ν. In general, at the inner edge of the nebula one takes $\tau_\nu \ll 1$ and then

$$W = \frac{1}{2}\left[1 - \sqrt{1 - \left(\frac{r_*}{r}\right)^2}\,\right] . \tag{4}$$

In this case the dilution coefficient takes on a geometrical meaning: it represents the ratio of the solid angle subtended by the central star at the point of observation to 4π. Then, when $r_*/r \ll 1$, we have, instead of (4),

$$W = \frac{1}{4}\left(\frac{r_*}{r}\right)^2 . \tag{5}$$

In PN the ratio r_*/r is of the order of 10^{-6}–10^{-7}. Therefore, the radiation density given by (2) shows the weakening of a factor of 10^{13} compared with the radiation density at the surface of the star. In spite of this, the composition of the spectrum is not changed, i.e., as was pointed out before, the emission maximum falls in the far ultraviolet region of the spectrum. This would not occur if the radiation in the nebula were in equilibrium. Actually, if we denote by T_1 the temperature that would be obtained in the nebula in the presence of thermodynamic equilibrium, the integrated radiation density can be written as $\rho = \alpha T_1^4$, where α is Stefan's constant. On the other hand, for the integrated radiation density, we have from (2) that $\rho = W\alpha T_*^4$. Equating both these values, we obtain $T_1 = T_* W^{1/4}$, which leads to a value of T_1 of the order of a few tens of degrees by taking T_* to be of the order of tens of thousands of degrees. This means that the maximum of the energy distribution over the spectrum would be displaced to the very far infrared band, which, as we have seen, does not occur.

Thus, we come to the first, principal conclusion: *PNs are objects in which the condition of thermodynamic equilibrium is not maintained.* Hence, the processes of radiation and absorption of the light photons into the nebula are not determined by thermodynamic equilibrium. In other words, the interaction between the radiation and matter must lead to a redistribution in frequency of the radiation in the direction of a more probable distribution, i.e. to a redistribution of energy over the spectrum that strengthens the long wavelength region and weakens the short wavelength region.

Quantitatively this process of redistribution of radiation within the nebula can be described by means of Rosseland's theorem.

2.2 Rosseland's Theorem

The essence of Rosseland's theorem is as follows: under nebular conditions, i.e. in the presence of dilute radiation, short wavelength radiation energy emitted by the star is transformed into long-wavelength radiation energy.

Consider an atom having three energy levels 1, 2 and 3 with energies $\varepsilon_1 < \varepsilon_2 < \varepsilon_3$. As a result of the absorption of a quantum of radiation, we can have both simple transitions related to resonant scattering of the type $1 \to 2 \to 1$ as well as more complex cyclic processes. The most important among them will be the mutually opposite processes

$$1 \to 2 \to 3 \to 1 \qquad \text{and} \qquad 1 \to 3 \to 2 \to 1\,.$$

In the first of these processes a sort of synthesis occurs when as a result of the absorption of two quanta of lower frequency ν_{12} and ν_{23} a quantum of higher frequency ν_{13} is emitted. In the second case, on the other hand, there occurs a process of subdivision of a quantum when, as a result of the absorption of one high frequency quantum ν_{13}, two quanta of lower frequency, ν_{12} and ν_{23}, are emitted. The problem consists in establishing which of these two cyclic transitions dominates under nebular conditions, i.e. at the strongly diluted radiation. To solve this problem let us estimate the ratio of the number of transitions $1 \to 2 \to 3 \to 1$ to the number of transitions $1 \to 3 \to 2 \to 1$ per unit time.

From elementary considerations one can derive the ratio of the numbers of transitions by these two cycles:

$$\frac{N_{1\to2\to3\to1}}{N_{1\to3\to2\to1}} = \frac{B_{12}\rho_{12}B_{23}\rho_{23}(A_{31} + B_{31}\rho_{13})}{B_{13}\rho_{13}(A_{32} + B_{32}\rho_{23})(A_{21} + B_{21}\rho_{12})}\,. \tag{6}$$

In order to simplify this expression one can use Einstein's relations:

$$A_{ki} = B_{ik}\frac{g_i}{g_k}\sigma_{ik}\,, \qquad B_{ki} = \frac{g_i}{g_k}B_{ik}\,, \qquad \sigma_{ik} = \frac{8\pi h\nu_{ik}^3}{c^3}\,, \tag{7}$$

where g_i and g_k are the statistical weights of the levels. In addition we put

$$\bar{\rho}_{ik} = \left(e^{h\nu_{ik}/kT_*} - 1\right)^{-1}, \tag{8}$$

and using (2) we can write

$$\rho_{ik} = W\sigma_{ik}\bar{\rho}_{ik}\,. \tag{9}$$

By means of the above definitions we can write instead of (6):

$$\frac{N_{1\to2\to3\to1}}{N_{1\to3\to2\to1}} = W\frac{\bar{\rho}_{12}\bar{\rho}_{23}(1+W\bar{\rho}_{13})}{\bar{\rho}_{13}(1+W\bar{\rho}_{12})(1+W\bar{\rho}_{23})}\,. \tag{10}$$

In the photospheres of the stars we have $W = 1$ and the ratio (10) is almost exactly equal to 1. In the gaseous envelopes of some stars $W < 1$, and therefore in those cases (10) will be smaller than unity. Finally, in nebulae where $W \sim 10^{-13}$, i.e. $W \ll 1$, and the factor $\bar{\rho}_{12}\bar{\rho}_{23}/\bar{\rho}_{13}$ is of the order of unity, we have

$$\frac{N_{1\to2\to3\to1}}{N_{1\to3\to2\to1}} \approx W\,, \tag{11}$$

i.e. the number of transitions of the type $1 \to 2 \to 3 \to 1$ is extremely small in comparison with the number of transitions of the type $1 \to 3 \to 2 \to 1$. This is Rosseland's theorem.

Thus, according to Rosseland's theorem, a nebula, which absorbs high frequency (ultraviolet) quanta emitted by the central star, transforms them into quanta of lower frequencies, and among them quanta of visible light. Therefore, since the number of ultraviolet quanta emitted by the central star is very high in comparison with the number of visual quanta, the total number of visual quanta emitted by the nebula will also be very high. As a result, the visual brightness of the nebula will considerably exceed the visual brightness of the central star.

Thus, one of the mysteries concerning the nature of the nebula – that the nebulae are brighter than their central stars – is explained. At the same time, Rosseland's theorem predicts the emission nature for PN spectra.

2.3 Fluorescence: Zanstra's Theory

The process of transformation of shortwave into longwave radiation, i.e. when the high frequency quanta are transformed into low frequency quanta, is nothing other than *fluorescence.*

Let us consider a hydrogen nebula that surrounds the nucleus, the central star. In spite of the small density of ionizing radiation – it is $1/W \sim 10^{13}$ times smaller than at the surface of the star – the degree of ionization in the nebula, as we shall see later, will be very high – of the order of one thousand (i.e. $n^+/n_1 \approx 1\,000$, where n^+ and n_1 are, respectively, the ion and neutral atom concentrations). This is explained by the small matter density in the nebula, i.e. the small electron concentration (the degree of ionization is inversely proportional to the electron concentration). Then it follows that almost all hydrogen atoms will be ionized. As far as the small number of neutral atoms is concerned, practically all of them will be in the ground state; excited atoms will be almost absent since the density of the radiation necessary for excitation is too small.

Thus, the atoms in the nebula will be practically in only two states: ground and ionized. If the mass of the nebula is sufficiently large it will be opaque at the frequencies absorbed by the ground state hydrogen (beyond the Lyman limit). It can be shown that this condition will be fulfilled for a density as low as 1 atom cm^{-3} if the radius of the nebula is of the order of 10 000 a.u. In this case the optical depth of the nebula τ_c at frequencies beyond the Lyman limit will be of the order of one. At the same time, the nebula will be completely transparent in the subordinate lines of the hydrogen – Balmer, Paschen, etc.

On the other hand, the ratio of the absorption coefficient of the neutral hydrogen atom in the first lines of the Lyman series – L_α, L_β and so forth – to the absorption coefficient of the same atom beyond the series limit is of the order of 10^4 up to 10^5. Therefore, the optical depth in these lines, under the assumed conditions ($\tau_c \approx 1$), is of the order of 10^4 to 10^5, i.e. very high. Let us see what will happen to the ultraviolet quanta absorbed by the nebula under these conditions. In what follows, by ultraviolet or L_c quanta we shall mean quanta whose frequencies are higher than that of the Lyman series limit.

An L_c quantum that reaches the nebula sooner or later will be absorbed by a hydrogen atom, as a result of which the atom will be ionized. This process is called *photoionization*. After a certain time the electron will be captured by a photon: in this case we have a process called *recombination*. Two possibilities exist: (1) the electron falls directly to the first (ground) level, and (2) the electron falls into one of the higher levels. In the first case an L_c quantum is emitted and everything returns to the beginning. In the second case the electron according to Rosseland's theorem has to make a series of cascade transitions or jumps, thus emitting quanta in the subordinate series which escape from the nebula since the nebula is completely transparent at the frequencies of these series. In the nebula the radiation is so diluted, and the material density is so low, that this chain of cascade transitions goes uninterrupted in the vast majority of cases. The last link in the cascade is the transition to the first level, accompanied by the emission of a quantum in the Lyman series.

Two cases are possible. Assume that the electron finds itself in the second level. This obviously must be proceeded by a transition of the atom from some higher state (discrete or continuous) to the second and hence by the emission of one, and only one, quantum in the Balmer series or Balmer continuum, which escapes from the nebula without any obstacle. Afterwards a $2 \rightarrow 1$ transition takes place with the emission of one L_α quantum. However, according to our assumptions, the optical depth of the nebula at the frequencies of the Lyman lines and, among them, of L_α is very high. Therefore, this quantum, after travelling a short distance inside the nebula, will be absorbed by a neutral hydrogen atom in the ground state. The atom is excited to the second state and again, owing to the lack of collisions, it will make a spontaneous transition to the ground state after a very short time (of the order of 10^{-8} sec), emitting a L_α quantum. Thus, under nebular conditions the L_α

quanta cannot be modified further: they will undergo many scatterings until finally they reach the outer boundary of the nebula and escape.

Assume now that the electron has been captured to the third state. This capture will be accompanied by the emission of a quantum beyond the Paschen limit, which then escapes from the nebula. Now the electron has two possibilities: it can move either directly to the first level with the emission of an L_β quantum or first to the second level and then to the first one with the emission of two quanta, H_α and L_α. In the first case the emitted L_β quanta will be absorbed again owing to the large optical depth of the nebula in the frequencies of the Lyman lines, and later on, it will again excite an atom to the third state. This process will continue until the second possibility is realized. Here the H_α quantum leaves the nebula and the L_α quantum remains and is scattered many times. The same reasoning can be applied to the cases where, instead of an L_α quantum, one has L_γ quanta, L_δ quantum and so forth.

Thus, one L_α quantum and one Balmer quantum result from each L_c quantum absorbed in the nebula. If the optical depth τ_c of the nebula is of the order of unity or higher, it will absorb all the L_c quanta emitted by the star per unit time. In that case the total number of Balmer quanta emitted by the nebula should be equal to the number of L_α quanta.

This is how the Balmer quanta emitted by the nebulae originate. The fluorescence mechanism just described was first proposed by Zanstra (1927). This mechanism can be extended to explain the luminosity of the PNs, not only in the hydrogen lines, but in the lines of other elements – helium, oxygen, nitrogen and so forth, as well.

2.4 Excitation of Forbidden Lines

The excitation of many emission lines, including the most intense ones, cannot be attributed to the fluorescence mechanism. Bowen (1927) was the first to show that in the nebulae there exists still another mechanism of excitation of emission lines quite different from fluorescence.

Bowen started by identifying the important nebular lines N_1 and N_2 (5 007 Å and 4 959 Å respectively). It turned out that they are due to forbidden transitions in double ionized oxygen. Another pair of nebular lines – 3 726 Å and 3 729 Å – is also due to forbidden transitions, but of singly ionized oxygen. In the same way, the lines 6 548 Å and 6 584 Å correspond to forbidden transitions in singly ionized nitrogen, and so forth. A schematic representation of these and other forbidden transitions is given in Fig. 2.1.

Forbidden lines are not observed under normal laboratory conditions. Therefore we must accept that in the nebulae there exist special conditions favouring the formation of such lines.

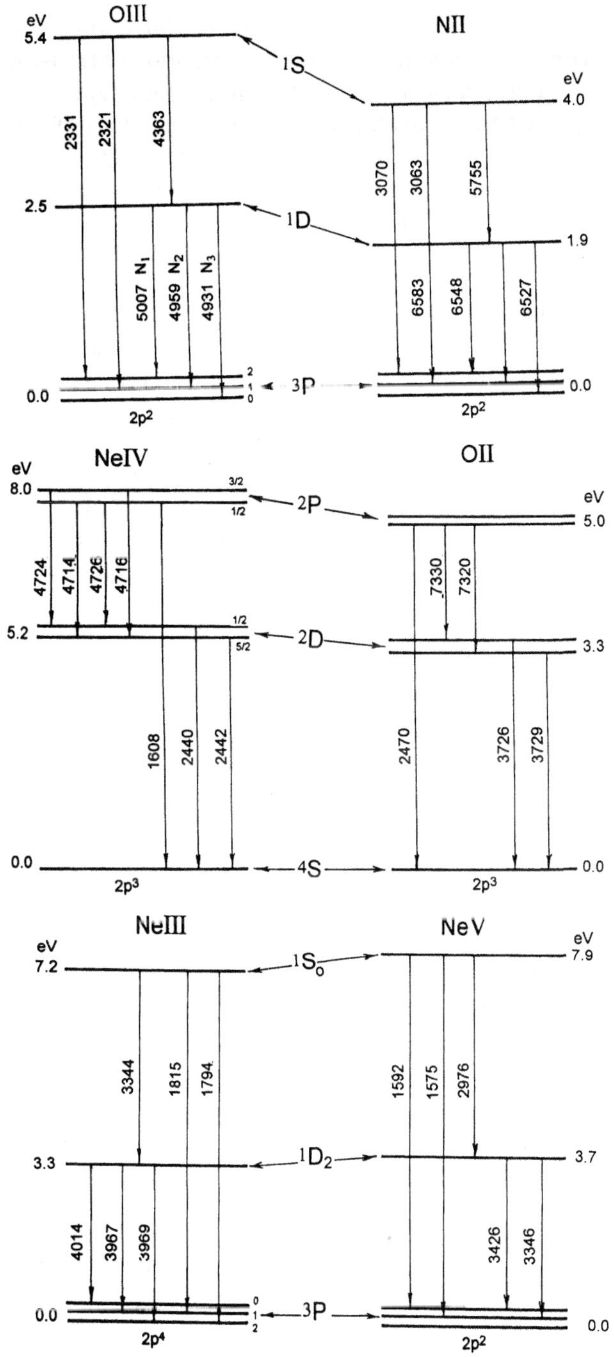

Fig. 2.1. Scheme of transitions for some forbidden lines of OIII and NII (configuration $2p^2$), OII and NeIV ($2p^3$), NeIII and NeV ($2p^4$ and $2p^2$)

First, recall that the usual "selection rules" for atomic transitions are related with certain approximations in the calculation of transition probabilities. In a first approximation, when only dipole radiation is taken into account, one finds that the probabilities of certain transitions should be equal to zero. Actually, with a more exact treatment, i.e. when quadrupole and other radiations are included, as well as the effects connected with the magnetic moment of both the electron and the nucleus, and so forth, one finds that the probabilities of those transitions are different from zero, although their values are a million or even a billion times smaller than the probabilities of allowed transitions.

Table 2.1 shows the Einstein probability coefficients for spontaneous transitions A_{ij} for a many forbidden lines, excited by ions OIII, NII, SIII, NeV (configuration $2p^2$), OII, NeIV, SiII ($2p^3$), NeIII, ArIII ($2p^4$), collected from different sources (see summary by Osterbrock, 1974). As we see, the values of A_{ij} for forbidden transitions are extremely small – less than unity – and for most interesting forbidden lines are of the order of 10^{-2}–10^{-4} s^{-1}. For the usual transitions these coefficients are of the order of 10^7–10^8 s^{-1}.

However, the fact that the magnitude of the transition probability is different from zero is still not a sufficient condition for the production of an observable line. The *intensity* of the forbidden line is also important. If it is not large enough, the forbidden line will still be unobservable. Evidently, a necessary condition for observation is the requirement that the intensity of the forbidden line be comparable to the intensity of the usual lines.

In connection with this one can note that if the forbidden lines are emitted by a transition whose upper level can also give a permitted line, then the intensity of the forbidden line can never, under any conditions, be comparable with the intensity of the permitted line since for the given upper level the permitted transitions responsible for the permitted line will occur a million times faster than the forbidden transitions. Thus one has the first condition:

> *For the emission of a forbidden line of observable intensity the upper level responsible for the forbidden transition has to be metastable.*

Even if they satisfy this condition, not all forbidden lines will be obscured. If several forbidden transitions are possible from a given metastable state then, obviously, those forbidden more strongly cannot give a sufficiently strong line. We can illustrate this point with the example of the OIII ion, for which the $^1D_2$ level is metastable. Below this level we have three levels which are very close to each other: $^3P_2$, $^3P_1$ and $^3P_0$. Therefore, in addition to the N_1 and N_2 lines which arise from transitions from the level to the levels $^3P_1$ and $^3P_1$, one could expect the emission of still another line, N_3, with wavelength 4932 Å originating from the transition $^1D_2 - {}^3P_0$ (see Fig. 2.1). However, as follows from the data in Table 2.1, the probability of this transition is ten thousand times smaller than the probability of the $^1D_2 - {}^3P_2$, which gives the line N_1. This means that N_3 line will be ten thousand times less intense than N_1, i.e. it will be observed only with great difficulty and in special conditions.

Table 2.1. Transition probabilities A_{ij} (s^{-1}), "collision strengths" $\Omega(ij)$ and excitation energies ε_{ij} (eV) for ions OIII, NII, NeV, SIII (configuration $2p^2$), OII, NeIV, SII ($2p^3$) and NeIII, ArIII ($2p^4$)

Ion	Transition	λ, Å	A_{ij}	$\Omega(ij)$	ε_{ij}
[OIII]	$^1D_2$–$^1S_0$	4363	1.6	0.58	2.84
	$^3P_1$–$^1S_0$	2321	2.3×10^{-1}	0.30	5.37
	$^3P_2$–$^1D_2$	5007 N_1	2.1×10^{-2}	2.50	2.45
	$^3P_1$–$^1D_2$	4959 N_2	7.1×10^{-3}	2.50	2.47
	$^3P_0$–$^1D_2$	4931 N_3	1.9×10^{-6}	2.50	2.49
[NII]	$^1D_2$–$^1S_0$	5755	1.1	0.39	2.15
	$^3P_1$–$^1S_0$	3063	3.4×10^{-2}	0.36	4.05
	$^3P_2$–$^1D_2$	6584	3.0×10^{-3}	2.99	1.89
	$^3P_1$–$^1D_2$	6548	1.0×10^{-3}	2.99	1.90
[NeV]	$^1D_2$–$^1S_0$	2976	2.6	0.19	4.12
	$^3P_1$–$^1S_0$	1575	4.2	0.20	7.88
	$^3P_2$–$^1D_2$	3426	3.8×10^{-1}	1.47	3.62
	$^3P_1$–$^1D_2$	3346	1.4×10^{-1}	1.47	3.71
[SIII]	$^1D_2$–$^1S_0$	6312	2.5	1.34	1.93
	$^3P_1$–$^1S_0$	3722	8.5×10^{-1}	0.75	3.34
	$^3P_2$–$^1D_2$	9532	6.4×10^{-2}	3.87	1.30
	$^3P_1$–$^1D_2$	9069	2.5×10^{-2}	3.87	1.37
[OII]	$^2D_{5/2}$–$^2P_{3/2}$	7320	1.5×10^{1}	0.78	1.70
	$^4S_{3/2}$–$^2P_{3/2}$	2470	6.0×10^{2}	0.45	5.03
	$^4S_{3/2}$–$^2D_{5/2}$	3729	4.2×10^{5}	1.47	3.33
	$^4S_{3/2}$–$^2D_{3/2}$	3726	1.8×10^{4}	1.47	3.33
[NeIV]	$^2D_{5/2}$–$^2P_{3/2}$	4714	4.0×10^{1}	0.61	2.63
	$^4S_{3/2}$–$^2P_{3/2}$	1602	1.3	0.43	7.75
	$^4S_{3/2}$–$^2D_{5/2}$	2425	5.9×10^{4}	1.14	5.12
	$^4S_{3/2}$–$^2D_{3/2}$	2423	5.6×10^{3}	1.14	5.12
[SII]	$^2D_{5/2}$–$^2P_{3/2}$	10320	2.1×10^{1}	5.15	1.20
	$^4S_{3/2}$–$^2P_{3/2}$	4069	3.4×10^{1}	2.72	3.05
	$^4S_{3/2}$–$^2D_{5/2}$	6716	4.7×10^{4}	5.66	1.85
	$^4S_{3/2}$–$^2D_{3/2}$	6731	1.8×10^{3}	5.66	1.84
[NeIII]	$^1D_2$–$^1S_0$	3342	2.8	0.20	3.71
	$^3P_1$–$^1S_0$	1815	2.2	0.17	6.84
	$^3P_2$–$^1D_2$	3869	1.7×10^{-1}	1.14	3.21
	$^3P_1$–$^1D_2$	3967	5.2×10^{-2}	1.14	3.13
[ArIII]	$^1D_2$–$^1S_0$	5192	3.1	0.84	2.40
	$^3P_1$–$^1S_0$	3109	4.0	0.59	3.99
	$^3P_2$–$^1D_2$	7139	3.2×10^{-1}	3.92	1.74
	$^3P_1$–$^1D_2$	7751	8.3×10^{-2}	3.92	1.60

The same is true for the 6 528 [NII] line. The line N_3[OIII] was actually discovered as early as 1955 by Aller and coauthors in the spectrum of the PN NGC 7 027 with an extremely small relative intensity: the ratio $I(N_3)/I(N_1)$ was of the order of 0.0005.

From the transition probabilities of forbidden lines shown in Table 2.1, it follows that the mean life of the atom or ion in a metastable state is quite large, i.e. of the order of a fraction of a second, several minutes, or even hours. If during this interval of time the excited atom undergoes a collision with another particle, for example, with a free electron, then the excitation of the metastable atom will be transformed into the kinetic energy of this particle (electron) instead of the radiation energy in the forbidden line.

Thus, the second condition follows:

The emission of a forbidden line requires relatively low matter density.

This condition may be formulated quantitatively in the following way. Denoting by n_2 the number of atoms (ions) per unit volume in the metastable state, the number of spontaneous transitions $2 \to 1$ responsible for the forbidden line will be per unit time $n_2 A_{21}$. The number of induced transitions $2 \to 1$ caused by collisions of the second kind is $n_2 n_e q_{21}(T_e)$, where $q_{21}(T_e)$ is the probability of the $2 \to 1$ transition by means of collisions. Then, the condition for the appearance of a forbidden line can be written in the following form:

$$n_2 A_{21} \gg n_2 n_e q_{21}(T_e) \, ,$$

or

$$n_e \ll \frac{A_{21}}{q_{21}(T_e)} \, , \tag{12}$$

where $q_{21}(T_e)$ is given in the form

$$q_{21}(T_e) = 8.63 \times 10^{-6} \frac{\Omega(1,2)}{\omega_2} T_e^{-1/2} \, , \tag{13}$$

where $\Omega(1,2)$ is the "collision strength", the numerical values of which are also given in Table 2.1, ω_2 is the statistical weight of the upper (metastable) level and T_e is the electron temperature of the nebula.

Condition (12) must be satisfied for an electron concentration and, consequently, for a material density in the nebula in order not to "put out" the forbidden line completely.

Denote the right hand side in (12) by $n_e(\mathrm{cr})$,

$$n_e(\mathrm{cr}) = \frac{A_{21}}{q_{21}(T_e)} \, , \tag{14}$$

calling it "critical concentration". When $n_e < n_e(\mathrm{cr})$, the forbidden line will be observed with full intensity. But this still does not mean that the line cannot appear when the electron concentration is of the order of the critical

Table 2.2. Critical electron concentrations n_e(cr) for some forbidden lines

Line	Upper level	n_e(cr) [cm^{-3}]	Line	Upper level	n_e(cr) [cm^{-3}]
3342 [NeIII]	$^1D_2$	8.1×10^8	4959 [OIII] N_2	$^1D_2$	1.6×10^5
6312 [SIII]	$^1D_2$	1.1×10^8	6584 [NII]	$^1D_2$	5.8×10^4
4363 [OIII]	$^1S_0$	3.2×10^7	6548 [NII]	$^1D_2$	1.9×10^4
3426 [NeV]	$^1D_2$	1.5×10^7	6731 [SII]	$^4D_{3/2}$	1.3×10^4
3346 [NeV]	$^1D_2$	5.5×10^6	6716 [SII]	$^2D_{5/2}$	5.8×10^3
10320 [SII]	$^2D_{3/2}$	1.9×10^6	3726 [OII]	$^2D_{3/2}$	5.7×10^3
5007 [OIII] N_1	$^1D_2$	4.9×10^5	3729 [OII]	$^4D_{5/2}$	2.0×10^3
2423 [NeIV]	$^2D_{3/2}$	2.3×10^5	4931 [OIII] N_3	$^1D_2$	0.4×10^2

concentration, i.e. when $n_e \sim n_e$(cr). Forbidden lines can be observed even when $n_e > n_e$(cr), but they will be weak. The quantitative calculations of the intensity of a forbidden line in the latter case [$n_e > n_e$(cr)] present a special problem which will be considered in Chap. 15.

Table 2.2 gives the critical electron concentrations n_e(cr) obtained with the help of (14) at $T_e = 10\,000$ K for the most interesting forbidden lines.

For the majority of PNs $n_e < 10^4$ cm^{-3}. Comparing this value with those in Table 2.2, we see that the requirement of small density, or more precisely of a small electron concentration, in PNs is fully satisfied for the N_1 and N_2 lines, as well as for the lines 6 312 [SIII], 4 363 [OIII], 6 548+6 584 [NII], and for the infrared doublet 10 336+10 370 [SII] and ultraviolet doublet 2 423+2 425 [NeIV]. However, the forbidden lines of the doublet 3 726+3 729 [OII] may be excited not at full strength; their excitation at full strength is possible only in low density PNs with $n_e \leq 10^3$ cm^{-3}. As far as the 4 363 [OIII] line is concerned, it can be observed even in the very dense PNs – with $n_e \sim 10^7$–10^8 cm^{-3} – and also during the early phase of the evolution of nova envelopes, long before the appearance of the nebular lines N_1 and N_2.

However, an atom in a metastable state can, under the influence of the external radiation, realize a transition to higher states with no emission in forbidden lines. Therefore we require that no photons will be absorbed during the lifetime of the metastable state. Thus we arrive at the third condition of the emission of forbidden lines, formulated by Eddington (1927):

The radiation density within the nebula must be small.

As we shall see in the following section this condition is also satisfied in PNs.

2.5 Accumulation of Atoms in Metastable Levels

The intensity E_{ik} of a forbidden line is proportional to the spontaneous transition probability A_{ik} and the number of atoms in the level n_i,

$$E_{ik} = n_i A_{ik} h\nu_{ik} \, .$$

The spontaneous transition probability from a metastable level, as we saw above, is extremely small. Nevertheless, the intensity of the forbidden lines observed in PNs is rather high. Obviously, this is possible only because of the large number of atoms in the metastable state. Thus, under nebular conditions the atom can remain in the metastable state for a long enough time (with respect to the possibility of a spontaneous transition downwards). Therefore, a huge number of atoms can be accumulated in such a state. The problem of the accumulation of atoms in a metastable state was first discussed by Ambartsumian (1933).

Consider an atom consisting of three levels, of which the middle one is metastable. Assuming that the excited atoms can only make radiative transitions, one can write the equilibrium conditions for the number of atoms per unit volume found in the first and third levels:

$$\begin{aligned} n_1 B_{12} \rho_{12} + n_1 B_{13} \rho_{13} &= n_2 A_{21} + n_3 A_{31} \, , \\ n_1 B_{13} \rho_{13} + n_2 B_{23} \rho_{23} &= n_3 A_{31} + n_3 A_{32} \, . \end{aligned} \tag{15}$$

Here we neglect induced transitions since the radiation field is strongly diluted ($W \ll 1$).

From (15) we get, eliminating n_3 and using the relations between the Einstein coefficients and also (9) for the quantity ρ_{ik},

$$\frac{n_2}{n_1} = W \frac{A_{21}(g_2/g_1)\bar{\rho}_{12} + (1-p)A_{31}(g_3/g_1)\bar{\rho}_{13}}{A_{21} + pA_{32}(g_3/g_2)W\bar{\rho}_{23}} \, , \tag{16}$$

where $p = A_{31}/(A_{31} + A_{32})$.

We have also the condition that the second state is metastable, i.e. $A_{21} \ll A_{31}, A_{32}$. Therefore, the first term in (16) can be neglected in comparison with the second. Each one of the two terms in the denominator can dominate depending on the circumstances. We have to consider both possibilities.

I. $W \ll A_{21}/A_{32}$. In this case the second term in the denominator can be neglected:

$$\frac{n_2}{n_1} = W(1-p)\frac{A_{31}}{A_{32}}\frac{g_3}{g_1}\bar{\rho}_{13} \approx W\frac{A_{31}}{A_{21}}\bar{\rho}_{13} \, , \tag{17}$$

where we have taken $(1-p)g_3/g_1 \approx 1$.

II. $W \ll A_{21}/A_{32}$. Now one can neglect the first term in the denominator:

$$\frac{n_2}{n_1} = \frac{g_3}{g_1}\frac{\bar{\rho}_{13}}{\bar{\rho}_{23}} \approx \bar{\rho}_{12} \, . \tag{18}$$

Consider, for comparison, the situation that would arise if the second level is not metastable, i.e. when the transition $2 \to 1$ is permitted. Denoting the Einstein coefficient for the permitted transition by A^0_{21}, one can write $A_{21} \approx A_{31}, A_{32}$, and from (16) one finds that

$$\frac{n_2^0}{n_1} = W\frac{g_2}{g_1}\,\bar{\rho}_{12} + W(1-p)\frac{A_{31}}{A_{21}^0}\,\frac{g_3}{g_1}\,\bar{\rho}_{13} \approx W\frac{A_{31}}{A_{21}^0}\,\bar{\rho}_{13}\,, \tag{19}$$

since both terms on the right hand side of (23) are of the same order.

Comparing (17) and (19) we see that when we have a metastable state in the first of the two cases considered, the ratio n_2/n_1 exceeds the value of this ratio for a normal excited state by a factor of the order of the ratio of the probabilities for the allowed and the forbidden transitions ($n_2/n_2^0 \approx A_{21}^0/A_{21}$). This means that the product of the number of atoms times the probability of spontaneous emission, i.e. the intensity of the forbidden line, is in this case of the same order as the intensity of the permitted line (since $n_2A_{21} \approx n_2^0A_{21}^0$).

In the second case of the two cases the ratio n_2/n_1 is determined approximately by Boltzmann's formula and, as shown by comparing (16) and (17), it is of the order of $1/W$ times greater than in the case of a normal excited state. It follows that, again in this case, there is a strong accumulation of atoms in the metastable state. However, the number of transitions from the metastable state is not as large as the number of transitions from the normal state. Indeed, we have

$$n_2A_{21} \approx n_1\bar{\rho}_{12}A_{21} \approx n_2^0\,\frac{A_{21}}{W} \ll n_2^0A_2^0\,. \tag{20}$$

This means that in case II the intensity of the forbidden line will be smaller compared with the intensity of the permitted lines.

The physical meaning of cases I and II is as follows. The first term in the denominator of (17) corresponds to the forbidden transition $2 \rightarrow 1$, and the second to the upward transition from the metastable state $(2 \rightarrow 1)$. In the first case the transitions $2 \rightarrow 1$ predominate and the forbidden line is strong. In the second case there is a relatively strong emission field which induces the transition $2 \rightarrow 3$ and the forbidden transition is weak.

Since in PNs W is of the order of 10^{-13}, almost all metastable states correspond to case I. A possible exception is the 2^3S state of helium, where the electron can remain for a long time (of the order of days and weeks). On the other hand, in envelopes of small radius – in the envelopes of Wolf–Rayet, Be and similar stars – almost all metastable states correspond to case II.

Thus, the third condition necessary for the appearance of a forbidden line – as discussed at the end of the previous section – can be expressed by the relation

$$W \ll A_{21}/A_{21}^0\,. \tag{21}$$

The ratio A_{21}/A_{21}^0 is of the order of 10^{-8} for the 4 363 [OIII] line, 10^{-10} for N_1 and N_2, and 10^{-12} for 3 727 [OII].

Equations (12) and (21) determine the upper limit of the matter and radiation densities. These conditions must be fulfilled irrespective of the particular mechanism of excitation of the atom.

2.6 The Role of Electron Collisions

Above, in Sect. 2.4, it was pointed out that several lines, and among them the important nebular lines N_1 and N_2, cannot arise as a result of the photoionization of ions and their subsequent recombinations. This conclusion is confirmed by the following facts.

Let us assume that the luminosity of the nebula in the N_1 and N_2 lines is due to stellar energy beyond the limit of the principal series of the doubly ionized oxygen. Then the number of photons emitted by the star beyond the limit of this series ($\lambda < 226$ Å) cannot be smaller than the number of photons emitted by the nebula in the N_1 and N_2 lines. In order to satisfy this condition we require a very high stellar temperature – of the order of a million degrees.

The following argument seems more convincing: since the ionization potentials of OIII and HeII are almost identical (54.5 and 54.2 eV) these two ions absorb stellar energy in the same region of the spectrum. Therefore, if the luminosity of the nebula in the lines N_1 and N_2 is due only to this energy (fluorescence), then the intensity of the lines N_1 and N_2 could not exceed that of the HeII lines, since the number of helium atoms is not smaller than the number of oxygen atoms. However, in a number of PNs the helium lines can hardly be seen (and in diffuse nebulae they disappear altogether), while the N_1 and N_2 lines appear with high intensity.

All these show that in the nebulae there exists an excitation mechanism other than recombination. Such a mechanism was suggested again by Bowen (1927, 1928) and, according to this, the luminosity of the nebulae in the *nebulium* lines is due to the *kinetic energy of the free electrons.*

The excitation potential of the metastable levels that give rise to forbidden lines, and among them the nebular lines N_1 and N_2, is not large – of the order of a few eV. For example, for the level $^1D_2$ of the ion O^{++} during the transitions from which the lines N_1 and N_2 were excited, this potential is equal to 2.5 eV (Fig. 2.1). For the lines 6 548 and 6 584 [NII] it is even smaller – 1.89 eV, and so forth. On the other hand, the residual energy of the ultraviolet photon after the photoionization of hydrogen is transformed into kinetic energy of the free electrons. This energy is sufficiently high to excite the OIII ions to the metastable state through inelastic collisions with the electrons. These collisions of course occur only rarely since the density of matter in the nebula is very low – low enough that the ion once in a metastable state will not undergo another collision with a free electron before making a spontaneous transition downwards. Nevertheless, the number of collisions is high enough to excite the ground state ions to the metastable levels since the time that the ion, for example, O^{++}, spends in the ground state is very large – of the order of a week.

Thus, collisions with free electrons raise the ions from the ground state to the metastable state at the expense of the kinetic energy of the electrons, and then spontaneous transitions downwards lead to the almost complete *ex-*

traction of the kinetic energy of their electrons in order to emit the forbidden lines.

Consider a two-level ion, 1 and 2, of which the higher (2) is metastable. Denote by $n_1q_{12}(T_e)$ the number of transitions from the the normal (1) to the metastable (2) level by electron collisions of the first kind; by $n_2q_{21}(T_e)$ the number of transitions from the metastable to the ground state by means of collisions of the second kind; and by n_2A_{21} the number of spontaneous transitions from the metastable to the ground state. Then the equilibrium condition yields

$$n_1q_{12} = n_2(q_{21} + A_{21}) \,. \tag{22}$$

The coefficients q_{12} and q_{21} depend on the concentration and density distribution of free electrons, and the relation between them is given by the relationship

$$q_{12} = q_{21}\frac{\omega_2}{\omega_1}\, e^{-\varepsilon_{12}/kT_e} \,, \tag{23}$$

where ε_{12} is the excitation energy of the metastable level, and q_{21} yields (13). Then from (22) and (23) we get:

$$\frac{n_2}{n_1} = \frac{q_{21}}{A_{21} + q_{21}}\frac{\omega_2}{\omega_1}\, e^{-\varepsilon_{12}/kT_e} \,. \tag{24}$$

If $q_{21} \gg A_{21}$, then (24) fits into Boltzmann's law for n_2/n_1; however, this inequality indicates a preponderance of induced over spontaneous transitions $2 \to 1$. Under such conditions the nebular lines would be weakened. Actually, as we have seen before, in the nebulae condition (12) is fulfilled and, hence, a predominance of spontaneous transitions over induced ones takes place. Therefore we have instead of (24)

$$n_2 = n_1\frac{q_{21}}{A_{21}}\frac{\omega_2}{\omega_1}\, e^{-\varepsilon_{12}/kT_e} \,. \tag{25}$$

For the intensity of the forbidden line we obtain

$$E_{21} = n_2A_{21}h\nu_{12} = n_1q_{21}\frac{\omega_2}{\omega_1}\, h\nu_{12} \,, \tag{26}$$

or, using (23),

$$E_{21} = n_1q_{12}h\nu_{12} \,, \tag{27}$$

i.e. *the number of forbidden transitions is equal to the number of excitations by electron collisions.*

Thus, the forbidden lines are excited by inelastic electron collisions.

2.7 The Deactivation Effect

However, the same electron collisions can lead to the weakening of the forbidden line in spite of the fact that at $n_e \sim n_e(\mathrm{cr})$ the electron collisions may occur not only via ions in ground state (1) but also via ions which are in metastable state (2). In the latter case the transitions $2 \rightarrow 1$ will never be radiative. As a result, a process of the weakening – *deactivation* – of forbidden line will take place.

If $n_2 A_{21}$ is the number of radiative transitions $2 \rightarrow 1$ and $n_2 n_e q_{21}(T_e)$ the number of non-radiative transitions $2 \rightarrow 1$ provoked by electron collisions, then we can write for the *deactivation factor* X

$$X = \frac{n_2 A_{21}}{n_2 A_{21} + n_2 n_e q_{21}(T_e)},$$

or

$$X = \left[1 + \frac{n_e}{n_e(\mathrm{cr})}\right]^{-1}. \tag{28}$$

If, in the limiting case, we have for the intensity of a forbidden line, according to (27), $E_{21} = n_1 q_{12} h\nu_{12}$, now, in view of the deactivation factor X, its intensity will be

$$E_{21} = n_1 q_{12}(T_e) h\nu_{12} X. \tag{29}$$

Hence

$$\begin{aligned} X &= 1 \quad \text{for} \quad n_e \ll n_e(\mathrm{cr}), \\ X &< 1 \quad \text{for} \quad n_e \sim n_e(\mathrm{cr}), \\ X &> 0 \quad \text{for} \quad n_e \gg n_e(\mathrm{cr}). \end{aligned} \tag{30}$$

Using the data of Table 2.2, we have the following relationships for deactivation factor X.

For the lines of [NeIII], [NeV] and [SIII],

$$X = \left[1 + 10^{-6} n_e\right]^{-1}; \tag{31}$$

for [OIII],

$$X = \left[1 + 2 \times 10^{-6} n_e\right]^{-1}; \tag{32}$$

for [NII] and [SII],

$$X = \left[1 + 1.5 \times 10^{-5} n_e\right]^{-1}; \tag{33}$$

and for [OII],

$$X = \left[1 + 5 \times 10^{-4} n_e\right]^{-1}. \tag{34}$$

In all cases, however, at extremely high electron concentrations, i.e. when $n_e > 10^6\ \mathrm{cm}^{-3}$, $X \rightarrow 0$, and hence forbidden lines cannot be excited.

2.8 Stratification of Radiation: The Phenomenon

Thus, the fact that in the nebula there exist at least two different excitation mechanisms for the emission lines – recombinational and collisional – should lead to different sizes of monochromatic images of the nebula in different lines, i.e. to the presence of stratification of radiation, as was mentioned in Chap. 1. Let us analyse this question following Bowen (1935) himself. However, we will be forced to refer to the basics of the theory of ionization and excitation of nebula discussed below in the first sections of Chap. 4.

Suppose we have a spherical nebula with a hot star at its center. Obviously, near the nucleus oxygen will be highly ionized, say, in the fourth stage of ionization. The ionized atoms will undergo recombinations with electrons, together with cascade transitions downwards from the upper states, and as a result different lines (only permitted) of the OIV spectrum will be emitted. Thus, in the central region of the nebula, in zone I, we shall observe such lines as 3 412 OIV, 3 737 OIV and so forth. In zone II the oxygen will be in the third ionization state via absorption of a photon with a wavelength shorter than 225 Å. The capture of an electron by an O^{+++} ion leads to the permitted lines in the spectrum of OIII. Therefore, the dimensions of the region of the nebula where the OIII lines are emitted, both permitted – 3 760 OIII, 5 592 OIII – and forbidden – 5 007 [OIII], 4 959 [OIII], 4 363 [OIII]. These considerations can be applied to the following zones: for each lower ionization state the dimensions of the zone where the permitted or forbidden lines

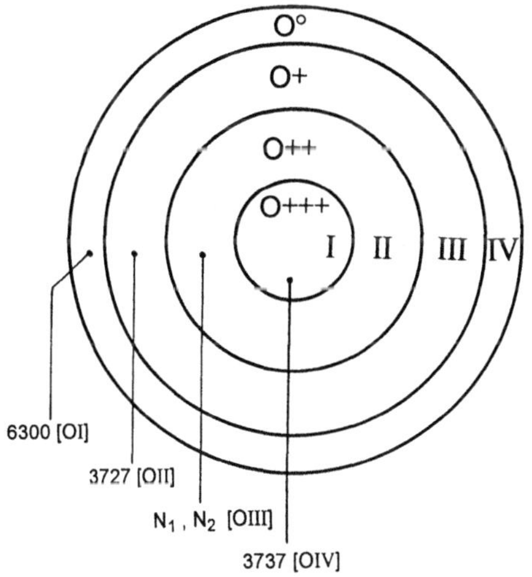

Fig. 2.2. Scheme illustrating the stratification of the radiation for a spherical planetary nebulae in the case of oxygen ions. I, II, III and IV are the radiation zones of the nebula in the lines O^{+++}, O^{++}, O^{+} and O^{0}, respectively. Examples of forbidden lines radiating in these zones are also given

of the spectrum of a given ion are emitted, increase (Fig. 2.2). Obviously, we shall also have a similar picture for the lines of other elements.

Thus, with the decrease of the ionization state (of every element) the sizes of the corresponding zones will increase. Now it can be understood, for example, why the sizes of the monochromatic images of PNs in the line 3 426 [NeV] are always smaller than the sizes of the zone in the the line 3 869 [NeIII]: because in the first case, NeV, the ionization potential (126 eV) is essentially larger than the ionization potential in the second case, NeIII (63 eV). Also, the sizes of zone He^{++} (ionization potential 54 eV) in the line 4 686 HeII must always be smaller compared with the sizes of the He^{+} zone (ionization potential 25 eV) in the line 4 471 HeI.

The smallest ionization potential is that of the hydrogen atom. Therefore, the largest image size should correspond to the hydrogen lines. Neutral oxygen has an ionization potential almost identical to that of hydrogen. However, because of the abundance of hydrogen atoms in a nebula the oxygen atoms are usually left with an insufficient fraction of the ionization energy. It follows, then, that the size of the region containing OI lines should be smaller than the size of the region corresponding the hydrogen lines. It is also clear that none of the forbidden lines can occupy a zone greater than that of the hydrogen lines.

As for the helium, the 4 686 HeII line is emitted when an electron is captured by a doubly ionized helium atom which has an ionization potential four times greater than that of hydrogen. Doubly ionized helium can be found only in the inner regions of the nebula, and therefore the size of the image in the 4 686 HeII line should be considerably smaller than the HII region.

2.9 Emission Lines of Neutral Atoms

In the spectrum of the gaseous nebulae we sometimes observe the forbidden lines of neutral atoms, particularly of OI, NI and MgI, excited by inelastic electron collisions. The best known among them are the lines 6 300+6 364 [OI], 5577 [OI], 5 198+5 200 [NI], 4 571 [MgI]. It must be noted, however, that these lines are observed usually in the spectrum of low excitation nebulae, in particular, in diffuse nebulae and in Herbig–Haro objects.

We have an extreme example of high a excitation PN: NGC 7 027, in whose spectrum these lines, though rather weak, are present. The temperature of the nucleus of this object is of the order of 15 000 K or higher, and, even in such conditions, the observed intensities of these lines of neutral elements are as follows ($H_\beta = 100$):

6 300 [OI]	20	5 198 [NI]	1.2;
6 364 [OI]	10	5 200 [NI]	0.3;
5 577 [OI]	1.5	4 571 [NI]	0.9.

Table 2.3. The atomic parameters for neutral oxygen and nitrogen (forbidden transitions) and magnesium (permitted transitions). The values of Ω_{ij} are given for $T_e = 10^4$ K

Line [Å]	Transition	A_{ij} [s^{-1}]	$\Omega(ij)$	e_{ij} [eV]
1 302.2 [OI]	$^3P_2 - {}^3S^0$	2.38×10^8	0.25	9.52
1 304.9 [OI]	$^3P_1 - {}^3S^0$	8.47×10^7	0.15	9.52
1 306.9 [OI]	$^3P_0 - {}^3S^0$	9.36×10^6	0.05	9.52
2 958.4 [OI]	$^3P_2 - {}^1S_0$	2.88×10^{-4}	0.0324	4.18
2 972.3 [OI]	$^3P_1 - {}^1S_0$	7.32×10^{-2}	0.0324	4.18
5 577.4 [OI]	$^1D_2 - {}^1S_0$	1.22	0.105	2.21
6 300.3 [OI]	$^3P_2 - {}^1D_2$	6.34×10^{-3}	0.266	1.97
6 363.8 [OI]	$^3P_1 - {}^1D_2$	2.11×10^{-3}	0.266	1.97
6 391.5 [OI]	$^3P_0 - {}^1D_2$	7.23×10^{-7}	0.266	1.97
5 197.9 [NI]	$^4S_{3/2} - {}^3D_{3/2}$	2.02×10^{-5}	0.194	2.39
5 200.4 [NI]	$^4S_{3/2} - {}^3D_{5/2}$	7.27×10^{-6}	0.290	2.39
4 562.5 MgI	$^1S_0 - {}^3P_2^0$	4.13×10^{-4}	1.6	2.72
4 571.1 MgI	$^1S_0 - {}^3P_1^0$	1.80×10^2	1.6	2.72

The peculiarity of the situation is also due to the presence in the spectrum of the same nebula NGC 7 027 of the emission lines belonging to the ions of very high ionization potential, higher than 100 eV – OV, NeV, CaVII, etc. The conclusion is clear: even the extremely high excitation nebulae are surrounded by a zone of neutral atoms, and those forbidden lines are excited most probably in the *transition* zones – from HII to HI; because for the generation of forbidden lines [OI], [NI] it is necessary to have also *free* electrons.

The atomic parameters for the most interesting lines of neutral atoms compiled by Mendoza (1983) and Osterbrock (1974) are given in Table 2.3. We shall return to the ultraviolet triplet 1 302, 1 305 and 1 307 [OI] in Chap. 7.

2.10 The Chemical Composition of Planetary Nebulae

In principle the determination of the chemical composition of the PN is simple: it can be obtained by comparing the theoretical and observed intensities of the emission lines in their spectra.

In the case when the forbidden lines are used for the determination of the relative abundance of chemical elements in PNs, the sequence of the calculations is as follows.

We have for the complete energy $E(A)$ emitted by the PN in a forbidden line λ_A belonging to some atom (ion) A

$$E_f(A) = n_e \, n_1\,(A)\, q_A\,(T_e)\, h\, \nu_A\, V_A\,, \tag{35}$$

where $n_1(A)$ is the concentration of atoms (ions) of type A in its ground state and V is the volume of the PN.

On the other hand, we have for the complete energy $E(H_\beta)$ emitted by the nebula in the line H_β of hydrogen (see Sect. 3.1):

$$E(H_\beta) = z_{42}\, A_{42}\, h\, \nu_\beta\, n_e\, n^+\, V_H\,, \tag{36}$$

where V_H is the volume of the PN illuminated in the hydrogen lines.

From (35) and (36) we find for the relative abundance $n_1(A)/n_H$ of atoms (ions) of type A, taking $n^+ = n_H$, that

$$\frac{n_1(A)}{n_H} = n_e \frac{E_f(A)}{E(H_\beta)} \frac{z_{42} A_{42}}{q_A(T_e)} \frac{\nu_\beta}{\nu_A} \left(\frac{d_H}{d_A}\right)^3, \tag{37}$$

where d_H and d_A are the apparent sizes of monochromatic images of the PN in the H_β and λ_A lines.

Substituting in (37) $z_{42} = 0.36 \times 10^{-20}$ at $T_e = 10\,000$ K (Table 3.5 Model B), $A_{42} = 8.37 \times 10^6$ s^{-1}, $\nu_\beta = 0.617 \times 10^{15}$ s^{-1}, as well as the value of $q_A(T_e)$ from (13), we obtain

$$\frac{n_1(A)}{n_H} = 5.00 \times 10^{11} \frac{\omega_i}{\nu_A \Omega(ij)} \frac{n_e}{T_e^{1/2}} \left(\frac{d_H}{d_A}\right)^3 \frac{E_f(A)}{E(H_\beta)}\,. \tag{38}$$

Here the ratios d_H/d_A and $E_f(A)/E(H_\beta)$ taken directly from observations. The rest parameters are known for a given ion A. The application of this relationship requires also the knowledge of the electron concentration n_e and electron temperature T_e for the PN under examination.

The situation is different, to some degree, in the case of the determination of the relative abundance of elements with the help of recombination lines. The determination of theoretical intensities of recombination lines for each atom or ion begins with the composition of the corresponding equations of stationarity of transitions analogous to those which were realized for hydrogen atoms (see the system of (10) in Chap. 3). By solving this system of equations, we obtain the relative number of atoms (ions) in different excitation levels, after which we get the energy emitted by the PN in one or the other line with the help of the formula as in (36) above. Combined with the equation for hydrogen (36), we obtain the theoretical ratio of concentration of these atoms (ions) in the nebula.

In spite of the simplicity of the procedure described above for the determination of the chemical composition of the PN, its practical application has associated with it a number of difficulties. First, not all atoms and ions are known with sufficient accuracy of the atomic parameters – the coefficients of spontaneous transitions and recombinations, the efficient cross-sections of collisions, etc. Second, the atoms of a given element can be distributed among

Table 2.4. Chemical composition of planetary nebulae. The relative abundance of 13 elements is given on the scale log H = 12 together with data for the Sun and the stars

Element	Planetary nebulae	Stars	Sun	Element	Planetary nebulae	Stars	Sun
H	12.00	12.00	12.00	Ar	7.0	6.80	6.0
He	11.0	10.93	10.8	Cl	6.9	5.60	5.5
C	8.8	8.52	8.62	Na	6.6	6.25	6.28
O	8.6	8.82	8.84	Ca	6.4	6.30	6.36
N	8.1	7.96	7.94	K	5.7	4.95	5.16
Ne	7.9	7.92	7.57	P	5.3	5.52	5.50
S	7.9	7.20	7.2	F	4.9	4.6	4.56

the different states of ionization and excitation, while we know their concentration only in one of these states. We have at our disposal a method for transition from one ionization state to another but as a rule in such a way that it is difficult to preserve the required reliability owing to the uncertainties connected with the knowledge of a number of parameters.

Despite these difficulties, the relative abundance of chemical elements has been determined for dozens of PNs. In Table 2.4, their average chemical composition is given, i.e. the relative abundance of 13 elements scaled by log H = 12 (e.g. Aller, 1978; Pottasch, 1984). For comparison data for the Sun are also given.

The most abundant element in PNs is hydrogen. Then comes helium, which is usually nearly 10% relative to hydrogen atoms. The third is carbon; then follows oxygen, nitrogen etc. The total amount of all the remaining elements in PNs, starting from carbon, is of the order of one percent.

In general, however, the chemical composition of PNs is the same as for the Sun, for stellar atmospheres and for the interstellar medium.

2.11 Continuous Absorption and Recombination Coefficients

There are a number of problems of the physics of gaseous nebulae where one has to know the numerical values of continuous absorption and recombination coefficients for the definite types of atoms and ions. Such data are necessary, for example, for calculating continuous spectra and the continuum, the intensities of emission lines and the ionization models of nebulae, and for the determination of the chemical composition and abundance of elements, etc.

The continuous absorption coefficient, α_ν, estimated for one atom or ion, can be represented in a form dependent on the frequency of the photon $\nu \geq \nu_0$:

$$\alpha_\nu = \alpha_{\nu_0}\left[\beta\left(\frac{\nu}{\nu_0}\right)^{-s} + (1-\beta)\left(\frac{\nu}{\nu_0}\right)^{-s-1}\right]. \tag{39}$$

Here ν_0 is the frequency of ionization for an atom or ion and α_{ν_0} is the absorption coefficient at the frequency of ionization $\nu = \nu_0$. The numerical values of the constants α_{ν_0}, β and s for a number of atoms and ions are given in Table 2.5 (e.g. Seaton, 1958; Henry, 1970; Osterbrock, 1974).

In the overwhelming majority of cases we can confine ourselves to considering only the first term in (47). Then we have for hydrogen atom

$$\alpha_\nu(\mathrm{H}) = 6.30\times10^{-18}\left(\frac{\nu}{\nu_0}\right)^{-3}\ \mathrm{cm}^2. \tag{40}$$

For the neutral and ionized helium we have

$$\alpha_\nu(\mathrm{He}) = 7.83\times10^{-18}\left(\frac{\nu}{\nu_0}\right)^{-2}\ \mathrm{cm}^2, \tag{41}$$

and

$$\alpha_\nu(\mathrm{He}^+) = \frac{1}{4}\alpha_\nu(\mathrm{H}) = 1.58\times10^{-18}\left(\frac{\nu}{\nu_0}\right)^{-3}\ \mathrm{cm}^2. \tag{42}$$

The recombination coefficient $C(T_\mathrm{e})$ of a free electron with an ion represents the contribution of two types of recombinations, radiative and dielectric:

$$C(T_\mathrm{e}) = C_\mathrm{rad}(T_e) + C_\mathrm{di}(T_e) \tag{43}$$

i.e. the efficiency of both types of recombination depends on the electron temperature of the medium T_e.

The derived expression for the radiative recombination coefficient has the following form:

$$C_\mathrm{rad}(T_\mathrm{e}) = A_\mathrm{rad}\left(\frac{10^4}{T_\mathrm{e}}\right)^{\eta}. \tag{44}$$

The numerical values of A_rad and η, collected from different sources for the most important ions for the physics of gaseous nebulae, are presented in Table 2.6 (e.g. Flower, 1968; Brown, 1971; Aldrovandi, Pequignot, 1973; Shull, Steeberg, 1982).

As for the dielectric recombination, for the overwhelming majority of ions it is efficient only at fairly high electron temperatures – of the order of 100 000 K and higher. In the case of PNs where T_e lies in the interval of 10 000–20 000 K, the dielectric recombination plays a noticeable role only for a limited type of ion, and in the remaining cases this coefficient is numerically too small – smaller than 10^{-15} cm^3 s^{-1}.

The dielectric recombination coefficient may be presented in a two-parameter form in the following manner (Burgess, 1965):

$$C_\mathrm{di}(T_\mathrm{e}) = A_\mathrm{di}(T_\mathrm{e}^{-3/2})\exp(-T_0/T_\mathrm{e}), \tag{45}$$

Table 2.5. The parameters of the continuous absorption coefficient in (39) for a number of atoms and ions

Initial condition	Final condition	ν_0 [10^{15} s^{-1}]	α_{ν_0} [10^{-18} cm^2]	β	s
$H^0(2S)$	$H^+(^1S)$	3.29	6.30	1.34	2.99
$He^0(^1S)$	$He^+(^2S)$	5.95	7.83	1.66	2.05
$He^+(^2S)$	$He^+(^1S)$	13.17	1.58	1.34	2.99
$C^0(^3P)$	$C^+(^2P)$	27.27	12.2	3.32	2.0
$C^+(^2P)$	$C^{+2}(^1S)$	5.91	4.60	1.95	3.0
$C^+(^1S)$	$C^{+3}(^2S)$	11.58	1.60	2.6	3.0
$C^+(^2S)$	$C^{+4}(^1S)$	15.60	0.68	1.0	2.0
$N^0(^1S)$	$N^+(^3P)$	3.51	11.4	4.29	2.0
$N^+(^3P)$	$N^{+2}(^1P)$	7.17	6.65	2.86	3.0
$N^{+2}(^2P)$	$N^{+3}(^1S)$	14.67	2.06	1.63	3.0
$N^{+3}(^1S)$	$N^{+4}(^2S)$	18.75	1.08	2.6	3.0
$N^{+4}(^2S)$	$N^{+5}(^1S)$	23.70	0.48	1.0	2.0
$O^0(^3P)$	$O^+(^4S)$	3.29	2.94	2.66	1.0
$O^0(^3P)$	$O^+(^2D)$	4.09	3.85	4.38	1.5
$O^0(^3P)$	$O^+(^2P)$	4.50	2.26	4.31	1.5
$O^+(^4S)$	$O^{+2}(^3P)$	8.51	7.32	3.84	2.5
$O^{+2}(^3P)$	$O^{+3}(^2P)$	13.30	3.65	2.01	3.0
$O^{+3}(^2P)$	$O^{+4}(^2S)$	18.73	1.27	0.83	3.0
$O^{+4}(^1S)$	$O^{+5}(^2S)$	27.56	0.78	2.6	3.0
$Ne^0(^1S)$	$Ne^+(^2P)$	5.22	5.35	3.77	1.0
$Ne^+(^2P)$	$Ne^{+2}(^3P)$	9.94	4.16	2.72	1.5
$Ne^+(^2P)$	$Ne^{+2}(^1D)$	10.72	2.71	2.15	1.5
$Ne^{+2}(^3P)$	$Ne^{+3}(^4S)$	15.42	1.80	2.28	2.0
$Ne^{+3}(^4S)$	$Ne^{+4}(^3P)$	25.52	3.11	1.96	3.0
$Ne^{+4}(^3P)$	$Ne^{+5}(^2P)$	30.60	1.40	1.47	3.0
$Mg^0(^2S)$	$Mg^+(^2S)$	1.15	7.94	1.0	2.67
$Mg^+(^2S)$	$Mg^{+2}(^1S)$	2.27	0.20	4.21	1.5
$Mg^{+2}(^1S)$	$Mg^{+3}(^2P^0)$	12.09	5.04	4.44	1.75
$Mg^{+3}(^2P^0)$	$Mg^{+4}(^3P)$	16.50	2.3	1	2
$Mg^{+4}(^3P)$	$Mg^{+5}(^4S^0)$	21.32	1.9	1	2.3
$S^0(^3P)$	$S^+(^4S^0)$	1.56	22.9	2.0	3
$S^+(^4S^0)$	$S^{+2}(^3P)$	3.52	8.2	1.0	6
$S^{+2}(^3P)$	$S^{+3}(^2P^0)$	5.26	1.16	6.38	3
$S^{+3}(^2P^0)$	$S^{+4}(^1S)$	7.14	2.95	4.53	2
$Si^0(^3P)$	$Si^+(^2P)$	1.23	39.2	4.42	5
$Si^+(^2P^0)$	$Si^{+2}(^1S)$	2.47	2.65	0.6	3
$Si^{+2}(^1S)$	$Si^{+3}(^2S)$	5.05	2.48	2.3	1.8
$Si^{+3}(^2S)$	$Si^{+4}(^1S)$	6.81	0.85	2.0	1

Table 2.6. Parameters of the coefficient of radiative recombination for a number of ions

Ion	A_{rad} [cm^3 s^{-1}]	η	α_n [cm^3 s^{-1}]	Ion	A_{rad} [cm^3 s^{-1}]	η	α_n [cm^3 s^{-1}]
HeI	4.3(−13)*	0.672	1.8(−13)	NeIII	4.4(−12)	0.675	1.2(−12)
CI	4.7(−13)	0.624	2.7(−13)	NeIV	9.2(−12)	0.668	2.7(−12)
CII	2.3(−12)	0.645	1.2(−12)	NeV	1.5(−11)	0.684	4.0(−12)
CIII	3.2(−12)	0.770	1.3(−13)	MgI	1.4(−13)	0.885	1.6(−15)
CIV	7.5(−12)	0.817	4.0(−13)	MgII	8.8(−13)	0.838	5.4(−15)
NI	4.1(−13)	0.608	1.8(−13)	MgIII	3.5(−12)	0.734	4.0(−13)
NII	2.2(−12)	0.639	1.0(−12)	MgIV	7.7(−12)	0.714	1.3(−12)
NIII	5.0(−11)	0.676	1.9(−12)	MgV	1.4(−11)	0.716	2.7(−12)
NIV	6.5(−12)	0.743	2.3(−13)	SiI	5.9(−13)	0.601	4.6(−13)
NV	1.5(−11)	0.850	6.5(−13)	SiII	1.0(−12)	0.786	1.3(−13)
OI	3.1(−13)	0.678	1.1(−11)	SiIII	3.7(−12)	0.693	1.0(−12)
OII	2.0(−12)	0.646	7.5(−13)	SiIV	3.5(−12)	0.821	2.0(−14)
OIII	5.1(−12)	0.666	1.9(−12)	SI	4.1(−13)	0.630	2.5(−13)
OIV	9.6(−12)	0.670	3.1(−12)	SII	1.8(−12)	0.686	9.4(−13)
OV	1.2(−12)	0.779	3.6(−13)	SIII	2.7(−12)	0.745	1.2(−13)
OVI	2.3(−11)	0.802	9.7(−13)	SIV	5.7(−12)	0.755	2.9(−13)
NeI	2.2(−13)	0.759	3.4(−14)	SV	1.2(−11)	0.701	2.5(−12)
NeII	1.5(−12)	0.693	3.5(−13)				

* 4.3(−13) means 4.3×10^{-13}.

where the numerical values of the parameters $A_{\rm di}$ and T_0 are given in Table 2.7 (Aldrovandi, Pequignot, 1973; see also Storey, 1981, 1994).

Note the absence in Table 2.7 of oxygen ions OI, OII etc.; as well as of neon ions NeI, NeII etc., the forbidden lines of which are widely seen in the spectra of PNs. This fact is explained by the extremely small values of $C_{\rm di}$ – of the order of 10^{-12} cm^3 s^{-1}.

The numerical values of the recombination coefficient $C_{\rm n}$ $(T_{\rm e})$ for hydrogen atoms are presented in Table 3.2 for a large interval of electron temperatures

Table 2.7. Parameters of the dielectric recombination coefficient for a number of ions

Ion	$A_{\rm di}$ [cm^3 s^{-1} K$^{3/2}$]	T_0 [10^4 K]	Ion	$A_{\rm di}$ [cm^3 s^{-1} K$^{3/2}$]	T_0 [10^4 K]
CI	6.9(−4)	11	MgI	1.7(−3)	5.1
CII	7.0(−3)	1.5	SiI	6.2(−3)	11
CIII	3.8(−3)	9.1	SiII	1.4(−2)	12
NI	5.2(−4)	13	SiIII	1.1(−2)	10
NII	1.7(−3)	14	SI	7.3(−3)	11
NIII	1.2(−2)	18	SII	4.9(−3)	12
NIV	5.5(−3)	11	SIII	9.1(−3)	13

and for the first fourteen levels. The values of the total recombination coefficient $C(T_e)$, i.e. for all levels of hydrogen atoms, are given in Table 3.3.

3. Emission Lines of Hydrogen and Helium

3.1 The Intensities of Balmer Lines of Hydrogen

In the overwhelming majority of cases the PNs are transparent to the radiation in the subordinate lines of hydrogen, in particular the Balmer series, and this circumstance opens the possibility of solving the important problem of the distribution of the atoms over the different excited levels by the simplest means. The final aim is to find out the relative intensity of the Balmer lines, or, as it is usually called, the *Balmer decrement.*

In the simplest case when the role of inelastic electron collisions can be ignored, the following two processes should be taken into account: (a) the ionizations only from the ground state of hydrogen; the ionizations from the subordinate levels can be neglected owing to the negligible number of atoms in excited levels; and (b) the recombinations of free electrons with protons at all excitation levels accompanied by downward cascades.

The starting point in calculating the population of the excited states is the condition of stationarity according to which the number of atoms that arrive in a given state must be exactly equal to the number of atoms leaving it.

Let us write the condition of stationarity for a level of the hydrogen atom, characterized by the principal quantum number n and azimuthal quantum number l. First we have to write down the total number of atoms arriving at the level nl, then the total number of atoms leaving the same level, and finally we equate the two expressions.

The total number of atoms arriving at the level nl is given by the following three quantities.

(a) The number of free electrons captured directly to the level nl:

$$N_e N^+ C_{nl}(T_e)\,,$$

where $C_{nl}(T_e)$ is the recombination coefficient of the electron for the level nl; N_e and N^+ are, respectively, the free electron and proton concentrations.

(b) The number of spontaneous transitions from higher discrete states $n'l'$ to the states nl:

$$\sum_{n'=n+1}^{\infty} \sum_{l'=0}^{n'-1} N_{n'l'} A_{n'l', nl} ,$$

where $N_{n'l'}$ is the number of atoms in the state $n'l'$ and $A_{n'l', nl}$ is the Einstein coefficient for the spontaneous transition $n'l' \to nl$.

(c) The number of transitions from the ground state to the level nl under the influence of radiation (radiative transitions):

$$N_1 B_{1, nl} \, \rho_{1, nl} ,$$

where $B_{1, nl}$ is the Einstein coefficient for the induced transition $1 \to nl$ and $\rho_{1,nl}$ is the radiation density corresponding to this transition.

Thus, the total number of atoms arriving at the level nl per unit time and per unit volume will be

$$N_e N^+ C_{nl}(T_e) + \sum_{n'=n+1}^{\infty} \sum_{l'=0}^{n'-1} N_{n'l'} A_{n'l', nl} + \mathrm{N}_1 B_{1, nl} \, \rho_{1,nl} \, . \tag{1}$$

On the other hand, the total number of atoms leaving the level nl will be

$$N_{nl} \sum_{n''=1}^{n-1} \sum_{l''=0}^{n''-1} A_{nl, n''l''} , \tag{2}$$

since from the state nl only spontaneous downward transitions are possible (to the level $n''l''$); N_{nl} is the number of atoms per unit volume in the state nl.

If we equate (1) to (2), according to our condition of the stationarity of all transitions, we get

$$\begin{aligned} N_e N^+ C_{nl}(T_e) + \sum_{n'=n+1}^{\infty} \sum_{l'=0}^{n'} N_{n'l'} A_{n'l', nl} + N_1 B_{1n, nl} \rho_{1, nl} = \\ N_{nl} \sum_{n''=1}^{\infty} \sum_{l''=0}^{n''-1} A_{nl, n''l''} , \end{aligned} \tag{3}$$

In the case of thermodynamic equilibrium the distribution of atoms over levels nl (or $n'l'$) is given by the relation

$$N_{nl} = \frac{2l+1}{n^2} N_n , \tag{4}$$

where N_n is the total number of atoms per unit volume in the level n, i.e.

$$N_n = \sum_{l=0}^{n-1} N_{nl} , \tag{5}$$

and $2(2l + 1)$ and $2n^2$ are, respectively, the weights of the levels nl and n.

The solution of (3) cannot satisfy (4). However, taking into account (4), we can simplify system (3) and obtain an approximate solution. Substituting (4) into (3) and summing over all the terms with a given value of n and all possible values of l (from $l = 0$ to $l = n + 1$), we get the following set of algebraic equations:

$$N_e N^+ C_n(T_e) + \sum_{n'=n+1}^{\infty} N_{n'} A_{nn'} + N_1 B_{1n} \rho_{1n} = N_n \sum_{n'=1}^{n-1} A_{nn'} , \tag{6}$$

where

$$C_n(T_e) = \sum_{l=0}^{n-1} C_{nl}(T_e) ,$$

$$A_{n'n} = \frac{1}{n^2} \sum_{l=0}^{n-1} \sum_{l'=0}^{n'-1} (2l' + 1) A_{n'l', nl} , \tag{7}$$

$$B_{1n} = \sum_{l=0}^{n-1} B_{1\,nl}; \qquad \rho_{1n} = \sum_{l=0}^{n-1} \rho_{1\,nl}$$

and N_n is given by (5).

In general, (6) is independent of the optical properties of the nebula. One often considers two nebular models. In the first one (case A), one assumes that the nebula is thin at the frequencies of the Lyman series. In this case the last member of the line on the left hand side of (6) will be equal to zero (so far as $\rho_{1n} = 0$) and the equation takes the form

$$N_e N^+ C_n(T_e) + \sum_{n'=n+1}^{\infty} N_{n'} A_{n', n} = N_n \sum_{n'=1}^{n-1} A_{n'} . \tag{8}$$

In the second case (case B) we assume that the nebula is thick in the Lyman lines; this corresponds to Zanstra's model. In this case all photons in the lines of the Lyman series emitted within the nebula will also be absorbed. In other words, the number of $n \to 1$ transitions must be exactly balanced by the number of $1 \to n$ transitions, i.e.

$$N_n A_{n1} = N_1 B_{1n}\, \rho_{1n} . \tag{9}$$

Therefore, in case B, (6) yields

$$N_e N^+ C_n(T_e) + \sum_{n'=n+1}^{\infty} N_{n'} A_{n'n} = N_n \sum_{n'=2}^{n-1} A_{n'n} . \tag{10}$$

Thus, in both cases we are led to a set of linear algebraic equations depending on the parameter $N_n / N_e N^+ = z_n$. Each set is solved independently to obtain the population of the energy levels of hydrogen (starting from the third). If we know the number z_n it is easy to find out the relative intensities of the emission lines. The numerical values of transition probability coefficients for hydrogen atom A_{ij}, where $i(=n')$ corresponds to the upper level and $j(=n)$ to the lower one, are given in Table 3.1 for a number of transitions $i \rightarrow j$ of practical interest.

Table 3.1. The values of transition probability coefficients A_{ij} for the hydrogen atom

	j						
i	1	2	3	4	5	6	7
2	4.67×10^8	–	–	–	–	–	–
3	5.54×10^7	4.39×10^7	–	–	–	–	–
4	1.27×10^7	8.37×10^6	8.94×10^6	–	–	–	–
5	4.10×10^6	2.52×10^6	2.19×10^6	2.68×10^6	–	–	–
6	1.64×10^6	9.68×10^5	7.74×10^5	7.67×10^5	1.02×10^6	–	–
7	7.53×10^5	4.37×10^5	3.34×10^5	3.03×10^5	3.24×10^5	4.50×10^5	–
8	3.85×10^5	2.20×10^5	1.64×10^5	1.42×10^5	1.38×10^5	1.55×10^5	2.26×10^5

Note that the third term on the left hand side of (6) is an inconvenient one because of the existence of an unknown parameter – the density of the radiation ρ_{1n} vanishes for both models A and B (though for different reasons). Formally (8) and (10) differ from each other only in that in case (A) the sum on the right hand side starts from $n = 1$ while in the second case (B) it starts from $n = 2$.

The energy emitted by the nebula in the Balmer series corresponding to the transition $n \rightarrow 2$ is equal to

$$E_{n2} = h\nu_{n2}A_{n2} \int N_n \, \mathrm{d}V \,, \tag{11}$$

where the integral extends over the entire volume of the nebula. However, $N_n = z_n N_\mathrm{e} N^+$, and if we assume that the electron temperature does not vary within the nebula, we obtain

$$E_{n2} = z_n A_{n2} \, h\nu_{n2} \int N_\mathrm{e} N^+ \, \mathrm{d}V \,. \tag{12}$$

This formula gives the intensity of the Balmer lines up to a certain factor. In a particular case, when we have a nebula that is homogeneous and completely ionized over the whole volume, then $N_\mathrm{e} = N^+$, and we have for the intensity of any Balmer line

$$E_{n2} = z_n \, A_{n2} \, h\nu_{n2} \, N_\mathrm{e}^2 \, \mathrm{d}V \,. \tag{13}$$

As we can see, numerically z_n depends only on T_e, and hence relationship (13) in essence reveals the dependence of the intensity of hydrogen lines of all series, including the Balmer one, on the electron concentration N_e: this dependence is universal for all cases and has the form

$$E_{nk} \sim N_e^2 \,. \tag{14}$$

Rewriting (13) for the line $H_\beta(n = 4)$ and dividing by (13), we get for the relative intensities of the Balmer lines (with respect to the intensity of H_β) i.e. for the *Balmer decrement*, the following expression:

$$\frac{E_{n2}}{E_{42}} = \frac{z_n}{z_4} \frac{A_{n2}}{A_{42}} \frac{\nu_{n2}}{\nu_{42}} \,, \qquad n = 3,\, 4 \,\ldots \,. \tag{15}$$

Equation (10) was first solved by Cillie (1936), who used the first 12 levels ($n = 3, 4, \ldots, 14$) and neglected the rest. Subsequent solutions of (8) and (10) were obtained by a number of authors aiming to improve the original results. Obviously, the quantity N_n, for example, with $n \to \infty$ should be greater than for $n = 14$ since the increase of the number of levels leads to the increase of the population of the given level N_n formed by cascade transitions from the higher levels.

Equations (8) and (10) were solved for a practically infinite number of levels, $n \to \infty$, in 1938 by Baker and Menzel. A dimensionless parameter b_n is introduced:

$$b_n = z_n \frac{(2\pi\mu k T_e)^{3/2}}{h^3} \frac{e^{-\kappa_n/kT_e}}{n^2} \,, \tag{16}$$

which indicates by what factor the value $N_n/N_e N^+$ in the nebula differs from the value of the same ratio $N_n/N_e N^+$ in the case of thermodynamic equilibrium at the temperature T_e. The obtained value of b_n is fairly close to unity and, as one would expect, $b_n \to 1$ as $n \to \infty$.

In the most general form, the problem of the relative intensities of hydrogen lines as well as the lines of neutral and ionized helium, and in particular of the Balmer decrement, has been analysed and solved comprehensively by a number of theoreticians (e.g. Pengelly, Seaton, 1964; Brocklehurst, 1971, 1972). The essence of the subsequent improvement of the theory depends on taking into account of a number of factors: the specification of the numerical value of the Kramers–Gaunt factor at the determination of the Einstein transition coefficients and the influence of the degeneracy of energetic levels of hydrogen atoms in relation to azimuthal quantum number n.

Now, we have quite a satisfactory solution of (8) and (10) for the infinitive levels ($n \to \infty$) taking into account the mentioned factors. During these calculations the most precise values for the recombination coefficient $C_n(T_e)$ for the hydrogen-like atoms are used:

$$C_e(T_e) = 5.197 \times 10^{-14}\, z\, \frac{\gamma^{1/2}}{n} X_n S_n(\gamma) \,, \tag{17}$$

Table 3.2. Values of the recombination coefficient $C_n(T_e)$ for hydrogen atoms (in 10^{-14} cm^3 s^{-1})

	T_e, K							
n	5 000	10 000	15 000	20 000	40 000	60 000	80 000	100 000
1	22.148	15.446	12.390	0.549	7.007	5.410	4.446	3.820
2	11.639	7.655	5.889	4.832	2.876	2.064	1.604	1.299
3	7.309	4.544	3.3521	2.672	1.469	1.005	0.761	0.597
4	5.010	2.955	2.117	1.639	0.755	0.569	0.421	0.333
5	3.641	2.050	1.423	1.098	0.547	0.350	0.250	0.202
6	2.724	1.490	1.011	0.766	0.372	0.238	0.173	0.134
7	2.093	1.112	0.741	0.560	0.264	0.169	0.120	0.092
8	1.655	0.866	0.570	0.431	0.194	0.125	0.088	0.072
9	1.338	0.670	0.446	0.326	0.150	0.092	0.066	0.052
10	1.103	0.550	0.354	0.259	0.116	0.072	0.052	
11	0.909	0.446	0.288	0.208	0.094	0.058		
12	0.767	0.372	0.237	0.172	0.072			
13	0.656	0.310	0.197	0.143	0.062			
14	0.556	0.263	0.168	0.119	0.052			

where

$$\gamma = 157\,890\,\frac{Z^2}{T_e}\,,$$

$$x_n = \frac{\gamma}{n^2}\,,$$

are tabulated by Seaton (1960). Table 3.2 has been compiled on the basis of these data, where the numerical values of the function $C_n(T_e)$ for the hydrogen atom ($Z = 1$) and for numbers of n up to $n = 14$ and for some magnitudes of T_e that are of astrophysical interest are given.

An asymptotic expression of the function $C(Z, T_e)$, the so-called complete recombination coefficient, has been derived as well, i.e. the recombination coefficient when the electrons are captured at all levels, from $n = 1$ up to $n = \infty$:

$$C(Z,T_e) = \sum_{n=1}^{\infty} C_n(Z, T_e) = 5.197 \times 10^{-14}\,\gamma\, Z(0.4288 + 0.5 \ln\gamma - 0.469\gamma^{-1/3})\,. \tag{18}$$

Table 3.3. Values of the function $C(T_e)$ for hydrogen in units of 10^{-13} cm^3 s^{-1}

T_e, K	5 000	10 000	15 000	20 000	40 000	60 000	80 000	100 000
$C(T_e)$	6.727	4.118	3.066	2.477	1.456	1.055	0.831	0.692

Table 3.4. Values of recombination coefficients $C_n(T_e)$ at levels S and P for hydrogen in units of 10^{-14} cm^3 s^{-1}

Level	T_e, K 5 000	10 000	20 000
$1\,^2S$	22.8	15.8	10.8
$2\,^2S$	3.37	2.34	1.60
$2\,^2P$	8.33	5.35	3.24

This formula gives the value of $C(Z,T_e)$ with an error smaller than 0.5 percent when $T_e/Z^2 < 10^5$ and of 3 percent when $T_e/Z^2 = 10^6$. Table 3.3 gives the numerical values of the function $C(T_e)$ for hydrogen for some values of T_e.

For the solution of some problems connected with the physics of gaseous nebulae it is necessary to know the exact values of the recombination coefficient $C_n(T_e)$ including the ground level, 1^2S, as well as both sublevels, 2^2S and 2^2P, of hydrogen; these data are given in Table 3.4 (Seaton, 1960).

The ultimate result of the solution of (8) and (10) is the determination of the numerical values of z_n for line values of n of hydrogen atoms, for both nebular models, A and B, and for three values of the electron temperature T_e: 5 000, 10 000 and 20 000 K; these data are given in Table 3.5. Then the relative intensities of the Balmer lines – the Balmer decrement – are obtained with the help of (15), as well as the Paschen decrement (transitions $n \to 3$): the final results are given in Table 3.6 (model A) and Table 3.7 (model B).

Table 3.5. Values of the quantity z_n in units of 10^{-20} cm^{-3}

n	T_e, K 10 000	20 000	10 000	20 0000
	Model A		Model B	
2	0.0321	0.0185	–	–
3	0.0797	0.0439	0.25	0.134
4	0.167	0.085	0.36	0.186
5	0.284	0.145	0.54	0.272
6	0.448	0.223	0.78	0.38
7	0.656	0.319	1.08	0.52
8	0.908	0.436	1.44	0.68
9	1.213	0.571	1.86	0.87
10	1.564	0.726	2.34	1.08
15	4.062	1.81	5.66	2.50
20	7.87	3.41	10.6	4.54
25	12.97	5.54	17.0	6.87
30	19.47	8.21	25.2	10.52

Table 3.6. Theoretical intensities of hydrogen lines in units of $H_\beta = 1.00$: Model A

Line	λ; Å	T_e, K 5 000	10 000	20 000
		Lyman series		
L_α	1 216	32.6	32.7	32.8
		Balmer series		
3–2H_α	6 563	3.10	2.86	2.69
4–2H_β	4 861	1.00	1.00	1.00
5–2H_γ	4 340	0.458	0.470	0.485
6–2H_δ	4 101	0.250	0.262	0.271
7–2H_ε	3 970	0.153	0.159	0.167
8–2H_8	3 889	0.102	0.107	0.112
9–2H_9	3 835	0.0717	0.0748	0.0785
10–2H_{10}	3 797	0.0522	0.0544	0.0571
		Paschen series		
4–3P_α	18 751	0.562	0.466	0.394
5–3P_β	12 818	0.241	0.216	0.196
6–3P_γ	10 938	0.1260	0.1179	0.1100
8–3P_8	9 545	0.0497	0.0474	0.0452
10–3P_{10}	9 014	0.0251	0.0239	0.0228
$4\pi J(H_\beta)/n_e n^+$ $\times 10^{-25}$ ergs cm^{-3} s^{-1}		1.54	0.83	0.42

Table 3.7. Theoretical intensities of hydrogen lines in units of $H_\beta = 1.00$: Model B

Line	Å	T_e, K 5 000	10 000	20 000	Line	Å	T_e, K 5 000	10 000	20 000
		Balmer series					Paschen series		
H_α	6 563	3.05	2.87	2.76	P_α	18 751	0.427	0.352	0.293
H_β	4 861	1.00	1.00	1.00	P_β	12 818	0.187	0.165	0.1463
H_γ	4 340	0.451	0.466	0.474	P_γ	10 938	0.0991	0.0906	0.0820
H_δ	4 101	0.249	0.256	0.262	P_8	9 545	0.0392	0.0367	0.0343
H_ε	3 970	0.153	0.158	0.162	P_{10}	9 014	0.0199	0.0185	0.0172
H_8	3 889	0.101	0.105	0.107	P_{15}	8 948	0.0057	0.0053	0.0050
H_{10}	3 797	0.051	0.053	0.054	P_{20}	8 395	0.0024	0.0022	0.0021
$4\pi J(H_\beta)/n_e n^+$, $\times 10^{-25}$ ergs cm^{-3} s^{-1}							2.20	1.24	0.66

For model A, the calculated intensities of the line L_α – the first and the only hydrogen line which can be generated in accordance with model B; the intensities of the rest of the lines of this series, – L_β, L_γ etc. – are equal to zero.

The numerical values of the parameter b_n determined with the help of (16) are as follows:

n	3	5	10	20	30
$b_n(T_e = 10\,000$ K)	0.089	0.233	0.448	0.571	0.618
$b_n(T_e = 20\,000$ K)	0.330	0.460	0.620	0.713	0.750

As we can see, the numerical values of b_n are noticeably different from unity, especially for low levels, and the divergence from thermodynamic equilibrium monotonically decreases, as is expected, while passing to high numbers of the series, $b_n \to 1$ at $n \to \infty$.

In various problems concerning the physics of nebulae, gaseous envelopes, stellar atmospheres, etc. one has to present the emission in one or another emission line in the units of the volume emission $J(H_\beta)$ in the hydrogen line H_β. The latter can be represented in the following form:

$$E_\beta = z_{42} N_2 N^+ A_{42} h\nu_\beta \,, \tag{19}$$

where z_{42} is, as follows from Table 3.5, a weak function of the electron temperature T_e. The latter dependence can be represented in the form

$$z_{42} = 0.36 \times 10^{-20} \left(\frac{10^4}{T_e}\right)^\alpha \,. \tag{20}$$

Within 3% accuracy the index $\alpha = 1$ in the interval of $T_e \sim 5\,000$–$20\,000$ K. Then the volume emission coefficient in the line H_β for a completely ionized medium, i.e. when $N^+ = N_e$, takes the following form, where we have also substituted $A_{42} = 8.37 \times 10^6$ s^{-1} (Table 3.1) and $h\nu_\beta = 4.08 \times 10^{-12}$ ergs:

$$E_\beta = 1.23 \times 10^{-25} (10^4/T_e) N_e^2 \text{ ergs cm}^{-3} \text{ s}^{-1} \,. \tag{21}$$

The volume emission E^0_β for one free electron in 1 cm^{-3} ($N_e = 1$ cm^{-1}) and at $T_e = 10\,000$ K yields

$$E^0_\beta = 1.23 \times 10^{-25} \text{ ergs cm}^{-3} \text{ s}^{-1} \,. \tag{21a}$$

The numerical values of the volume emission $J(H_\beta)$ in the line H_β as a function of T_e and in units of $n_e n^+/4\pi$ are also given in Tables 3.6 and 3.7 for both models A and B.

3.2 The Role of Collisions

It is not difficult to guarantee the negligible role of electron collisions on the Balmer decrement of gaseous nebulae. This is explained by three arguments. First, the mean energy of the escaped electron is approximately two times smaller at the star's temperature of 60 000 K than the energy of the excitation levels. Second, the sufficient part of the energy of free electrons goes on the excitation of the forbidden lines. Third, at non-elastic collisions of free electrons with hydrogen atoms the principal part of the energy is still at the expense of the excitation of Balmer lines.

However, all these considerations need a detailed quantitative analysis. Now the same system of stationarity of transitions (6) must be written bearing in mind also the transitions between different levels under the electron collisions. As a result, the initial system of equations takes the following form (for model B):

$$N_e N^+ C_n(T_e) + \sum_{n'=n+1}^{\infty} N_{n'} A_{n'n} + N_e \sum_{l'=l\pm1}^{n-1} N_{nl'nl} = N_n \left[\sum_{n'=2}^{n-1} A_{nn'} + N_e \sum_{l'=l\pm1}^{n-1} q_{nl\,nl'} \right] , \tag{22}$$

where $q_n = q(T_e)$ is the coefficient of collisional transitions. In (22) the impacts of the first kind have been taken into account (from down to up) as well as the impacts of the second kind (from up to down) without radiation.

Unlike (6), in (22) the electron concentration N_e is also present. Hence, if in the case of (6) the decrement depends on the electron temperature only (through recombination coefficient of $C_n(T_e)$), now the resulting decrement will depend, at least formally, on both T_e and N_e.

By solving (22), we get the populations of atoms at different levels of hydrogen, and then it is not difficult to obtain, with the help of relationships of type (15), the relative intensities of the lines, i.e. the decrement itself for any series. The results are presented in Table 3.8 for three values of T_e, 5 000, 1 000 and 20 000 K, and three values of N_e, 10^2, 10^4 and 10^6 cm^{-3} (Brocklehurst, 1972).

The results presented in Table 3.8 should be considered as of particular interest, since the dependence of relative intensities of the lines on N_e, though it exists, is extremely weak, however.

Thus, among the three possible factors which can influence the steepness of the Balmer decrement, at least one – the electron collisions – should be abandoned completely. However, there are still two factors – interstellar extinction and self-absorption in the nebula.

Table 3.8. Calculated Balmer decrement for hydrogen, for both recombinations and electron collisions. Model B. $H_\beta = 1.00$. $N_e = 10^2,\ 10^4,\ 10^6\ cm^{-3}$

	$T_e = 5\,000$ K		$T_e = 10\,000$ K			$T_e = 20\,000$ K	
Line	10^2	10^4	10^2	10^4	10^6	10^2	10^4
H_α	3.03	3.00	2.86	2.85	2.81	2.74	2.74
H_β	1.00	1.00	1.00	1.00	1.00	1.00	1.00
H_γ	0.459	0.480	0.469	0.469	0.471	0.476	0.476
H_δ	0.252	0.253	0.259	0.259	0.262	0.264	0.264
H_ε	0.154	0.155	0.159	0.159	0.163	0.163	0.163
H_8	0.102	0.102	0.105	0.105	0.110	0.107	0.107

3.3 Comparison with Observations

The remarkable result of the theory of the origin of hydrogen emission lines should be considered the complete independence of the theoretical Balmer decrement on the most important physical parameters of the nebula – the electron temperature T_e and the electron concentration N_e, as well as on the nucleus temperature T_* and the factor of dilution W. Owing to this, the Balmer decrement turns out to be a universal parameter which characterizes the nature of the PN itself.

At the same time, the "pure" or "universal" Balmer decrement can reach the observer suffering distortions under the inner and outer factors – self-absorption in the first case and interstellar extinction in the second. However, similar distortions yield to the quantitative analysis, and in each case the initial Balmer decrement can be reconstructed with sufficient reliability.

However, the theoretical "pure" Balmer decrement has a defect: its absolute value and, equally, the steepness are practically the same for both models A and B (Tables 3.6 and 3.7). In other words, by the magnitude or steepness of the Balmer decrement one cannot distinguish the optically thin, in the lines of Lyman series, nebula (model A) from the optically thick one (model B).

In Table 3.9, the observed Balmer decrements are given for a dozen PNs. For the first half of them for which the parameter of interstellar extinction c is less than 0.1, the directly observed decrements at least for these objects must be distorted by interstellar absorption to the smallest degree. However, this does not mean that the decrement is not already distorted within the nebula owing to self-absorption. Nevertheless, by comparing the observations, which are not corrected for any effect, with the theory (the last line in Table 3.9) we conclude that on average the agreement is quite good.

The next six PNs are influenced by an interstellar extinction to various degrees characterized by the values of c from 0.4 up to 2.5. However, their

Table 3.9. Balmer decrements for a group of planetary nebulae: directly observed (first six objects) and corrected for interstellar extinction (next six objects). $H_\beta = 1.00$

Planetary nebula	c	H_α	H_γ	H_δ
NGC 2392	0.02	2.91	0.473	0.260
NGC 3242	0.02	3.04	0.449	0.306
NGC 4361	0.01	2.79	0.464	0.284
NGC 6210	0.10	3.01	0.432	0.260
NGC 6720	0.10	3.06	0.463	0.247
Vy 1–2	0.02	2.96	0.456	0.289
IC 4846	0.44	2.91	0.505	0.244
NGC 6567	0.71	2.95	0.515	0.249
NGC 7027	1.25	2.86	0.522	0.261
M 1–5	1.43	2.94	0.485	0.273
M 1–67	1.68	3.13	0.549	0.275
M 3–35	2.54	2.80	0.507	0.230
Mean		2.95	0.485	0.265
Theory		2.85	0.469	0.259

corrected decrements turn out to be practically indistinguishable from each other, and on average are in complete accordance with the theory.

Thus one of the important achievements of the physics of PNs, photoionization or, as it is often called, the recombination theory of the excitation of hydrogen emission lines, is reliably confirmed by observations.

3.4 The Balmer Decrement: Observations

Theory thus predicts one and the same relative intensity for the hydrogen lines: $H_\alpha : H_\beta : H_y = 2.85 : 1 : 0.47$ (model B).

What do the observations show? Values $H_\alpha/H_\beta < 2.85$ have never been observed within the errors of measurements. The cases $H_\alpha/H_\beta = 2.85$, i.e. in complete accord with the theory, are extremely rare. The overwhelming majority of PN observations give $H_\alpha/H_\beta > 2.85$. This can be formulated as follows: the steepness of the observed Balmer decrement is larger compared with the steepness of the theoretical decrements, and the higher the degree of steepness, the stronger this inequality will be fulfilled.

In Table 3.10, a sample of PNs with increasing values of observed intensities of the first Balmer line, H_α, is given (Barker, 1978, 1980, 1982). As we can see, in some cases the ratio H_α/H_β differs from its theoretical value (2.85) for its larger values – up to 5–10, even up to 10 or 20 times. It should be mentioned, however, that the probability of finding PNs with a given value of H_α/H_β is too great to be constant. The largest number of PNs is expected with the observed ratio H_α/H_β in the interval of 3–4.

Table 3.10. Balmer decrement ($H_\beta = 1.00$) for selected planetary nebulae with increasing value of the intensity of the first line of the series H_α of hydrogen and increasing magnitude of the parameter c of interstellar extinction

Planetary nebula	H_α	H_γ	H_δ	c	$E(\alpha - \beta)$
NGC 4361	2.79	0.46	0.28	0.011	–0.023
NGC 3242	3.04	0.45	0.31	0.023	0.070
IC 2149	3.42	0.43	0.21	0.272	0.298
Hu 2–1	4.15	0.42	0.22	0.451	0.408
M 2–50	4.73	0.42	0.13	0.705	0.550
M 1–74	5.64	0.38	0.21	0.78	0.741
M 3–27	6.53	0.48	0.22	1.08	0.900
NGC 7027	7.41	0.34	0.15	1.25	1.037
M 1–5	8.85	0.29	0.14	1.43	1.230
M 1–67	11.40	0.30	0.13	1.68	1.505
He 2–76	14.5			2.11	1.76
He 2–84	18.9			1.46	2.85
M 3–35	20.1	0.21	0.07	2.54	2.122
Th 3–4	23.2			2.73	2.27
He 2–143	31.4			3.12	2.60
He 2–145	52.6*			3.79	3.16
Theory	2.85	0.47	0.26	0.00	0.000

* Possible HII region.

The relative intensities of the lines H_γ and H_δ are also given in Table 3.10. The derivations of their intensities as is expected appear towards the lower side of their theoretical values (last line of the table). In general the increase of H_α is accompanied by the monotonic fall of the strength of the lines H_γ, H_δ, etc. On average this is true, though the monotonic fall of H_γ is somewhat distorted (we do not mention the object M 3-27, which will be discussed separately in Chap. 15).

The main reason for these deviations of the observed Balmer decrements is established to be selective interstellar extinction. Extinction is large at short wavelengths and smaller at long wavelengths. Therefore the observed ratio H_α/H_β becomes larger the larger the absorption or the larger the distance of the nebula.

The observed intensities of emission lines must be corrected for interstellar selective extinction. The essence of this procedure is as follows. It is useful to characterize the complete interstellar extinction along the path from the observer to the given object by the so-called *extinction parameter* c:

$$c = \log \frac{I(H_\beta)}{F(H_\beta)} , \tag{23}$$

where $I(H_\beta)$ is the possible flux in the absence of the extinction and $F(H_\beta)$ is the observed flux distorted by extinction. In stellar magnitudes the extinction in the line will be $A_\beta = 2.5\ c$.

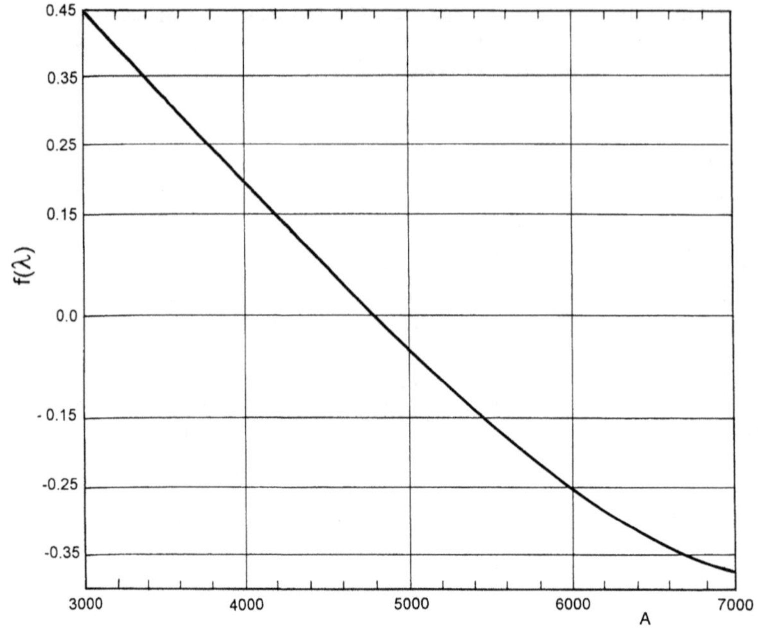

Fig. 3.1. Numerical values of the interstellar extinction function $f(\lambda)$ as a function of wavelength for the optical region (3 000–7 000 Å)

Once we know the magnitude of c, the value of the flux $I(\lambda)$ in the line λ corrected for the interstellar extinction can be obtained in the following manner:

$$I(\lambda) = F(\lambda)\,10^{c[1+f(\lambda)]}\,, \tag{24}$$

where $F(\lambda)$ is the observed flux in the line λ and $f(\lambda)$ is the *reddening function* the numerical values of which depend on λ and are determined by special observations. For the first eight lines of the Balmer series of hydrogen the values of $f(\lambda)$ are as follows:

Line	H_α	H_β	H_γ	H_δ	H_ε	H_8	H_9	H_{10}
λ, Å	6 563	4 861	4 340	4 101	3 970	3 889	3 835	3 797
$f(\lambda)$	−0.334	0	0.114	0.168	0.204	0.223	0.239	0.247

The dependence of $f(\lambda)$ on λ is given for the wavelength interval 3 000–7 000 Å in Fig. 3.1, drawn according to the data of various observations (see Nandy et al., 1975; Savage, Mathis, 1979; Stasinska et al., 1993). Note that the function $f(\lambda)$ is negative in the region $\lambda > 4\,861$ Å.

Rewriting (24) for the line H_β and dividing again by (24), and in view of the condition $f(\lambda_\beta) = 0$ for the line H_β, we obtain

$$\frac{I(\lambda)}{I(H_\beta)} = \frac{F(\lambda)}{F(H_\beta)}\,10^{cf(\lambda)}\,. \tag{25}$$

With the help of this expression one can obtain the corrected values of the relative intensities of any line, $I(\lambda)/I(H_\beta)$, if the observed ratio $F(\lambda)/F(H_\beta)$ of the line and the numerical value of c are known for the given nebula. As for the numerical values of the extinction function $f(\lambda)$, they are one and the same for all nebulae.

For the well known PN NGC 7027, for example, we have $H_\alpha/H_\beta = 7.41$, $c = 1.25$ (Table 3.10) and $f(H_\alpha) = -0.334$. With these data we get from (25) $I(H_\alpha)/I(H_\beta) = 2.83$, which is in complete agreement with the theoretical model B.

The reddening of the nebula by interstellar extinction should be characterized also by the *the colour excess* $E(\alpha - \beta)$ for the Balmer lines in the following manner:

$$E(\beta - \alpha) = 2.5 \log \frac{I(H_\beta)}{I(H_\alpha)} \frac{F(H_\beta)}{F(H_\alpha)}, \tag{26}$$

or substituting $I(H_\alpha)/I(H_\beta) = 2.85$, we get

$$E(\beta - \alpha) = -1.137 + 2.5 \log \frac{F(H_\alpha)}{F(H_\beta)}, \tag{27}$$

In the absence of extinction we have $F(H_\alpha)/F(H_\beta) = I(H_\alpha)/I(H_\beta) = 2.85$, and hence $E(\alpha - \beta) = 0$. The numerical values of $E(\alpha - \beta)$ for our sample of PNs are given in the last column of Table 3.10.

3.5 The Decrement Diagram

The Balmer decrement is characterized not only by the ratio H_α/H_β, but also by H_γ/H_β. In an ideal case when only the interstellar excitation is taken into consideration, the divergence of H_α/H_β and H_γ/H_β from their theoretical values must correspond to one and the same value of the extinction parameter c. However, in the majority of cases these divergences for one and the same PN correspond to the various values of c. But this is not logical, and the reason for this divergence should be searched for elsewhere, namely, in the additional source of absorption besides interstellar extinction. The *self-absorption* within the nebula can be such a source. The self-absorption can arise only in optically thick, in the line H_α, nebula, i.e. when $\tau(H_\alpha) > 1$. So, the concept of the influence of self-absorption on the Balmer decrement arises.

The most obvious way to reveal the dependence of the Balmer decrement on both factors – interstellar extinction and self-absorption within the nebula – is to build the *decrement diagram* as shown in Fig. 3.2 (Miller, Mathews, 1972; Barker, 1978). By the lower inclined line IS the dependence between $F(H_\alpha)/F(H_\beta)$ and $F(H_\gamma)/F(H_\beta)$ is shown to correspond only for the first effect – interstellar extinction at the same c for all three lines – H_α, H_β and

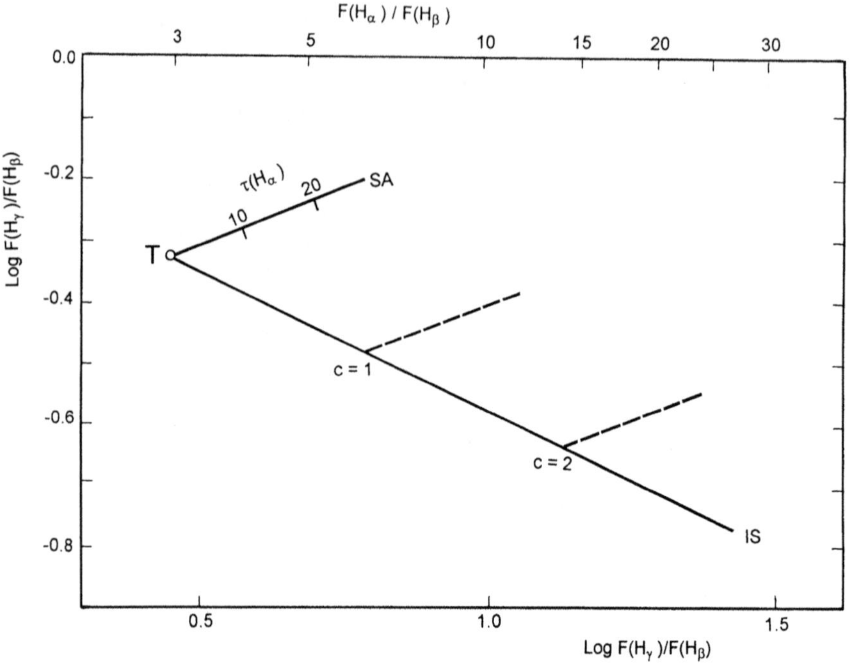

Fig. 3.2. Balmer decrement diagram – the graphical dependence of $\log F(\mathrm{H}_\gamma)/F(\mathrm{H}_\beta)$ on $\log F(\mathrm{H}_\alpha)/F(\mathrm{H}_\beta)$. The lower line ($IS$) relates to the effect of interstellar extinction: the points corresponding to the values of extinction $c = 1$ and $c = 2$ are fixed. The upper slope (SA) and parallel broken lines are related to the effect of self-absorption within the nebula: the points corresponding to the value of $\tau(\mathrm{H}_\alpha) = 10$ and 20 are also fixed. The circle on the left (point T) represents the theory at $c = 0$ $\tau(\mathrm{H}_\alpha) = 0$ and $T_\mathrm{e} = 10\,000$ K (model B)

H_γ. Each point on this line corresponds to the following theoretical values of the decrement:

$$F(\mathrm{H}_\alpha)/F(\mathrm{H}_\beta) = I(\mathrm{H}_\alpha)/I(\mathrm{H}_\beta)\,10^{-f(\lambda_\alpha)c} = 2.85 \times 10^{+0.334c}\,,$$

$$F(\mathrm{H}_\gamma)/F(\mathrm{H}_\beta) = I(\mathrm{H}_\gamma)/I(\mathrm{H}_\beta)\,10^{-f(\lambda_\alpha)c} = 0.47 \times 10^{-0.114c}\,,$$

or

$$\log F(\mathrm{H}_\alpha)/F(\mathrm{H}_\beta) = 0.455 + 0.334c\,, \tag{28}$$

$$\log F(\mathrm{H}_\gamma)/F(\mathrm{H}_\beta) = -0.328 - 0.114c\,. \tag{29}$$

In Fig. 3.2 the positions of the points with $c = 1$ and $c = 2$ are shown. The circle on the left (point T) corresponds to the value of the Balmer decrement at $c = 0$ and $T_\mathrm{e} = 10\,000$ K of model B, i.e. $F(\mathrm{H}_\alpha)/F(\mathrm{H}_\beta) = 2.85$ and $F(\mathrm{H}_\gamma)/F(\mathrm{H}_\beta) = 0.47$ (Table 3.8).

The upper inclined line AS in Fig. 3.2 is related to the second effect – self-absorption in the nebula; the points corresponding to the values of $\tau(\mathrm{H}_\alpha) = 10$ and 20 are marked. In fact, the line AS should be continued

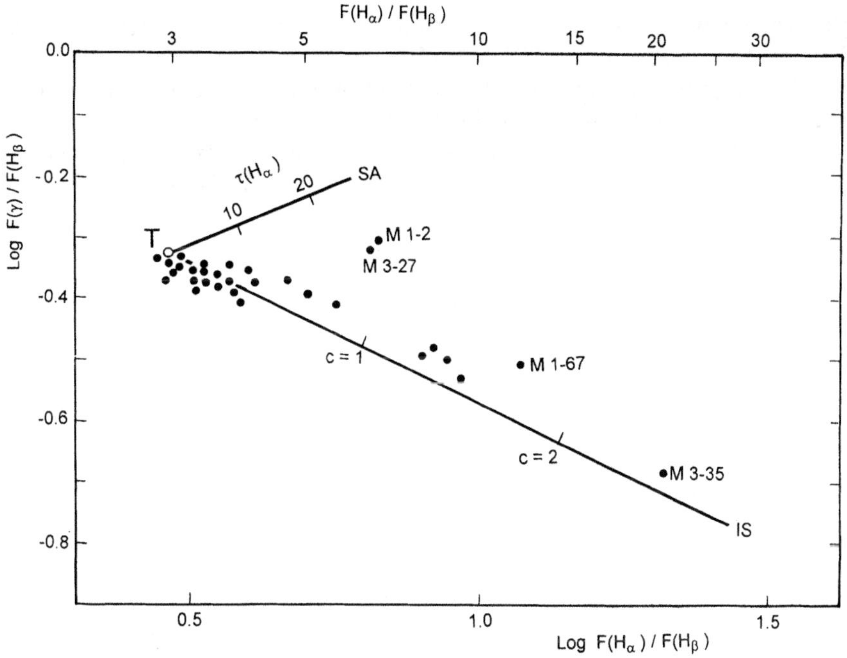

Fig. 3.3. A group of PNs on the Balmer decrement diagram (*dots*). The largest self-absorption, with $t(\mathrm{H}) = 15$, is for M 1-2 and M 3-27; the largest interstellar extinction, with $c = 2.5$, is for M 3-35

from any point of the lower declined line of the interstellar extinction IS; for two points, $c = 1$ and $c = 2$, these lines are shown by broken lines.

3.6 Planetary Nebulae on the Decrement Diagram

The positions of PNs with reliable measurements of the Balmer decrement $F(\mathrm{II}_\alpha)/F(\mathrm{II}_\beta) \sim F(\mathrm{H}_\gamma)/F(\mathrm{H}_\beta)$ are marked by dots on the decrement diagram, Fig. 3.3, using Barker's observations (1978). As we can see, the divergence of the directly observed decrement from the theoretical position – point T – is caused mainly by the first effect – interstellar extinction.

At the same time we also observe the scatter in the observed points around the line IS, interstellar extinction, the evidence of the action also of the second effect – self-absorption. However, judging by the fact that in the majority of cases the observed points are scattered over both sides of the line IS and the same mean value, we conclude that the effect of self-absorption does not play an essential role, if its existence is real and is not caused by errors of measurement. This remark does not refer, however, to the PNs M 1-2, M 3-27 and M 1–67 for which both effects are evident: the interstellar extinction in the amount of $c = 0.50$ and $c = 1.50$ and self-absorption in $\tau(H_\alpha) \approx 15$ or 10.

The large role of the interstellar extinction up to $c \approx 2.5$ is observed for the PN M 3-35 (the limiting point on the right in Fig. 3.3). Note that all four mentioned PNs are anomalous, i.e. they are star-like and presumably anomalously dense, with an electron concentration of about 10^6–10^8 cm^{-3} (see Chap. 15).

3.7 The Self-absorption Effect

The majority of emission lines – the lines of Balmer, Paschen and the highest series of hydrogen and the lines of neutral and ionized helium, all without exception forbidden lines, etc., leave the nebula without any obstacle owing to its transparency to these lines. Hence, there is not a problem concerning the transfer of radiation in these lines.

The situation is different in the case when the nebula is opaque to some line; then both processes – the scattering and the absorption of the photons – should be taken into account. Two extreme cases were examined above: (a) the nebula is transparent to the lines of all series of hydrogen – Lyman, Balmer, etc. (model A), (b) the nebula is opaque to the lines of the Lyman series but is transparent to the lines of the Balmer series (model B).

In addition, one can have a situation when a nebula that is completely opaque in the Lyman lines, as in model B, is at the same time partially opaque for the Balmer series of hydrogen. In this case the Balmer lines cannot leave the nebula without obstruction.

If the nebula is partially opaque, say, in the line H_β, then only a part of the H_β photons resulting from fluorescence can leave the nebula unperturbed, while the remaining H_β photons would be absorbed, sooner or later; as a result the atoms must jump to the level $n = 4$. There is a certain probability that these initial H_β photons will vanish completely via cascade transitions $4 \to 3 \to 2 \to 1$ with the consequent emission of P_α, H_α and L_α photons. As a result, the number of H_β photons emerging from the nebula will decrease, while the number of H_α photons will increase, leading to the rise of the ratio H_α/H_β in comparison with "pure" model B. Then the position of the point H_α/H_β will move to the right, towards larger values of H_α/H_β on the decrement diagram (Fig. 3.2).

Also, if the nebula is opaque, even partially, in the line of H_γ, then after the absorption of this photon we shall have a definite probability of its death, if the atom will perform cascade transitions, say, $5 \to 3 \to 2 \to 1$, with the emission of photons P_β, H_α and L_α. In this case new H_α photons will appear which again will lead to the increase of H_α/H_β. However, H_γ/H_β will also increase in comparison to its value in model B so long as, first, the number of H_β photons is small and, second, the opacity in the line H_γ is smaller than in the line H_β (the absorption coefficient in the lines $\sim \lambda^2$ most probably is $\sim \lambda^2 A_{ij}$). As a result, the corresponding point H_γ/H_β on the decrement diagram will move upwards on the ordinate.

Thus, when $\tau(H_\alpha) > 1$, $\tau(H_\beta) > 1$, etc. the geometrical place of points with the coordinates of H_α/H_β and H_γ/H_β can be presented on the decrement diagram by a track directed upwards and to the right relative to the initial point corresponding to model B, i.e. with the coordinates of $H_\alpha/H_\beta = 2.85$ and $H_\gamma/H_\beta = 0.47$.

The essence of the "effect of self-absorption" lies just in the discrepancy in the values of H_α/H_β and H_γ/H_β compared with model B and in the appearance of the positively inclined line on the decrement diagram.

This is the situation from the qualitative point of view. As for the quantitative interpretation of the problem, this results in the solution of the linear equations of the stationarity of transitions with the addition of terms associated with the "effect of self-absorption". As a result, one has the upper inclined line obtained in 1969 by Cox and Mathews, as shown in Figs. 3.2 and 3.3; from the left side it starts from $\tau(H_\alpha) = 0$ with the two marked points on $\tau(H_\alpha) = 10$ and 20.

3.8 Helium Emission Lines

The starting point to obtain the intensities of helium emission lines, both neutral and ionized, is the same equations of stationarity of transitions ((10) or, taking into account electron collisions, (22)). However, now the corresponding atomic parameters for HeI and HeII – the Einstein coefficients of spontaneous transitions, "collision strengths", etc. – should be used in these equations. In the case of HeII, as a hydrogen-like atom, this operation is realized easily, i.e. by substitution of $Z = 2$ in the corresponding expressions for HI. The situation is more complicated with HeI. The intensities for HeI and HeII were determined for both models A and B, with the role of electron collisions being ignored and being taken into account. These collisions, as one might expect, are negligible.

The predicted intensities for the first five or six lines of HeII of Pickering ($n \to 4$) and Pfund ($n \to 5$) series for model B without taking into account the electron collisions are given in Table 3.11 in accord with Pangelly (1964). The same for the triplet and singlet recombination lines of HeI, in model B, and with the account being taken of electron collisions, are given in Table 3.12 (Brocklehurst, 1972; Robbins, 1970).

In contrast with the hydrogen lines, the relative intensities of HeI and HeII lines, and especially of HeI lines, are noticeably sensitive to the electron temperature T_e. In particular, the steepness of the decrements of all series falls with the increase of T_e. The calculations reveal the most intense line of HeI: it appears to be the pseudoresonance infrared line 10 830 HeI (Table 3.12). The next is the subordinate line 5 876 HI, and also 3 889 HeI.

Table 3.11. Calculated intensities of recombination lines of HeII in units of the intensity J of the line 4 686 HeII, i.e. J(4 686HeII) = 1. Model B

Transition	λ [Å]	T_e, K 5 000	10 000	20 000	400 000
		Pickering series ($n \to 4$)			
5–4	10 123	0.295	0.274	0.256	0.237
6–4	6 560	0.131	0.135	0.135	0.135
7–4	5 411	0.0678	0.0734	0.0779	0.0799
8–4	4 859	0.0452	0.0469	0.0506	0.0527
9–4	4 542	0.0280	0.0315	0.0345	0.0365
10–4	4 339	0.0198	0.0226	0.0249	0.0262
		Pfund series ($n \to 5$)			
6–5	11 626	0.108	0.0956	0.0856	0.0758
7–5	9 344	0.0547	0.0539	0.0513	0.0474
8–5	8 236	0.0320	0.0331	0.0327	0.0311
10–5	7 178	0.0144	0.0144	0.0160	0.0157
$\varepsilon(4\,686) \times 10^{-25}$ ergs cm^{-3} s^{-1}		2.50	1.25	0.60	0.28

While comparing the theory with observations, one should mention that for various reasons not all HeI and HeII lines have been observed with enough reliability in the PN spectra. Among the HeII lines, those of 5 411 HeII and 4 542 HeII are often observed, both being from the Pickering series. For fifty PNs the observed relative intensities of two lines of the Pickering series of ionized helium, 5 411 HeII and 4 542 HeII, were on average two times larger than their predicted values (intensities in units of J(4 686HeII) = 1.00):

Table 3.12. Calculated intensities of recombination lines of HeI with electronic collisions ($n_e = 10^4$ cm^{-3}) in units of J(4 471) = 1. Model B

Line [Å]	T_e, K 5 000	10 000	20 000	Line [Å]	T_e, K 5 000	10 000	20 000
	Triplet lines				Singlet lines		
5 876	3.01	2.76	2.58	6 678	0.867	0.791	0.731
4 026	0.459	0.474	0.487	4 922	0.276	0.274	0.271
3 820	0.251	0.264	0.274	5 016	0.512	0.588	0.689
7 065	0.243	0.328	0.477	3 965	0.199	0.234	0.279
10 830	3.96	4.42	5.01				
3 889	1.90	2.26	2.79				
3 187	0.75	0.92	1.16				
$\varepsilon(4\,471), \times 10^{-27}$ ergs cm^{-3} s^{-1}					9.31	4.84	2.34

	$J(5\,411\text{HeII})$	$J(4\,542\text{HeII})$
Observational mean	0.137	0.072
Theoretical mean	0.073	0.032

The situation is different for HeI lines. For three lines – 5 876 HeI, 6 678 HeI and 4 922 HeI – the mean observed values were corrected for interstellar extinction for a sample of PNs and turned out to be in astonishing accordance with theory ($J(4\,471\text{HeI}) = 1.00$):

	$J(5\,876\text{HeI})$	$J(6\,678\text{HeI})$	$J(4\,922\text{HeI})$
Observational mean	2.76	0.72	0.23
Theoretical mean	2.76	0.79	0.27

In the last lines of Tables 3.11 and 3.12 the specific emission coefficients $\varepsilon(4\,471)$ and $\varepsilon(4\,686)$ for 4 471 HeI and 4 686 HeII are given (for $T_e = 10\,000$ K). The volume emission coefficients in these lines can be written in the form

$$\varepsilon(4\,471\ \text{HeI}) = 4\pi\varepsilon(4\,471)\,\text{Ne}\,\text{N}\,\text{He}^{+})\,,$$

$$\varepsilon(4\,686 = 4\pi\varepsilon(4\,686)\,\text{Ne}\,\text{N}\,\text{He}^{++})\,.$$

However, there is one more series of HeII lines which was never discussed, i.e. the Balmer series of ionized helium (transitions $n \to 2$); this series is represented by the lines 1 640 (He_α), 1 215 (He_β), 1 084 (He_γ), etc. The theoretical Balmer decrement of this series must be exactly the same as for the Balmer decrement of HI. The observational aspect is more complicated. The second line of this series, 1 215 HeII, is blended by the powerful hydrogen line L_α = 1 216 HI of the geocorona; and the fourth line 1 025 HeII is blended with the hydrogen line L_β = 1 026 HI. Therefore, in fact the observations concern mainly the line of the Balmer series – 1 640 HeII.

The line 1 640 HeII is observed in the spectra of almost all PNs as the most powerful line in the ultraviolet region of their spectra (see Chap. 7). Its intensity is usually compared with the intensity of 4 686 HeII – the first line of the Paschen series of ionized helium. The theoretical intensities of both lines are known (see Table 3.6, with account taken of z_n):

$$\frac{I(1\,640\ \text{HeII})}{I(4\,686\ \text{HeII})} = \frac{I(6\,563\ \text{HeI})}{I(18\,751\ \text{HeI})} = \left|\begin{array}{l} 7.14 \text{ at } T_e = 5\,000\ \text{K} \\ 8.15 \text{ at } T_e = 10\,000\ \text{K} \\ 9.41 \text{ at } T_e = 20\,000\ \text{K} \end{array}\right.$$

The observed ratio $F(1\,640\ \text{HeII})\ /F(4\,686\ \text{HeII})$ as a rule strongly differs from its theoretical value, the reason being the interstellar absorption. That is why this ratio is used for the determination of numerical values of the extinction parameter c for each nebula separately.

The line 1 640 HeII is completely of recombination origin and has an important role in the physics of PNs. Owing to the fact that photons within the energy interval from 54 eV up to $\sim$ 300 eV are absorbed by HeII atoms, the ratio 1 640 HeII / 1 216 HI carries information on the energy distribution in the Lyman continuum of hydrogen and ionized helium.

3.9 Recombination Lines of HeII in the Ultraviolet

All Balmer lines of ionized helium, 1 640 HeII and shorter, and almost all lines of the Paschen series, with the exception of the first two lines, 4 686 HeII and shorter, lie in the ultraviolet region, shorter than 3 000 Å, and hence these lines may be observed only beyond the Earth's atmosphere. The interpretation of these observations cannot be carried out without knowing the calculated values of the intensities of the lines of HeII. Such calculations have been carried out by Seaton (1978).

First, it is not difficult to guarantee that the luminescence of real nebulae occurs in accordance with model B, i.e. when $\tau(\mathrm{L}_\alpha) \gg 1$, not only in hydrogen lines but also in the lines of ionized helium.

Let us start from the Lyman series of HeII. All lines of this series are four times shorter than the hydrogen lines of the same series, namely 304 HeII = L_α(HeII), 256 HeII = L_β(HeII), etc., and the series boundary is at 228 Å. According to model B, all radiation coming from the central star in the frequencies of the ionization of helium, i.e. shorter than 228 Å, should be transformed via fluorescence into the lines and continuum of the Balmer, Paschen and higher series, and the Lyman series will be represented by only one line – 304 L_α(HeII). However, these lines can be detected only in space conditions and only for the nearby PNs.

The procedure for calculation is the same as for hydrogen lines. The energy, $J(\lambda)$, emitted in the given line of ionized helium per unit volume is given by the expression

$$J(\lambda) = \frac{1}{4\pi}\,\alpha(nn')\,\mathrm{Ne}\,\mathrm{N}(\mathrm{He}^{++})\,h\,\nu\,, \qquad (31)$$

where $\alpha(n \rightarrow n')$ is the effective recombination coefficient of free electrons with doubly ionized helium, $\mathrm{N(He^{++})}$ – the concentration of $\mathrm{He^{++}}$ ions.

The value of $\alpha(n \rightarrow n')$ depends on the electron concentration weakly but decreases with the increase of electron temperature T_e nearly linearly as follows from the following data (in units of 10^{-14} cm^3 s^{-1}):

T_e, K	5 000	10 000	20 000
$\alpha(3 \rightarrow 2)$ 1 640 HeII	143.8	81.0	44.2
$\alpha(4 \rightarrow 3)$ 4 686 HeII	69.2	34.9	16.9

In Table 3.13 the calculated intensities of the emission lines of HeII of the Balmer and Paschen series are given at $N_e = 10^4$ cm^{-3} and two values of T_e: 10 000 and 20 000 K.

Table 3.13. Calculated intensities of HeII lines of the Balmer ($n-2$) and Paschen ($n-3$) series in units of $J(4686) = 100$

n	Line [Å]	T_e, K 10 000	T_e, K 20 000	n	Line [Å]	T_e, K 10 000	T_e, K 20 000
	Balmer series				Paschen series		
3	1 640	660	745	4	4 686	100	100
4	1 215	201	246	5	3 204	40.3	45.2
5	1 085	90	113	6	2 734	20.5	23.9
6	1 025	49.1	61.8	7	2 512	12.2	14.4
8	972	19.8	25.0	8	2 386	8.0	9.5
10	949	10.0	12.7	10	2 253	4.1	4.8

From the data in Table 3.13 the following conclusions can be derived.

(a) For HeII lines, the steepness of the Balmer decrement is much larger than that of the Paschen decrement.

(b) The ratio $H_\alpha(1640)/P_\alpha(4686)$ for HeII varies from 6.6 up to 7.5, i.e. the first line of the Balmer series 1 640 HeII is 7–8 times as powerful as the first line of the Paschen series 4 686 HeII.

(c) The Balmer decrement of HeII is steeper than the Balmer decrement of HI.

(d) HeII decrements are relatively sensitive to T_e and less sensitive to N_e. Hence, the Balmer and Paschen decrements cannot be an indicator of electron concentration in PNs.

Thus the informativeness of HeII emission lines for the physics of PN is much higher compared with the hydrogen lines.

3.10 The Optical Depth of Nebulae at the Frequency of Helium Ionization

The emission line 4 686 HeII, the analog of the hydrogen line H_β, is observed in the spectra of PNs, occasionally, while 1 640 HeII, the analog of the hydrogen line H_α, is always observed. This circumstance may be used to solve one of the important problems of the physics of PN, i.e. for the determination of the complete optical depth t of PNs at the frequencies of the ionization of helium, i.e. shorter than 228 Å.

Denote by $\phi(H_\beta)V$ the short wavelength (L_c) photons emitted by the central star of a PN in the frequency interval located within the limits of the ionization of hydrogen, i.e. from ν_0 up to $4\nu_0$; they are transformed via fluorescence to H_β photons. Denote by $\phi(1\,640)$ the photons ionizing He^+, i.e. those within the interval from $4\nu_0$ up to ∞, transformed to the photons 1 640 HeII. Then we can write for the fluxes of radiation escaping from the nebula

$$F(H_\beta) = \phi(H_\beta)\, h\,\nu_\beta\, 4\pi r_*^2 (1 - e^{-\tau_c}) \frac{2\pi}{c^2} \left(\frac{kT_*}{h}\right)^3 \int_{x_0}^{4x_0} \frac{x^2 dx}{e^x - 1}, \quad (32)$$

$$F(1\,640\ \mathrm{HeII}) = \phi(1\,640)\, h\,\nu_{1640}\, 4\pi r_*^2 (1 - e^{-t_c}) \frac{2\pi}{c^2} \left(\frac{kT_*}{h}\right)^3 \int_{4x_0}^{\infty} \frac{x^2 dx}{e^x - 1}, \quad (33)$$

where $x_0 = h\nu_0/kT_*$, ν_* is the frequency of the ionization of hydrogen, $4\nu_0$ is the frequency of the ionization of helium, r_* is the radius of the PN nucleus, τ_c is the optical depth of the nebula at frequencies of the L_c radiation.

From (32) and (33) we have

$$\frac{F(1\,640\,\mathrm{HeII})}{F(H_\beta)} = 0.377 \frac{\phi(1\,640)}{\phi(H_\beta)} \frac{1 - e^{-t_c}}{1 - e^{-\tau_c}} Q(T_*)\,, \quad (34)$$

where the following notations are used:

$$Q(T_*) = \int_{4x_0}^{\infty} \frac{x^2 dx}{e^x - 1} \Big/ \int_{x_0}^{4x_0} \frac{x^2 dx}{e^x - 1}. \quad (35)$$

The numerical values of $\phi(H_\beta)$ and $\phi(1\,640$ HeII) can be obtained by using the recombination theory of PN luminescence. The calculations give, at $T_e = 10\,000$ K, $\phi(H_\beta) = 0.118$ and $\phi(1\,640) = 0.40$. Then we have

Table 3.14. Optical depths, τ_c and t_c, at the frequencies of the ionization of hydrogen and helium, for a sample of planetary nebulae

Planetary nebula	T_*, K	$\frac{F(1\,640\ \mathrm{HeII})}{F(H_\beta)}$	τ_c	t_c
NGC 7009	100 000	0.43(1)	0.25	0.27
NGC 6572	60 000	0.052(2)	0.14	0.21
NGC 2371	100 000	5.89(3)	0.10	2.1
NGC 6720	130 000	4.21(4)	0.48	>1
NGC 6905	70 000	1.12(5)	0.11	>1
NGC 7027	165 000	3.92(6)	<1	<1
NGC 7662	100 000	3.07(7)	0.23	>1

(1)Perinotto, Benvenuti, 1981b. (2)Flower, Penn, 1981. (3)Pottasch et al., 1981. (4)Barker, 1982. (5)Feibelman, 1982b. (6)Grewing et al., 1978b. (7)Harrington et al., 1978b, Storey, 1994.

$$\frac{F(1\,640)}{F(\mathrm{H}_\beta)} = 10.0\,Q(T_*)\,\frac{1 - e^{-t_c}}{1 - e^{-\tau_c}}\,. \tag{36}$$

There are two possibilities. First, if one can somehow obtain the genuine value of τ_c (e.g. via two-photon emission, Chap. 9), then one can obtain from (36) the numerical value of t_c. Second, in the complete absence of data, one can obtain the numerical value only of the upper limit of τ_c, taking $t_c \gg 1$.

The method just described for the determination of τ_c or t_c is applied to a sample of PNs with known values of the nucleus temperature and with known observed fluxes in both the lines 1 640 HeII and H_β (Table 3.14).

For three objects, NGC 7 009, NGC 6 572 and NGC 2 371, the numerical values of τ_c are known; they were obtained by the method of $2q$ emission (Chap. 9). For these one can determine numerical values of t_c directly. In the remaining cases only the upper limits of τ_c are given if the nebula is optically thick ($t_c > 1$) at the frequencies of helium ionization. In one case (NGC 7 027) the PN is optically thin at both frequencies, i.e. $\tau_c < 1$ and $t_c < 1$; then the ratio can be obtained from (36) for $(1 - e^{-t_c})/(1 - e^{-\tau_c}) \approx t_c/\tau_c = 1.2$, ($Q(165\,000) = 0.33$ is used).

Some conclusions:

(a) The optical depth of a PN at helium ionization frequencies, t_c, is always larger than τ_c, the optical depth at hydrogen ionization frequencies.
(b) Nebulae are not always optically thin at both frequencies, i.e. when $\tau_c < 1$ and $t_c < 1$. Very often the nebulae, being optically thin at the frequencies of hydrogen ionization ($\tau_c < 1$), remain optically thick also at the frequencies of helium ionization ($t_c > 1$).

4. Ionization Equilibrium of Planetary Nebulae

4.1 Ionization of Nebulae

The starting point for the study of the ionization structure of planetary nebulae, and gaseous nebulae in general, is the ionization formula. The atoms can be ionized in two ways: under direct radiation of the central star, and by collisions with free electrons. In the first case one has photoionization, in thc second, collisional ionization. First, let us consider photoionization.

The formula for the calculation of the degree of ionization in a nebula under the action of radiation can be deduced in the usual way, i.e. from the equilibrium condition, when the number of atoms that reach the continuum through photoionization must be equal to the number of atoms that leave the continuum as a result of the recombination of ionized atoms with free electrons.

The characteristic feature of the conditions in gaseous nebulae is extremely low degree of excitation of all atoms, including hydrogen. Therefore we can restrict ourselves to the examination of photoionization, which occurs only from the ground states of atoms. Then, if we call n_1 the number of atoms per unit volume in the ground state, the number of acts of ionization per unit time will be

$$n_1 \int_{\nu_0}^{\infty} k_{1\nu} \frac{\rho_\nu c}{h\nu} \, \mathrm{d}\nu \,, \tag{1}$$

where $k_{1\nu}$ is the coefficient of continuous absorption for one atom in the ground state, ν_0 is the ionization frequency and ρ_ν is the density of ionizing radiation at a given point of the nebula.

Recombinations with free electrons can occur at all levels. The total number of recombinations will be

$$n_e n^+ \sum_1^{\infty} C_n(T_e)$$

or

$$\frac{n_e n^+}{p} C_1(T_e) \,, \tag{2}$$

where p is the fraction of captures directly to the first level, i.e.

$$p = \frac{C_1(T_e)}{\sum_1^\infty C_n(T_e)} . \tag{3}$$

Application of the equilibrium condition gives

$$n_1 W \int_{\nu_0}^{\infty} k_{1\nu}\, e^{-\tau_\nu}\, \frac{\rho_\nu^* c}{h\nu}\, \mathrm{d}\nu = \frac{n_e n^+}{p}\, C_1(T_e) , \tag{4}$$

where we have taken $\rho_\nu = W \rho_n u^* e^{-\tau_\nu}$, and where ρ_ν^* is the radiation density at the surface of the star at the frequency ν, τ_ν is the optical depth of the nebula at the frequency ν beyond the Lyman limit, and W is the coefficient of dilution.

From (4) we have

$$\frac{n^+}{n_1} n_e = \frac{Wp}{C_1(T_e)} \int_{\nu_0}^{\infty} k_{1\nu}\, e^{-\tau_\nu}\, \frac{\rho_\nu^* c}{h\nu}\, \mathrm{d}\nu . \tag{5}$$

In fact this is the ionization equation for a nebula. However, it can be considerably simplified by using the relation that exists between the quantities $k_{1\nu}$ and $C_1(T_e)$, and in fact the function ρ_ν^* may be approximated with high accuracy using Planck's formula at the effective temperature T_*, and also that the law of distribution of velocities in an electron gas is given by the Maxwellian law at the electron temperature T_e. We also assume that $k_{1\nu} \sim \nu^{-3}$ for all types of atom, although this relationship is different for various atoms (e.g. $k_{1\nu} \sim \nu^{-3}$ for hydrogen). Taking account of all these factors, and also after integrating (5), we can write for the final ionization formula for the nebula

$$\frac{n_e}{n_1} = p\, \frac{g^+}{g_1}\, W \sqrt{\frac{T_e}{T_*}}\, \frac{2(2\pi\mu k T_*)^{3/2}}{h^3}\, e^{-\frac{h\nu_0}{kT_*}}\, e^{-\tau_c} , \tag{6}$$

where τ_c is the mean value of the optical depth (distance) from a given layer in the nebula at the frequencies of the Lyman series of hydrogen, n^+, n_e and n_1 are the concentrations of ionized atoms, free electrons and neutral atoms (in the ground state) respectively, and μ is the electron mass.

The degree of ionization of atoms (ions) at thermodynamic equilibrium is determined, as is known, by the Saha formula. Equation (6) derived for ionization in nebulae differs from the Saha formula by a factor $p\,(g^+/g_1)\, W \sqrt{T_e/T_*}$ on the right hand side.

The factor $p(g^+/g_1)$ is usually close to unity. Therefore the final ionization formula for a nebula, at least in its inner layers, i.e. when $\tau_c \approx 0$, takes the form

$$\frac{n^+}{n_1} n_e = W \sqrt{\frac{T_e}{T_*}}\, \frac{2(2\pi\mu k T_*)^{3/2}}{h^3}\, e^{-\frac{h\nu_0}{kT_*}} . \tag{7}$$

Thus the main difference from the Saha formula is the presence of the factor W on the right hand side. In nebulae this factor is extremely small, of the order of 10^{-14}. Nevertheless, the degree of ionization n^+/n_1 reaches a very high value, about 1 000 at the usual temperatures of the nuclei ($\sim$ 50 000–100 000 K). This is explained by the small concentration of free electrons, about 10^3 cm^{-3} (Chap. 7). Because of the low electron concentration alone, the degree of ionization turns out to be larger than unity ($n^+/n_1 > 1$) at temperatures $T_* \sim 20\,000$ K and $T_* \sim 33\,000$ K for helium.

We will deal with the ionization formula (7) in subsequent sections and chapters.

4.2 Ionization and Excitation by Electron Collisions

Free electrons which appear in nebulae as a result of the photoionization of atoms, primarily hydrogen, possess enough kinetic energy for the excitation and ionization of atoms by inelastic impacts. The efficiency of the utilization of such an energetic resource – the kinetic energy of free electrons – is higher the smaller the excitation potential of atoms and ions is. Many atoms and ions (OI, OII, OIII, NI, NII, etc.) possess excitation levels with excitation potentials of the order of 2–5 eV, i.e. they are located not far from the ground state. In the case of hydrogen, however, the excitation potential of the nearest level – resonance ($2S$) – is located rather high, at 10.2 eV from the ground state ($1S$). This circumstance only determines the low efficiency of electron collisions in the "life" of hydrogen atoms in nebular conditions.

Denote by $\alpha_c(T_e)$ the probability coefficient of ionization of the hydrogen atom from the ground state owing to collisions. Then, for the number of ionizations Z_{ion} per unit time and per unit volume we have

$$Z_{\mathrm{ion}} = n_1 n_{\mathrm{e}} \alpha_{\mathrm{c}}(T_{\mathrm{e}}) \,, \tag{8}$$

with

$$\alpha_{\mathrm{c}}(T_{\mathrm{e}}) = \int_{v_0}^{\infty} Q(v)\, v f(v)\, \mathrm{d}v \,, \tag{9}$$

where $Q(v)$ is the effective collisional cross-section of the electron with a neutral hydrogen atom, $f(v)$ is Maxwell's function for the electron velocities, and v_0 is the minimum velocity of the free electron necessary to detach the bound electron.

The theoretical values of $\alpha_c(T_e)$ as a function of electron temperature were obtained in various approximations in the 1950s (Chamberlain, 1953; Geltman, 1956). The effective collisional ionization cross-section $Q(v)$ can be represented with a high degree of accuracy by a linear dependence on E/E_0 in the following form:

$$Q(v) = 0.56\,\pi\alpha_0^2 \left(\frac{E}{E_0} - 1\right) = 0.56\,\pi\,\alpha_0^2 \left[\left(\frac{v}{v_0}\right)^2 - 1\right] \,, \tag{10}$$

where $\pi\alpha_0^2 = 0.88 \times 10^{-16}$ cm^2. Substituting (10) into (9) and integrating, we find for the function $\alpha_c(T_e)$ that

$$\alpha_c(T_e) = 2.24\,\pi\,\alpha_0^2 \left(\frac{kT_e}{2\pi\mu}\right)^{1/2} \left(1 + \frac{2kT_e}{\kappa_0}\right) e^{-\kappa_0/kT_e}\,, \tag{11}$$

where κ_0 is the ionization potential from the ground state of hydrogen (= 13.6 eV). Table 4.1 gives the quantity $\alpha_c(T_e)$ computed according to this formula for a number of values of T_e.

Table 4.1. Values of the function $\alpha_c(T_e)$ (in units of $\pi\alpha_0^2 = 0.88 \times 10^{-16}$ cm^2) and the degree of ionization of hydrogen n^+/n_1 caused by electron collisions

T_e, K	$\alpha_c(T_e)$	n^+/n_1
10 000	5.16	0.001
15 000	1.17×10^3	0.34
20 000	2.23×10^4	8
40 000	1.98×10^6	1200

The number of recombinations Z_{rec} of free electrons with the proton to all energetic levels per unit volume and per unit time yields:

$$Z_{rec} = n^+ n_e \sum_{n=1}^{\infty} C_n(T_e) = n_e n^+ C(T_e)\,. \tag{12}$$

Under equilibrium conditions we obviously must have $Z_{ion} = Z_{rec}$. Then, from (8) and (12), we find that

$$\frac{n^+}{n_1} = \frac{\alpha_c(T_e)}{C(T_e)}\,. \tag{13}$$

This is the ionization equation or hydrogen ionization formula when the ionization occurs through inelastic collisions with electrons. It is interesting to see that in this case the degree of ionization does not depend on the electron concentration. Recall that in the case of photoionization the degree of ionization is inversely proportional to the electron concentration (9). The values of n^+/n_1 calculated using (13) for the same values of T_e are given in the last column of Table 4.1.

The electron temperatures in PNs are usually of the order of 10 000 K and in all cases less than 20 000 K. Therefore the degree of ionization by electron collisions will be $n^+/n_1 < 1$. In the case of photoionization we have $n^+/n_1 \sim 1\,000$. Hence, it follows that the collisional ionization of hydrogen has practically no role to play in PNs.

4.3 The Degree of Excitation

The order of magnitude of the degree of excitation of the hydrogen atoms in a nebula can be obtained from the approximate formula derived for the detailed equilibrium between the induced transitions $1 \to n$ under the action of the radiation of the central star and the spontaneous transitions $n \to 1$:

$$N_1 W \rho_{1n} B_{1n} = N_n A_{n1} , \tag{14}$$

where $W\rho_{1n}$ is the density of the induced radiation in the inner layers of the nebula. Bearing in mind that

$$\rho_{1n} B_{1n} = \frac{g_n}{g_1} \bar{\rho}_{1n} A_{n1} ,$$

from (14) we find for the degree of excitation that

$$\frac{N_n}{N_1} \approx W \bar{\rho}_{1n} = \frac{W}{e^{h\nu_{1n}/kT_*} - 1} , \tag{15}$$

where T_* is the temperature of the central star. When $T_* = 50\,000$ K and $W \approx 10^{-13}$ we get for the lower levels ($n = 2, 3, \ldots$)

$$\frac{N_n}{N_1} \sim 10^{-14} . \tag{16}$$

The exact solution of the problem of the degree of excitation can be obtained in the manner described in Chap. 3, i.e. with the help of the quantity z_n, which describes the ratio of the number of atoms at the level n to the multiplier $N_e N^+$: $z_n = N_n / N_e N^+$. Since the numerical values of z_n have been determined above (Table 3.5) one can use the magnitude of multiplier $N_e N^+ = N_n / z_n$ in the ionization formula (7):

$$\frac{N_n}{N_1} = z_n W \frac{2(2\pi \mu k)^{3/2}}{h^3} T_* T_e^{1/2} e^{-\kappa^+/kT_*} , \tag{17}$$

where $\kappa^+ = h\nu_0$ is potential of ionization. Substituting the numerical values of constants, we get

$$\frac{N_n}{N_1} = 4.81 \times 10^{15} z_n W T_* T_e^{1/2} e^{-158\,000/T_*} . \tag{18}$$

This is the formula for determination of the degree of excitation in nebular conditions for hydrogen atoms. Note that it differs from the Boltzmann formula

$$\frac{N_n}{N_1} = \frac{g_n}{g_1} e^{(\epsilon_1 - \epsilon_n)/kT_*} ,$$

which gives the distribution of atoms over the excited levels in the condition of thermodynamic equilibrium. The major contrast is the existence on the right hand side of (18) of the dilution coefficient W, i.e. of an extremely small quantity. This fact once more confirms the conclusion that PNs are objects where thermodynamic equilibrium is not established completely.

The numerical values of N_n/N_1 for levels 2–5 of hydrogen, obtained with the help of (18) for both models A and B, and at $T_* = 50\,000$ K, $T_e = 10\,000$ K and $W = 10^{-13}$, and with the values of z_n from Table 3.5, are as follows:

n	2	3	4	5
Model A	0.3×10^{-13}	0.8×10^{-13}	2.8×10^{-13}	4.5×10^{-13}
Model B	–	2.5×10^{-13}	3.6×10^{-13}	5.4×10^{-13}

As we can see, both formulae (15) and (18) give one and the same value for the degree of excitation of hydrogen atoms in nebulae, i.e. $N_n/N_1 \sim 10^{-13}$–10^{-14}. This is an extremely small quantity. Under special conditions (see Chap. 9) the second state of hydrogen ($n = 2$) may be an exception to this rule. It is interesting that the degree of excitation of a nebula increases when we go to higher levels, just the opposite of what follows from the Boltzmann formula.

The conclusion regarding the extremely low degree of ionization of hydrogen atoms should be considered one of the most important properties of the physics of gaseous nebulae. For comparison, in the solar atmosphere ($T_* = 5\,700$ K) the degree of excitation of the second level is much higher, about 10^{-9}, and even higher, 10^{-5}, in the photospheres of type A stars ($T_* \sim 100\,000$ K).

4.4 The Balmer Decrement for Collisions

Here we shall consider the problem of the excitation of real hydrogen atoms by electron impacts. The number of acts of excitation at the level n per unit volume and per unit time, will be $N_1\ N_e\ \alpha_n(T_e)$, where $\alpha_n(T_e)$ is analogous to (9) with the replacement of v_0 by v_n – the least electron velocity necessary to move the atom from the ground state ($n = 1$) up to level n.

One can obtain the distribution of atoms over the different excited states when the excitation is purely due to collisions from the following equilibrium condition:

$$N_n \sum_{k=2}^{n-1} A_{nk} = \sum_{k=n+1}^{\infty} N_k A_{kn} + N_1 N_e \alpha_n(T_e)\,. \tag{19}$$

Solving this system by known numerical values of $\alpha_n(T_e)$, we obtain, first, the degree of the excitation; this turns out to be one or two orders smaller ($N_n/N_1 \sim 10^{-15}$) than we obtained in the case of photoionization (Chap. 3). Then we calculate the Balmer decrement: its values are as follows (for model B):

	H_α	H_β	H_γ	H_δ	H_ε
$T_e = 10\,000$ K	5.76	1.00	0.291	0.136	0.076
$T_e = 20\,000$ K	4.79	1.00	0.347	0.169	0.097

As we can see, in the case of the excitation of hydrogen by only electron collisions, the Balmer decrement is steeper than in the case when the excitation is due to photoionization only (Table 3.7).

When the excitation is due to both process – photoionization and electron collisions – the resulting Balmer decrement can be represented as a linear combination of both types of Balmer decrement, i.e.

$$\frac{E_{n2}}{E_{42}} = (1-x)\left(\frac{E_{n2}}{E_{42}}\right)_{\mathrm{R}} = x\left(\frac{E_{n2}}{E_{42}}\right)_{\mathrm{C}}, \tag{20}$$

where x depends on N_1/N_e and T_e.

In all cases, the role of the electron collisions on the Balmer decrement of PNs does not appear to be large.

4.5 The Ionization Structure of Planetary Nebulae

To solve the problem of the ionization structure of PNs one needs to find out the law of variation of the degree of ionization within the nebula for a given distance from its central star, the source of the ionizing radiation. The initial data are as follows:

(a) The temperature of the central star T_e.
(b) The luminosity of the central star L_* or its radius R_*.
(c) The chemical composition of the nebula (usually limited by taking account of the eight elements – H, He, O, Ne, C, N, Fe, Si) and the relative abundance of the elements, i.e. the ratios X/H.
(d) The inner and outer radii of the nebula.
(e) The concentration of hydrogen atoms N_H and the law of its variation with distance from the centre.
(f) The atomic parameters of the elements under consideration.

Below, only the main principle behind the solution of the problem will be described together with some final conclusions.

The equation of transfer of radiation of intensity I_ν and frequency ν of a continuous spectrum and at a distance r from the centre of a nebula is written in the form

$$\frac{dI_\nu}{dr} = -k_\nu I_\nu + j_\nu, \tag{21}$$

where j_ν is the volume emission coefficient, k_ν is the effective volume continuous absorption coefficient, determined by summary absorption of different ions X^i, X^{i+1} etc. at a frequency ν:

$$k_\nu = \sum N(X^i)\, s_\nu(X^i), \tag{22}$$

where the summation is realized over all ions X^i (X is the element, i is the degree of ionization, $i = 0$ is the neutral atom) and s_ν is the coefficient of continuous absorption at a frequency ν and on one ion X^i. The dependence $s_\nu \sim f(\nu)$ is determined by (39), and is given in Table 2.5, where the numerical values of the interpolation parameters are given for the main atoms and ions.

To (21) one should add the equation of ionization equilibrium, which equates the numbers of ionization and recombination events:

$$4\pi\, N(X^i) \int_{\nu_0}^{\infty} \frac{J_\nu}{h\nu} S_\nu(X^i) d\nu = N_e\, N(X^{i+1})\, C(X^i)\,, \tag{23}$$

where J_ν is the mean intensity of the ionizing radiation at a given point within the nebula:

$$J_\nu = \frac{1}{4\pi} \int I_\nu d\omega\,. \tag{24}$$

The unknown quantities J_ν and $N(X^i)$ together with k_ν should be obtained as a function of distance by the joint solution of a large number of equations of (21) and (23) type written for all elements and ions.

The procedure just described for the determination of the ionization structure became common once the early models of nebula had been constructed in the 1960s (Harrington, 1969; Fowler, 1969). The integration and summation of all the equations, particularly of (23), are performed using numerical methods.

Two examples of predicted ionization structures for the hydrogen, helium and oxygen of a certain model of nebula are shown in Fig. 4.1. Here the curves of the variation of the relative abundance of H^0, H^+, He^0, He^+, He^{++}, as well as for O^{+4}, O^{+3}, O^{+2}, O^+ and O^0 are given as functions of the distance r from the centre of the nebula.

Two conclusions follow from these curves:

(a) There is a variation of the number of ions with distance, both the increase and the decrease occuring, in general, rather sharply.
(b) A clear sequence of zones with diminishing ionization is seen, e.g. the O^{+4} zone follows those of O^{+3}, O^{+2}, etc., with negligible overlapping between the zones.

As for the dimensions of the zones, they are quite different; for some zones (He^+, O^{+2}, C^{+3}, N^{+3}) their sizes are comparable with the nebula's radius, for others (O^+, Ne^+) they are much smaller than the nebula's radius.

4.6 Thermal Equilibrium of Nebulae

The exact value of the electron temperature in PNs can be obtained from the condition of detailed thermal balance, e.g. the equality between the thermal energy of a nebula E_t and the energy of cooling E_{cul}.

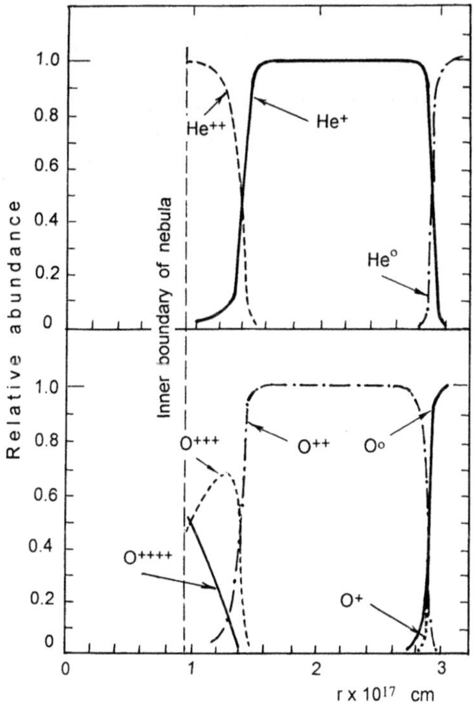

Fig. 4.1. The ionization structure for a planetary nebula model – the distribution of ions of H, He and O along the radius of the nebula r. $T_* = 100\,000$ K, $r_{\rm in} = 0.031$ pc, $r_{\rm out} = 0.095$ pc and $n_e = 7\,000$ cm^{-3}

The heating of a nebula is determined by the photoionization of hydrogen atoms, and its energy, $E_{\rm ph}$(II), is completely spent on heating the nebula, i.e. $E_t = E_{\rm ph}$(H).

The cooling of the nebula occurs by means of four processes: (a) recombination of free electrons with protons, $L_{\rm R}$; (b) free–free transitions of free electrons or thermal bremsstrahlung, $L_{\rm FF}$; (c) radiation in the emission lines, which are mainly forbidden, $L_{\rm C}$; (d) inelastic collisions of free electrons with hydrogen atoms, $L_{\rm H}$. Therefore, we have for the total energy of cooling

$$E_{\rm cul} = L_{\rm R} + L_{\rm FF} + L_{\rm C} + L_{\rm H}\,. \tag{25}$$

For stationary nebulae we have $E_{\rm ph}({\rm H}) = E_{\rm cul}$, or

$$E_{\rm ph}({\rm H}) = L_{\rm R} + L_{\rm FF} + L_{\rm C} + L_{\rm H}\,. \tag{26}$$

Denoting by J_ν the intensity of the ionizing radiation reached at a given point within the nebula from the central star, one can write for the number of absorption events, or equivalently for the number of ionization events $N_{\rm ion}$, per unit time per unit volume

$$N_{\text{ion}} = N_1 \int_{\nu_0}^{\infty} \frac{J_\nu}{h\nu} s_\nu \, \mathrm{d}\nu \,, \tag{27}$$

where N_1 is the concentration of neutral hydrogen atoms and s_ν is the continuous absorption coefficient per hydrogen atom.

On the other hand, the complete energy $E_{\text{ph}}(\text{H})$ gained via photoionization by free electrons per unit time and per unit volume will be

$$E_{\text{ph}}(\text{H}) = N_1 \int_{\nu_0}^{\infty} \frac{J_\nu}{h\nu} h(\nu - \nu_0) s_\nu \, \mathrm{d}\nu \,. \tag{28}$$

The ratio of (28) to (27) is obviously the mean energy per free electron $\bar{\varepsilon}$

$$\bar{\varepsilon} = \frac{\displaystyle\int_{\nu_0}^{\infty} \frac{J_\nu}{h\nu} h\,(\nu - \nu_0)\, s_\nu \, \mathrm{d}\nu}{\displaystyle\int_{\nu_0}^{\infty} \frac{J_\nu}{h\nu} s_\nu \, \mathrm{d}\nu} \,. \tag{29}$$

Thus, $\bar{\varepsilon}$ depends on the temperature of the central star T_*. In the case when the radiation of the central star can be approximated with the Planck law, i.e. when $J_\nu(T_*) = B_\nu(T_*)$, we have from (29), taking into account also that $s_\nu \sim \nu^{-3}$,

$$\bar{\varepsilon} = kT_* \, A(T_*) \,, \tag{30}$$

where $A(T_*)$ is a quantity close to unity and weakly dependent on T_* and which at $\tau_c \ll 1$ has the form

$$A(T_*) = \frac{\displaystyle\int_{x_0}^{\infty} \frac{\mathrm{d}x}{e^x - 1}}{\displaystyle\int_{x_0}^{\infty} \frac{\mathrm{d}x}{x(e^x - 1)}} - x_0 \,, \tag{31}$$

where $x_0 = h\,\nu_0 / kT_*$.

However, at $\tau_c \ll 1$ one should also take into account photoionization under the action of the diffuse component of the ionizing radiation, formed within the nebula via fluorescence:

$$A(T_*) = \frac{\displaystyle\int_{x_0}^{\infty} \frac{x^3 \mathrm{d}x}{e^x - 1}}{\displaystyle\int_{x_0}^{\infty} \frac{x^2 \mathrm{d}x}{e^x - 1}} - x_0 \,. \tag{32}$$

As the calculations show, for a broad range of T_* (40 000 – 200 000 K) the numerical values of $A(T_*)$ change rather weakly: 1–0.5 in the case of $\tau_c \ll 1$ and 1–2 in $\tau_c \gg 1$. Therefore to a good approximation one can adopt $A(T_*) = 1$. Then we shall have the following values for the mean energy of free electrons $\bar{\varepsilon}$ (in eV):

T_*, K	40 000	60 000	80 000	100 000	150 000	200 000
$\bar{\varepsilon}(\tau_c \ll 1)$	2.9	4.0	4.9	5.9	7.9	8.8
$\bar{\varepsilon}(\tau_c \gg 1)$	5.0	8.4	12.1	15.9	25.9	34.8

For other forms of the function J_ν, the integration in (29) should be carried out by numerical methods.

Thus, we have for the total recombination rate per unit volume

$$N_e N^+ \sum_1^\infty C_n(T_e) = N_e N^+ C(T_n) ,$$

which is, at the same time, the number of electrons appearing as a result of photoionization. Therefore we get for the total kinetic energy of heating the nebula

$$E_{ph}(H) = N_e N^+ C(T_e)\,\bar{\varepsilon} = N_e N^+ C(T_e)\,kT_x A(T_*) , \tag{33}$$

where $C(T_e)$ is the effective coefficient of recombination of all electrons at all levels of the hydrogen atom.

One also has to obtain E_{cul}, i.e. the sum of the four types of cooling energies on the right hand side of (26).

Recombinations. The amount of energy lost by electrons with velocities within the interval v and $v+dv$ during recombination at all levels of hydrogen is equal to

$$N_e N^+ f(v)\,v \frac{mv^2}{2}\,dv \sum_1^\infty \sigma_n(T_e) , \tag{34}$$

where $\sigma_n(T_e)$ is the cross-section of electron–proton recombination at the level n.

The total energy L_R lost by all electrons with arbitrary velocities up to ∞ as a result of recombination will be

$$L_R = N_e N^+ kT_e\,\beta_A(T_e) , \tag{35}$$

where

$$\beta_A(T_e) = \frac{1}{kT_e} \int_0^\infty \sum_1^\infty \sigma_n(T_e) \frac{mv^2}{2} f(v)\,v\,dv , \tag{36}$$

and $f(v)$ is the Maxwellian velocity distribution function. The numerical values of the function $\beta_A(T_e)$ dependent on T_e are as follows (e.g. Hummer, Seaton, 1963):

T_e, K	5 000	10 000	15 000	20 000
$\beta_A(T_e)$, $\times 10^{-13}$	5.42	3.23	2.40	1.88

Thermal Bremsstrahlung (Free–Free Transitions). Free electrons can also lose energy without recombination, i.e. via radiation during close encounters with one another as well as with protons, i.e. at free–free transitions, known as *thermal bremsstrahlung.* The spectrum of this radiation is continuous over all frequencies, from $\nu = 0$ up to $\nu = \infty$. The intensity of radiation at the frequency ν is given by the expression

$$F_\nu = N_e N^+ \frac{2^5 \pi e^6}{3^{3/2} h m c^3} \left(\frac{2\pi k}{m} \right)^{1/2} Z^2 T_e^{-1/2} \exp\left(-\frac{h\nu}{kT_e} \right) , \tag{37}$$

i.e. it depends only on the electron temperature T_e.

The total losses of energy by bremsstrahlung will be

$$L_{FF} = \int_0^\infty F_\nu \, d\nu = \\ N_e N^+ \frac{2^5 \pi e^6}{3^{3/2} h m c^3} \left(\frac{2\pi k}{m} \right)^{1/2} Z^2 = 1.42 \times 10^{-27} N_e N^+ T_e^{1/2} , \tag{38}$$

where $Z = 1$.

Electron Collisions: Radiation in Forbidden Lines. In the simplest case, the collision of an electron with an ion leads to the excitation of this ion to the nearest metastable level. The reverse downward transition is accompanied by emission of a photon escaping from the nebula without any difficulty. Thus, a part of the kinetic energy of free electrons is lost by means of excitation in forbidden lines. The loss of energy in this process is

$$L_c(k) = N_e N_1(k) \, q_{12} h\nu_{12} , \tag{39}$$

or, taking account of the deactivation effect,

$$L_c(k) = N_2(k) A_{21} \, h\nu_{12} = N_e N_1(k) \, q_{12} h\nu_{12} \frac{1}{1 + N_e q_{21}/A_{21}} , \tag{40}$$

where $N_1(k)$ is the concentration of ions of type k, and $q_{12} = q(k, T_e)$ is the coefficient of probability of collisional transition $1 \to 2$ known as a "collisional rate".

The complete energy losses by electron collisions on excitation of all forbidden lines will be

$$L_c = \sum_k L_c(k) = N_e \sum_k N_1(k) \, q_{21}(k) h\nu_k , \tag{41}$$

where k refers to all possible forbidden lines.

The problem, however, can be sufficiently simplified in spite of the fact that the summary relative power of all forbidden lines for given nebula, i.e. the ratio $\sum \mathrm{Neb}/E(\mathrm{H}_\beta)$, is known from direct observations. Then from (41) we have

$$L_c = E(H_\beta) \frac{\sum \text{Neb}}{E(H_\beta)} = z_{42} N_e N^+ A_{42} h\nu_\beta \frac{\sum \text{Neb}}{E(H_\beta)}, \tag{42}$$

where $E(H_\beta)$ is the volume emission coefficient in the line H_β.

Substituting in (42) $z_{42} = 0.36 \times 10^{-20}(10^4/T_e)$ (Chap. 3) we finally get

$$L_c = 1.23 \times 10^{-21} N_e N^+ T_e^{-1} \frac{\sum \text{Neb}}{E(H_\beta)}. \tag{43}$$

Collisions with Hydrogen Atoms. The neutral hydrogen can perform cooling through inelastic collisions with electrons. It affects, in particular, the collisional excitation of resonance level 2^2P with energy 10.2 eV and the subsequent transition to the ground state, via emission of an L_α-photon. Simultaneously the excitation of the metastable level 2^2S of neutral hydrogen with the same energy 10.2 eV occurs with the emission at downward transitions of of two photons of arbitrary frequencies – so-called $2q$ or two-photon emission (see Fig. 9.1).

So, we can write for the energy losses of neutral hydrogen by the collisional mechanism

$$L_H = N_e N_1 q_H(T_e) h\nu_\alpha, \tag{44}$$

where

$$q_H(T_e) = \sum_n q_H(1\,^2S \rightarrow n) \tag{45}$$

and $q_H(1\,^2S \rightarrow n)$ is the coefficient of collisional excitation of neutral hydrogen from its ground state $1\,^2S$ up to levels $n = 2, 3 \ldots$ including all sublevels of thin structure, and at a given T_e. The largest value of q_H, however, occurs for the transition at two resonance levels $2\,^2S$ and $2\,^2P$ ($n = 2$); their numerical values are given in the following table, together with the values of $q_H(T_e)$, being the sum of all five transitions – two at $n = 2$ and three at $n = 3$, which is enough for practical purposes (e.g. Burke et al., 1967):

T_e, K	$q_H(1\,^2S - 2\,^2S)$	$q_H(1\,^2S - 2\,^2P)$	$q_H(T_e)$
10 000	0.94×10^{-13}	1.6×10^{-13}	2.70×10^{-13}
15 000	3.8×10^{-12}	7.0×10^{-12}	12.5×10^{-12}
20 000	2.5×10^{-11}	5.0×10^{-11}	9.3×10^{-11}

The value of N_1 in (44) depends on the degree of hydrogen ionization N^+/N_1; therefore it is convenient to rewrite (44) in the form

$$L_H = 1.63 \times 10^{-11} N_e N^+ \frac{N_1}{N^+} q_H(T_e). \tag{46}$$

The curves of all four types of energy losses – L_R, L_{FF}, L_C and L_H – are shown in Fig. 4.2 as a function of the electron temperature T_e.

Now the condition of thermal equilibrium takes the following form, if we substitute (46), (43), (38), (32) and (30) into (26):

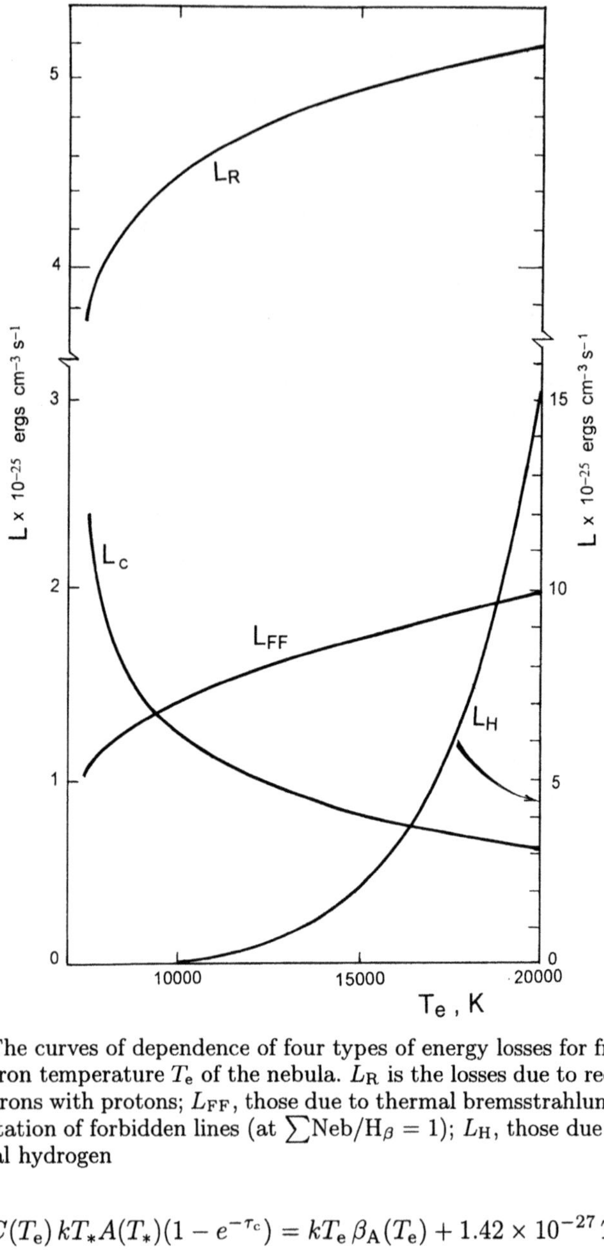

Fig. 4.2. The curves of dependence of four types of energy losses for free electrons at the electron temperature T_e of the nebula. L_R is the losses due to recombination of free electrons with protons; L_{FF}, those due to thermal bremsstrahlung; L_C, those due to excitation of forbidden lines (at $\sum \mathrm{Neb}/\mathrm{H}_\beta = 1$); L_H, those due to collisions with neutral hydrogen

$$C(T_e)\, kT_* A(T_*)(1 - e^{-\tau_c}) = kT_e\, \beta_A(T_e) + 1.42 \times 10^{-27}\, T_e + 1.23 \times 10^{-21}\, T_e^{-1} \frac{\sum \mathrm{Neb}}{E(\mathrm{H}_\beta)} + 1.63 \times 10^{-11} \frac{N_1}{N^+} q_H(T_e)\,. \quad (47)$$

This expression, in fact, interrelates the three parameters – T_*, T_e and τ_c. An important advantage of this expression is the absence of the electron concentration N_e, if, obviously, the ratio N_1/N^+ in the last term on right hand side is not taken into account. However, the degree of ionization in nebulae N^+/N_1 is a quantity that is more or less stable, namely, of the order of 1 000 in high excitation nebulae and 100 in low excitation ones. In this way we preserve the three-parameter structure of (47).

If the electron temperature T_e is assumed to be a known quantity, (47) turns out to be a dependence between the temperature of the central star T_* and the optical depth of the nebula τ_c. If T_* is also obtained via some alternative way, then (47) should determine τ_{ct} – a crucial parameter for any nebula. The values of τ_c obtained in such a way for a sample of PNs are given in Table 4.2.

Rather remarkable is the distribution of the energy losses of free electrons according to different "canals" of losses, as shown in Table 4.3 for a sample of PNs. Judging by these data, the main part of the energy losses, nearly 75%, of the energy of electrons is lost upon the excitation of forbidden and pseudoforbidden lines (L_C). The next are the losses for recombination processes (L_R), i.e. during radiation in the lines and continuum of various series – 16%. The losses due to thermal bremsstrahlung (L_{FF}), i.e. due to the radiation in the continuum, as well as losses due to the collisions with neutral hydrogen atoms (L_H), are of the same order, about $\sim$ 4–5%.

Among the essential results of the present section should be considered the establishment of the numerical value of the parameter η – the part of the energy of free electrons spent on the excitation of all forbidden lines: it turns out to be equal to $\eta = 0.75$. However, the intensity of $N_1 + N_2$ lines are nearly half of the power of all forbidden lines. Therefore the excitation of $N_1 + N_2$ requires nearly one third of the energy of the free electrons, i.e. $\eta(N_1 + N_2) \approx 0.3$.

The condition of thermal equilibrium enables us to determine the exact interrelations between the various parameters of the nebula and its nucleus. Among them are the depth τ_c and the nucleus temperature T_*. However,

Table 4.2. The optical depth τ_c obtained from the condition of thermal equilibrium for a sample of planetary nebulae

Planetary nebula	Excitation class p	T_e, K	T_*, K	$\sum$ Neb/H$_\beta$	τ_c
NGC 2371	12	15 000	100 000	25	0.48
NGC 6572	7	11 500	60 000	16	0.62
NGC 6720	10	10 500	130 000	26	0.30
NGC 6853	10	13 000	144 000	24	0.26
NGC 7009	7	15 400	100 000	19	0.40
NGC 7027	10	16 000	165 000	60	0.56
NGC 7293	7	13 000	123 000	26	0.34
NGC 7662	10	14 000	100 000	32	0.60

Table 4.3. Distribution (as a percentage) of the energy of free electrons lost by the processes: L_R – recombination, L_{FF} – thermal bremsstrahlung, L_C – excitation of forbidden lines, L_H – collisions with hydrogen atoms

Planetary nebula	Percentage			
	L_R	L_{FF}	L_C	L_H
NGC 2867	13	5	80	2
NGC 2371	17	6	70	7
NGC 6302	8	3	80	9
NGC 6720	12	5	83	0.3
NGC 6853	16	5	76	3
NGC 7009	20	8	61	12
NGC 7027	9	3	82	6
NGC 7293	15	5	77	3
IC 418	20	7	73	0
IC 2149	28	6	66	0
Mean	16	5	75	4

the large uncertainties in the value of τ_c in each concrete case makes highly uncertain the determination of T_*.

The correct way to determine of the nuclei temperatures is to search in the direct spectra of the nuclei in the far ultraviolet – 1 000–2 000 Å.

4.7 Stratification of Radiation: Theory

As a phenomenon, the stratification of radiation has been considered from the observational point of view in Sects. 1.6 and 3.11. Below we will consider its theoretical interpretation as a consequence of specific conditions of luminescence of different atoms and ions in nebular conditions.

Two examples are shown in Fig. 4.3 illustrating the locations of the predicted zones with various degrees of ionization of He, O and Ne; as can be seen, the radii of zones for one and the same ion, O or Ne, are nearly identical. Note that practically all oxygen, helium and neon is only in the ionized state corresponding to the given zone.

Each of these zones can generate only definite lines – forbidden lines N_1 and N_2[OIII] or 4 363 [OIII] in zone O^{++}, the doublet 3 726 + 3 729 [OII] in zone O^+, forbidden lines 1 584 [NeV], 3 885 [NeV] or 3426 [NeV] in zone Ne^{++++}, etc. The same is the case with recombination lines: the lines 1 640 HeII and 4 686 HeII can be excited only in zone He^{++}, and 5 876 HeI and 4 471 HeI, also recombination lines, in zone He^+, etc. This means that the dimensions of the monochromatic images must differ greatly from one another: we must have the smallest sizes in the case of lines with high excitation potentials and the largest sizes for the lines with low potentials.

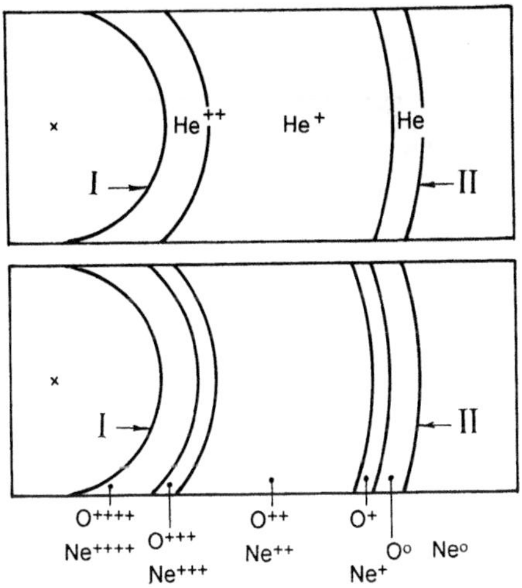

Fig. 4.3. The stratification of the radiation in PNs. The predicted positions of ionization zones for He, O and Ne are shown for a nebula with $T_e = 100\,000$ K, $n_e = 7\,000$ cm^{-3}. I and II are the inner and outer boundaries of the nebula; the cross denotes its centre

A decisive role in the stratification phenomenon is played by the temperature of the central star. One can see this with the example of the helium zones. Above, we have seen that the boundaries of the ionization zones are rather sharp. This being so, one can speak, say, of the radius r_1 of the ionized zone of hydrogen HII.

The values of $r_1 = r_1$(HII) can be obtained from elementary considerations, e.g. by equating the total number of photons emitted by the star in the Lyman continuum with recombination by hydrogen atoms in the zone HII, i.e.

$$\int_{\nu_0}^{\infty} \frac{L_\nu}{h\nu}\,\mathrm{d}\nu = \frac{4\pi}{3} r_1^3 n_\mathrm{H} n_\mathrm{e} C_\mathrm{B}(\mathrm{H}^+)\,, \tag{48}$$

where L_ν is the luminosity of the central star with a temperature T_* at frequency ν, and C_B is the coefficient of recombination of a free electron with a proton.

Analogously one can write for the zones of singly and doubly ionized helium with radii r_2(HeI) and r_3(HeII), respectively,

$$\int_{\nu_0}^{\infty} \frac{L_\nu}{h\nu}\,\mathrm{d}\nu = \frac{4\pi}{3} r_2^3 n(\mathrm{He}^+) n_\mathrm{e}' C_\mathrm{B}(\mathrm{He}^+)\,, \tag{49}$$

$$\int_{4\nu_0}^{\infty} \frac{L_\nu}{h\nu}\,\mathrm{d}\nu = \frac{4\pi}{3} r_3^3 n(\mathrm{He}^{++}) n_\mathrm{e}'' C_\mathrm{B}(\mathrm{He}^{++})\,, \tag{50}$$

where we have taken into account (through $n_\mathrm{e}'$and n_e'') that among the suppliers of free electrons within the zones He^+ and He^{++} is not only the hydrogen ($n_\mathrm{e} = n_\mathrm{H}$), but also helium.

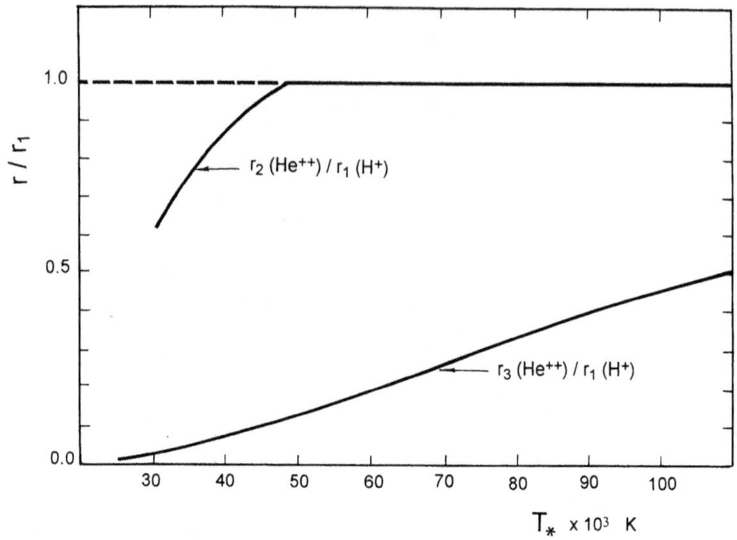

Fig. 4.4. Curves of the dependence of the ratios r_2/r_1 and r_3/r_1 of ionization zones for He^{++} and H^+ on the temperature of the central star T_*. Both zones He^+ and He^{++} are combined at $T_* = 45\,000$ K. The radius of the zone He^{++} is smaller than that of the zone H^+, even at $T_* = 100\,000$ K

If the radiation of a star is given by the Planck law, we have from (48–50) for the ratio of the radii of these zones r_2/r_1 and r_3/r_1

$$\left(\frac{r_2}{r_1}\right)^3 = \frac{f(x_2)}{f(x_0)}\frac{n(\mathrm{H}^+)}{n(\mathrm{He}^+)}\frac{C_\mathrm{B}(\mathrm{H}^+)}{C_\mathrm{B}(\mathrm{He}^+)}\left[1+\frac{N(\mathrm{He})}{N(\mathrm{H})}\right]^{-1}, \tag{51}$$

$$\left(\frac{r_3}{r_1}\right)^3 = \frac{f(4x_0)}{f(x_0)}\frac{n(\mathrm{H}^+)}{n(\mathrm{He}^{++})}\frac{C_\mathrm{B}(\mathrm{H}^+)}{C_\mathrm{B}(\mathrm{He}^{++})}\left[1+2\frac{N(\mathrm{He})}{N(\mathrm{H})}\right]^{-1}, \tag{52}$$

where

$$f(x) = \int_x^\infty \frac{z^2\,\mathrm{d}z}{e^z-1}, \tag{53}$$

$x_0 = h\,\nu_0/kT_*$, $x_1 = h\,\nu_1/kT_*$ and ν_2 is the frequency of the first ionization of helium.

The results of calculations carried out at $T_\mathrm{e} = 10\,000$ K and He/H $= 0.1$ are presented in Fig. 4.4 in the form of curves of dependence of r_2/r_1 and r_3/r_1 on the star's temperature T_*. Note that the ratios of the radii of zones do not depend on the electron concentration of nebula n_e. In Fig. 4.4, the nebula's size is limited by the radius where $r_2(\mathrm{He}^+)/r_1(\mathrm{H}^+) = 1$.

As follows from these curves, up to $T_* < 45\,000$ K the radius of the zone of singly ionized helium r_2 is smaller compared with the radius r_1 of the ionized hydrogen zone. As for the zone He^{++}, the ratio $r_3(\mathrm{He}^{++})/r_1(\mathrm{H}^+) < 1$ at $T_* = 120\,000$ K. Even at $T = 80\,000$ K the radius of the zone He^{++} is two times smaller than the radius of H^+, and three times smaller at $T_* = 120\,000$ K.

4.8 Bowen Resonance Fluorescence

The wavelength of the resonance line of ionized helium L_α HeII303.78 Å accidentally almost exactly coincides with the wavelength of 303.80 Å of the transition $^3P_2 - n^3P_2^0$ of doubly ionized oxygen O^{++}. The latter, absorbing the photons of L_α HeII, can be raised from the ground state $^3P_{2,1,0}$ to the level $^3P_2^0$. Subsequent downward cascade transitions are, first, to the levels ^3P, ^3S, ^3P, then to $^3P^0$ and, finally to the initial state $^3P_{2,1,0}$. These transitions are accompanied by emission of over twenty lines in the optical band, half of which are relatively strong (3 444, 3 133, 3 341, etc.). This is the essence of Bowen's fluorescence mechanism.

Another example of Bowen resonance fluorescence occurs in the case of neutral oxygen OI in which the transition from the ground state $2p\ ^3P$ to the excitation level $3d\ ^3D^0$ is accompanied by the absorption of a photon with a wavelength of 1 025.77 Å, quite near the 1 025.72 L_β line of hydrogen. The difference $\Delta\lambda = 0.05$ Å is within the limits of the Doppler width at $T_e = 10\,000$ K. The recombinations and consequent cascade transitions at normal fluorescence result in the transfer of a definite number of atoms to the level $3d\ ^3D^0$, the subsequent downward transitions leading to the appearance of two infrared photons, 11 287 OI and 8 446 OI, as well as three ultraviolet lines, 1 302, 1 305 and 1 306 OI.

The described mechanism of Bowen resonance fluorescence has interesting applications in various problems concerning the physics of gaseous nebulae.

4.9 Excitation of the Resonance Level of Hydrogen by Electron Collisions

The hydrogen atom has a metastable level $2S$ with energy of excitation of 10.2 eV. The lifetime of an atom at this level is very long – about 0.1 s. Therefore, at sufficiently high electron energies an accumulation of atoms at the level $2S$ can take place, i.e this level will have an anomalously high population. To leave this level, $2S$, the atom should emit two photons of arbitrary frequencies (Fig. 9.1).

In a stationary state we have

$$n_1 n_e \alpha_{2s(T_e)} = n_2 A_{2S,\,1S}\,, \tag{54}$$

where the left hand side gives the number of transitions $1 \to 2$ due to collisions; the function $\alpha_{2s}(T_e)$ is similar to (9) and is related to the transition $1S - 2S$. We have also for spontaneous transition $2S - 1S$, $A_{2s,1s} = 8.227\ \mathrm{s}^{-1}$; calculations carried out for $T_e = 20\,000$ K give

$$\frac{n_{2S}}{n_1} \approx 10^{-7}\,, \tag{55}$$

which considerably exceeds – by six or seven orders – the degree of excitation of the ordinary levels of hydrogen.

5. Temperature of Nebular Nuclei

5.1 Statement of the Problem

The nuclei of PN may well be the hottest among the known stars. This conclusion can be drawn from the large apparent brightness of the nebulae as well as from the almost complete invisibility of the nuclei together with their dwarf-like sizes, and from the presence in the spectrum of nebulae of the emission lines belonging to highly ionized elements – HeII, CIV, NV, NeV, OV, etc.; for their ionization a powerful source of radiation in the wavelength region shorter than 100 Å is necessary.

To reveal the observational indications of the high temperatures of the nuclei, one has to establish quantitative correlations between these temperatures and certain directly observable quantities. Thus, the problem of developing methods for the determination of nuclei temperatures has arisen.

As a result, the theory of determining of the temperatures of nebular nuclei turns out to be one of the most developed and classical branches of theoretical astrophysics. It appears that the methods developed for PNs can also be applied to a wide class of objects possessing gaseous envelopes – non-stationary stars of various types, Wolf–Rayet and P Cygni stars, novae, supernovae, the interstellar medium and even galactic nuclei and quasars.

5.2 The Recombination Method: Zanstra's First Method

This method is based on the theory of the hydrogen line emission of PNs. As we have seen in Chap. 2, a nebula reemits a photon in certain lines of the Balmer series of hydrogen for each L_c quantum radiated by the central star and absorbed by the nebula. If N_c is the number of L_c photons emitted per second in all directions by the star and if the nebula is able to absorb all these quanta, then the total number of Balmer quanta radiated by the nebula will also be N_c. Consequently, if one might somehow compute the number of Balmer quanta emitted by the nebula per unit time, then one could determine the rate of ultraviolet radiation emitted by the central star and hence its colour temperature corresponding to Planck's law.

The number of L_c quanta radiated by the central star per second in the frequency interval from ν to $\nu + d\nu$ is

$$dN_c = 4\pi r_*^2 \frac{2\pi}{c^2} \frac{\nu^3 d\nu}{e^{h\nu/kT_*} - 1}, \tag{1}$$

where T_* and r_* are the temperature and radius of the central star.

The rate of emission of Balmer photons by the nebula is

$$N_{\mathrm{Ba}} = 4\pi r_*^2 \frac{2\pi}{c^2} \int_{\nu_0}^{\infty} \frac{(1 - e^{-\tau_\nu})\, \nu^2\, d\nu}{e^{h\nu/kT_*} - 1}, \tag{2}$$

where τ_ν is the optical thickness of the nebula at some frequency ν beyond the Lyman limit, and ν_0 is the frequency of the Lyman limit.

Now we have to determine from observations the same quantity N_{Ba}, i.e. the total number of Balmer photons radiated by the nebula, and equate this with (2), after which we can determine the required temperature of the nucleus.

Let E_i be the rate of the emission of photons in the ith Balmer line by the whole nebula. This quantity can be determined in arbitrary units from measurements of spectrograms of the nebula. Let $(\partial E_*/\partial \nu)_i$ be the rate of total radiation of the nucleus per unit frequency, just outside the ith Balmer line. This quantity also can be obtained in the same units from measurements of the spectrograms of the nucleus of the nebula. Thus we have the dimensionless ratio

$$A_i = \frac{E_i}{\nu_i (\partial E_*/\partial \nu)_i} \tag{3}$$

for all Balmer lines accessible to observation.

On the other hand, we have

$$\left(\frac{\partial E_*}{\partial \nu}\right)_i = 4\pi r_*^2 \frac{2\pi h \nu_i^3}{c^2} \frac{1}{e^{h\nu/kT_*} - 1}. \tag{4}$$

Therefore the total number of Balmer photons emitted by the nebula yields

$$\begin{aligned} N_{\mathrm{Ba}} &= \sum_i \frac{E_i}{h\,\nu_i} = \frac{1}{h} \sum_i A_i \left(\frac{\partial E_*}{\partial \nu}\right)_i = \\ &4\pi r_*^2 \frac{2\pi}{c^2} \sum_i \frac{\nu_i^3 A_i}{e^{h\nu_i/kT_*} - 1}, \end{aligned} \tag{5}$$

where the summation runs over all lines of the Balmer series and over the Balmer continuum.

Equating (2) to (5) we obtain

$$\int_{\nu_0}^{\infty} \frac{(1 - e^{-\tau_\nu}\, \nu^2\, d\nu)}{e^{h\nu/kT_*} - 1} = \sum_i \frac{x_i^3\, A_i}{e^{h\nu_i/kT_*} - 1}. \tag{6}$$

Let us define

$$x = \frac{h\nu}{kT_*}, \qquad x_i = \frac{h\nu_i}{kT_*}, \qquad x_0 = \frac{h\nu_0}{kT_*}. \tag{7}$$

Then, taking into account that $\tau_\nu \sim \nu^{-3}$, instead of (6) we will have

$$\int_{x_0}^{\infty} \frac{\left[1 - e^{-\tau_0\left(\frac{x_0}{x}\right)^3}\right] x^2 \, dx}{e^x - 1} = \sum_i \frac{x_i^3 \, A_i}{e^{x_i} - 1} , \tag{8}$$

where τ_0 is the optical thickness of the nebula at the frequency ν_0. This formula can be somewhat simplified by using the optical thickness τ_c averaged over the Lyman continuum frequencies:

$$\left(1 - e^{-\tau_c}\right) \int_{x_0}^{\infty} \frac{x^2 \, dx}{e^x - 1} = \sum_i \frac{x_i^3 \, A_i}{e^{x_i} - 1} . \tag{9}$$

This is the precise formula for determination of the temperature T_* for the nucleus of a nebula; the total optical depth beyond the Lyman limit equals τ_c.

Assuming in (9) that $\tau_c \gg 1$, i.e. if the nebula absorbs all L_c radiation of the nucleus, we arrive at Zanstra's formula (1930):

$$\int_{x_0}^{\infty} \frac{x^2 \, dx}{e^x - 1} = \sum_i \frac{x_i^3 \, A_i}{e^{x_i} - 1} . \tag{10}$$

As will be shown below, for the majority of PNs the condition $\tau_c \gg 1$ is not satisfied. Consequently (10), generally speaking, gives the lower limit of the temperature of the nucleus, which enters into x_0 and x_i.

Equation (9) can be solved numerically and graphically. For this and similar problems, we give in Table 5.1 the numerical values of the integrals

$$J_2(x_0) = \int_{x_0}^{\infty} \frac{x^2 \, dx}{e^x - 1} , \tag{11}$$

$$J_3(x_0) = \int_{x_0}^{\infty} \frac{x^3 \, dx}{e^x - 1} , \tag{12}$$

for various values of x_0. In calculations of $J_2(x_0)$ and $J_3(x_0)$ at $x_0 > 12$ the following expressions may be useful:

$$\begin{aligned} \int_{x_0}^{\infty} \frac{x^2 \, dx}{e^x - 1} &= \left(x_0^2 + 2x_0 + 1\right) e^{-x_0} , \\ \int_{x_0}^{\infty} \frac{x^3 \, dx}{e^x - 1} &= \left(x_0^3 + 3x_0^2 + 6x_0 + 6\right) e^{-x_0} . \end{aligned} \tag{13}$$

Zanstra applied (10), which is valid for $\tau_c \gg 1$, to three PNs – NGC 6543, NGC 6572 and NGC 7009 – and found for the temperatures of their nuclei 39000, 40000 and 55000 K respectively. However, at least two of these nebulae, NGC 6543 and NGC 7009, have double shells and consequently $\tau_c < 1$. As regards NGC 6572, a stellar-type nebula, it has still not been shown to have double shells. Therefore the values of T_* for at least two of

Table 5.1. Numerical values of integrals necessary for the determination of the temperatures of the nuclei of planetary nebulae

x_0	$\int_{x_0}^{\infty} \frac{x^2 \mathrm{d}x}{e^x-1}$	$\int_{x_0}^{\infty} \frac{x^3 \mathrm{d}x}{e^x-1}$	x_0	$\int_{x_0}^{\infty} \frac{x^2 \mathrm{d}x}{e^x-1}$	$\int_{x_0}^{\infty} \frac{x^3 \mathrm{d}x}{e^x-1}$	x_0	$\int_{x_0}^{\infty} \frac{x^2 \mathrm{d}x}{e^x-1}$	$\int_{x_0}^{\infty} \frac{x^3 \mathrm{d}x}{e^x-1}$
0	2.4041	6.4939	3.6	0.6120	3.1187	7.1	0.05497	0.4604
0.1	2.3993	6.4936	3.7	0.5764	2.9890	7.2	0.05096	0.4316
0.2	2.3854	6.4915	3.8	0.5425	2.8620	7.3	0.04722	0.4045
0.3	2.3634	6.4859	3.9	0.5103	2.7379	7.4	0.04375	0.3790
0.4	2.3342	6.4756	4.0	0.4797	2.6169	7.5	0.04052	0.3550
0.5	2.2986	6.4596	4.1	0.4506	2.4991	7.6	0.03752	0.3323
0.6	2.2574	6.4369	4.2	0.4230	2.3846	7.7	0.03473	0.3110
0.7	2.2113	6.4069	4.3	0.3969	2.2735	7.8	0.03214	0.2909
0.8	2.1610	6.3691	4.4	0.3721	2.1659	7.9	0.02974	0.2720
0.9	2.1071	6.3232	4.5	0.3487	2.0618	8.0	0.02751	0.2543
1.0	2.0502	6.2691	4.6	0.3266	1.9612	8.1	0.02544	0.2376
1.1	1.9909	6.2068	4.7	0.3057	1.8641	8.2	0.02352	0.2220
1.2	1.9296	6.1364	4.8	0.2860	1.7706	8.3	0.02174	0.2073
1.3	1.8669	6.0580	4.9	0.2675	1.6806	8.4	0.02009	0.1936
1.4	1.8031	5.9719	5.0	0.2500	1.5941	8.5	0.01856	0.1807
1.5	1.7387	5.8784	5.1	0.2336	1.5110	8.6	0.01715	0.1686
1.6	1.6740	5.7781	5.2	0.2181	1.4313	8.7	0.01584	0.1572
1.7	1.6093	5.6714	5.3	0.2036	1.3550	8.8	0.01463	0.1466
1.8	1.5449	5.5587	5.4	0.1899	1.2819	8.9	0.01350	0.1367
1.9	1.4810	5.4406	5.5	0.1771	1.2120	9.0	0.01246	0.1274
2.0	1.4179	5.3176	5.6	0.1650	1.1453	9.2	0.01061	0.1105
2.1	1.3558	5.1903	5.7	0.1537	1.0817	9.4	0.00903	0.09580
2.2	1.2949	5.0593	5.8	0.1432	1.0210	9.6	0.00767	0.08296
2.3	1.2353	4.9252	5.9	0.1333	0.9632	9.8	0.00652	0.07176
2.4	1.1771	4.7884	6.0	0.1241	0.9081	10.0	0.00554	0.06202
2.5	1.1204	4.6495	6.1	0.1154	0.8557	10.2	0.00470	0.05354
2.6	1.0653	4.5091	6.2	0.1073	0.8060	10.4	0.00399	0.04619
2.7	1.0119	4.3676	6.3	0.09976	0.7588	10.6	0.00338	0.03918
2.8	0.9603	4.2256	6.4	0.09271	0.7140	10.8	0.00286	0.03428
2.9	0.9105	4.0835	6.5	0.08612	0.6715	11.0	0.00242	0.02949
3.0	0.8624	3.9417	6.6	0.07998	0.6312	11.2	0.00205	0.02536
3.1	0.8162	3.8007	6.7	0.07425	0.5931	11.4	0.00173	0.02178
3.2	0.7717	3.6608	6.8	0.06891	0.5571	11.6	0.00146	0.01870
3.3	0.7291	3.5224	6.9	0.06393	0.5230	11.8	0.00124	0.01604
3.4	0.6883	3.3856	7.0	0.05929	0.4908	12.0	0.00104	0.01375
3.5	0.6493	3.2510						

these objects cannot be considered correct. The calculations show that the temperatures of nuclei when $\tau_c < 1$ are much higher than those obtained with $\tau_c \gg 1$. For NGC 6 543, NGC 6 572 and NGC 7 009, the nucleus temperatures T_*, obtained from (9) at different values of τ_c, are as follows:

τ_c	$\gg 1$	1	0.1	0.01
NGC 6 543	39 000	43 000	80 000	210 000
NGC 6 572	40 000	45 000	83 000	212 400
NGC 7 009	55 000	80 000	130 000	$> 300\,000$

In principle, this method of determination of temperatures of nuclei may also be applied to other lines formed via recombination, e.g. the lines of neutral and ionized helium (He I, He II). Such attempts have been made and they have led to still higher values of the nuclei temperatures. Thus, for example, for the nucleus of NGC 7 009 Zanstra obtained $T_* = 70\,000$ K using the lines of ionized helium instead of $T_* = 55\,000$ K with hydrogen. This provided evidence that in fact the optical depth τ_c of the nebula beyond the Lyman limit in hydrogen is smaller than unity.

5.3 The "Nebulium Method": Zanstra's Second Method

Unlike the lines of hydrogen and helium, forbidden lines are not formed by recombination processes. These lines are excited by means of collisions of the first kind of ions with free electrons, produced primarily by photoionization of hydrogen. In other words, the energy in forbidden lines is drawn from the kinetic energy of free electrons.

The energy of free electrons is gained from the ultraviolet radiation of the star. Part of the latter is spent in removing electrons from atoms, while the remaining part is transferred to the electrons as kinetic energy. Finally part of the energy of the stellar ultraviolet radiation will be reemitted by the nebula in the form of forbidden lines. Thus, if we can compute the fraction of the ultraviolet energy which escapes via forbidden lines as a function of the star's temperature, and moreover if we can determine from observations the energy radiated by the nebula in forbidden lines, then we will be able to estimate the temperature T_* of the nucleus. This is the essence of the "nebulium method".

The relationship analogous to (9) can be derived in this case:

$$\eta(1 - e^{-\tau_c}) \int_{x_0}^{\infty} \frac{(x - x_0)\, x^2\, dx}{e^x - 1} = \sum_{\text{Neb}} \frac{x_i^4 A_i}{e^{x_i} - 1}, \tag{14}$$

where η is the part of the kinetic energy used for the excitation of forbidden lines, and A_i is given by (3). In (14) the unknown parameter is the temperature T_* of the nucleus which enters in x_0 and x_i. The parameters τ_c and η are considered known.

The recombination method has some advantage with respect to the nebulium method: in the former case the number of unknown parameters was only one, τ_c, while in the latter case two parameters, τ_c and η, are unknown. Ideally, if both methods could be applied to the same object (nebula), the simultaneous solution of (9) and (14) would enable one to estimate the value of η.

In particular, when $\eta = 1$ and $\tau_c \gg 1$, we obtain from (14) Zanstra's second formula:

$$\int_{x_0}^{\infty} \frac{x^3\,dx}{e^x - 1} - x_0 \int_{x_0}^{\infty} \frac{x^2\,dx}{e^x - 1} = \sum \frac{x_i^4\, A_i}{e^{x_i} - 1} . \tag{15}$$

This formula again gives only the lower limit to the temperature of the nucleus, so long as in a real nebula $\tau_c < 1$ and $\eta < 1$. The optical depth τ_c has already been discussed above. As regards η, the method of determination was described earlier (Sect. 4.6); though η is smaller than unity, its mean value is equal to 0.75, i.e. close to unity, and hence its role for T_* is small. The largest influence on the value of T_* is τ_c; in this respect (14) and (9) are equivalent.

The application of (15) for the nuclei of PNs already discussed above, NGC 6543, NGC 6572 and NGC 7009, gives for T_*: 37000, 38000 and 50000 K, i.e. the same values that were obtained via Zanstra's first method. This fact, however, is not enough to conclude that the results obtained by two methods are not the lower limits but are close to the true temperatures; the reason is the similar uncertainty in both formulae (9) and (14), so that their results can be incorrect to the same extent.

Moreover, the "nebulium method" is not an independent method. The difference between the recombination method and the "nebulium method" is formal. In principle, both methods can be replaced by a single one if one considers simultaneously the balance between the ultraviolet radiation of the star beyond the Lyman limit and the total energy radiated by the nebula in the lines of both hydrogen (recombination) and "nebulium" (forbidden).

In addition, Zanstra's methods have a general procedural defect: for their application one must obtain spectrograms of both the nebula and its central star. The latter is not always possible owing to the weakness of central stars.

5.4 The HeII/HI Method: Ambartsumian's Method

For certain pairs of quantum states of hydrogen and ionized helium the ratio of Einstein coefficients for spontaneous transitions are constant and equal to $Z^4 = 16$. This is the case, for example, for the transitions corresponding to the lines H_β and 4686 HeII, H_γ and 4512 HeII etc. Assume that: (a) there are enough He^+ ions in the nebula that all radiation lying beyond the ionization limit $4\nu_0$ of singly ionized helium is absorbed, i.e. $\tau_c(\text{HeII}) > 1$; (b) radiation lying in the frequency interval ν_0–$4\nu_0$, is completely absorbed by hydrogen alone, i.e. $\tau_c > 1$. Then the following relationship can be derived for the determination of the temperature of the nebula's nucleus (Ambartsumian, 1932):

$$\frac{\displaystyle\int_{4x_0}^{\infty} \frac{x^2\,dx}{e^x - 1}}{\displaystyle\int_{x_0}^{4x_0} \frac{x^2\,dx}{e^x - 1}} = \frac{A_{43}}{A_{42}} \frac{\lambda_{4686}}{\lambda_{H_\beta}} \frac{I_{4686}}{I_{H_\beta}} = 0.90\,\frac{I_{4686}}{I_{H_\beta}} . \tag{16}$$

For a stricter statement of the problem, however, one has to proceed from the real number of recombination acts at the initial level of the excitation of the lines 4 686 HeII and 4 861 HI. In this case we shall have, instead of (16), the following relationship for the nebula, which is optically thick to the L_c radiation, i.e. $\tau_c \gg 1$ (model B):

$$\frac{\displaystyle\int_{4x_0}^{\infty} \frac{x^2\,dx}{e^x - 1}}{\displaystyle\int_{x_0}^{4x_0} \frac{x^2\,dx}{e^x - 1}} = \frac{\alpha(H_\beta)}{\alpha_{4686}} \frac{\alpha_B(He^+)}{\alpha_B(H^0)} \frac{\nu(4861)}{\nu(4686)} \frac{I_{4686}}{I_{H_\beta}} . \tag{17}$$

For the optically thin nebula, $\tau_c \ll 1$, model A, α_B in (17) must be replaced by α_A .

The final result for the case $\tau_c \gg 1$ (model B) and $T_e = 10\,000$ K is as follows (Harman, Seaton, 1966):

$$\frac{\displaystyle\int_{4x_0}^{\infty} \frac{x^2\,dx}{e^x - 1}}{\displaystyle\int_{x_0}^{4x_0} \frac{x^2\,dx}{e^x - 1}} = 0.615\,\frac{I_{4686}}{I_{H_\beta}} , \tag{18}$$

where I_{4686} and $I_{H\beta}$ are the observed intensities of the emission lines 4 686 HeII and H_β in the spectrum of PNs.

For $\tau_c \ll 1$ (model A) the numerical coefficient at $I_{4686}/I_{H\beta}$ is replaced by 0.527.

The first assumption, τ_c (HeII) $\gg 1$, is justified for those PNs for which the dimensions of monochromatic images in the line 4 686 HeII are smaller than those of monochromatic images in the hydrogen lines (NGC 2 165, NGC 3 242, NGC 6 818, NGC 7 662 etc.). However, the second assumption, $\tau_c \gg 1$, is not valid for the majority of PNs. Consequently the temperatures of the nuclei determined in this way will be the upper limits of their true values.

Finally, there is the third category of nebulae for which both conditions are not fulfilled, i.e. for it τ_c (HeII) < 1 and $\tau_c < 1$. These are highly evolved ring-like PNs of gigantic sizes, e.g. NGC 7 094, K 1-14, K 1-16, A 7, A 36, A 73 etc.; their diameters are of the order of a parsec and larger.

Thus the nuclei temperatures determined by the HeII/HI method will be in the majority of cases only the upper limits of their true values.

The application of the HeII/HI method to the three PNs mentioned above, NGC 6 543, NGC 6 572 and NGC 7 009, yields the temperatures for their nuclei of 74 100, 92 300 and 103 000 K, i.e. two times larger than we had from Zanstra's methods. As we shall see later (Sect. 5.8), for the majority of PNs nearly 90% – the HeII/HI method results in temperatures for nuclei beyond 100 000 K.

When $\tau_c < 1$ and τ_c (HeII) > 1, (18) should be written in the form

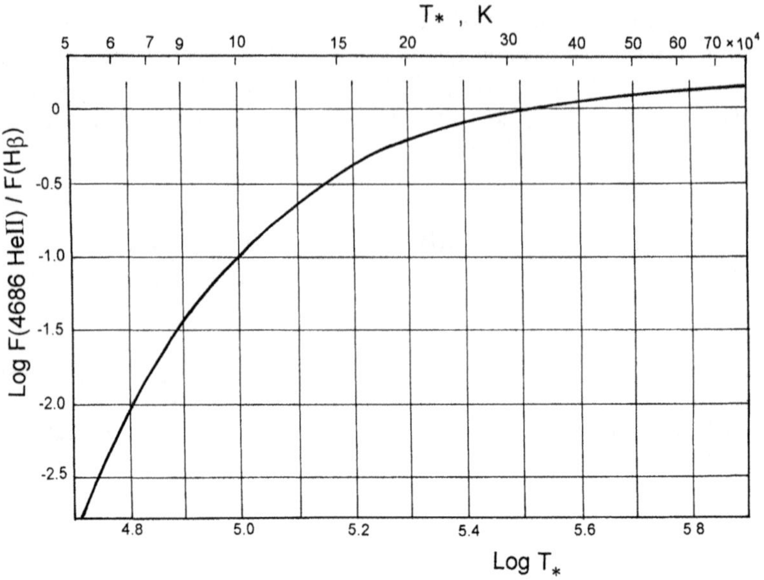

Fig. 5.1. The HeII/HI method for determining nebular nuclei temperatures T_*: the dependence of T_* on $I(4686\ \mathrm{HeII})/F(\mathrm{H}_\beta)$

$$\frac{\displaystyle\int_{4x_0}^{\infty} \frac{x^2\,\mathrm{d}x}{e^x - 1}}{\displaystyle\int_{x_0}^{4x_0} \frac{x^2\,\mathrm{d}x}{e^x - 1}} = 0.615\,(1 - e^{-\tau_c})\,\frac{I_{4686}}{I_{\mathrm{H}_\beta}}\,. \tag{19}$$

The advantage of the HeII/HI method is that applying it the spectrogram of the nebula only is sufficient.

The temperatures of nuclei determined by this method decrease as τ_c decreases. In the case of Zanstra's methods we had the contrary: the temperature of the nucleus increased as τ_c decreased. Therefore, one might suppose that there must exist some value of τ_c for which the temperatures of the nucleus of a given nebula determined according to both Ambartsumian's and Zanstra's methods coincide with each other. In Fig. 5.1 the curve of dependence of the ratio $I_{4686\mathrm{HeII}}/I_{\mathrm{H}_\beta}$ on T_* is given, according to (19).

As a matter of fact, in (19) we have one equation with two unknown parameters – T_* and τ_c. For their determination it is necessary to have one more equation containing T_* and τ_c. One of Zanstra's formulae can be used for this purpose. Combining, therefore, (9) and (19), and excluding the term $(1 - e^{-\tau_c})$, we arrive at the following precise formula for the determination of the temperature of the nucleus of an optically thin nebula, i.e. $\tau_c < 1$ but τ_c (HeII) > 1:

$$\frac{\displaystyle\int_{x_0}^{\infty}\frac{x^2\,\mathrm{d}x}{e^x-1}\int_{4x_0}^{\infty}\frac{x^2\,\mathrm{d}x}{e^x-1}}{\displaystyle\int_{x_0}^{4x_0}\frac{x^2\,\mathrm{d}x}{e^x-1}} = 0.615\,\frac{I_{4686}}{I_{\mathrm{H}_\beta}}\sum_i \frac{x_i^3\,A_i}{e^{x_i}-1}\,, \tag{20}$$

or, with sufficient accuracy,

$$\int_{4x_0}^{\infty}\frac{x^2\,\mathrm{d}x}{e^x-1} = 0.615\,\frac{I_{4686}}{I_{\mathrm{H}_\beta}}\sum \frac{x_i^3\,A_i}{e^{x_i}-1}\,. \tag{21}$$

This is, in fact, Zanstra's formula written for ionized helium – a kind of hydrogen recombination method, with the difference that the sum of the right hand side of (21) is related to hydrogen lines, while only the intensity of 4 686 HeII is used among helium lines.

Determining from (21) the correct value of T_*, we can, with the help of (9) or (19), obtain τ_{c}. In this manner we have a method for determining the optical thickness for at least those PNs for which τ (HeII) > 1, i.e. for which the dimensions of monochromatic images in 4 686 HeII are smaller than in hydrogen lines.

Equation (21) was applied to two PNs, NGC 6 572 and NGC 7 009, revealing the line 4 686 HeII and with known values of A_i. The two key parameters, T_* and τ_{c}, for these nebulae, obtained in such manner, are as follows:

	I(4 686 HeII)/H_β	T_*, K	τ_{c}
NGC 6 572	0.13	75 000	0.14
NGC 7 009	0.14	94 000	0.40

As can be seen, the true temperatures of the nuclei of PNs are rather high and significantly exceed the values obtained by Zanstra's methods.

5.5 The Method of Fluxes: Seaton's Method

The absolute luminosity of the star-nucleus of a nebula Λ depends on its temperature T_* and apparent magnitude m_{s}. Assuming that the radiation of the star is represented by Planck's law, the following dependence between T_* and Λ can be derived:

$$\Lambda(S) = 2.590\times 10^{-4}\,T_*^4\left(e^{29\,600/T_*}-1\right)10^{\delta}\,, \tag{22}$$

where

$$\delta = -0.4\,m_s + 0.14c - 4.70 \tag{23}$$

and $0.14c$ is the difference in interstellar extinction between the line 4 861 H_β and the effective wavelength 4 365 Å (nearly in the B band) for which m_{s} is determined.

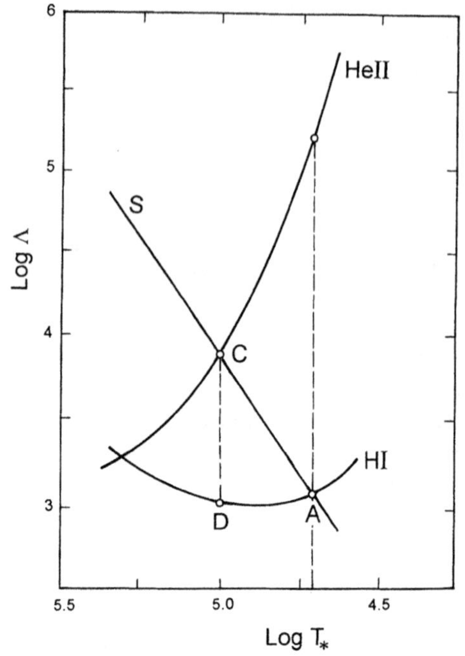

Fig. 5.2. The determination of nuclei temperatures T_* by the "method of fluxes". The symbol Λ denotes the luminosity of the nebula's nucleus (see text)

The same magnitude Λ, however, can be represented through the absolute fluxes in the emission lines of $F(\mathrm{H}_\beta)$ and $F(4\,686\ \mathrm{HeII})$. The corresponding dependence on T_* has the following form, when $\tau_c(\mathrm{HeII}) > 1$ but $\tau_c < 1$:

$$\Lambda(\mathrm{HI}) = 5.89 \times 10^{11} \frac{F(\mathrm{H}_\beta)}{1 - e^{-\tau_c}} \left[\int_{x_0}^{4x_0} \frac{x^2\,\mathrm{d}x}{e^x - 1} \right]^{-1} , \tag{24}$$

$$\Lambda(\mathrm{HeII}) = 3.64 \times 10^7\, T_* F(4\,686\ \mathrm{HeII}) \left[\int_{x_0}^{4x_0} \frac{x^2\,\mathrm{d}x}{e^x - 1} \right]^{-1} . \tag{25}$$

In three equations, (22), (24) and (25), there are two unknown variables, T_* and τ_c; however, if the condition $\tau_c \gg 1$ is used in (24), then one will have only one unknown, T_*.

Obviously, the three curves of type $\Lambda \sim T_*$ defined by these formulae – of the star (S), of the line H_β (HI) and of the line 4 686 HeII (HeII) – should intersect at a single point. However, if $\tau_c < 1$, then the intersection will not occur at a single point, but at two – at the points C and A, as shown in Fig. 5.2. The point of intersection of curves S and HeII gives us the magnitude of T_*, after which we obtain τ_c with the help of (24). Note that the magnitude of $1 - e^{-\tau_c}$ corresponds to the section CD in Fig. 5.2.

This is the essence of Seaton's method (Harman, Seaton, 1966). The application of this method requires knowledge of the absolute fluxes of the radiation of the nebula in H_β and 4 686 HeII, as well as the brightness of the nucleus m_s at the effective wavelength 4 365 Å.

For low excitation nebulae, with no HeII in their spectra, one of the emission lines of neutral helium, e.g. 4 471 HeI or 5 876 HeI, can be used. Then analogously (24) and (25) we will have

$$A(\text{HeII}) = 1.36 \times 10^8\, T_* F(4\,471) \left[\int_{x_1}^{\infty} \frac{x^2\, dx}{e^x - 1} \right]^{-1} , \tag{26}$$

where $x_1 = 1.807 x_0$. In this case the temperature T_* is obtained by the point of intersection of three curves – S, HI and HeI. Once we have the T_*, it is not difficult to obtain the numerical value of τ_c for any nebula, as described in the previous section.

This method has been applied for 42 PNs with available photoelectric measurements of fluxes in H_β and 4 686 HeII in absolute units. The temperatures T_* of the nuclei obtained for them along with their optical depths τ_c are given in Table 5.2. From these data it can be seen that 13 nebulae, or ~ 30%, are completely or almost completely opaque.

Table 5.2. Nuclei temperatures T_* and the optical depths at the frequencies of hydrogen ionization τ_c for a sample of planetary nebulae obtained by Seaton's method

Planetary nebula	T_* K	τ_c	Planetary nebula	T_* K	τ_c
NGC 40	34 000	$\gg 1$	NGC 6 751	76 000	0.08
NGC 650-1	182 000	1.77	NGC 6 772	122 000	0.49
IC 351	91 000	0.10	NGC 6 778	85 000	0.27
NGC 1 501	72 500	0.05	NGC 6 781	91 000	0.23
NGC 1 525	74 000	0.10	NGC 6 804	72 500	0.03
IC 418	42 600	$\gg 1$	BD+30	45 700	0.007
NGC 2 022	91 000	0.06	NGC 6 818	195 000	0.96
IC 2 149	49 000	$\gg 1$	NGC 6 826	69 000	0.28
NGC 2 371	100 000	0.08	NGC 6 853	132 000	1.43
NGC 2 392	67 500	0.02	NGC 6 881	69 000	0.04
NGC 3 242	93 000	0.14	NGC 6 891	56 300	0.12
NGC 3 587	105 000	0.51	NGC 6 894	31 600	0.43
IC 3 568	31 500	$\gg 1$	NGC 6 905	102 000	0.20
NGC 6 058	72 500	0.04	NGC 7 008	31 000	0.09
NGC 6 210	50 000	$\gg 1$	NGC 7 009	81 000	0.30
NGC 6 309	95 300	0.09	NGC 7 026	97 500	0.36
NGC 6 439	72 500	0.11	NGC 7 139	97 500	0.60
NGC 6 445	182 000	3.80	IC 5 217	74 000	0.18
NGC 6 543	66 000	1.36	NGC 7 354	102 000	0.22
NGC 6 572	61 600	1.13	NGC 7 662	100 000	0.19

5.6 The [OIII]/[OII] Method

In this method one utilizes the ratio of the intensity of the green doublet of doubly ionized oxygen $N_1 + N_2$ [OIII] to that of the violet doublet of singly ionized oxygen 3727 [OII]. Obviously, the ratio $E(N_1+N_2)/E(3727)$ will increase with the temperature of the nucleus, because at high temperatures the number of O^{++} ions will be larger compared with the number of O^+ ions.

The theoretical relations for the determination of the temperature T_* of the nuclei of PNs corresponding to the known ratios $E(N_1+N_2)/E(3727)$ have been derived in the following form (Gurzadyan, 1955):

$$\frac{E(N_1+N_2)}{E(3727)} = q(T_e)\frac{W}{n_e}T_e \exp\left(-\frac{414000}{T_*}\right) , \tag{27}$$

where n_e is the electron concentration, W is the dilution coefficient and

$$q(T_e) = \frac{2(2\pi\mu k)^{3/2}}{h^3}T_e^{1/2} = 4.80 \times 10^{17}\ \mathrm{cm^{-3}\ K^{-1}} \tag{28}$$

at $T_e = 10000$ K.

Substituting into (27) the obvious expression for W

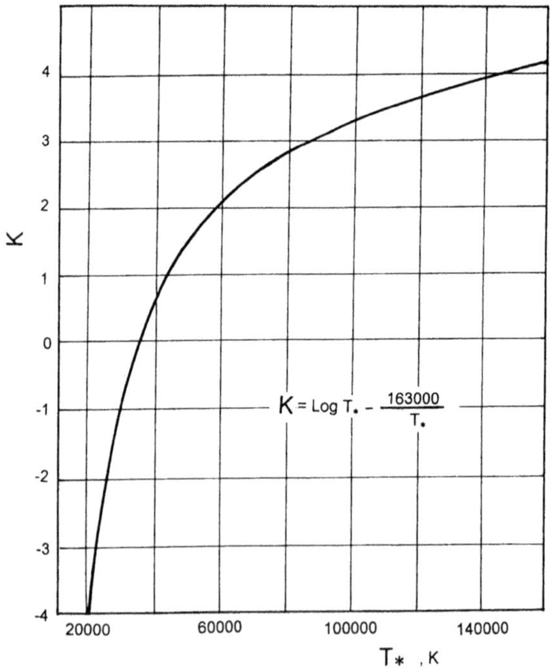

Fig. 5.3. The [OIII]/[OII] method for the determination of the temperature T_* of a nebula's nucleus showing the dependence of T_* on K (see (30))

Table 5.3. Temperatures of the nuclei of planetary nebulae obtained by the [OIII]/[OII] method. n_e is in units of $\times 10^3$ cm^{-3} and T_* in K

Nebula	$\frac{E(N_1+N_2)}{E(3727)}$	n_e	T_*	Nebula	$\frac{E(N_1+N_2)}{E(3727)}$	n_e	T_*
NGC 40	0.12	1.44	31 000	NGC 6826	48.5	2.0	39 000
NGC 650-1	2.95	1.25	94 000	NGC 6833	43	5.0	62 000
NGC 1535	75	3.50	60 000	NGC 6853	3.4	0.28	76 000
NGC 2022	88	1.00	73 000	NGC 6884	113	7.1	160 000
NGC 2371	7.3	1.0	57 000	NGC 6886	6.8	5.6	69 000
NGC 2392	11.2	5.62	43 000	NGC 6891	29.5	2.5	31 000
NGC 2440	20	2.81	122 000	NGC 6894	3.6	0.85	80 000
NGC 2452	10.3	1.25	103 000	NGC 6905	147	4.0	140 000
NGC 3132	5.1	0.63	36 000	NGC 7008	32	10.0	120 000
NGC 3242	163	14.1	78 000	NGC 7009	89	2.24	60 000
NGC 3587	4	0.13	66 000	NGC 7026	15.5	25	90 000
NGC 6058	136	2.5	78 000	NGC 7027	154	19	150 000
NGC 6210	38	5.6	49 000	NGC 7293	1.7	0.14	80 000
NGC 6445	5.7	0.8	115 000	NGC 7662	155	3.2	67 000
NGC 6543	38	3.55	50 000	IC 351	125	5.0	78 000
NGC 6567	12	3.2	58 000	IC 418	0.61	15.8	37 000
NGC 6572	16	5.0	48 000	IC 1747	34.6	3.2	69 000
NGC 6629	19	2.0	47 000	IC 2003	26.5	12.6	59 000
NGC 6720	4.7	0.63	63 000	IC 2149	2.0	2.0	43 000
NGC 6741	12.2	2.5	60 000	IC 2165	24.4	2.8	74 000
NGC 6751	5.9	25	68 000	IC 3568	104	5.0	69 000
NGC 6772	5.2	0.2	92 000	IC 4593	8.3	2.24	29 000
NGC 6778	6.9	0.62	67 000	IC 4634	22.8	6.2	102 000
NGC 6781	3.7	0.63	81 000	IC 4846	60	8.9	69 000
NGC 6790	58	12.6	51 000	IC 4997	45	125	72 000
NGC 6803	34	5.0	55 000	IC 5117	59	79	68 000
NGC 6807	40	33.0	54 000	IC 5217	56	8.9	69 000
NGC 6818	19.4	3.2	75 000	J 320	69	7.9	67 000

$$\log W = 7.00 - 0.4m_* - 2\log D'' - \frac{14\,700}{T_*} \tag{29}$$

we obtain

$$K = \log T_* - \frac{163\,000}{T_*}\,, \tag{30}$$

where

$$K = \log \frac{E(\mathrm{N_1+N_2})}{E(3\,727)} - 0.4m_* + 2\,\log D'' - \log n_e - 10.7\,. \tag{31}$$

Here m_* is the photographic magnitude of the nucleus and D'' the angular size of the nebula in arc seconds. Having from observations the numerical value of ratio $E(\mathrm{N_1+N_2})/E(3\,727)$ and determining n_e by one of existing methods, we obtain from (31) first, the value of K, and then the temperature

T_* from (30). To simplify the calculations, the curve of $K \sim T_*$ is given in Fig. 5.3.

In Table 5.3, the nuclei temperatures obtained by the "[OIII]/[OII] method" are given for more than fifty PNs from known values of the ratio $E(\mathrm{N_1{+}N_2})/E(3\,727)$, corrected for interstellar extinction.

The next step is the consideration of a more realistic model of a nebula when the variation of the degree of ionization along the thickness of the nebula should be taken into account, as well as the role of the cooling of ions. For such a model one has (e.g. Köppen, Tarafder, 1978)

$$\frac{E(\mathrm{N_1 + N_2})}{E(3\,727)} \sim \left(\frac{R_*}{n_\mathrm{e}}\right)^{1/3} T_\mathrm{e}^{1/2} \exp\left(-\frac{h(\nu_1 - \nu_2)}{kT_*}\right), \tag{32}$$

i.e. the dependence of the ratio of intensities of lines both on the radius of a nebula's nucleus and on the electron concentration are extremely weak. Nevertheless for the majority of cases (30) and (32) give almost the same results.

Thus, in spite of a number of uncertainties, the results obtained by the methods described above enable us to conclude that the temperatures of the nuclei of PNs are rather high, within the limits 60 000–100 000 K.

5.7 Temperatures of High Excitation Nebula Nuclei

The criterion for including a PN within the category of high excitation objects is usually the existence of the sufficiently strong emission line 4 686 HeII or 1 640 HeII in its spectrum. For such a purpose both lines are equivalent, in view of the fact that the ratio of their intensities, $F(1\,640)/F(4\,686)$, is nearly a constant and practically independent of T_e and n_e; at $T_\mathrm{e} = 10\,000$ K and $n_\mathrm{e} = 10^4\ \mathrm{cm}^{-3}$ it equals 6.50 (Seaton, 1978).

The relative intensity of 4 686 HeII (in units $\mathrm{H}_\beta = 1.00$) rarely exceeds unity even for extremely excited nebulae, since the ratio $F(4\,686\mathrm{HeII})/F(\mathrm{H}_\beta)$ is less than unity. In Table 5.4, a compilation of 80 high excitation nebulae is listed for which $F(4\,686\mathrm{HeII})/F(\mathrm{H}_\beta) > 0.15$. The total intensities of both lines, $\mathrm{N_1 + N_2}$ [OIII], are also given (e.g. Barker, 1978; Perinotto, Benvenuti, 1981; Kaler, 1982; Feibelman, 1982; Harrington et al., 1982; Perinotto et al, 1994). All intensities are corrected for interstellar excitation. At the end of this table, 15 PNs of medium excitation for which $F(4\,686\mathrm{HeII})/F(\mathrm{H}_\beta) <$ 0.15, are also given.

In Table 5.4, a tendency for the observed intensities of 4 686 HeII to decrease in parallel with the rise of the intensity of the lines $\mathrm{N_1 + N_2}$ can be noticed. This becomes clearer when the relationship between the intensity $F(4\,686$ HeII) and the ratio of $F(\mathrm{N_1{+}N_2})/F(4\,686$ HeII) is considered. Graphically this relationship is presented in Fig. 5.4 for the data in the second and fourth columns of Table 5.4 (Gurzadyan, 1988). Scatter of the points

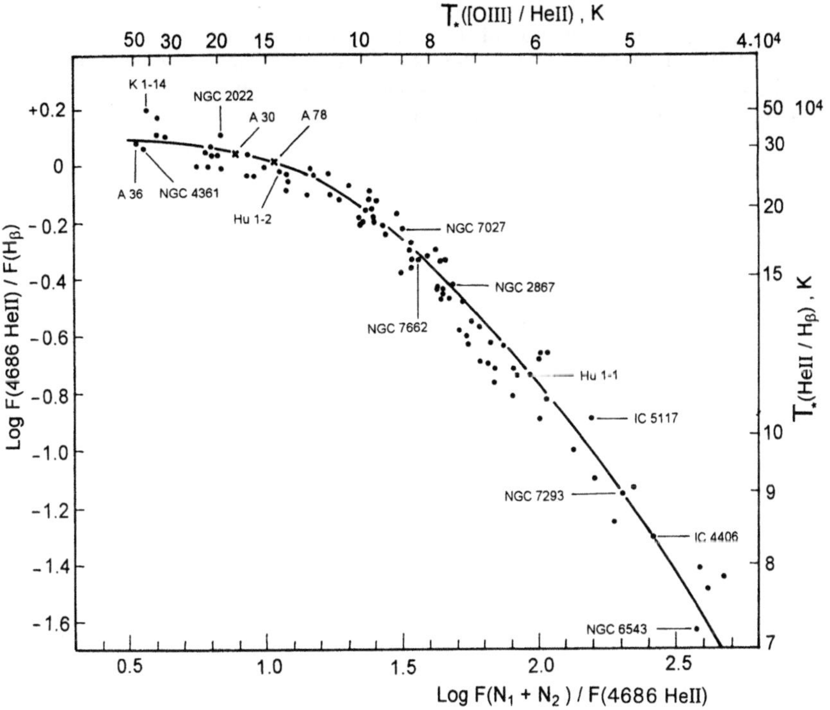

Fig. 5.4. Empirical dependence between the relative intensity $F(4\,686$ HeII) and the ratio of intensities $F(N_1{+}N_2)/F(4686$ HeII) for high excitation (4 686 HeII/$H_\beta > 0.15$) planetary nebulae according to the data of Table 5.4

is smaller and hence reveals the existence of a strong correlation between $F(4\,686$ HeII) and $F(N_1{+}N_2)/F(4686$ HeII). The corresponding empirical formula has the form

$$R_{\rm He} = -0.04 + 0.48\,R_{\rm N} - 0.42\,R_{\rm N}^2\,, \tag{33}$$

where $R_{\rm He} = \log F(4686{\rm HeII})/F({\rm H}_\beta)$ and $R_{\rm N} = F({\rm N}_1 + {\rm N}_2)/F(4\,686\ {\rm HeII})$; see Fig. 5.4.

Thus the possibility of the existence of a positive correlation between the mentioned lines can be predicted. The problem is to derive this relationship in quantitative form.

For the intensities of the lines 4 686 HeII and $N_1 + N_2$ emitted by the nebula we have

$$F(4\,686{\rm HeII}) \sim n_{\rm e}\, n({\rm He}^{++})\, V\,({\rm He}^{++})\,, \tag{34}$$

$$F({\rm N}_1 + {\rm N}_2) \sim n_{\rm e}\, n({\rm O}^{++})\, V\,({\rm O}^{++})\,, \tag{35}$$

where $n({\rm He}^{++})$ and $n({\rm O}^{++})$ are concentrations of the ions He^{++} and O^{++} in their zones, and $V({\rm He}^{++})$ and $V({\rm O}^{++})$ are the volumes of these zones.

Table 5.4. Observed intensities of the emission lines 4 686 HeII and $N_1 + N_2$ [OIII] ($H_\beta = 1.00$), corrected for interstellar extinction, in the spectra of high excitation planetary nebulae, together with the values of the temperatures of their nuclei obtained by the HeII/HI and [OIII]/[OII] methods

Planetary nebula	$F(4686)$	$F(N_1+N_2)$	$\log \frac{F(N_1+N_2)}{F(4686)}$	T_*, K HeII/HI	T_*, K [OIII]/[OII]
NGC 246	1.32	6.9	0.72	367 000	243 000
NGC 650-1	0.61	16.2	1.42	186 000	94 000
NGC 1 360	0.99	9.7	0.99	263 000	152 000
NGC 1 501	0.76	16.1	1.33	211 000	102 000
NGC 1 514	0.18	14.9	1.92	115 000	65 300
NGC 1 535	0.18	14.8	1.91	115 000	64 000
NGC 2 022	1.00	9.1	0.96	266 000	212 000
NGC 2 346	0.23	12.7	1.74	124 000	72 000
NGC 2 371-2	0.94	13.8	1.18	250 000	123 000
NGC 2 392	0.63	13.8	1.34	188 000	100 000
NGC 2 438	0.44	13.0	1.49	153 000	88 000
NGC 2 440	0.76	17.9	1.37	211 000	102 000
NGC 2 452	0.57	15.3	1.43	178 000	93 000
NGC 2 610	0.99	5.6	0.75	263 000	233 000
NGC 2 792	0.94	15.5	1.22	250 000	116 000
NGC 2 818	0.80	11.4	1.15	210 000	125 000
NGC 2 867	0.39	18.9	1.68	150 000	75 000
NGC 3 211	0.68	20.8	1.48	197 000	89 000
NGC 3 242	0.37	16.3	1.64	146 000	79 000
NGC 3 587	0.19	12.9	1.83	117 000	69 200
NGC 3 918	0.51	21.5	1.62	164 000	77 000
NGC 4 361	1.15	4.1:	0.55	303 000	–
NGC 5 307	0.28	15.7	1.75	132 000	72 000
NGC 6 058	0.75	13.6	1.26	209 000	110 000
NGC 6 210	0.034	13.9	2.61	76 700	45 000
NGC 6 302	0.70	16.9	1.38	199 000	97 000
NGC 6 309	0.70	16.2	1.36	199 000	100 000
NGC 6 326	0.35	15.3	1.64	145 000	74 000
NGC 6 369	0.22	23.1	2.02	122 000	60 200
NGC 6 439	0.17	12.55	1.87	112 000	64 600
NGC 6 445	0.54	18.5	1.53	172 000	85 000
NGC 6 537	0.97	14.2	1.16	257 000	125 000
NGC 6 543	0.023	8.5	2.57	71 600	45 000
NGC 6 563	0.15	11.9	1.90	110 000	63 800
NGC 6 565	0.13	12.7	2.00	105 000	61 000
NGC 6 572	0.13	14.4	2.04	105 000	58 200
NGC 6 644	0.22	23.1	2.03	122 000	60 000
NGC 6 720	0.65	14.5	1.35	190 000	100 000
NGC 6 741	0.47	21.0	1.65	160 000	77 000
NGC 6 751	0.34	15.4	1.66	143 000	77 600
NGC 6 772	0.34	15.0	1.64	141 000	78 000
NGC 6 778	0.17	10.2	1.78	112 000	70 800
NGC 6 781	0.24	15.8	1.83	126 000	68 000

Table 5.4. (continued)

Planetary nebula		$F(4686)$	$F(N_1+N_2)$	$\log \frac{F(N_1+N_2)}{F(4686)}$	T_*, K HeII/HI	T_*, K [OIII]/[OII]
NGC	6790	0.037	17.1	2.67	78500	43000
NGC	6803	0.04	15.4	2.58	79400	45000
NGC	6804	0.81	9.5	1.07	221000	137000
NGC	6818	0.76	18.9	1.40	210000	95000
NGC	6826	0.57	10.2	1.25	178000	126000
NGC	6853	0.42	15.0	1.49	153000	88000
NGC	6881	0.66	15.2	1.39	193000	96000
NGC	6884	0.21	21.1	2.00	120000	60000
NGC	6886	0.47	21.0	1.65	160000	77000
NGC	6891	0.02	10.7	2.73	70000	35600
NGC	6894	0.25	13.4	1.73	127000	73000
NGC	6894	0.08	12.7	2.20	92300	53700
NGC	6905	0.17	13.8	1.83	112000	68000
NGC	7008	0.89	10.7	1.08	237000	135000
NGC	7009	0.14	14.0	2.00	107000	60200
NGC	7026	0.26	12.1	1.70	129000	70000
NGC	7027	0.60	19.2	1.50	184000	87000
NGC	7094	1.16	7.0	0.78	309000	267000
NGC	7139	0.19	10.8	1.75	117000	70800
NGC	7293	0.10	13.3	2.12	87700	56200
NGC	7354	0.63	15.5	1.39	186000	96000
NGC	7662	0.47	16.9	1.56	160000	84000
IC	289	0.97	6.6	0.84	257000	193000
IC	351	0.48	18.2	1.58	162000	81000
IC	1747	0.27	16.3	1.78	131000	70000
IC	2003	0.65	14.3	1.34	190000	100000
IC	2165	0.50	16.7	1.52	165000	86000
IC	2448	0.38	15.9	1.62	148000	79000
IC	4642	1.14	7.0	0.79	302000	210000
IC	5117	0.13	20.3	2.19	105000	54000
IC	5217	0.074	16.1	2.34	91200	50000
A	15	1.28	5.1	0.60	355000	350000
A	20	1.49	6.0	0.60	457000	360000
A	30	1.52	5.5	0.56	473000	420000
A	33	0.81	11.6	1.15	221000	126000
A	36	1.18	4.0:	0.53	316000	–
A	43	0.94	11.1	1.07	250000	137000
A	50	0.34	16.8	1.71	141000	71000
A	51	0.90	8.0	0.95	243000	160000
A	72	1.11	0.3	0.82	295000	200000
A	75	0.81	18.9	1.37	221000	98000
H	3-29	1.12	9.7	0.93	298000	165000
Hu	1-1	0.18	16.5	1.96	115000	62000
Hu	1-2	0.95	10.7	1.05	251000	140000
J	900	0.47	15.9	1.53	160000	85000
K	1-14	1.60	5.8	0.56	562000	420000

Table 5.4. (continued)

Planetary nebula		$F(4\,686)$	$F(\mathrm{N_1{+}N_2})$	$\log \frac{F(\mathrm{N_1+N_2})}{F(4686)}$	T_*, K	
					HeII/HI	[OIII]/[OII]
K	1-16	1.26	5.4	0.63	343 000	320 000
K	3-27	0.92	7.9	0.93	245 000	165 000
K	3-63	0.79	13.4	1.23	216 000	115 000
M	1-1	1.00	6.2	0.79	266 000	212 000
M	1-7	0.16	18.0	2.05	111 000	58 200
Me	2-1	0.86	16.9	1.30	229 000	105 000

The values of $V(\mathrm{He^{++}})$ and $V(\mathrm{O^{++}})$ can be obtained from the condition of equality between the numbers of ionizing photons emitted by the central star and the total number of recombinations,

$$\int_{\nu_1}^{\infty} \frac{L_\nu}{h_\nu}\, \mathrm{d}\nu = n_\mathrm{e} n(\mathrm{He^{++}})\, \alpha_\mathrm{B}(\mathrm{He^{++}})\, V\,(\mathrm{He^{++}})\,, \tag{36}$$

$$\int_{\nu_2}^{\infty} \frac{L_\nu}{h_\nu}\, \mathrm{d}\nu = n_\mathrm{e} n(\mathrm{O^{++}})\, \alpha_\mathrm{B}(\mathrm{O^{++}}) \left[1 - \frac{V(\mathrm{O^{+++}})}{V(\mathrm{O^{++}})}\right]\,, \tag{37}$$

where L_ν stands for the luminosity of the nuclei, and ν_1 and ν_2, for the ionization frequencies of $\mathrm{He^{++}}$ and $\mathrm{O^{++}}$, respectively. From (37) the volume of ionization zone $\mathrm{O^{+++}}$ is deduced.

Assuming the radiation of the central star is Planckian at the temperature T_*, and taking T_e = const. for all nebulae, we obtain

$$\frac{F(\mathrm{N_1 + N_2})}{F(4\,686\mathrm{HeII})} = C\delta_1 \left[\frac{J_2(x_1)}{J_2(x_2)} - 0.53\right]\,, \tag{38}$$

where $x_1 = 408\,000/T_*$, $x_2 = 631\,000/T_*$, and the function $J_2(x)$ is given by (11). The multiplier δ_1 characterizes the efficiency of absorption by the nebula of the radiation ionizing the helium and oxygen, i.e.

$$\delta_1 = \frac{1 - e^{-\tau(\nu_2)}}{1 - e^{-\tau(\nu_1)}}\,, \tag{39}$$

where $\tau(\nu_1)$ and $\tau(\nu_2)$ are the total optical depths of the nebula at the frequencies of helium ($\mathrm{H^{++}}$) and oxygen ($\mathrm{O^{++}}$), respectively.

The condition $\tau(\nu_1) = \tau(\mathrm{He^{++}}) > 1$ is fulfilled for the majority of nebulae. Therefore, the following values of δ_1 are possible depending on the star's temperature T_*:

$\delta_1 = 1$ at $\tau(\mathrm{He^{++}}) > 1$ and $\tau(\mathrm{O^{++}}) > 1$, T_* – small,
$\delta_1 < 1$ at $\tau(\mathrm{He^{++}}) > 1$ and $\tau(\mathrm{O^{++}}) < 1$, T_* – large,
$\delta_1 < 1$ at $\tau(\mathrm{He^{++}}) < 1$ and $\tau(\mathrm{O^{++}}) \ll 1$, T_* – very large.

In deducing (38), the condition $\nu(\mathrm{O}^{+++}) \approx \nu(\mathrm{He}^{++})$ is used, and

$$\frac{V(\mathrm{O}^{+++})}{V(\mathrm{O}^{++})} = \frac{\alpha_{\mathrm{B}}(\mathrm{O}^{++})}{\alpha_{\mathrm{B}}(\mathrm{O}^{+++})}\frac{J_2(x_2)}{J_2(x_1)} = 0.53\,\frac{J_2(x_2)}{J_2(x_1)}\,.$$

As for the coefficient of proportionality C in (38), its numerical value can be obtained using this formula even for a single point on the curve on Fig. 5.4 with known values of $F(\mathrm{N}_1 + \mathrm{N}_2)/F(4\,686\ \mathrm{HeII})$ and T_*. If we assume that $\delta_1 \approx 1$ can be fulfilled at $T_* = 80\,000$ K and $R_{\mathrm{N}} = \log F(\mathrm{N}_1 + \mathrm{N}_2)/F(4\,686\ \mathrm{HeII}) = 1.60$, we will obtain $C = 5.50$. Then taking $\delta_1 = 1$ we will have the following expression for the determination of the *lower* limit of T_*:

$$\frac{F(\mathrm{N}_1 + \mathrm{N}_2)}{F(4\,686\mathrm{HeII})} = 5.50\left[\frac{J_2(x_1)}{J_2(x_2)} - 0.53\right]. \qquad (40)$$

To determine readily the T_* with the help of (40), the curve of dependence of T_* on R_{N} is given in Fig. 5.5.

Equation (40) represents a semi-empirical dependence of the ratio $F(\mathrm{N}_1 + \mathrm{N}_2)/F(4\,686\ \mathrm{HeII})$ only from one parameter, the star's temperature T_*. In a stricter formulation of the problem, we must have, instead of a one-parameter formula, a much more complicated expression depending not only on T_* but also on other parameters of both the nebula and its nucleus.

With the aid of (40) we can obtain the lower limit of the temperature T_* of the nebulae, located on the curve of Fig. 5.4 or close to it. In this case

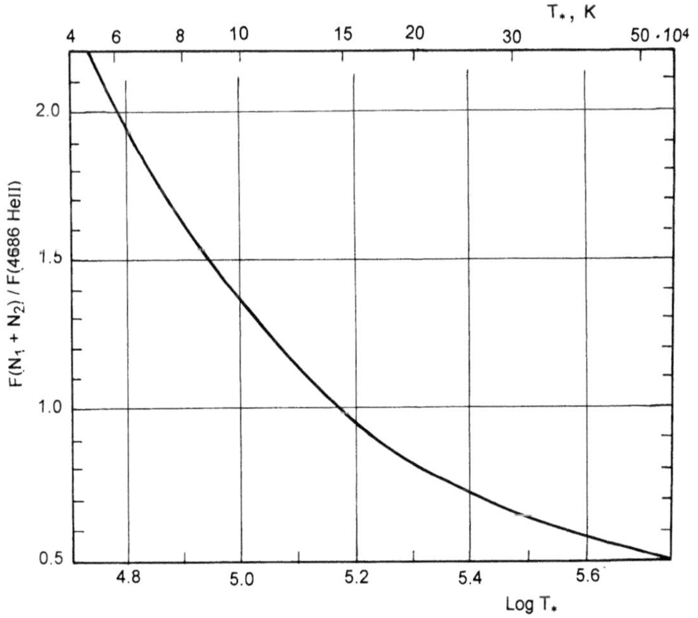

Fig. 5.5. Theoretical dependence between $\log T_*$ and $\log R_{\mathrm{N}}$ (40)

one should speak of the statistical determination of T_* with the help of the $R_N = \log F(N_1+N_2)/F(4\,686\ \text{HeII})$ method according to the following scale:

T_*, K	500 000	300 000	150 000	100 000	80 000	60 000	40 000
R_N	0.52	0.65	1.00	1.35	1.60	2.00	2.80

The temperature scale derived in such a way is given in the sixth column of Table 5.4, as well as in Fig. 5.4 along the upper abscissa.

The temperatures of the nuclei of PN can also be determined by the "HeII/HI" method, i.e. via the observed ratio $F(4\,686\ \text{HeII})/F(\text{H}_\beta)$. Then, taking into account that both the zones H^+ and He^{++} usually do not coincide with each other, we have

$$\frac{F(4\,686\text{HeII})}{F(\text{H}_\beta)} = 1.626\,\delta_2\,\frac{J_2(4x_0)}{J_2(x_0)}\,, \tag{41}$$

where $x_0 = 158\,000/T_*$ and

$$\delta_2 = \frac{1 - e^{-\tau(\text{He}^{++})}}{1 - e^{-\tau(\text{H}^+)}}\,, \tag{42}$$

where $\tau(\text{H}^+)$ and $\tau(\text{He}^{++})$ are the optical depths of nebulae at the frequencies of ionization of hydrogen and helium.

The radius of the zone He^{++} is always smaller than that of the H^+. Only at very high temperatures, about 10^6 K, can both zones coincide. For the majority of nebulae the condition $\tau(\text{He}^{++}) > 1$ and $\tau(\text{H}^+) < 1$ is fulfilled, so that $\delta_2 > 1$. Therefore using (41) we can obtain for $\delta_2 = 1$ only the *upper* limit of the star's temperature T_* with the help of the relationship

$$\frac{F(4\,686\text{HeII})}{F(\text{H}_\beta)} = 1.626\,\frac{J_2(4x_0)}{J_2(x_0)}\,. \tag{43}$$

The temperature scale derived by the HeII/H_β method and using (41), is shown as the ordinate on the right hand side in Fig. 5.4, as well as in the fifth column of Table 5.4.

Thus, we have two equations, (40) and (41), describing the temperature of T_* from the ratio of the intensities of various emission lines. Formally, if one eliminates T_* from these equations, a relationship of (33) type might be obtained. However, to do that one has, first, to reveal the behaviour of factors of δ_1 and δ_2 while moving to nebulae with high magnitudes of T_*.

Some conclusions:

(a) The upper and lower limits of nuclei temperatures in some cases, however, can be close to each other (NGC 7 094, NGC 2 610, A 15, A 30, etc.), on average they differ by 1.5–2 times (NGC 3 918, NGC 6 302, NGC 6 537, NGC 7 027, A 50, A 75 etc.).

(b) Nebulae with $T_* > 100\,000$ K form about 90% of the total, with $T_* > 200\,000$ K – 37%, and with $T > 300\,000$ K – nearly 10% of the total number of analysed objects.

(c) The existence of nebulae with hot nuclei – higher than 300 000 K – is certain, being confirmed by both methods. Such objects are NGC 2 022, NGC 7 094, NGC 4 361, NGC 246, NGC 2 610, etc. as well as K 1–14, K 1-15, A 15, A 20, A 36 etc. Two "helium" nebulae – A 30 and A 78 – are also among them. Though absent in Table 5.4, they are conditionally plotted in Fig. 5.4 (crosses) owing to their ratios $F(\mathrm{N_1+N_2})/F(4\,686)$, which are equal to 8.0 and 10.5, respectively. For them $F(\mathrm{H}_\beta) = 0$, and hence $F(4\,686\ \mathrm{HeII})/F(\mathrm{H}_\beta) \to \infty$. Temperatures T_* for these nebulae estimated by various methods are about 880 000 K and 38 000 K (Jacoby, Ford, 1983).

(d) The existence of PNs with nuclei temperatures of the order of 400 000 K and even 500 000 K seems probable. This can be considered the main conclusion one can arrive at by statistical analysis of the dependence of $F(\mathrm{N_1+N_2})/F(4\,686) \sim T_*$ shown in Fig. 5.4.

5.8 Nuclei Temperatures of Low Excitation Nebulae

To the low excitation PN category are assigned those objects in the spectrum of which emission lines of high potential ions – HeII, [NeIV], [NeV], NV – are absent. Morphologically it includes all types of PN – planet-like NGC 40, IC 418, Hu 2–1, two-envelope NGC 6 543, NGC 6 826, superdense IC 4 997, with low electron temperature BD+30°3 639 and diffuse nebulae.

The spectra of low excitation PNs are represented mainly by strong forbidden lines [OIII], [OII], [NII] as well as by Balmer emission lines of hydrogen. Hence, to determine the temperatures of their nuclei it is necessary to have a method based on the possibility of using those lines. Such a method has been proposed by Stoy (1933) (for more details see Preite-Martinez and Pottasch, 1983) based on the assumption that the ratio of total energy emitted by a nebula in forbidden lines to the energy emitted in one of the hydrogen lines must depend on the nucleus temperature T_*. Then the following relationship may be derived

$$\frac{F(f)}{F(\mathrm{H}_\beta)} = 1.38 \times 10^{-14} \frac{\displaystyle\int_{\nu_0}^{\infty} F_\nu(T_*)(\nu - \nu_\mathrm{e})\,\frac{\mathrm{d}\nu}{\nu}}{\displaystyle\int_{\nu_0}^{\infty} F_\nu(T_*)\,\frac{\mathrm{d}\nu}{\nu}}\,, \tag{44}$$

where $F(f)$ is the total observed flux in all forbidden lines, $F_\nu(T_*)$ is the flux of the radiation of central star, ν_0 is the ionization frequency of hydrogen, and the frequency ν_e is determined from the condition $h\nu_\mathrm{e} = h\nu_0 + \frac{3}{2}kT_\mathrm{e}$.

If the radiation of nucleus is approximated with Planck's law from (44) one can write

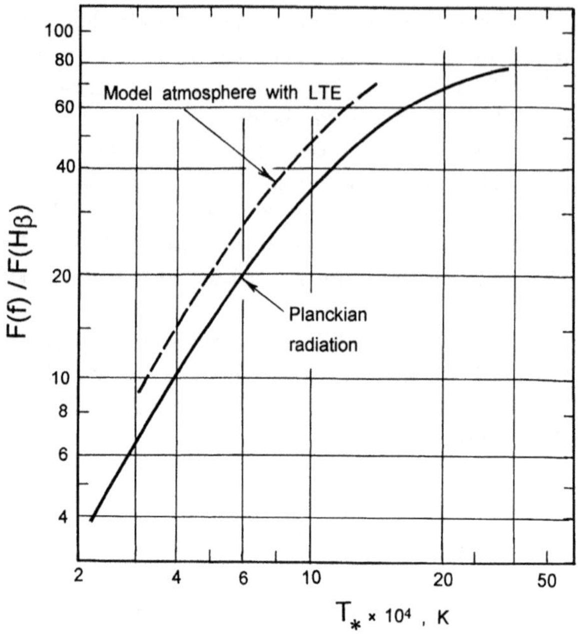

Fig. 5.6. Dependence of the temperature of the central star T_* on the observed ratio $F(f)/F(\mathrm{H}_\beta)$ (Stoy's method)

$$\frac{F(f)}{F(\mathrm{H}_\beta)} = 2.88 \times 10^{-5}\, T_* \frac{\displaystyle\int_{x_0}^{\infty} \frac{(x - x_0)x^2}{e^x - 1}\, \mathrm{d}x}{\displaystyle\int_{x_0}^{\infty} \frac{x^2}{e^x - 1}\, \mathrm{d}x}\,, \tag{45}$$

where the searched for temperature T_* appears both directly and through $x_0 = h\nu_0/kT_*$. For model atmospheres the integration in (44) is performed numerically.

In Fig. 5.6 the curves of the dependence of $F(f)/F(\mathrm{H}_\beta)$ on T_* are given for both the Planckian law (solid line) and the Hummer–Mihalas (1970) model atmosphere (broken line).

In Table 5.5, the temperatures of the nuclei for a sample of low excitation class PNs, obtained by this method, are given for both the model atmosphere and Planckian radiation cases. As can be seen, in all cases both values of T_* almost coincide with each other, and therefore (45) can be sufficient for the application of Stoy's method. Note that all these considerations concern

Table 5.5. Temperatures of the nuclei of a sample of low excitation class (p) planetary nebulae obtained by Stoy's method for a model atmosphere and for Planckian radiation of the nucleus

Planetary nebula	p	T_*, K Model	T_*, K Planck	Planetary nebula	p	T_*, K Model	T_*, K Planck
NGC 40	1	32 000	32 000	IC 418	1	33 000	34 000
NGC 6 210	4	45 000	58 000	IC 2 149	2	34 000	37 000
NGC 6 543	5	39 000	42 000	IC 4 593	6	38 000	42 000
NGC 6 567	4	44 000	56 000	J 320	3	45 000	57 000
NGC 6 629	5	40 000	45 000	Hu 2–1	1	34 000	37 000
NGC 6 891	4	42 000	49 000	BD +30°3639	1	31 000	30 000

nebulae that are optically thick at the frequencies of L_c radiation ($\tau_c > 1$). The transition to the case $\tau_c < 1$ can be realized in the same manner, as done for the other methods above.

6. Electron Temperature. Electron Concentration

6.1 Statement of the Problem

The electron temperature and electron concentration are like conductors of the physical processes in any plasma, including the strongly ionized medium of gaseous nebulae.

The relative abundance of any ion in a nebula is determined first of all by the effective temperature of its central star. If the abundance of ions of a given type in a nebula is sufficiently high, the strength of their emission lines is determined exclusively by the electron temperature T_e and electron concentration n_e of the medium, i.e. nebula. If this is so, the opposite statement should be true also: the emission lines are the most reliable and in fact the only tools for investigating nebulae. With favourable conditions such an investigation can be carried out at almost any point within the nebula owing to the transparency of the medium for practically all emission lines. As a result, one has a unique possibility of constructing a detailed picture of the distribution of electron concentration within the nebula at various distances from its centre.

As distinct from celestial objects, which allow the application of dynamical methods of determination of their masses owing to the dominating role of gravitational forces, in gaseous nebulae the latter have practically no role. Therefore, for example, the problem of determining of such an important parameter as the mass of the nebula is solved only by considerations based on the knowledge of electron concentration and electron temperature, along with the geometric parameters of the nebula.

Early methods of determining the electron temperature and electron concentration already showed the astonishing constancy of the electron temperature for all PNs. The same, though to a lesser degree, was true of the electron concentration. However, the diversity in other parameters, such as the absolute dimensions of PNs, their shapes, internal structure, morphological peculiarities, evolutionary stage, – is rather large. This unique stability and homogeneity, of two key physical parameters – the electron temperature and electron concentration – should be considered one of the specific properties of PNs which distinguish them from other types of galactic object.

Our knowledge of the problem of the electron temperature and electron concentration in gaseous nebulae has a rather solid background. It has reached its present state by the elaboration and application of a series of original and efficient methods, which though developed specifically for gaseous nebulae, have entered firmly into the arsenal of astrophysics as powerful tools for analysing and deciphering the processes occurring in various cosmic objects. The present chapter is devoted to the description of these methods and their applications.

6.2 The Electron Temperature

Upon the photoionization of the hydrogen atom the remaining energy of the photon is transmitted to the ejected electron according to the relation $mv^2/2 = h\nu - h\nu_0$, where $h\nu$ is the energy of the photon and $h\nu_0$ is the ionization energy of hydrogen. This quantity, $mv^2/2$, varies for different acts of photoionization. The mutual collisions of electrons lead to their Maxwellian velocity distribution. In this case one can speak of the mean energy of free electrons and, therefore, on the mean temperature of the electron gas, called the *electron temperature* of the medium. Obviously, as the temperature of the nucleus emitting ultraviolet ionizing photons increases, the average remaining energy of the electrons and, consequently, the electron temperature will increase as well. However, this is so at a first glance only.

Formally, the dependence of the electron temperature T_e of the nebula on the temperature of the nucleus T_* can be estimated easily, proceeding from the following two conditions: (a) the condition of stationarity – the rate of photoionization must be equal to the rate of recombination; (b) the condition of radiative equilibrium – the amount of energy spent for the photoionization of hydrogen must be equal to the amount emitted via recombinations.

The application of both these conditions leads to the relation $f_1(T_e) = f_2(T_*)$, from which T_e can be determined for a given T_*. From the calculations for an optically thin hydrogen nebula the following values for T_e were obtained:

T_*, K	T_e, K	T_*, K	T_e, K
5 000	5 000	40 000	34 000
10 000	9 500	80 000	57 000
20 000	18 000	100 000	92 000

However, all these considerations concern the case when the loss of electron energy occurs only through recombination of hydrogen, i.e. the case of hydrogen cooling only.

In the conditions of gaseous nebulae, as we have seen in the previous chapter, there exists an even more powerful mechanism for the loss of energy by free electrons, namely, the excitation of forbidden lines. This leads to a still larger decrease in the electron temperature. There remains only one reliable way to solve this problem: to find a procedure to establish the electron temperature for each nebula directly from observations. The electron temperature estimated in this way becomes an independent parameter characterizing the physical conditions in the nebula.

Below we shall describe the methods which are used to determine the electron temperature of the nebulae.

The [OIII] Method. Ions, such as O^{++}, N^{+} etc., have not only a $^1D_2$ metastable level from which nebular lines originate, but also a higher metastable level $^1S_0$ which transits to $^1D_2$, emitting another forbidden line. For example, for O^{++} the excitation potentials from the ground state ^3P to levels ^1D and ^1S are 2.5 eV and 5.3 eV, respectively (Fig. 2.1). So, the transition from the level $^1S_0$ to the $^1D_2$ level leads to the emission of the forbidden line 4 363 [OIII], and that from $^1D_2$ downward to the levels $^3P_{1,2}$ creates the N_1 and N_2 lines. Obviously, as the electron temperature of the nebula increases, the relative number of ions in the $^1S_0$ state compared with the number in $^1D_0$ will increase, and consequently 4 363 [OIII] will become brighter compared with the lines N_1 and N_2.

Thus, it is possible to estimate the electron temperature of the nebula from the observed intensities of 4 363 [OIII] and $N_1 + N_2$ [OIII] (Ambartsumian et al., 1958). The problem, therefore, is to derive a theoretical dependence of the electron temperature T_e on the ratio of the intensities of the lines $N_1 + N_2$ [OIII] and 4 363 [OIII], i.e. on $E(N_1 + N_2)/E(4\,363)$.

Consider the problem in its general formulation. Let n_1, n_2 and n_3 denote the concentrations of O^{++} ions in the ^3P, ^1D and ^1S levels, respectively. Below, these levels will be designated by the numbers 1, 2 and 3. Rewrite the conditions of stationarity for levels 1 and 2, taking into account the electron collisions of the first kind (b_{ji}), collisions of the second kind (a_{ij}) and spontaneous transitions (A_{ij}):

$$\begin{aligned} n_1(b_{12} + b_{13}) &= n_2(A_{21} + a_{21}) + n_3(A_{31} + a_{31})\,, \\ n_2(A_{21} + a_{21} + b_{23}) &= n_1 b_{12} + n_3(A_{32} + a_{32})\,, \end{aligned} \tag{1}$$

where b_{ji} and a_{ij} are the probabilities of collisional transitions of the first and second kinds, respectively, and A_{ij} is the Einstein coefficient for a spontaneous radiative transition. From (1) we obtain

$$\frac{n_2}{n_1} = \frac{b_{12}(A_{31} + a_{31}) + (b_{12} + b_{13})(A_{32} + a_{32})}{b_{23}(b_{12} + b_{13}) + b_{13}(A_{21} + a_{21})}\,. \tag{2}$$

The coefficients $b_{ji} = q_{ji}$ and $a_{ij} = q_{ij}$ are related by the expression:

$$q_{ji} = q_{ij}\,\frac{\omega_i}{\omega_j}\,e^{-(\varepsilon_i - \varepsilon_j)/kT_e}\,, \tag{3}$$

and q_{ij} is given by the formula

$$q_{ij} = 8.63 \times 10^{-6}\,\frac{\Omega(ij)}{\omega_i}\,n_e T_e^{-1/2}\,, \tag{4}$$

where $\Omega(\mathrm{ji})$ is the "collision strength" and ω_i is the statistical weight of the ith level. For example, for the levels ${}^1S$, ${}^1D$ and ${}^3P$, ω_i equals 1, 5 and 9, respectively.

For the intensities of the lines $N_1 + N_2$ [OIII] and 4 363 [OIII], we have

$$E(\mathrm{N}_1 + \mathrm{N}_2) = n_2 \left[A({}^1D_2 \rightarrow {}^3P_2) h\nu_{\mathrm{N}_1} + A({}^1D_2 \rightarrow {}^3P_1)\, h\nu_{\mathrm{N}_2}\right] = n_2 A_{21} h\nu_{12}\,, \tag{5}$$

$$E(4\,363) = n_3 A_{32}\, h\nu_{23}\,, \tag{6}$$

where ν_{N_1}, ν_{N_2} and ν_{23} are the frequencies of the lines N_1, N_2 and 4 363 Å, respectively.

From (2–6) we find that

$$\frac{E(\mathrm{N}_1 + \mathrm{N}_2)}{E(4\,363)} = F(n_\mathrm{e},\, T_\mathrm{e})\, e^{33\,000/T_\mathrm{e}}\,, \tag{7}$$

where, if we also substitute the numerical values of A_{ij} and $\Omega(\mathrm{ji})$ for the ion O^{++} from Table 2.1,

$$F(n_\mathrm{e}, T_\mathrm{e}) = 0.0753 \frac{1 + 2.67\times10^5 T_\mathrm{e}^{1/2}/n_\mathrm{e} + (0.04 + 1.27\times10^4 T_\mathrm{e}^{1/2}/n_\mathrm{e})\, e^{-33\,000/T_\mathrm{e}}}{1 + 2.30\times10^3 T_\mathrm{e}^{1/2}/n_\mathrm{e} + 0.09 e^{-33\,000/T_\mathrm{e}}}\,. \tag{8}$$

For PNs one usually has $n_\mathrm{e} \sim 10^3 - 10^4$ cm^{-3}, $T_\mathrm{e} \sim 10\,000$ K. Therefore, with sufficient accuracy we can obtain from (8)

$$\frac{E(\mathrm{N}_1 + \mathrm{N}_2)}{E(3\,727)} = 0.0753 \frac{1 + 2.67 \times 10^5\, T_\mathrm{e}^{1/2}/n_\mathrm{e}}{1 + 2.30 \times 10^3\, T_\mathrm{e}^{1/2}/n_\mathrm{e}}\, e^{33\,000/T_\mathrm{e}}\,. \tag{9}$$

At relatively small values of the electron concentration $n_\mathrm{e}(< 10^3$ cm$^{-3})$ the right hand side of (9) appears to be less dependent on n_e and T_e, and in this case, $n_\mathrm{e} = 0$, we find from (8) that $F(0,\, T_\mathrm{e}) = 8.74$. Then (7) takes the form

$$\frac{E(\mathrm{N}_1 + \mathrm{N}_2)}{E(4\,363)} = 8.74\, \mathrm{e}^{33\,000/T_\mathrm{e}}\,. \tag{10}$$

Thus, for not too high values of the electron concentration, $n_\mathrm{e} < 10^3$ cm^{-3}, we have via (10) the possibility of determining the electron temperature T_e by means of the ratio $E(\mathrm{N}_1{+}\mathrm{N}_2)/E(4\,363)$, obtained from direct observations. In the remaining cases one must employ (7) and (9), for which it is necessary to determine n_e beforehand by any procedure.

The [NII] Method. The electron temperature can also be estimated with the help of other ions. This is especially relevant to the forbidden lines of ionized nitrogen 6 548 + 6 584 [NII] and 5 755 [NII]. The corresponding relation has the form (Aller, 1956, p. 193)

$$\frac{E(6\,548 + 6\,584)}{E(5\,755)} = 1.625 \times 10^{-2} \frac{1 + 1.94 \times 10^5\, T_\mathrm{e}^{1/2}/n_\mathrm{e}}{1.03 + 3.20 \times 10^2\, T_\mathrm{e}^{1/2}/n_\mathrm{e}}\, e^{25\,000/T_\mathrm{e}} \tag{11}$$

or, for small values of n_e,

$$\frac{E(6\,548 + 6\,584)}{E(5\,755)} = 9.85\, e^{25\,000/T_e}\,. \tag{12}$$

Analogous formulae may be derived for the forbidden lines of other ions, e.g. [NeIII], [NeV]. Equations (9–10) and (11–12) had found wide application. With the help of these formulae for the lines [OIII] and [NII] the electron temperatures for more than a hundred PNs have been obtained.

Usually the ratio $E(N_1+N_2)/E(4\,363)$ for the majority of PNs is on average of the order of 100, the ratio $E(6\,548 + 6\,584)/E(5\,755)$ of the order of 60. This is mainly due to the small intensity of 4 363 [OIII] in the former case, and of 5 755 [NII] in the latter, about 0.2 and 0.1, respectively (on the scale of $H_\beta = 100$). Therefore, one should be cautious of measurements of the 4 363 [OIII] and 5 755 [NII]. In the case of the lines 6 548 and 6 584 [NII] the problem of their separation from the strong hydrogen line 6 563 HI arises; the latter lies between both components of [NII]. Note that the electron temperature itself is not too sensitive to the ratio of intensities of these lines owing to the fact that their ratios appear as a logarithm.

The first application of (10) and (12) yields the same value of T_e, on average about $T_e \sim 10\,000$ K.

Estimations of T_e with the help of (9–10) and (11–12) including 107 PNs have been performed by Kaler (1986); see Table 6.1. Although some discrepancies in the T_e, obtained by both [OIII] and [NII] methods, are worth discussion, the main results are the same: T_e[OIII] = 10 200 K and T_e[NII] = 10 300 K, on average.

The "Balmer Continuum" Method. The continuous spectrum beyond the Balmer limit, $\lambda < 3\,646$ Å (Balmer continuum), is assumed to arise from recombinations, and its intensity is determined by the expression

$$I_\lambda = ce^{-hc/\lambda k T_e}\,.$$

Comparing the intensities at two wavelengths λ_1 and λ_2, we obtain

$$\log \frac{I_{\lambda_1}}{I_{\lambda_2}} = 0.434\,\frac{hc}{kT_e}\left(\frac{1}{\lambda_1} - \frac{1}{\lambda_2}\right)$$

or

$$T_e = 0.629\,\frac{1/\lambda_1 - 1/\lambda_2}{\log(I_{\lambda_1}/I_{\lambda_2})} \tag{13}$$

By means of this formula one can determine, in principle, the electron temperature of the nebula T_e from the observed value of the ratio of intensities at two points of the continuous spectrum for wavelengths shorter than 3 646 Å.

This method, despite its simplicity, gives poor results for a number of reasons: (a) one cannot always obtain or measure spectra of the Balmer continuum with sufficient accuracy; (b) the observed energy distribution in the

Table 6.1. Electron temperatures T_e of planetary nebulae obtained by the [OIII] and [NII] methods

Planetary	T_e, K		Planetary	T_e, K	
nebula	[OIII]	[NII]	nebula	[OIII]	[NII]
NGC 40	–	8 040	NGC 6 853	11 260	10 130
NGC 650–1	10 660	9 520	NGC 6 884	10 890	13 750
NGC 1 535	11 210	–	NGC 6 886	12 170	10 600
NGC 2 022	14 380	–	NGC 6 894	–	9 900
NGC 2 346	13 640	–	NGC 6 905	12 130	–
NGC 2 392	13 750	10 140	NGC 7 009	12 270	13 370
NGC 2 438	10 870	–	NGC 7 026	9 050	9 510
NGC 2 440	13 930	10 260	NGC 7 027	12 220	14 150
NGC 2 452	11 040	9 490	NGC 7 293	11 360	10 810
NGC 2 610	17 170	–	NGC 7 662	11 430	11 310
NGC 2 792	14 820	–	IC 351	12 300	–
NGC 2 818	14 800	11 380	IC 418	8 420	8 450
NGC 2 867	11 240	9 180	IC 1 297	10 820	8 340
NGC 3 132	10 640	9 060	IC 1 747	10 130	–
NGC 3 211	14 320	–	IC 2 003	11 850	13 180
NGC 3 242	11 370	10 060	IC 2 149	9 730	8 900
NGC 3 587	10 910	–	IC 2 165	13 850	11 740
NGC 3 918	12 490	9 280	IC 2 448	13 120	–
NGC 4 361	19 370	–	IC 2 501	9 870	9 940
NGC 5 307	12 680	–	IC 2 553	10 100	11 700
NGC 5 315	9210	10 200	IC 2 621	13 200	9 990
NGC 5 873	13 070	–	IC 3 568	10 400	–
NGC 5 882	9 280	8 290	IC 4 191	10 500	12 600
NGC 6 058	13 200	–	IC 4 406	10 500	9 400
NGC 6 210	9 570	9 650	IC 4 593	8 900	11 400
NGC 6 302	15 610	16 380	IC 4 634	9 700	11 200
NGC 6 309	11 250	9 850	IC 4 776	8 600	13 900
NGC 6 537	15 720	16 760	IC 4 846	10 000	9 700
NGC 6 543	8 230	9 680	IC 5 117	11 600	12 500
NGC 6 567	10 950	15 180	IC 5 217	11 400	11 500
NGC 6 572	10 170	11 880	BD+30°	–	8 600
NGC 6 644	12 600	13 620	Cn 3–1	22 200	6 500
NGC 6 720	11 200	10 180	H 1-55	–	5 300
NGC 6 751	10 330	7 910	H 4-1	11 900	10 300
NGC 6 765	11 700	8 740	He 2-5	11 000	11 500
NGC 6 778	7 950	7 730	He 2-7	12 700	12 900
NGC 6 790	11 780	16 700	He 2-108	9 700	–
NGC 6 803	9 630	10 160	He 2-131	–	7 100
NGC 6 804	12 900	–	Hu 1-1	12 900	10 200
NGC 6 818	12 820	11 410	Hu 1-2	17 100	11 600
NGC 6 826	10 560	12 240	M 1-74	9 500	12 100
NGC 6 833	13 160	20 900	M 2-9	8 700	9 000
Hu 2–1	8 900	12 600	M 2-50	12 200	–
J 320	12 400	11 600	M 3-1	12 300	9 900
J 900	12 100	10 600			
M 1-1	15 500	–			

Table 6.1. (continued)

Planetary nebula	T_e, K [OIII]	T_e, K [NII]	Planetary nebula	T_e, K [OIII]	T_e, K [NII]
M 1-4	12 800	–	M 3-35	9 000	11 200
M 1-5	9 100	11 700	M 4-18	–	6 800
M 1-14	10 500	9 800	Me 2-1	13 100	11 600
M 1-41	–	10 900	Me 2-2	10 800	13 200
M 1-67	–	7 400	Vy 1-2	9 700	7 700

Balmer continuum must be corrected for interstellar extinction; (c) the continuous spectrum of PNs consists of several components, including the Balmer continuum itself; it is hard to estimate the role of recombination of hydrogen in each of them (see Chap. 9).

The determination of T_e by the ratio of intensities in some part of the continuous spectra to the intensity of any emission line of hydrogen, say, of H_β can be considered to be the like the methods described. For example, any expression of the type

$$\frac{I(3\,646^-) - I(3\,646^+)}{E(H_\beta)} \tag{14}$$

must depend on T_e. Combining the data in Table 3.6 and Table 9.3 (Chaps. 3 and 9), the theoretical values of the ratio (14) can be determined depending on T_e and then, by comparison with observations, one can obtain T_e. This method also has found a limited application.

6.3 Cooling of the Inner Regions of a Nebula

For a long time it was believed that PNs are isothermal configurations. This was based on the a priori assumption that the efficiency of the cooling mechanism by various ions is the same within the nebula. However, the peculiarities of the expansion and structure of nebulae indicated the invalidity of that viewpoint. The early estimations (Gurzadyan, 1957) of the electron temperatures in the central parts of a PN had already led to unexpected results – they turned out to be essentially two or three times higher than the usual electron temperatures, $T_e \sim 10\,000$ K, in the outer parts of the nebula. This conclusion was confirmed soon after by Osaki (1962).

As we have seen, the cooling of a nebula involves such ions as O^+, O^{++}, N^+, S^+, S^{++}, etc.; monochromatic images of nebulae in the lines of these ions have the largest dimensions. Moreover, the lowest metastable state of each ion has a rather low excitation potential. Therefore, the electron temperatures determined from forbidden lines of these ions will be characteristic for outer regions of the nebula and, at the same time, they will represent the minimum temperatures in the nebula.

The situation is different in the central parts of nebulae where these elements will be in three, four or more times ionized states and consequently can perform the cooling of the gas. Therefore, with the increase of degree of ionization the excitation potential to the nearest metastable level ε will increase as well. So, for eight elements – C, N, O, Ne, S, F, Fe, Cl, – the dependence of the mean value of ε on the state of ionization is as follows:

Degree of ionization	II	III	IV	V	VI
$\bar{\varepsilon}$, eV	2.7	3.6	6.0	8.4	10.7

In the outer parts of nebulae the elements exist in singly and doubly ionized states, with the mean potential of the lowest metastable level of the order of $\varepsilon_{\rm I} \approx 2.5$ eV. In the central regions of nebulae the elements are at higher degrees of ionization with higher excitation potentials, $\varepsilon_{\rm II} \sim 8-10$ eV. Since $\varepsilon_{\rm II} > \varepsilon_{\rm I}$, one can readily conclude that the electron temperature in the central region of PNs, $T_{\rm eII}$, must noticeably exceed that of the outer regions, e.g. $T_{\rm eII} > T_{\rm eI}$.

We conclude, thus, that *PNs cannot be isothermal.*

6.4 The Temperature Gradient in a Nebula

Consider now the following problem: estimate the ratio $T_{\rm eII}/T_{\rm eI}$ for a PN. One can solve this problem by writing the condition of conservation of energy for the free electrons both for the inner and outer regions of the nebula and then comparing the results – magnitudes of $T_{\rm eII}$ and $T_{\rm eI}$.

We assume that free electrons gain their energy from the photoionization of hydrogen and lose it in four ways:

(a) By recombinations of electrons with protons and by free–free transitions. The corresponding rate is denoted by E_1.
(b) By recombinations of electrons with He^+ ions and free–free transitions of electrons in the field of He^+ ions (E_2).
(c) By inelastic collisions of electrons with neutral hydrogen (E_3).
(d) By excitation of forbidden lines of various atoms and ions (E_4).

If $E_{\rm e}$ is the rate of energy gain by free electrons, then the condition of conservation of energy takes the form

$$E_{\rm e} = E_1 + E_2 + E_3 + E_4 \,. \tag{15}$$

The energy spent at the recombination and free–free transitions of electrons can be represented in the form

$$E_1 = n_{\rm e} n^+ \left[F_r(T_{\rm e}) + F_{\rm ff}(T_{\rm e})\right] = n_{\rm e} n^+ \, F_1(T_{\rm e}) \,, \tag{16}$$

where n_e and n^+ are the concentrations of free electrons and protons, respectively. The function $F_1(T_e)$ can be given explicitly through known functions (e.g. Seaton, Osaki, 1962).

Analogously one can write for singly ionized helium with concentration $n(He^+)$

$$E_2 = n_e n(\mathrm{He}^+)\, F_2(T_e)\,. \tag{17}$$

As for the determination of E_3, the energy spent in inelastic collisions of electrons with ground state hydrogen, we will limit ourselves to accounting for the collisional excitation of the second level only, with the probability coefficient $\alpha_2(T_e)$:

$$E_3 = n_1 n_e\, \alpha_2(T_e)\, h\nu_\alpha\,, \tag{18}$$

where n_1 is the concentration of neutral hydrogen atoms in the ground state. Substituting the value of n_1 from the ionization formula (4.7), we get

$$E_3 = n_e n^+ \frac{n_e}{W\varphi(T_*) T_e^{1/2}} F_3(T_e)\,, \tag{19}$$

where W is the dilution coefficient. The functions $\varphi(T_*)$ and $F_3(T_e)$ are defined as follows:

$$\varphi(T_*) = \frac{2(\pi\mu k T_*)^{3/2}}{h^3} e^{-h\nu_0/kT_*}\,,$$
$$F_3(T_e) = \alpha_2(T_e) h\nu_\alpha\,, \tag{20}$$

where ν_0 is the hydrogen ionization frequency and T_* is the nucleus temperature.

The numerical values of the functions $F_1(T_e)$, $F_2(T_e)$ and $F_3(T_e)$ as functions of T_e are as follows:

T_e, K	10 000	15 000	20 000	30 000
$F_1(T_e)$, $\times 10^{-24}$	0.64	0.74	0.82	0.93
$F_2(T_e)$, $\times 10^{-24}$	0.56	0.65	0.72	0.80
$F_3(T_e)$, $\times 10^{-24}$	0.02	1.24	9.80	78.0

In the case of $F_2(T_e)$, it was assumed that the relative abundance of helium atoms is He/H = 0.2, and that the degree of ionization of helium is He^+/He = 0.9. For other ratios the numerical values of $F_2(T_e)$ should be multiplied by He/H and He^+/He.

The dependence of the function of $\varphi(T_*)$ on the nucleus temperature T_* is as follows:

T_*, K	40 000	60 000	80 000	100 000	120 000
$\varphi(T_*)$, $\times 10^{-20}$	0.10	0.38	0.80	1.40	1.90

Finally, for the free electron energy spent for the excitation of forbidden lines, we have

$$E_4 = n_e \sum_i n_i q_{ij} h\nu_{ij} = n_e n^+ \sum_i \frac{n_i}{n^+} q_{ij} h\nu_{ij}\,, \tag{21}$$

where n_i is the concentration of ions exciting a given forbidden line, q_{ij} is the probability coefficient for collision-induced transitions and ν_{ij} is the frequency of a given line. As for the ratio n_j/n^+, it is easy to see that it must be practically one and the same for the outer and inner regions of the nebula.

Using (4) for coefficients q_{ij}, which are no longer constants throughout the nebula, since they depend on the mean excitation energy ε of the metastable level, which is larger in the inner regions, we finally get

$$E_4 = Cn_e n^+ \frac{1}{T_e^{1/2}} e^{-\bar{\varepsilon}/kT_e} , \tag{22}$$

where C is a constant for a given nebula and depends on its chemical composition, and $\bar{\varepsilon}$ is the mean value of the excitation energy of the metastable levels. For the outer parts of a nebula $\bar{\varepsilon} = \bar{\varepsilon}_{\rm I} \approx 2.5$ eV, and for the inner parts $\bar{\varepsilon} = \bar{\varepsilon}_{\rm II} \approx 8$−−10 eV.

Finally, for the rate of electron heating E_e by photoionization, we have

$$E_e = \bar{\varepsilon} n_e n^+ C(T_e) , \tag{23}$$

where $C(T_e)$ is the summary recombination coefficient for all levels of hydrogen and $\bar{\varepsilon}$ is the mean electron energy. The latter depends on the nucleus temperature and is given in the following form:

$$\bar{\varepsilon} = kT_* A(T_*) , \tag{24}$$

where $A(T_*)$ is a function of T_*, appearing for both cases $\tau_c \gg 1$ and $\tau_c \ll 1$ (Sobolev, 1986).

Substituting (24) into (23) we find that

$$E_e = n_e n^+ K(T_*, T_e) , \tag{25}$$

where

$$K(T_*, T_e) kT_* A(T_*) C(T_e) . \tag{26}$$

Numerical values of the function $K(T_*, T_e)$ depending on T_* and for several values of T_e are given in Table 6.2 for both cases $\tau_c \ll 1$ and $\tau_c \gg 1$.

Substituting (26), (22), (19), (17) and (16) into (15), we finally obtain

$$K(T_*, T_*) = F_1(T_e) + F_2(T_e) + \frac{n_e}{W\varphi(T_e)T_e^{1/2}} F_3(T_e) + CT_e^{-1/2} \exp\left(-\frac{\bar{\varepsilon}}{kT_e}\right) . \tag{27}$$

This relation gives the dependence of the electron temperature T_e on the nucleus temperature T_*, on the electron concentration n_e, on the dilution coefficient W and on the chemical composition C. However, above all, it represents the dependence of T_e on $\bar{\varepsilon}$. Therefore the electron temperature will change sufficiently within a nebula, since $\bar{\varepsilon}$ varies strongly within the nebula: note that $\bar{\varepsilon}$ is in the exponent.

Table 6.2. Numerical values of $K(T_*, T_e)$ as a function of the electron temperature T_e of the nebula and the nucleus temperature T_* for both cases $\tau_c \ll 1$ and $\tau_c \gg 1$

T_e	T_*, K					
[K]	50 000	80 000	120 000	50 000	80 000	120 000
		$\tau_c \ll 1$			$\tau_c \gg 1$	
10 000	2.3	3.5	5.0	4.5	8.0	13.8
15 000	1.7	2.4	3.3	3.4	6.0	10.5
20 000	1.3	1.9	2.2	2.7	4.8	8.3
30 000	0.9	1.4	1.9	1.9	3.9	6.0

$K(T_*, T_e)$ – in units of 10^{-24} ergs cm^{-6}

Let us apply (27) for a model of a PN with $T_* = 50\,000$ K, $n_e = 10^4$ cm^{-3}, $W = 10^{-14}$, in order to determine the difference between the electron temperatures in the inner and outer regions.

To do this, one must rewrite (27) first, for the outer parts of the nebula, where $T_e = 10\,000$ K and $\varepsilon = 2.5$ eV, and where the only unknown parameter is C. Determining the latter we find 0.995×10^{-21} at $\tau_c \ll 1$ and 2.89×10^{-21} for $\tau_c \gg 1$. Since C is constant for the given nebula, substituting its value into (27) and also $\varepsilon_{II} = 8$ eV, we obtain T_{eII} for the central parts of the nebula:

$$T_{eII} = 16\,000 \text{ K for } \tau_c \ll 1,$$
$$T_{eII} = 21\,000 \text{ K for } \tau_c \gg 1.$$

Here it was assumed that the inner radius of the nebula is half of the outer one, hence the dilution coefficient for the inner region is $4W$.

In such a manner one can obtain T_{eII} for other values of ε_{II} and T_*. The results of calculations are given in Table 6.3.

Table 6.3. The electron temperature T_e as a function of the temperature of the central star T_* for various values of ε_{II}: 2.5, 6, 8 and 10 eV

	T_*	ε_{II}, eV			
	[K]	2.5	6	8	10
$\tau_c \ll 1$	50 000	10 000	15 500	16 600	17 000
	80 000	10 000	17 000	20 000	24 000
	100 000	10 000	17 800	22 200	27 500
$\tau_c \gg 1$	50 000	10 000	18 000	21 000	22 500
	80 000	10 000	19 800	25 000	30 500
	100 000	10 000	21 500	27 000	32 400

As we see, the electron temperatures in the central parts of the nebula are significantly, 1.5–2 times, higher than in the outer regions. The difference in electron temperatures increases with the temperature of the nucleus.

In such a way, one arrives at the concept of the inevitable gradient of the electron temperature within a PN.

The application of (27) to NGC 6543, NGC 7009, NGC 7293 and NGC 7662 yields the following results for the electron temperatures in their central parts T_{eII}:

	T_*, K	W	n_e cm^{-3}	T_{eI}, K	T_{eII}, K
NGC 6543	46 000	1×10^{-14}	1.7×10^4	9 000	17 300
NGC 7009	57 000	3×10^{-15}	1.5×10^4	12 800	18 000
NGC 7293	70 000	1×10^{-16}	1.2×10^2	11 000	19 000
NGC 7662	56 000	3×10^{-15}	1.0×10^4	11 400	20 000

Finally, in Table 6.4 the fractions of the energy of free electrons spent in various processes are given for the same PN. Calculations were made for both the outer (I) and inner (II) regions of nebulae.

Table 6.4. Fraction of free electron energy (in %) spent in various processes in the outer (I) and inner (II) regions of nebulae

Planetary	Recombinations Hydrogen		Recombinations Helium		Forbidden Lines		Collisions with hydr.	
nebula	I	II	I	II	I	II	I	II
NGC 6543	24	48	21	42	55	10	0	0
NGC 7009	33	47	30	41	36	6	1	6
NGC 7293	22	45	19	39	59	14	0	2
NGC 7662	38	50	34	44	23	3	5	3

Some conclusions follow:

(a) The largest fraction of free electron energy is spent in the recombination of hydrogen and helium.
(b) Inelastic collisions with hydrogen atoms have some role only in the central regions (II) of a nebula; in the outer regions they have almost no effect.
(c) The fraction of free electron energy spent in the excitation of forbidden lines is several times larger in the outer regions (I) of a PN than in the inner ones.

Note that the high electron temperatures in the central parts of nebulae imply high gas pressure and lower density, i.e. conditions supporting the formation of ring-like PNs.

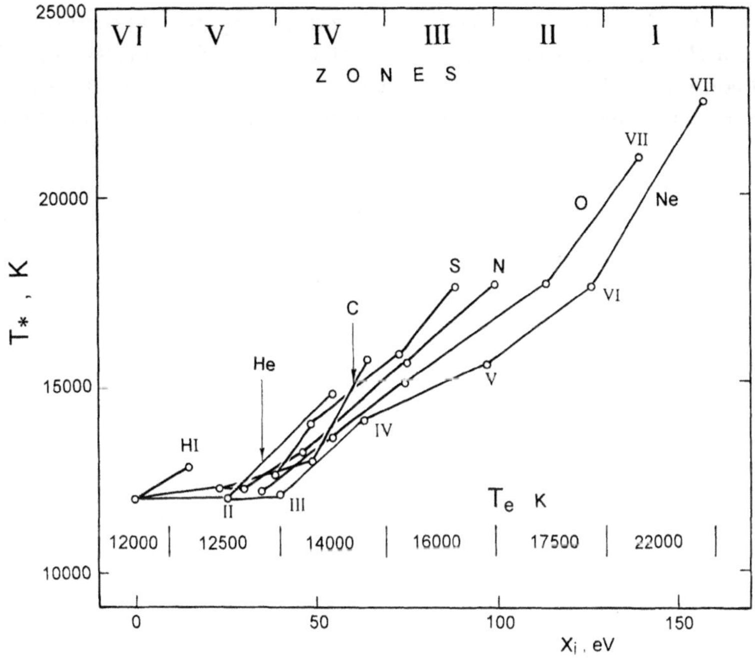

Fig. 6.1. Semi-empirical gradient of electron temperature T_e when passing from the central parts of planetary nebula NGC 3 918 (zones I–II) to its periphery (zones V–VII)

6.5 The Temperature Gradient Within the Nebula: Observations

A PN of a high excitation class, NGC 3 918 ($p = 9$), was the subject of thorough investigations at various wavelengths – ultraviolet, optical, infrared and radio, both via ground-based and space-based facilities, including the *International Ultraviolet Explorer* (*IUE*). Numerous emission lines have been recorded, from extreme ultraviolet, 1 176 CIII, up to the near infrared, 9 537 [SIII], and far infrared, 18.7 μm SIII, lines of neutral atoms, 6 300 + 6 365 [OI], and of ions at the VI–VII degree of ionization (Clegg et al., 1987). Based on these data a model of this nebula was constructed ($T_* = 140\,000$ K), and some of its parameters were reconstructed including the gradient of the electron temperature within the nebula. Figure 6.1 shows the dependence of T_e (ordinate) on the ionization potential κ (abscissa).

The origin of high potential ions, with $\phi_i > 100$ eV, is related to the central parts of the nebula; ϕ_i decreases as one moves outward. Therefore, Fig. 6.1 is nothing other than a semi-empirical law of variation of T_e from the centre of a nebula to its periphery, i.e. we have the sought-for gradient of T_e.

In practice, however, to determine the electron temperatures of the central parts of a PN, it is necessary to use the relative intensities of certain pairs of emission lines of highly ionized ions with the potential of ionization larger than 60 eV. The lines [NeIV] with the parameters given in Table 2.2 satisfy this condition. Then one can derive

$$\frac{E(2\,440+2\,442)}{E(1\,608)} = 2.97\,e^{30\,500/T_e}\,, \tag{28}$$

$$\frac{E(2\,440+2\,442)}{E(4\,720)} = 12.67\,e^{30\,500/T_e}\,. \tag{29}$$

In (29) under the 4 720 one must have in mind the summary intensities of four lines: 4 714 + 4 716 + 4 724 + 4 726 [NeIV].

The lines [NeIV] were detected and measured in the spectra of NGC 3 918 and NGC 7 027. For NGC 3 918 we have ($H_\beta = 1.00$): $E(2\,425) = 1.32$, $E(1\,608) = 0.066$, $E(4\,720) = 0.026$. The data for NGC 7 027 are collected in Table 7.8. Then we obtain from the above formulae the following temperatures T_{eII} for the central parts of these nebulae

	T_{eI}, K	T_{eII}, K
NGC 3 918	10 900	16 000
	10 900	21 800
NGC 7 027	12 400	23 300

The main hope must be associated with the lines [NeV]. Using the data in Table 2.1, one can derive the following expressions for the forbidden lines [NeV]:

$$\frac{E(3\,426+3\,346)}{E(1\,575+1\,592)} = 5.46\,e^{47\,800/T_e}\,, \tag{30}$$

$$\frac{E(3\,426+3\,346)}{E(2\,976)} = 16.9\,e^{47\,800/T_e}\,. \tag{31}$$

The lines 1 575 + 1 592 [NeV] were discovered in the spectrum of NGC 3 918 with a relative intensity of 0.056. The values $E(3\,426) = 0.821$ and $E(3\,346) = 0.378$ were measured as well. Then we obtain from (30) $T_{eII} =$ 34 800 K, i.e. two times higher than was obtained above.

One more ratio can be derived for the line doublets of [NeIII] (see Table 2.3):

$$\left.\begin{aligned}\frac{E(3\,869+3\,967)}{E(1\,815+1\,794)} &= 12.4\,e^{43\,000/T_e}\,,\\ \frac{E(3\,869+3\,967)}{E(3\,343)} &= 18.0\,e^{43\,000/T_e}\,.\end{aligned}\right. \tag{32}$$

Thus, the most efficient for the diagnosis of the central parts of a PN seem to be the forbidden lines, and, above all, [NeIV] and [NeV].

6.6 The Electron Concentration

A general estimate of the electron concentration in PNs can be reached with the help of the ionization formula (4.7), if $\tau_c \sim 1$ for the nebula is assumed. One can obtain from the ionization formula, substituting $n_e \approx n^+$, $T_* = 50\,000$ K, $T_e = 10\,000$ K and $W = 10^{-14}$,

$$\frac{n_e^2}{n_1} \approx 10^6 . \tag{33}$$

We have $\tau_c = \kappa_c n_e R_n = 1$, where $\kappa_c = 6.3 \times 10^{-18}$ cm^2 is the absorption coefficient per neutral hydrogen atom at the Lyman limit and R_n is the radius of the nebula. Adopting $R_n = 10\,000$ a.u. $= 1.5 \times 10^{17}$ cm we find that $n_1 \sim 1$ cm^{-3} and consequently $n_e = n^+ \approx 10^3$ cm^{-3}. These results indicate the high degree of ionization in PNs: $n^+/n_1 \sim 1\,000$.

There are a number of methods for determining the electron concentration of PNs from observational data.

The Surface Brightness Method (H Method). This method was proposed long time ago by Menzel and Aller (1941) and is based on the idea that the surface brightness of a nebula in the hydrogen lines depends on its electron concentration.

The energy emitted by the nebula in a line of the Balmer series of hydrogen is

$$E_{ki} = z_k \, A_{ki} \, h\nu_{ik} \, n_e^2 \, V , \tag{34}$$

where $z_k = n_k / n_e n^+$ and V is the volume of the nebula:

$$V = \frac{4\pi}{3} R_n^3 . \tag{35}$$

From (34) we have

$$n_e = \left(\frac{E_{ki}}{z_k \, A_{ki} h\nu_{ik} \, V} \right)^{1/2} . \tag{36}$$

Proceeding from the luminosity E_{ki} of the nebula in a given line to the surface brightness S_{ki} in the same line by means of the relation $S_{ki} = E_{ki}/4\pi \, R_n^2$, and using (35), we find instead of (36) that

$$n_e = \left(\frac{3 S_{ki}}{z_k \, A_{ki} R h\nu_{ik}} \right)^{1/2} . \tag{37}$$

Denote by H_{ki} the surface brightness of a nebula in the given line expressed in stellar magnitudes per square arc minute. The relation between S_{ki} and H_{ki} has the form

$$S_{ki} = 0.84 \times 10^3 (2.512)^{-\mathrm{H}_{ik}} . \tag{38}$$

Representing z_k in the form $z_k = b_k z_k^0$, where z_k^0 is the value of z_k for thermodynamic equilibrium, and applying (37) to the line H_β, we find that

$$n_e = 3.18 \times 10^{11} \frac{T_e^{3/4}}{(Rb_4)^{1/2}} e^{-\kappa_4/kT_e} (2.512)^{-H_\beta/2} . \tag{39}$$

From here the electron concentration in the nebula by known surface brightness H_β in the line H_β and the nebula's radius R_n, can be obtained.

If H_β is replaced by H_n – the surface brightness of the nebula in the continuous spectrum beyond the Balmer limit, we will have

$$n_e = 8.54 \times 10^{11} \frac{T_e^{3/4}}{R^{1/2}} (2.512)^{-H_n/2} . \tag{40}$$

The luminosity of the nebula in the line H_β yields 1/5 (6–7%) of the total luminosity in the visible spectrum, and converting from luminosity to apparent stellar magnitude m_n, we obtain still another relation for the electron concentration:

$$n_e = 7.58 \times 10^8 \frac{\varphi^{-3/2}}{r^{1/2}} 10^{-0.2\, m_n} , \tag{41}$$

where φ is the angular diameter of nebula in arc seconds and r is the distance to the nebula.

Equations (39), (40) and (41) were applied to a number of PNs. It was shown that the electron concentrations in PNs vary between not too large limits and on average are of the order of $10^3 -- 10^4$ cm^{-3}.

The H_β-line Flux Method. We have for the volume emission coefficient ε_β in the line H_β

$$4\pi\varepsilon_\beta = 1.24 \times 10^{-25} n_e n^+ \left(\frac{10^4}{T_e}\right)^{0.85} \text{ergs cm}^{-3}\text{s} . \tag{42}$$

If F_β is the observed flux of emission in the line H_β, then

$$4\pi r^2 F_\beta = 4\pi\varepsilon_\beta V = 4\pi\varepsilon_\beta \frac{4\pi}{3} R_n^3 . \tag{43}$$

From the last two expressions we obtain, taking $n_e = n^+$ and substituting $R_n = r\varphi/206\,265''$,

$$n_e = 4.60 \times 10^2 O \left[\frac{F_\beta}{(r\varphi)^3}\right]^{1/2} \left(\frac{T_e}{10^4}\right)^{0.425} \text{cm}^{-3}, \tag{44}$$

or, if we represent F_β in units of 10^{-11} ergs cm^{-2}s^{-1} and r in kiloparsecs,

$$n_e = 2.65 \times 10^4 \left[\frac{F_\beta}{(r\varphi)^3}\right]^{1/2} \left(\frac{T_e}{10^4}\right)^{0.425} \text{cm}^{-3} . \tag{45}$$

In all formulae φ is expressed in arc seconds.

The Doublet 3 727 [OII] Method. The metastable levels $^2D_{3/2}$ and $^2D_{5/2}$ of the O^+ ion in the ground configuration $2p^3$ are located quite close to each other and at 3.32 eV above the ground level $4S_{3/2}$. The well-known violet doublet 3 729 [OII] and 3 726 [OII] is emitted at transitions $^2D_{5/2} - ^4S_{3/2}$ and $^2D_{3/2} - ^4S_{3/2}$ (Fig. 2.2). As distinct from the N_1 and N_2 lines, the ratio of intensities of which does not depend on the conditions in the nebula, the ratio of intensities of 3 729 [OII] and 3 726 [OII] does depend on n_e and T_e.

At low densities in the nebula the ratio of intensities of these lines will be equal to the ratio of the effective cross-sections, and hence $E(3\,729)/E(3\,726)=$ 1.47.

At high densities this ratio is determined by the condition of thermodynamic equilibrium: $E(3\,729)/E(3\,726) = 0.45$.

At intermediate densities this ratio will be between the above values; the corresponding relationship has the form (e.g. Saraph, Seaton, 1974)

$$\frac{E(3\,729)}{E(3\,726)} = 1.5\,\frac{1+3.26\,x}{1.02+(10.4+4.0\,e^{-19\,500/T_e})\,x} \tag{46}$$

or approximately, for $T_e \sim 10\,000$ K,

$$\frac{E(3\,729)}{E(3\,726)} = 1.5\,\frac{1+3.26\,x}{1.02+10.5\,x}\,, \tag{47}$$

where

$$x = 10^{-2}\,\frac{n_e}{T_e^{1/2}}\,. \tag{48}$$

For high values of n_e, in practice, for $n_e > 5\times10^4$ cm^{-3}, we have

$$\frac{E(3\,729)}{E(3\,726)} \to 0.45 \text{ at } n_e \to \infty\,,$$

while for low n_e, i.e. for $n_e < 50$ cm^{-3}, we obtain

$$\frac{E(3\,729)}{E(3\,726)} \to 1.47 \text{ at } n_e \to 0\,.$$

The dependence of $E(3\,729)/E(3\,726)$ on n_e is shown in Fig. 6.2 for $T_e =$ 10 000 K and $T_e = 20\,000$ K. We see that the dependence of $E(3\,729)/E(3\,726)$ on T_e is rather weak. Therefore, for the determination of the electron concentration by (47) it is enough to have a rough estimation of T_e.

Seaton (1954) was the first to apply this method to several PNs. Subsequently, Osterbrock (1960) measured the ratio $E(3729)/E(3726)$ and obtained electron concentrations for a number of PNs (Table 6.5).

Electron concentrations determined by the doublet [OII] appear to be significantly larger than those obtained by the H method. By this method the electron concentrations of low density and low excitation PNs were successfully determined, thereby giving an idea of the lower limit of electron concentrations in PNs. For example, in the densest regions of the giant PN

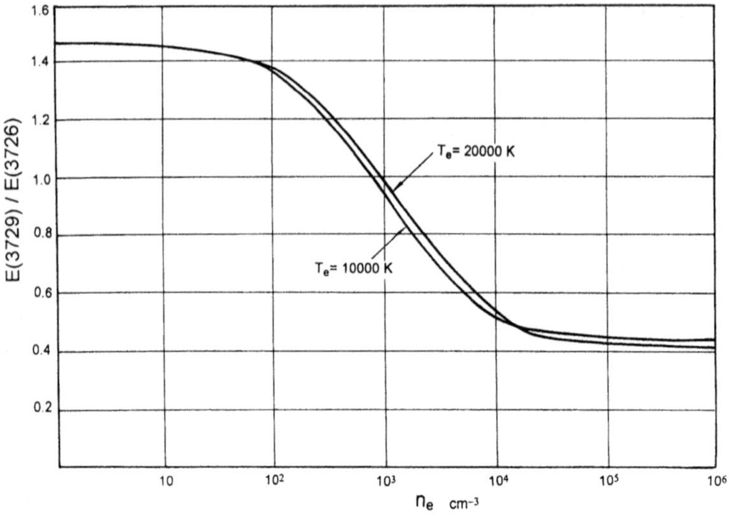

Fig. 6.2. Dependence of the ratio of $E(3\,729)/E(3\,726)$ on the electron concentration m_e at two values of the electron temperature T_e, 10 000 K and 20 000 K

in Aquarius, NGC 7 293, the electron concentration yields 140–180 cm^{-3}; for the bipolar nebula NGC 3 587 it was found to be 150 cm^{-3}, in certain regions even reaching 70 cm^{-3}.

The doublet [OII] method is simple and convenient to apply. Its disadvantage is the necessity of high dispersion spectrographs in order to separate the lines 3 729 Å and 3 726 Å while measuring their intensities.

The Doublet [SII] Method. The analogue of the violet doublet 3 727 [OII] is the red doublet of singly ionized sulphur 6 731 + 6 717 [SII]; this doublet is emitted at the same transitions from the same levels $^2D_{3/2,5/2}$ to $^4S_{3/2}$. The dependence of the ratio $E(6\,731)/E(6\,717)$ on n_e and T_e in this case is given by a formula analogous to (47) (e.g. Eissner et al., 1969; Canto et al., 1980):

$$\frac{E(6\,731)}{E(6\,717)} = 1.5\,\frac{1 + 0.35\,x}{1 + 0.96\,x}\,. \tag{49}$$

The dependence of this ratio on n_e and T_e is is basically the same as in the case of violet doublet $E(3\,729)/E(3\,726)$.

The red doublet [SII] method has also found wide application, as has that of the violet doublet [OII]. In the majority of cases both methods give nearly the same results for n_e (except for IC 4 997 and IC 418).

PNs with electron concentrations measured by these two methods, [OII] and [SII], are collected in Table 6.5.

The [OIII]–[NII] Method. We have already mentioned this method before in connection with the problem of determining the electron temperature of PNs. Combining (10) and (11), we obtain a system of equations, the left side

Table 6.5. Electron concentrations n_e (in units of 10^3 cm^{-3}) in planetary nebulae obtained by the violet doublet [OII] and red doublet [SII] methods

Planetary nebula	[OII]	[SII]	Planetary nebula	[OII]	[SII]	Planetary nebula	[OII]	[SII]
NGC 40	1.44	1.12	NGC 6537	3.14	0.79	NGC 7048	–	1.78
NGC 650-1	1.26	0.56	NGC 6543	3.55	0.79	NGC 7293	0.14	1.41
NGC 1501	–	2.00	NGC 6565	0.40	–	NGC 7662	3.16	4.47
NGC 1535	3.55	–	NGC 6567	3.16	–	IC 351	5.01	–
NGC 2022	1.00	0.16	NGC 6572	5.01	2.82	IC 418	15.8	0.89
NGC 2346	3.16	–	NGC 6629	2.00	–	IC 1747	3.16	5.01
NGC 2371	–	2.00	NGC 6644	6.30	–	IC 2003	–	5.62
NGC 2392	5.62	1.12	NGC 6720	0.16	1.58	IC 2149	2.00	5.62
NGC 2440	2.82	0.45	NGC 6741	2.51	2.82	IC 2165	2.82	4.47
NGC 2452	1.26	3.98	NGC 6778	–	1.58	IC 3568	5.00	–
NGC 2474	–	0.25	NGC 6781	0.16	–	IC 4593	2.00	1.58
NGC 2818	–	0.79	NGC 6790	12.6	25.1	IC 4642	0.63	–
NGC 2867	–	2.51	NGC 6803	5.01	2.24	IC 4776	4.47	–
NGC 2899	–	0.79	NGC 6818	3.16	2.84	IC 4846	8.91	5.01
NGC 3132	0.16	-	NGC 6826	2.00	3.55	IC 4997	126.0	5.00
NGC 3242	14.1	3.98	NGC 6853	0.28	–	IC 5117	–	5.02
NGC 3587	0.13	0.25	NGC 6884	7.07	7.94	IC 5217	8.91	7.34
NGC 3918	2.82	3.16	NGC 6886	5.60	8.91	BD+30°	7.08	2.82
NGC 4361	10.0	–	NGC 6891	2.52	–	Hu 1-1	–	2.24
NGC 6058	2.51	–	NGC 6905	–	1.00	M 1-2	–	5.62
NGC 6210	5.60	3.55	NGC 7008	10.0	–	M 1-1	2.00	–
NGC 6302	7.90	12.6	NGC 7009	2.24	5.00	MeWe 1-4	–	15.8
NGC 6309	3.16	2.82	NGC 7026	–	7.94	Me 2-1	1.26	1.58
NGC 6445	0.79	2.00	NGC 7027	5.00	10.0			

of which are determined by observations, while on the right side the quantities T_e and n_e appear. From the joint solution of these equations one can obtain T_e and n_e.

In spite of its simplicity, this method has limited application. It gives reliable results for comparatively dense nebulae. For intermediate concentrations ($n_e \sim 10^4$ cm^{-3}) the method becomes sensitive to errors in the measurements.

The Balmer Discontinuity Method: The D Method. The magnitude of the discontinuity in the continuous spectrum at the limit of the Balmer series depends on T_e and n_e. In particular, the higher the electron concentration of the nebula is, the smaller is the magnitude of the discontinuity $D = \log I(3\,646^-)/I(3\,646^+)$. However, D is particularly sensitive to the electron temperature. Therefore if one can establish T_e by alternative methods, then it is possible to determine n_e. The method is efficient when T_e is low and n_e is not too high. The corresponding calculations were performed by Seaton (1955) for a number of PN. The results obtained are in marginal accord with those of other methods. The main problem is the difficulty of the measuring of the real intensities of continuous spectrum just on the Balmer discontinuity.

The CIII Method. An interesting pair of ultraviolet lines of doubly ionized carbon turned out to be quite sensitive to the electron concentration. Both these lines, 1 906.68 CIII] (= 1 907 Å) and 1 908.73 CIII] (= 1 909 Å), are semi-forbidden ones. The sensitivity of the ratio of the intensities of both emission lines, $F(1\,907)/F(1\,909)$, to the electron concentration n_e is its highest between the limits $10^3 < n_e < 10^6$ cm^{-3}, which just corresponds to normal and dense PNs. The dependence of this ratio on the electron temperature T_e is weak, and T_e can be obtained by other methods. In Fig. 6.3 the dependence of the ratio of $F(1\,907)/F(1\,909)$ on n_e is shown for certain values of T_e, from 5 000 K up to 90 000 K.

In Table 6.6 the values of n_e obtained with the help of this method and based on *IUE* high resolution recordings are given for a sample of PNs (Feibelman et al., 1980, 1988). The observed ratios of $F(1\,907)/F(1\,909)$ were within the limits from 0.03 (IC 4 997) up to 1.40 (IC 351, IC 5 217). For comparison, in the last column of Table 6.6, the values of n_e, determined by the forbidden lines [OIII] and [OII] (Aller et al., 1980), are also given (see Table 6.5).

In Fig. 6.4, the *IUE* recordings of the profiles of these lines are given for four PN. In two cases, NGC 3 242 and Hu 1–2, these profiles are split, an obvious indication of the expansion of the nebulae, in the first case with

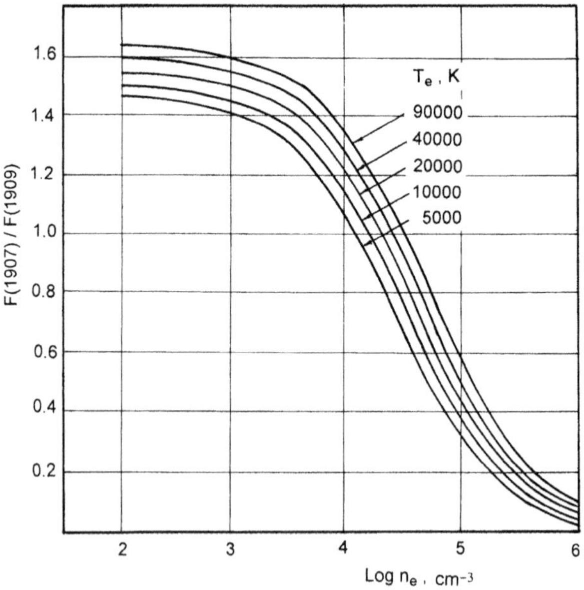

Fig. 6.3. Calculated curves of the dependence of the intensity ratio $F(1907)/F(1909)$ CIII] on the electron concentration n_e for five values of the electron temperature T_e

Table 6.6. Electron concentrations obtained by the CIII method – the ratio of the intensities of lines $F(1\,907)/F(1\,909)$ for a sample of planetary nebulae

Planetary nebula	T_e, [K]	$\frac{F(1\,907)}{F(1\,909)}$	$\log n_e$ cm^{-3} [CIII]	[OIII],[OII]
NGC 3 211	13 700	1.11	4.1	3.44
NGC 3 242	11 100	1.09	4.2	3.90
NGC 6 572	10 800	0.80	4.5	4.14
NGC 6 644	12 000	0.93	4.0	3.2:
NGC 6 818	18 000	1.19	4.1	3.84
NGC 6 891	10 000	1.08	4.2	3.86
NGC 7 009	10 500	1.14	4.2	3.88
NGC 7 027	13 000	0.6	4.8	4.3
IC 351	12 700	1.40	3.4	4.26
IC 418	9 100	0.93	4.4	4.27
IC 1 297	11 000	1.0:	4.4	–
IC 2 165	12 200	1.39	3.5	3.52
IC 3 568	12 400	1.16	4.1	3.02
IC 4 997	18 000	0.03	5.9	5.17
IC 5 217	11 000	1.40	3.4	3.78
J 900	11 900	1.37	3.5	3.27
Hu 1–2	17 000	1.25	3.9	3.65

velocity 39 km s^{-1}, in the second, 34.5 km s^{-1}. In both cases the nebulae are hollow in the centre and an asymmetry in their shapes can also be noticed.

In the case of NGC 6 572 no splitting seems to exist; however, the faint prominence from the long wavelength side of both profiles can still be interpreted as an expansion of this strongly asymmetrical nebula with a velocity ~ 70 km s^{-1}.

For IC 4 997, first, the splitting of 1 909 CIII is absent, and, second, the short wavelength component 1 907 CIII is strongly depressed, thus indicating a high electron concentration. Both facts confirm the conclusions derived from ground-based observations.

Note that the electron concentrations obtained by the CIII method are systematically higher than obtained by other methods.

6.7 The Masses of Planetary Nebulae

The mass of a PN $\mathcal{M}_n$ can be determined by its known electron concentration n_e and linear size:

$$\mathcal{M}_n = \frac{\pi}{6} D^3 n_H m_H \,, \tag{50}$$

where D is the diameter of the nebula, n_H is the concentration of hydrogen atoms, and m_H is the mass of the hydrogen atom. Adopting $n_H \approx n_e \sim$

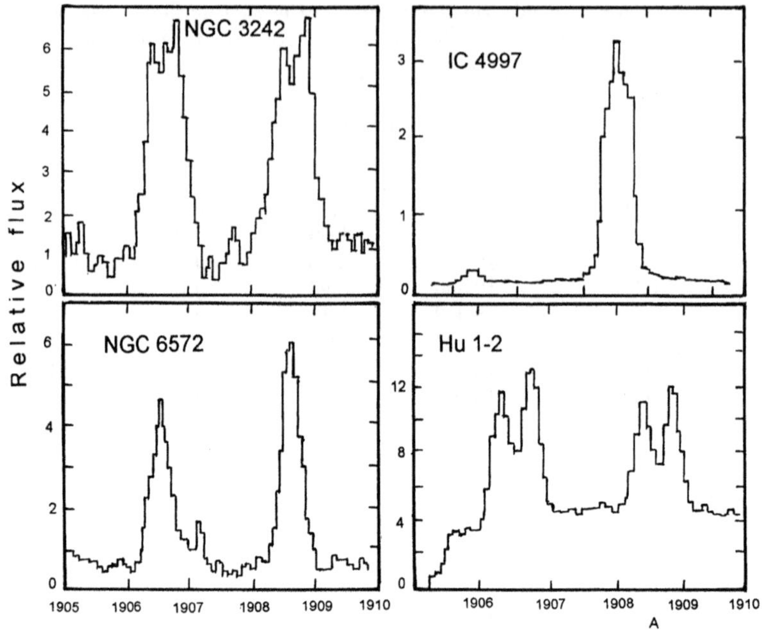

Fig. 6.4. *IUE* recordings of profiles of lines 1 907 III] and 1 909 CIII] in the spectra of the planetary nebulae NGC 3 242, NGC 6 572, IC 4 997 and Hu 1-2. Splitting of the profiles of NGC 3 242 and Hu 1-2 indicates that they are expanding, with a velocity of 39 and 35 km s^{-1}, respectively. The strong depression in the line 1 907 CIII] and the absence of splitting of 1 909 CIII] in the case of IC 4 997 indicates the compact structure of this nebula and its high electron concentration

$10^3 - 10^4$ cm^{-3} and $D = 20\,000$ a.u., we obtain

$$\mathcal{M}_{\rm n} \sim (0.01 - 0.1)\mathcal{M}_{\odot} \,. \tag{51}$$

Another more precise method of determining a nebula's mass is based on (3.12), containing the energy radiated by a nebula in the hydrogen lines:

$$E_{\rm n2} = z_{\rm n}\, A_{n2}\, h\nu_{\rm n2}\, n_{\rm e} n^+\, V \,. \tag{52}$$

Adopting $n_{\rm e} = n^+$, we get

$$n_{\rm e} = \left(\frac{E_{\rm n2}}{z_{\rm n}\, A_{\rm n2}\, h\nu_{\rm n2}\, V} \right)^{1/2} \,. \tag{53}$$

As long as

$$n^+ = n_{\rm e} = \frac{\mathcal{M}_{\rm n}}{m_{\rm H}\, V} \,,$$

we will have

$$\mathcal{M}_{\rm n} = m_{\rm H} \left(\frac{E_{n2}\, V}{z_{\rm n}\, A_{n2} h\nu_{12}} \right)^{1/2} \,. \tag{54}$$

If we pass from E_{n2} to the total visual luminosity L by the relation $E_{n2} = \gamma_{n2} L$, we get instead of (54):

$$\mathcal{M} = C\sqrt{LV}\,. \tag{55}$$

This formula was derived by Ambartsumian. The multiplier C can be assumed almost constant for all PNs as long as γ_{n2} varies weakly from one nebula to another:

$$C = m_{\mathrm{H}} \left(\frac{\gamma_{n2}}{z_n A_{n2} h\nu_{n2}} \right)^{1/2} = \mathrm{const.} \tag{56}$$

When, for example, the nebula is observed in the line H_β and the ratio N_2/H_β is equal to 3, we have $\gamma_{24} = 1/12$, and then $C = 1.5 \times 10^{12}$ (CGS).

Judging by the observational dispersion of the luminosity L and volume V, the masses of PNs can differ by at least over ten times.

Note that (55) gives the total mass of a nebula, planetary or diffuse, only in the case $\tau_c < 1$. In the case when $\tau_c > 1$, this formula renders only the luminous mass, i.e. ionized part, of the nebula. The mass of the neutral hydrogen can be obtained only by means of radio astrophysical methods, i.e. by observations in the 21 cm radio line.

One more relation can be derived for determining the nebula's mass $\mathcal{M}_n$ in the case when the flux F_β in absolute units is known.

Taking into account the contribution of helium, we have

$$\mathcal{M}_n = n_e m_{\mathrm{H}} \left(1 + 4\frac{\mathrm{He}}{\mathrm{H}} \right) V\,. \tag{57}$$

Combining (42) and (43) and adopting He/H = 0.1, we have

$$\frac{\mathcal{M}}{\mathcal{M}_\odot} = 2.35 \times 10^2 F_\beta \frac{r^2}{n_e} \left(\frac{T_e}{10^4} \right)^{0.85} = 1.16 \times 10^{-31} F_\beta \frac{r^2}{n_e} \left(\frac{T_e}{10^4} \right)^{0.85}, \tag{58}$$

when F_β is in units of ergs $\mathrm{cm}^{-2}\mathrm{s}^{-1}$, the nebula's radius, r, in cm, and

$$\frac{\mathcal{M}}{\mathcal{M}_\odot} = 11.20 F_\beta \frac{r^2}{n_e} \left(\frac{T_e}{10^4} \right)^{0.85}, \tag{59}$$

when F_β is in the units of 10^{-11} ergs $\mathrm{cm}^{-2}\mathrm{s}^{-1}$, r is in kpc, and n_e is determined by (45). Substituting n_e from (45) into (59,) we finally get

$$\frac{\mathcal{M}_n}{\mathcal{M}_\odot} = 1.40 \times 10^{-20} (r\varphi)^{3/2} F_\beta^{1/2} \left(\frac{T_e}{10^4} \right)^{0.425}. \tag{60}$$

The application of (59) for instance for the two PNs NGC 6 567 and NGC 7 293, the visible sizes of which differ over 100 times, yields for their masses 0.009 $\mathcal{M}_n$ and $0.30\mathcal{M}_\odot$, respectively (Pottasch, 1980).

For a stricter formulation of the problem one has to take into account also the influence of the so-called "filling factor" ε, i.e. the degree of divergence of the nebula from homogeneity. However, as the analysis shows, the influence of this factor is not too large, as long as $n_e \sim \varepsilon^{-1/2}$ and $\mathcal{M} \sim \varepsilon^{1/2}$, and even at $\varepsilon = 0.3$ the discrepancies in the obtained masses do not exceed a factor of two.

7. Planetary Nebulae in the Ultraviolet

7.1 The Structure of a Nebula's Spectra in the Ultraviolet

Historically the first ultraviolet spectral observations of PNs were performed by the space observatory *Orion-2*, when the first ultraviolet spectrogram of a PN, IC 2 149, was obtained (Gurzadyan, 1974), and by the *Aerobee* rocket, when the first short wavelength spectrogram of NGC 7 027 was registered (Bohlin et al., 1975). Systematic ultraviolet spectral observations of PNs have been connected, since 1978, with the space observatory satellite *International Ultraviolet Explorer, IUE* (Boggess et al., 1978).

Since 1991 the third short wavelength region has appeared, the so-called, extreme ultraviolet, including the wavelength region shorter than $\sim$ 600 Å. A space astrophysical observatory satellite, the *Extreme Ultraviolet Explorer*, with a grazing-incidence telescope and variable line-spaced grating spectrometers aimed to conduct an all-sky survey, as well as performing detailed studies of individual sources of extreme ultraviolet emission, including PNs (Bowyer, Malina, 1991; Bowyer, 1994). In 1994 the *First Extreme Ultraviolet Explorer Source Catalog* was published (Bowyer et al., 1994) including 410 point sources, of which 372 have plausible optical, ultraviolet, or X-ray identifications. Six PNs, NGC 246, NGC 1 360, NGC 3 587, NGC 4 361, K 1-16 and LoTr 5 (the latter located exactly at the north Galactic pole) were detected in the shortest EUV band, 58–174 Å; the emission most probably is originating from the PNs' central stars (Fruscione et al., 1996).

In general all emission lines and continuous spectra in the wavelengths between 912 Å and 3 000 Å are attributed to the ultraviolet range. For various reasons it is convenient to divide this range into two parts: *the near ultraviolet*, 2 000–3 000 Å, and *the far ultraviolet*, 1 000–2 000 Å.

The composition and structure of the spectra of PNs in each of these regions are considerably stable. The intensities themselves are not stable and can vary within sufficient limits if one passes from one nebula to another, depending on the excitation class.

The typical structure of the spectra of PNs in the far ultraviolet, 1 000–2 000 Å (more precisely, 912–2 000 Å), is presented in Fig. 7.1. The strongest line absolutely dominating within this region is the semi-forbidden doublet

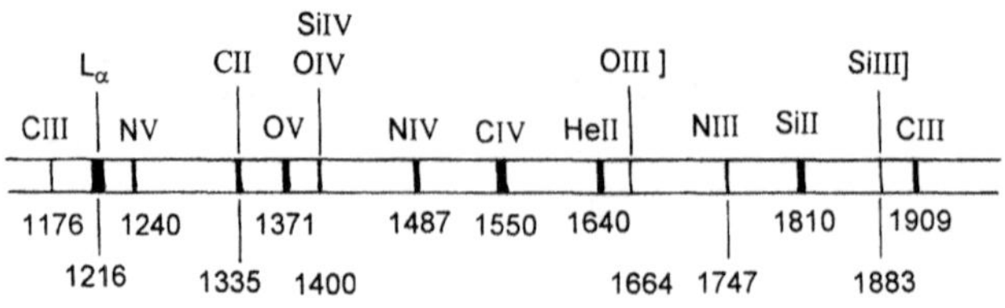

Fig. 7.1. Structure of the spectra of planetary nebulae in the far ultraviolet region, 1 000–2 000 Å

of triply ionized carbon 1 448 + 1 451 CIV (= 1 450 CIV). This doublet is present in the spectrum of all PNs almost without exception; more precisely it is absent in PNs of the lowest excitation class, $p \sim 12$. The intensity of this doublet strongly depends on the excitation class; it is very strong in general for classes $p = 9$ and higher. For classes $p \sim 10$ and higher the strength of this doublet is of the same order as of the doublet $N_1 + N_2$ [OIII], and in some cases (NGC 7 027 ($p = 10$) and M 1-1 ($p = 12$)) it is even stronger than $N_1 + N_2$ and is 10 to 30 times stronger in comparison with H_β. Rarely, 1 550 CIV can be fainter than H_β (NGC 7 009, NGC 3 242).

The next is the fluorescence (recombination) line 1 640 HeII which is again quite strong in PNs of high excitation classes as shown in the following table:

	p	N_1+N_2	1 550 CIV	1 640 HeII
M 1-1	12	7.6	18.2	7.1
NGC 7 027	10	18.0	31.7	7.7
NGC 7 662	10	17.0	14.0	6.3

In middle excitation class PNs ($p \sim 7$—–9) the doublet 1 550 CIV dominates over the line 1 640 HeII. The line 1 640 HeII is stronger than H_β almost always, with the exception again in PNs of excitation classes of $p \sim 1$—–2, where the lines of ionized helium are completely absent.

Then comes the semi-forbidden doublet 1 906–1 909 CIII] (= 1 908 CIII). In the high excitation PNs the strengths of these lines, 1 640 HeII and 1 908 CIII, are of the same order; however, in the low excitation nebulae, 1 640 HeII dominates over the doublets of 1 550 CIV and 1 908 CIII.

Among the next lines, depending on the excitation class of the PN, are the other lines in the far ultraviolet: low excitation, 1 335 CII, 1 810 SiII, and high excitation, 1 176 CIII], 1 400 OIV, 1 487 NIV, 1 644 OIII], 1 747 NIII, 1 883 SiIII]. In the far ultraviolet the highest excitation lines are the doublet 1 238–1 242 NV and the line 1 371 OV; however, they are rarely detected only in high excitation PNs.

All these lines, their mutual combinations, the absolute and relative intensities, even only the fact of their existence or absence, are sources of exceptional importance for the investigation of PNs. Note that the properties of lines of the far ultraviolet, their intensities and profiles are quite sensitive to the physical and kinematic parameters of a nebula and its nucleus.

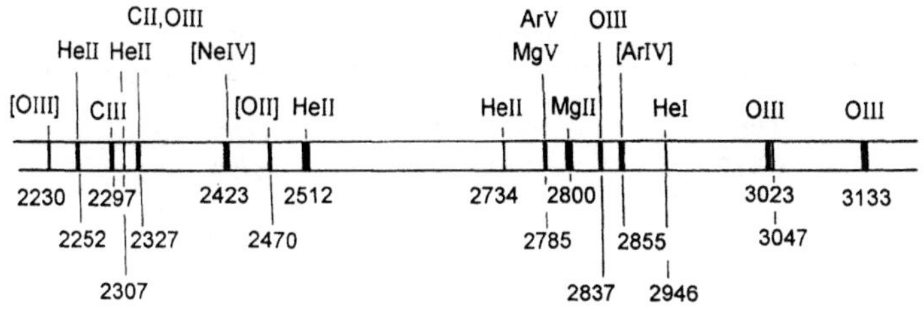

Fig. 7.2. Structure of planetary nebula spectra in the near ultraviolet region, 2 000–3 000 Å

The L_α line of hydrogen (= 1 216 Å) might also become a source of special importance. However the powerful emission of the geocorona completely dominates over the L_α line of any nebula.

Besides the emission lines, the far ultraviolet is represented also by the continuous spectrum. This includes the continuum of different ions, thermal bremsstrahlung of free electrons, the scattering of the ultraviolet radiation of the nucleus on the dust particles of the nebula and, of course, the two-photon emission of hydrogen, the spectra of which starts from 1 216 Å and reaches a maximum at 1 400 Å.

The structure of a nebula's spectra in the near ultraviolet (2 000–3 000 Å) is shown in Fig. 7.2. Here the strongest is the forbidden doublet 2 433+2 416 [NeIV] in high excitation nebulae (NGC 6 905) and lines 2 325+2 329 CII + OIII, as well as the doublet 2 800 MgII mainly in low and intermediate excitation nebulae (IC 4 997). The high potential lines 2 784 [MgV] and 2 785 [ArV], and also 2 855 [ArIV], are observed rarely and only in high excitation nebulae; these lines are completely absent in the spectra of low excitation nebulae. The remaining lines with moderate and low intensities belong mainly to ions OII, OIII, CII, CIII and are permanent satellites of the spectra for almost all excitation class nebulae. Among the lines of this group the doublet 3 023+3 047 OIII and line 3 133 OIII are usually distinguished. The ionized helium is represented by the lines 2 252 HeII, 2 307 HeII, 2 511 HeII, 2 734 HeII, and neutral helium by 2 946 HeI, 3 188 HeI.

The composition of the continuous spectra in the near ultraviolet is the same as in the far ultraviolet. In the region longer than 2 500 Å and up to 3 100 Å the Balmer continuum dominates (recombinations at the $n = 2$ level of hydrogen).

7.2 Resonance Lines in the Ultraviolet

The resonance lines are the strongest lines, without exception, in all atoms and ions, and therefore they are the most informative. The detection of resonance lines is the aim of any observations.

In the optical region, longer than 3 000 Å, we have a limited number of resonance lines of atoms and ions. The most important resonance lines are:

4 227 CaI	5 890 NaI	6 708 LiI
6 572 CaI	5 896 NaI	5 200 NI
7 665 KI	3 934 CaII	5 198 NI
7 669 KI	3 968 CaII	4 871 MgI
4 642 KI		

The lines CaI and LiI, the doublet KI and CaII in the spectra of PNs are completely absent. Among the rest of the lines, 4 571 MgI and the doublet 5 200 NI are extremely weak and are observed quite rarely. This is the reason why during the process of deciphering and interpreting the enormous amount of information concerning the physics of gaseous nebulae the resonance lines in the optical region are not always an attractive choice.

In the optical region, the *resonance forbidden* lines are the only exception, though not numerous, nevertheless being a source of information of greatest importance. These lines are

6 300 [OI]	4 959 [OIII]	6 548 [NII]
6 363 [OI]	5 007 [OIII]	6 584 [NII]
3 726 [OII]	3 967 [NeIII]	6 716 [SII]
3 729 [OII]	3 969 [NeIII]	6 730 [SII]

With the exception of these resonance and resonance-forbidden lines, the resonance lines of all the rest of the atoms and ions are located in the ultraviolet region – shorter than 3 000 Å and reaching well below the 912 Å, up to the hard ultraviolet (~ 100 Å) and X-rays (~ 10 Å). The number of such lines is enormous! In all cases to present here a list of even the most important ultraviolet resonance lines – single, doublet, triplet – is practically impossible. Below, the most studied ultraviolet resonance lines which have or might have some interest for the physics of gaseous nebulae are listed:

1 216 HI	2 139 NII	1 526 SiII
1 302 OI	992 NIII	1 817 SiII
1 304 OI	990 NIII	1 808 SiII
1 306 OI	1 243 NV	2 344 SiII
1 038 OVI	1 239 NV	2 334 SiII
1 032 OVI	2 852 MgI	1 892 SiIII
2 441 NeIV	2 796 MgII	1 206 SiIII
2 438 NeIV	2 803 MgII	1 403 SiIV
1 336 CII	2 750 MgV	1 394 SiIV
1 334 CII	2 892 MgV	1 741 NiII
1 908 CIII	2 669 AlII	1 751 NiII
1 909 CIII	1 671 AlII	2 343 FeII
1 548 CIV	1 863 AlIII	2 382 FeII
1 551 CIV	1 855 AlIII	2 599 FeII
2 143 NII	1 533 SiII	

The upper limit of wavelengths in this list is at $\sim 3\,000$ Å and is determined by the opacity of the Earth's atmosphere. The short wavelength boundary is quite definite, 912 Å, and is determined by the opacity of the interstellar medium, i.e. exceptionally by the absorption of interstellar neutral hydrogen in the Lyman continuum, shorter than 912 Å. Indeed, the optical thickness $\tau_c = knr$ in frequencies of the Lyman continuum, where the absorption coefficient for hydrogen is $k = 6.5 \times 10^{-18}$ cm^2 and the concentration of neutral hydrogen atoms is $n \sim 1$ cm^{-3}, yields 20, 200 and 2 000 at distances of 1, 10 and 100 pc. Even taking $n \sim 0.1$ cm^{-3} we will have $\tau_c > 1$. Therefore the band shorter than 912 Å and up to ~ 500 Å, in the direction of high galactic latitude up to ~ 700 Å, will be inaccessible for observations even of nearby stars: the only cosmic object observable at $\lambda < 912$ Å is the Sun.

Theoretically, among all resonance lines of atoms and ions, the most powerful should be L_α, the first line of the Lyman series of hydrogen, 1 216 HI. However, for the above mentioned reason, i.e. the practical impossibility of its revealing itself on the powerful background of the L_α emission of Earth's geocorona, we have no registration of this line emitted by a PN. The geocorona extends quite a long way, up to tens and hundreds of thousands of kilometres from the Earth. Therefore the L_α line can be recorded – and is in fact well recorded – only for extragalactic objects such as quasars, mainly owing to their large Doppler shift.

The resonance lines of neutral and ionized helium, the next element by its abundance, 584 HeI and 304 HeII, are found in the region of the hard ultraviolet. The opacity of the interstellar medium has a maximum at 912 Å, then it falls gradually according to a $\sim \lambda^{-3}$ law, so that at wavelengths shorter than 500 Å and even 600 Åit again becomes accessible for observations of objects located, however, not too far away.

The above list of ultraviolet resonance lines should be considered a basic one, especially in view of the fact that no resonance line of the optical region has a major role to play in the physics of gaseous nebulae. In fact, all the physics of gaseous nebulae has been built based on the forbidden optical lines, as well as on a few lines of subordinate series of hydrogen and some other elements. The possibility of operating with ultraviolet lines drastically changes the situation. This example once more reveals the motivation behind persistent efforts to develop space astrophysics.

7.3 The Macrostructure of the Spectra of Nebulae in the Ultraviolet

All PNs are identical or almost identical with respect to the main composition of emission lines both in the near (2 000–3 000 Å) and far (1 000–2 000 Å) ultraviolet. However, they are not identical and even differ essentially according to the macrostructure of spectra in the same regions. The determining

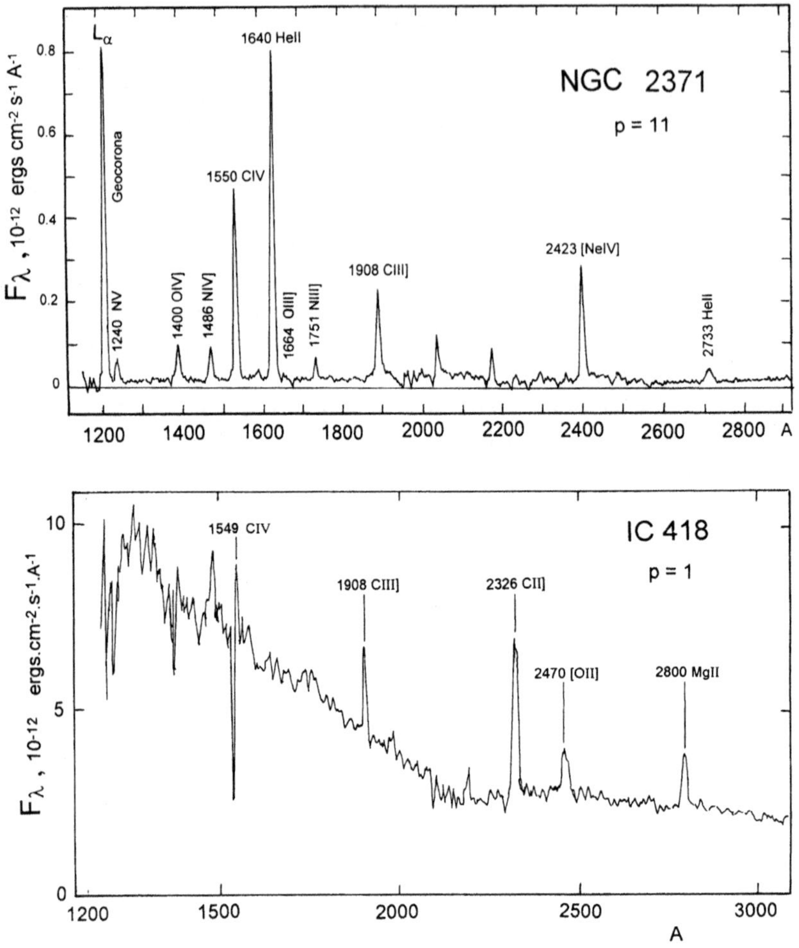

Fig. 7.3. *IUE* recordings of the spectra of two planetary nebulae, of high excitation – NGC 2371 ($p = 11$) – and low excitation – IC 418 ($p = 1$). In the second case the high excitation lines 1 240 NV, 1 550 CIV and 1 640 HeII are absent (in IC 418 the line 1 550 CIV with P Cyg type profile belongs to the central star of the nebula)

factor is the strength of the lines; some lines can be strong in the spectra of certain PNs, more fainter in others and completely absent in the still others. In this case the problem is of the dependence of the macrostructure of the spectra in the ultraviolet on the excitation class of a nebula.

The strongest high potential forbidden lines are observed as a rule in the spectra of high excitation PNs. In low excitation PNs, the high potential emission lines, in general, are absent.

The marked dependence of the macrostructure of short wavelength spectra on the excitation class of a nebula can be seen in Fig. 7.3. Here *IUE* recordings of spectra of two PNs of limiting excitation classes, p, are given, i.e. high, with $p = 11$, NGC 2 371 (see Pottasch, 1984), and low, with $p = 1$, IC 418 (Harrington et al., 1980). In the first case, NGC 2 371, the most intense are the high excitation lines 1 550 CIV and 1 640 HeII, then the forbidden line 2 423 [NeIV], and finally the semi-forbidden line 1 908 CIII]. If the nebula is of a high excitation class, the doublet 1 240 NV of four-fold ionized nitrogen would be seen literally in the vicinity of the L_α line of the geocorona, as it is seen in Fig. 7.3.

The situation is completely different in the case of IC 418. Note that here the continuous spectrum is related to the central star of the nebula, and the depression at $\sim$ 2 200 Å is due to interstellar extinction. The line 1 549 CIV with a P Cyg type profile also belongs to the central star, evidence proving the outflow of gaseous matter. In fact, the doublet 1 550 CIV in emission is absent in the spectra of this class of objects. The lines 1 640 HeII and 1 240 NV are also absent. The main lines for these ($P \sim 1$) objects are, thus, 1 908 CIII], 2 326 CII], 2 470 [OII]. The low excitation line 2 800 MgII is also present, though it was absent in the case of NGC 2 371.

The data in Table 7.1 give some evidence for the behaviour of the emission lines 1 550 CIV and 1 640 HeII independent of the excitation class p. Note that up to the class $p = 9$ these lines, 1 550 CIV and 1 640 HeII, are even stronger than $H_\alpha(= 2.85\ H_\beta)$.

Both lines 1 550 CIV and 1 640 HeII appear and disappear in the spectra of PNs in unison, e.g. in low excitation PNs, of classes 1–3, such as IC 418, M 2–9, BD+30°363 no traces of these lines are discovered.

There are PNs in which 1 240 NV is the strongest line in the region of 1 200–2 000 Å, even stronger than the lines 1 550 CIV and 1 640 HeII; in Fig. 7.4 is shown the *IUE* recording of one such nebulae, SMC P67 in the Small Magellanic Cloud (Aller et al., 1987). Such objects are rare.

Typically in the spectrum of high excitation PN both categories of emission lines coexist, the high excitation, such as 1 240 NV or 2 440 [NeV], and

Table 7.1. Behaviour of the intensities (in units of $H_\beta = 1.00$) of the strongest emission lines 1 550 CIV and 1 640 HeII as a function of the excitation class of the nebula p

Planetary nebula	p	$N_1 + N_2$	1 550 CIV	1 640 HeII
M 1–1	12	7.6	18.2	7.1
Me 2–1	11	18.6	11.4	5.8
NGC 7 027	10	18.6	31.7	7.7
NGC 3 918	9	21.4	5.1	3.0
NGC 6 905	8	11.5	1.6	1.4
NGC 7 009	7	14.0	0.1	0.1
IC 3 568	3	15.3	0.3	0.1
BD+30°	1	0.07	0	0

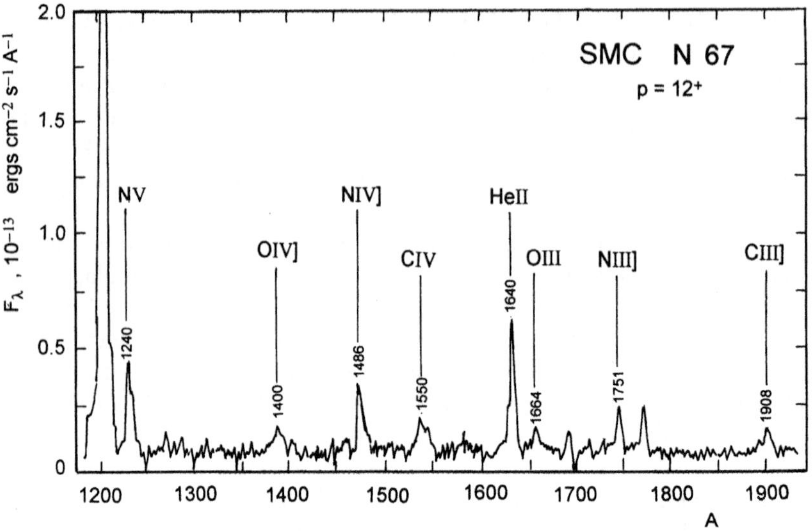

Fig. 7.4. *IUE* recording of the short wavelength spectrum of a planetary nebula, SMC P67, in the Small Magellanic Cloud, in which the emission doublet 1 240 NV is the strongest one in the region 1 200–2 000 Å

low excitation ones, such as the triplet 1 204 [OI] or doublet 2 800 MgII. Such a hybrid occurs, for example, in the nebulae NGC 7 027 and NGC 6 741, indicating the marked stratification of radiation zones within the nebulae.

In Table 7.2, a list of the ten most intense emission lines in the spectra of eight PNs is given including both the ultraviolet and optical regions. The strongest among emission lines is nevertheless the forbidden line N_1, 5 007

Table 7.2. The strongest emission lines in spectra of a sample of planetary nebula intensities are corrected for interstellar extinction

Line λ,Å	NGC 3 242	NGC 3 918	NGC 6 720	NGC 6 741	NGC 6 853	NGC 7 009	NGC 7 027	NGC 7 662
1 550 CIV	0.49	5.14	1.58	3.65	2.11	0.09	31.7	13.92
1 640 HeII	1.62	2.95	4.21	2.75	4.69	1.35	7.7	6.28
1 908 CIII]	2.14	5.40	2.14	7.19	3.69	0.61	22.9	4.23
2 422 [NeIV]	0.14	1.32	0.71	1.85	0.77	0.04	1.4	2.01
3 426 [NeV]	0.03	0.80	0.14	0.54	0.16	0.00	1.25	0.34
3 869 [NeIII]	1.10	1.20	0.94	1.50	1.15	1.16	1.10	0.75
4 861 H_β	1	1	1	1	1	1	1	1
4 959 [OIII]	4.51	5.30	4.42	5.17	3.99	3.93	4.79	3.06
5 007 [OIII]	13.53	15.60	14.09	14.53	12.64	12.66	13.15	9.53
6 563 H_α	2.78	2.46	3.38	2.85	2.77	2.85	2.62	2.84
p	9	9	10	9	10	7	10	10

Table 7.3. Ultraviolet emission line fluxes $F(\lambda)$, corrected for interstellar extinction, and the ratio $F(\lambda)/H_\beta$ for NGC 7 027, according to *IUE* observations (in units of 10^{-10} ergs cm^{-2} s^{-1})

Line	$F(\lambda)$	$\frac{F(\lambda)}{F(H_\lambda)}$	Line	$F(\lambda)$	$\frac{F(\lambda)}{F(H_\lambda)}$
1 406 OIV	0.8	1.78	2 734 HeII	0.9	0.24
1 487 NIV	0.5	0.92	2 783 [MgV]	1.0	0.21
1 549 CIV	20.4	31.7	2 800 MgII	1.6	0.33
1 608 [NeIV]	1.1	1.7	2 836 OIII	0.7	0.13
1 640 HeII	5.2	7.7	2 853 [ArIV]	0.2	0.035
1 663 OIII	0.9	1.3	2 928 [MgV]	0.2	0.029
1 750 NIII	0.8	1.3	3 024 OIII	0.4	0.053
1 909 CIII]	10.8	22.9	3 047 OIII	1.5	0.17
2 321 [OIII]	0.9	2.0	3 133 OIII	8.3	0.77
2 424 [NeIV]	1.3	1.4	3 203 HeII	2.2	0.16
2 470 OII	0.4	0.38	3 345 [NeV]	5.5	0.30
2 511 HeII	0.2	0.15	4 861 H_β	62.5	1.00

[OIII], though eventually it coincides by strength with those of 1 550 CIV and 1 908 CIII].

In Table 7.3 the ultraviolet emission line fluxes corrected for interstellar extinction for NGC 7 027 are given according to the first *IUE* results (Grewing et al., 1978).

7.4 The Doublet 1 550 CIV in Nebulae

The resonance doublet of triply ionized carbon 1 548.2 + 1 550.8 CIV (= 1 550 CIV) belongs to the same isoelectronic sequence as the doublets 2 796 + 2 803 MgII, 3 935 + 3 969 CaII, 1 393 + 1 403 SiIV, etc., i.e. the components of doublets originate in transitions from the nearest levels $^2P^0_{3/2}$ and $^2P^0_{1/2}$ with an excitation energy equal to 8.00 eV, down to the ground state $^2S_{1/2}$ (Fig. 7.5).

The levels $^2P^0_{3/2}$ are not metastable, and therefore, in contrast to the forbidden lines of the doublet 3 726 +3 729 [OII], when the ratio of intensities of components depends in fact only on electron concentration (see Fig. 6.2), in the case of the permitted doublet 1 550 CIV this ratio depends not on electron concentration but on the coefficient $g_i A_{ij}$. For 1 550 CIV we have: $A_{31} = 2.65 \times 10^8$ s^{-1}, $A_{21} = 2.63 \times 10^8$ s^{-1}, $g_3 = 4$, $g_2 = 2$ (Mendoza, 1983). Therefore, for the ratio of intensities of this doublet we have

$$\frac{I(1\,548\ CIV)}{I(1\,951\ CIV)} = 2.008\,.$$

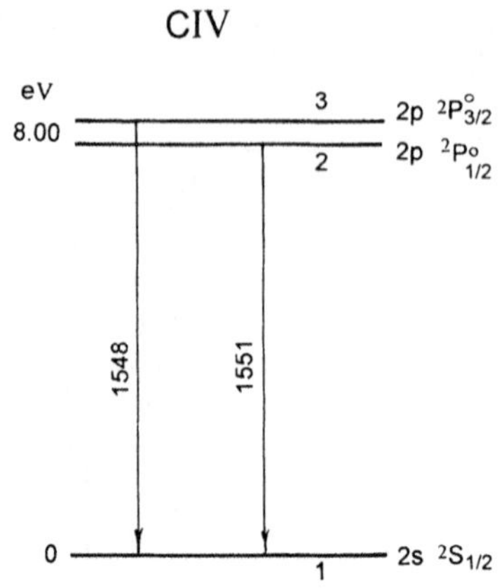

Fig. 7.5. Scheme of energy levels of CIV

Here the possible influence of electron collisions of the second kind is not taken into account. The observations give, however, values as a rule smaller than 2, i.e. within 1.3–1.7. According to some suggestions, this modification is caused by selective interstellar absorption or by circumstellar dust.

So far as the transitions ${}^2P^0_{3/2,1/2}-{}^2S_{1/2}$ are permitted, then in principle the doublet 1 550 CIV can have a fluorescent origin, i.e. after the recombination of C^{++++} with free electrons and the cascade to the ground state is performed ${}^2S_{1/2}$, with the escape of 1 550 CIV photons. However, it can be shown that in conditions of gaseous nebulae the origin of the doublet 1 550 CIV in emission is determined, in fact, purely by inelastic electron collisions of the first kind, i.e. collisional transitions from the ground state ${}^2S_{1/2}$ up to the resonance levels ${}^2P^0_{3/2,1/2}$ with spontaneous transitions and the escape of 1 550 CIV photons.

We can write for the volume emissivity $\varepsilon(\text{CIV})$ in the doublet 1 550 CIV (in ergs cm^{-3} s^{-1})

$$\varepsilon(\text{CIV}) = n_e \text{C}^{+3} q_{12}(T_e) h\nu_{1550} = 4.91 \times 10^{-16} n_e \text{C}^{+3} T_e^{-1/2} \exp\left(-\frac{92\,750}{T_e}\right). \tag{1}$$

As for the next strongest emission line, the resonance line 1 640 HeII, completely of fluorescent origin, we have for its volume emissivity

$$\varepsilon(\text{HeII}) = n_e \text{He}^{++}\, \alpha(\text{He}^{++})\, h\nu_{1640}\,, \tag{2}$$

where $\alpha(\text{He}^{++})$ is the effective recombination coefficient to the level $n = 3$ of the ion He^{++} at the transition $3 \to 2$, which gives the line 1 640 HeII.

With a sufficient approximation, $\alpha(T_e) \sim T_e$. Then one has (e.g. Osterbrock, 1974)

$$\alpha(\text{HeII}) = 11.98 \times 10^{-13}(10/T_e) = 1.198 \times 10^{-8} T_e\,. \tag{3}$$

We can write from (1) and (2) for the ratio of the intensities of the lines 1 550 CIV and 1 640 HeII

$$\frac{I(\mathrm{CIV})}{I(\mathrm{HeII})} = \frac{\varepsilon(\mathrm{CIV})}{\varepsilon(\mathrm{HeII})} = 3.38 \times 10^3 \frac{\mathrm{C}^{+++}}{\mathrm{He}^{++}} T_\mathrm{e}^{1/2} \exp\left(-\frac{92\,750}{T_\mathrm{e}}\right) . \tag{4}$$

An analogous expression can be derived for the ratio of intensities of the emission lines 1 550 CIV and H_β of hydrogen. For the latter we have (3.21). Therefore we get for the ratio $I(\mathrm{CIV})/I(\mathrm{H}_\beta)$

$$\frac{I(\mathrm{CIV})}{I(\mathrm{H}_\beta)} = 3.99 \times 10^5 \frac{\mathrm{C}^{+++}}{\mathrm{He}^{+}} T_\mathrm{e}^{1/2} \exp\left(-\frac{92\,750}{T_\mathrm{e}}\right) . \tag{5}$$

The observed values of the relative intensities of the emission lines 1 550 CIV and 1 640 HeII are known for a number of PNs in our Galaxy as well as in both Magellanic Clouds (e.g. Feibelman et al., 1985; Aller et al, 1987; Clegg et al., 1987 etc.). Note, first, that the intensities of the line 1 550 CIV are within the limits from $\sim$ 30 (NGC 7 027), $\sim$ 20 (M 1–2), up to 10, 5, 2, 1 and even to zero (M 2–9, BD+30°3639). Second, a correlation is discovered between the ratio of intensities 1 550 CIV/1 640 HeII and the intensity of the doublet 1 550 CIV itself, though its interpretation is not quite clear. More interesting, however, is the fact that the declinations of the lines of these relationships are essentially different for Galactic PNs and nebulae in the Magellanic Clouds. For example, for the Galactic PNs we have

$$\frac{I(\mathrm{CIV})}{I(\mathrm{HeII})} = 0.13\, I(\mathrm{CIV}) \tag{6}$$

and for PNs in the Magellanic Clouds

$$\frac{I(\mathrm{CIV})}{I(\mathrm{HeII})} = 2.33\, I(\mathrm{CIV}) , \tag{7}$$

i.e. the contrast in the declination of both correlation lines is about 20 times. Does this reflect a difference between physical properties between these two groups of PNs?

From observed ratios I(CIV)/I(H_β) and with the help of (5), the abundance of $\mathrm{C}^{+++}/\mathrm{H}^{+} \approx \mathrm{C/H}$ was obtained for a sample of PN. In majority of PNs this ratio yields $\sim 10^{-4}$–10^{-5}, i.e. in accord with the universal abundance of carbon in stellar atmospheres and on the Sun. However, in some cases, e.g. NGC 7 027, NGC 6 741 and IC 1 297, the ratio C/H is about $\sim 1.5 \times 10^{-3}$, i.e. higher by an order of magnitude. This anomalous abundance of carbon is one of the mysterious results concerning the chemical composition of PNs.

7.5 Ultraviolet CIII Lines in Nebulae

In the short wavelength PN spectra two lines of doubly ionized carbon have been discovered. One, 1 176 <CIII>, is a pseudoresonance (Chap. 8) and is excited at the transition $^3P -^3 P^0$ (Fig. 7.6), and the second, 1 909 CIII, a resonance but in fact semi-forbidden, is excited at the intercombination transitions $^3P_0$–^1S. The line 1 176 Å represents a sextet with the components within the interval 1 174.93–1 176.37 Å with a summary coefficient of spontaneous transition $A = 4.11 \times 10^9$ s^{-1} (e.g. Loulergue, Nussbaumer, 1974), while 1 909 Å is considered a doublet, 1 908.74 + 1 906.68 CIII, only formally, since the second component ($A = 5.22 \times 10^{-3}$ s^{-1}) is much fainter, up to five orders, than the first one ($A = 96.8$ s^{-1}).

The line 1 909 CIII] has been discovered and measured in the spectra of a number of PNs, and its intensity is comparable and even higher than that of H_β. The pseudoresonance line 1 176¡CIII> has been registered, e.g. for NGC 7 009 and NGC 7 662. The observed ratio of intensities, $E(1\,909)/E(1\,176)$, corrected for interstellar extinction is equal to 2.9 and 7.5 for NGC 7 009 (Perinotto, Benvenuti, 1981) and NGC 7 662 (Harrington et al., 1982), respectively.

Traditionally the ratios of the intensities of various emission lines of gaseous nebulae are being represented through the physical parameters of the nebulae electron temperature and electron concentration. However, in the case of the two above mentioned lines of CIII the results are in strong

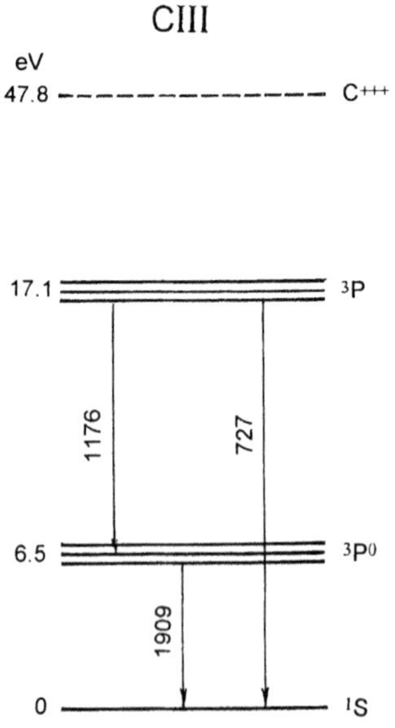

Fig. 7.6. Energy levels of CIII. Transitions at which the pseudoresonance line 1 176 <CIII> and semi-forbidden line 1 909 CIII] are excited are shown

divergence with the observations. This is the so-called *paradox of the CIII lines.*

Denote the levels ^1S, $^3P^0$ and ^3P, respectively, by 1, 2 and 3, and let n_1, n_2 and n_3 be the concentrations of ions C^{+++} on these levels. The basic data concerning the transitions between these levels are as follows:

	λ,Å	ε_{ij}, eV	A_{ij}, s^{-1}	Ω_{ji}
$^3P^0 - ^1S$	1 909	6.46	0.97×10^2	1.09
$^3P - ^3P^0$	1 176	10.64	4.11×10^9	7.89
$^3P - ^1S$	727	17.10	3.83×10^{-2}	0.0113

From the condition of stationarity of transitions for the levels 1 and 2 and on the assumption that all upward transitions are realized only under electron collisions, we get for the ratio of populations in the second and third levels of C^{++}

$$\frac{n_2}{n_3} = \frac{b_{12}}{b_{13}} \frac{A_{32}}{A_{21}} , \tag{8}$$

and then for the ratio of the intensities of lines 1 909 CIII] and 1 176<CIII>

$$\frac{E(1\,909)}{E(1\,176)} = \frac{n_2}{n_3} \frac{A_{21}}{A_{32}} \frac{h\nu_{12}}{h\nu_{23}} = 0.615 \frac{b_{12}}{b_{13}} = 55\, e^{118300/T_e} . \tag{9}$$

At $T_e = 15\,000$ K we find from here $E(1\,909)/E(1\,176) \sim 10^5$ which is at least four (!) orders larger compared with the observations.

Clearly this large discrepancy indicates that the excitation of these lines cannot be caused by a collisional mechanism. Presumably, a certain phenomenon that is not important in the case of other emission lines acquires a special role for C^{++} ions.

Two peculiarities have been noticed. First, the level P with the excitation potential 17.1 eV, and the initial one for the excitation of the line 1 176 <CIII>, is located comparatively not too far from the level of ionization of C^{+++} (47.8 eV). The ratio of both potentials is 2.8. Therefore, one cannot neglect the radiative transitions from the levels ^1S, $^3P^0$ and ^3P directly to the state of ionization, caused by *the action of ionized radiation* of the central star.

The second peculiarity is connected with the fact of the high temperatures of the nuclei of these PNs, higher than 100 000 K. This means that the maximum of their spectra lies at $\sim$ 300 Å. To realize transitions from the states ^1S and ^3P directly to the state of ionization for C^{++}, one needs wavelengths of 259 Å and 300 Å respectively. This chance coincidence of both wavelengths – of the maximum of the nucleus spectra and of the "working" photons for the ionization of C^{++} ions – should lead to an essential increase in the rate of radiative transitions of type $1 \rightarrow 4$, $2 \rightarrow 4$, etc. (4 is the state of ionization of C^{++}), and, hence, the necessity of taking them into account while writing the equation of stationarity of transitions.

As we shall see below, this extraordinary situation combined with individual parameters of atomic transitions provide specific conditions for the excitation of lines 1 909 CIII] and 1 176 <CIII>.

Our aim, therefore, is to obtain the theoretical ratio $E(1\,909)/E(1\,176)$, as a function of the parameters of the nebula and its central star, within the framework of the described scheme of transitions.

The equation of the stationarity of transitions for levels 1 and 2 should be written based on the four-level scheme for C^{++}, so that level 4 is identified with the ionization state C^{+++}. The following processes should be taken into account: (a) transitions $1 \to 2$ and $1 \to 3$, as well as $2 \to 3$, are realized only by electron collisions (b_{ij}); (b) transitions $1 \to 4$ and $2 \to 4$ (ionization) are realized under the action of the radiation of the central star (P_J); (c) transitions $3 \to 1$ and $2 \to 1$ are realized spontaneously (A_{ji}) and by electron collisions of the second kind (a_{ji}); (d) transitions $4 \to 1$ and $4 \to 2$ are identified with the recombination of ions C^{+++} directly at levels 1 and 2 (C_i).

The equations of the stationarity of transitions for levels 1 and 2 under these conditions should be written in the following form:

$$n_1(b_{12} + b_{13} + P_1) = n_2(A_{21} + a_{21}) + n_3(A_{31} + a_{31}) + n_e n^{+++} C_1 \,, \qquad (10)$$

$$n_2(b_{23} + P_2 + A_{21} + a_{21}) = n_1 b_{12} + n_3(A_{32} + a_{32}) + n_e n^{+++} C_2 \,, \qquad (11)$$

where C_1 and C_2 are the coefficients of recombination directly to levels 1 and 2, P_1 and P_2 are the rates of ionization from levels 1 and 2, and are given by the expressions

$$P_1 = W \int_{\nu_1}^{\infty} k_1(\nu) \, \frac{4\pi B_\nu(T_*)}{h\nu} \, d\nu \,, \qquad (12)$$

$$P_2 = W \int_{\nu_2}^{\infty} k_2(\nu) \, \frac{4\pi B_\nu(T_*)}{h\nu} \, d\nu \,. \qquad (13)$$

From (10) and (11) we obtain

$$\frac{n_2}{n_1} = \frac{[(A_{32} + a_{32})(b_{12} + b_{13} + P_1) + b_{12}(A_{31} + a_{31})]}{b_{12}(b_{23} + P_2) + (b_{13} + P_1)(b_{23} + P_2 + A_{21} + a_{21})} + \frac{\dfrac{n_e n^{+++}}{n_3} \, [b_{13}(C_1 + C_2) + C_2(b_{13} + P_1)]}{b_{12}(b_{23} + P_2) + (b_{13} + P_1)(b_{23} + P_2 + A_{21} + a_{21})} \,. \qquad (14)$$

Neglecting higher order terms in (14), we have for the ratio

$$\frac{n_2}{n_3} = \frac{b_{12}}{P_1} \frac{A_{32}}{A_{31}} \,. \qquad (15)$$

Comparing this ratio with (8) we see an essential difference: the rate of the transitions $1 \to 3$ by the collisional mechanism b_{13} is replaced by the rate of transitions P_1 (ionization) from the ground state (1) to ionization (4).

Now, the ratio of intensities of the lines 1 909 CIII and 1 176 <CIII> should be represented in the following form, instead of (9):

$$\frac{E(1\,909)}{E(1\,176)} = \frac{n_2}{n_3}\frac{A_{21}}{A_{32}}\frac{\nu_{12}}{\nu_{23}} = 0.615\,\frac{b_{12}}{P_1}\,. \tag{16}$$

We have for b_{12}

$$b_{12} = 8.73 \times 10^{-6} n_e T_e^{-1/2} \exp\left(-\frac{79\,800}{T_e}\right). \tag{17}$$

We have also for the function of P_1 from (18)

$$P_1(T_*) = WT_*\,\frac{2k}{hc^2}\,k_1\nu_1^2 \exp\left(-\frac{552\,000}{T_*}\right). \tag{18}$$

Substituting (17) and (18) into (16), we finally get (Gurzadyan, 1987)

$$\frac{E(1\,909)}{E(1\,176)} = 5.76 \times 10^{-10}\,\frac{n_e T_e^{-1/2}}{WT_*}\,\exp\left[+\frac{552\,000}{T_*}\left(1 - 0.144\,\frac{T_*}{T_e}\right)\right]. \tag{19}$$

As the calculations show, at $T_* = 100\,000$ K, $T_e = 15\,000$ K, $W = 10^{-13}$ and $n_e = 10^4$ cm^{-3}, the ratio $E(1\,909)/E(1\,176)$ is of the order of 10, i.e. three to four orders smaller than was obtained by Nussbaumer and Schild (1979) when only collisional transitions were considered.

The temperatures of the nuclei of the PNs mentioned above are equally high, $T_* = 100\,000$ K. The electron concentrations are also same for them: $n_e \sim 10^4$ cm^{-3}. Then we obtain with the help of (19) for the ratio $E(1\,909)/E(1\,176)$ depending on T_e:

T_e, K	10 000	12 500	15 000	17 500	20 000
$E(1\,909)/E(1\,176)$	0.51	2.22	5.70	11.7	19.2

Thus, the observed ratios of $E(1\,909)/E(1\,176)$ equal to 2.9 and 7.5 for NGC 7 009 and NGC 7 662 can be explained by electron temperatures of about 13 000 K and 16 400 K, respectively.

Note that in (19) the ratio $E(1\,909)/E(1\,176)$ depends not on n_e and W separately, but on their ratio n_e/W. Adopting, therefore, the ratio $p = n_e/W$ as an independent parameter, we can write, instead of (19):

$$\frac{E(1\,909)}{E(1\,176)} = 5.76 \times 10^{-10}\,\frac{p}{T_*}\,T_e^{-1/2} \exp\left[\frac{552\,000}{T_*}\left(1 - 0.144\,\frac{T_*}{T_e}\right)\right]. \tag{20}$$

Having the ratio $E(1\,909)/E(1\,176)$ for some fixed value of p, say, $p = 10^{16}$ cm^{-3}, one can obtain the ratio for any p with the help of the obvious relation

$$\frac{E(1\,909)}{E(1\,176)} = 10^{-16}\,p\left|\frac{E(1\,909)}{E(1\,176)}\right|_0, \tag{21}$$

where the subscript "0" indicates that this ratio is at $p_0 = 10^{16}$ cm^{-3}.

7.6 The Ultraviolet Doublet 2 800 MgII

The possibility of the excitation of the ultraviolet doublet 2 800 MgII in emission in the conditions of gaseous nebulae was already predicted theoretically already in the early 1960s, long before the epoch of space observations (Code, 1960; Aller, 1961; Gurzadyan, 1965).

The first detection of the doublet 2 800 MgII in emission was performed in 1978, when the high resolution short wavelength spectral recording of PN NGC 7 027 was obtained by *IUE*, with clearly visible k and h lines of 2 800 MgII, along with other emission lines (Fig. 7.5, Table 7.3). Figure 7.3. shows the *IUE* recording of PN IC 418 with the visible 2 800 MgII doublet.

The detection of the doublet 2 800 MgII with nearly the same intensities in the spectra of two representatives of PNs, one, IC 418, of extremely low excitation class ($p = 1$), the other, NGC 7 027, of quite high excitation class ($p = 10$), is a remarkable fact. How can these two extremes be combined?

These facts can be understood if one recalls that in the spectrum of NGC 7 027 the low potential lines 6 300 [OI] and 6 364 [OI] were registered earlier, with strengths only three times weaker than the H_β line. These facts can be explained if we assume that the nebula is enveloped by a dense zone of neutral hydrogen HI, while the forbidden lines of [OI] are excited in an intermediate narrow transition zone between those of HII to HI. Thus the 2 800 MgII also have to be excited within this zone. This model can also explain the existence of forbidden lines of four-times ionized magnesium, 2 783 [MgV] and 2 928 [MgV] – they are excited in the central, strongly ionized HII zone.

In strong contrast to this is the absence of the emission doublet 2 800 MgII in the spectra of PNs, e.g. in the PN NGC 7 662, a nebula of the same excitation class as NGC 7 027. Neither in the first UV recording of NGC 7 662 by *Aerobee* (Bohlin et al., 1978), nor later by *IUE* (e.g. Pena, Torres-Peimbert, 1981; Harrington et al., 1982), were any reliable signs of this doublet detected.

Among dozens of PNs observed by *IUE*, the absence of the doublet 2 800 MgII in emission has been confirmed for more than half of those objects of all excitation classes – high (NGC 2 371, 2 818, 6 302, 6 720), intermediate (NGC 3 242, 5 315, 6 905, 7 009) and low (BD +30°3 639). Apparently, we are dealing with the phenomenon of *depletion* – the condensation of magnesium atoms upon dust particles within the nebula, if the existence of the dust particles in PNs can be considered a confirmed fact (Chap. 10).

The sharpness of the problem can be illustrated with the case of the high excitation nebula NGC 3 242 ($p = 9$). In its spectrum, along with the strong lines of 1 640 HeII and 1 550 CIV, traces of 6 300 [OI] and 6 364 [OI] have been discovered too, as in IC 418, a low excitation PN. Hence, there was reason to expect the existence of the doublet 2 800 MgII in the spectrum of this PN as well. However, it is absent in both *IUE* recordings (Perinotto, Benvenuti, 1981; Barker, 1986). This fact leads to a new and perhaps even more complicated problem, namely, how one can combine the simultaneous existence in its optical spectrum of four emission lines – 4 572 MgI, 4 562 MgI,

Table 7.4. Emission line fluxes (in units of 10^{-12} ergs cm^{-2} s^{-1}) and the ratio $F(\lambda)/F(H_\beta)$, corrected for interstellar extinction, for IC 4 997

Line [λ, Å]	$F(\lambda)$	$\frac{f(\lambda)}{F(H_\beta)}$ [λ, Å]	Line [λ, Å]	$F(\lambda)$	$\frac{f(\lambda)}{F(H_\beta)}$ [λ, Å]
1 240 NV	0.74	0.10	1 909 CIII	21.56:	2.09:
1 335 CII	0.25	0.03	2 321 [OIII]	2.12	0.15
1 305 OI	0.73	0.09	2 470 OII	1.65	0.11
1 550 CIV	3.37	0.34	2 800 MgII	2.67	0.16
1 640 HeII	4.5	0.40	2 836 OIII	0.15	0.01
1 663 OIII	5.35	0.51	2 852 MgI	0.35	0.02
1 750 NIII	2.38	0.21	2 946 HeI	0.32	0.02:
1 817 NeIII	0.23	0.02	3 047 OIII	0.22:	0.01
1 892 SiIII	0.37	0.03	4 861 H_β	32.3	1.00

4 481 MgI and 4 391 MgII, all of subordinate series of neutral and ionized magnesium, with the absence of the strongest – resonance – line, 2 800 MgII.

Clearly in this case the depletion cannot be the only reason for such a situation. This doublet is likely to be an extremely sensitive indicator characterizing some peculiarities of the physical conditions in gaseous nebulae in general.

Table 7.5. Summary of observations of the doublet 2 800 MgII in emission in planetary nebulae (up to 1987). The ratio $F(2\,800\ \mathrm{MgII})/F(H_\beta)$ is corrected for interstellar extinction

Planetary nebula	Extinction class, p	$\frac{F(2\,800\mathrm{MgII})}{F(H_\beta)}$	Reference
NGC 7 027	10	0.33	Grewing et al., 1978b
IC 418	1	0.23	Harrington et.al., 1980
NGC 7 662	10	0.003	Harrington et.al., 1982
NGC 2 440	10	0.04:	Shields et al., 1981
NGC 6 572	7	< 0.01	Torres-Peimbert et.al., 1981
NGC 2 867	9	0.06	Aller et al., 1981b
IC 4 997	8	0.16	Feibelman 1982b
M 1–2	12	1.08	Feibelman 1984
M 2–9	1	1.28	Feibelman 1984
NGC 6 853	10	0.125–0.275	Barker 1984
NGC 6 741	9	0.625	Aller et al., 1986
NGC 3 918	9	< 0.01	Clegg ct al., 1987
NGC 6 302	10	0.09	Aller et al., 1980
NGC 6 826	9	0.16:	Aller, Keyes 1980

The case of IC 4997, a PN of an intermediate excitation class ($p = 8$), is also notable. Its *IUE* recording (Fig. 7.6) and the emission line flux data (Table 7.4, Feibelman, 1982c), the strong lines 1 550 CIV and 1 640 HeII, the absence of the line 1 240 NV – all are unusual for an object of that excitation class. At the same time, the doublet 2 800 MgII in emission is clearly seen with an intensity comparable with that of 1 550 CIV. Moreover, the resonance emission line of neutral magnesium, 2 852 MgI, is detected: I(2 852MgI)/I(2 800 MgII) = 0.13. Note that this PN was noted also for its high density ($n_e \sim 10^6$ cm^{-3}, see Chap. 15), as well as for time variations of the emission line 4 363 [OIII] (Chap. 18).

The emission line of neutral magnesium, 2 852 MgI, is also observed for the PN M 1-2, I(2 852 MgI)/I(2 800 MgII) = 0.44 (Feibelman 1982c); this object is again a superdense one, $\sim 10^7$ cm^{-3} (Chap. 15).

The data on the doublet 2 800 MgII in PN spectra are given in Table 7.5; objects with no evidence of it are twice as numerous.

7.7 The Doublet 2 800 MgII and the Physics of Gaseous Nebulae

What essential information can the doublet k and h MgII bring us with regard to the physics of PNs? What problems of a practical or theoretical character can be solved with the help of this doublet?

The doublet 2 800 MgII, analogous to doublet 1 550 CIV, is excited in the conditions of gaseous nebulae almost exclusively by electron collisions; the contribution of recombinations is negligible.

Analogous to (6), one can write for the volume coefficient of emission in the lines k and h MgII at their collisional excitation

$$I(\text{MgII}) = n_e \text{Mg}^+ \, q_{12}(T_e) \, h\nu_{2\,800} \,, \qquad (22)$$

where $q_{12}(T_e)$ is gives by (3) with the substitution $\Omega/\omega_1 = 8.25$ and $\varepsilon =$ 4.4 eV for Mg^+. Taking into account also (9), we have

$$\frac{I(\text{MgII})}{I(\text{H}_\beta)} = 2.05 \times 10^5 \, \frac{\text{Mg}^+}{\text{H}^+} \, T_e^{1/2} \exp\left(-\frac{51\,400}{T_e}\right) , \qquad (23)$$

where Mg^+/H^+ is the relative abundance of magnesium ions Mg^+ in the PN. Note that (23) for the relative intensity of magnesium emission does not depend on the electron concentration and is determined only by the electron temperature T_e.

Adopting $T_e = 100\,00$ K, we obtain from (23)

$$\frac{\text{Mg}^+}{\text{H}^+} = 0.83 \times 10^{-5} \, \frac{I(\text{MgII})}{I(\text{H}_\beta)} \qquad (24)$$

or

$$\frac{Mg^+}{H^+} \sim (0.08 - 0.8) \times 10^{-5}$$

at $I(\text{MgII})/I(H_\beta) \sim 0.1$–$1.0$ (Table 7.5).

The universal abundance of Mg/H is of the order of $(2-3) \times 10^{-5}$. Then, the observed intensity of the doublet 2 800 MgII in a PN might be explained if only 3–30% of magnesium atoms were in a singly ionized condition, Mg^+. For NGC 7 027 ($T_e = 12\,400$ K) this fraction is even less: $\sim$ 2%.

In contrast to CIV, the third level of Mg^+, $3d\ ^2D$ (Fig. 7.7), is not far from the ground state (8.8 eV), so that the electron collisions can be quite efficient. If so, then it is not difficult to predict the mechanism of emission of subordinate lines 2 790 MgII and 2 798 MgII.

In the theory of luminescence of gaseous nebulae all emission lines are divided into two groups depending on the mechanism of their excitation. The first group, *permitted lines*, are excited by fluorescence. The second group, *forbidden lines*, are excited by electron inelastic collisions. Now the third group of emission lines arises, i.e. *permitted by their nature but forbidden by their excitation.* This last group represents a new category of emission lines for the physics of gaseous nebulae. Although, as it turns out, other resonance lines are also excited by electron inelastic collisions, such as the doublet 1 550 CIV, 1 909 CIII, 2 326 CII etc., the "leadership", however, belongs to the magnesium lines. Indeed, the ionized magnesium has the lowest excitation energy of resonance levels among all resonance ultraviolet lines of ions of various elements.

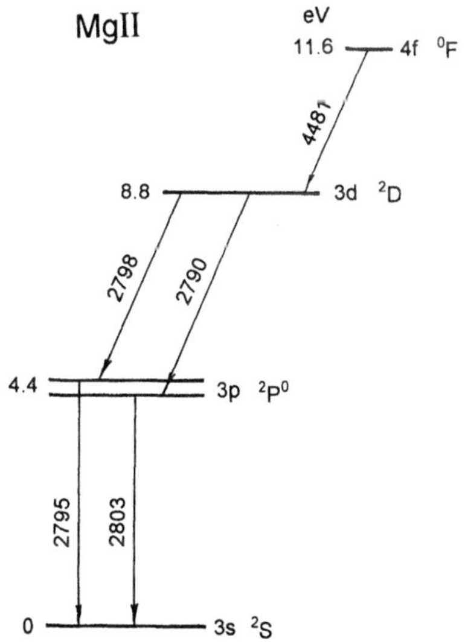

Fig. 7.7. Scheme of energy levels of Mg^+. The most important transitions are shown. Numbers near the levels denote the excitation energy in eV

The singly ionized magnesium has a number of other peculiarities, for example:

- it has no metastable levels;
- it cannot excite forbidden lines;
- it cannot excite pseudoresonance lines (Chap. 8).

Therefore all its "power" is in permitted transitions, and primarily in the doublet k and h MgII.

The serious exploitation of the doublet k and h MgII in the problems of physics of gaseous nebulae is still in its preliminary stage.

7.8 Ionization of Magnesium in Nebulae

In PNs with remarkably high temperatures of the central stars, the fraction of magnesium atoms in high states of ionization – higher than Mg^+ – can be quite large. Therefore, the problem of the distribution of magnesium atoms over various degrees of ionization is of particular interest.

First, one has to derive the formula of magnesium ionization proceeding in the condition of stationarity between the acts of ionization and recombination. As a result, we have for the rth stage of ionization

$$\frac{N^r}{N^{r-1}} n_e = W P_r(T_e) T_* \exp\left(-\frac{\chi_r}{kT_*}\right) , \tag{25}$$

where χ is the ionization potential up to rth stage, and

$$P_r(T_e) = \frac{2k}{hc^2} \frac{\nu_r^2 S_{\nu_r}}{\alpha_r(T_e)} \left(\frac{T_e}{10^4}\right)^{-\eta} , \tag{26}$$

or on substituting the numerical values of constants and taking also $(T_e/10^4)^{-\eta} \approx 1$ at $\eta = 0.28$ and $T_e = 10\,000$ K,

$$P_r(T_e) = 4.6 \times 10^{-11} \frac{\nu_r^2 S_{\nu_r}}{\alpha_r(T_e)} . \tag{27}$$

In these expressions N^r is the concentration of r-fold ionized magnesium; s_{ν_r} is the coefficient of continuous absorption of the ion N^r on the ionization frequency ν_r; $\alpha_r(T_e)$ is the recombination coefficient of ions N^{r+1} with the electron; W is the dilution coefficient. In the derivation of (25), $s_\nu \sim \nu^{-2}$ is assumed.

The next step is to determine the relative abundance of ions in various states of ionization, N^r/N_0, where N_0 is the total concentration of magnesium atoms independent of their degree of ionization.

The starting point for the solution of this problem is the condition

$$N_1 + \sum_1^r N^{+r} = N_0 \,. \tag{28}$$

Dividing by N^+ we get

$$\frac{N_1}{N^+} + \sum_1^r \frac{N^{+r}}{N^+} = \frac{N_0}{N^+} \,, \tag{29}$$

where the values N^{+r}/N^+ are taken from (25), and N_1/N^+ from the following relationship:

$$\frac{N^+}{N_1} n_e = W P_1(T_e) T_* \exp\left(1 \frac{\chi^+}{kT_*}\right) \,, \tag{30}$$

and $P_1(T_e)$ is given by (27) with $r = 1$.

Rewrite (29) explicitly for $r = 4$. In this case the problem is reduced to the solution of a simple algebraic equation for W in the following form:

$$\frac{abcd}{n_e^4} W^4 + \frac{abc}{n_e^3} W^3 + \frac{ab}{n_e^2} W^2 - \frac{a}{n_e} \left(\frac{N_0}{N^+} - 1\right) W + 1 = 0 \,, \tag{31}$$

where

$$a = P_1 T_* \exp\left(-\frac{88\,700}{T_*}\right) \,, \tag{32}$$

$$b = P_2 T_* \exp\left(-\frac{114\,260}{T_*}\right) \,, \tag{33}$$

$$c = P_3 T_* \exp\left(1 - \frac{929\,000}{T_*}\right) \,, \tag{34}$$

$$d = P_4 T_* \exp\left(1 - \frac{1\,267\,400}{T_*}\right) \,. \tag{35}$$

Here, the numerical value of N_0/N^+ is determined directly from (23) by replacing Mg^+/H^+ with the following expression:

$$\frac{Mg^+}{H^0} = \frac{N^+}{H^+} = \frac{N^+}{N_0}\frac{N_0}{H^+} = \frac{N^+}{N_0}\frac{Mg}{H} \,, \tag{36}$$

where $H^+ = H$, and Mg/H is the universal abundance of magnesium. Then we have from (23):

$$\frac{N_0}{N^+} = 2.0 \times 10^5 \left[\frac{I(\mathrm{MgII})}{I(\mathrm{H}_\beta)}\right]^{-1} \frac{\mathrm{Mg}}{\mathrm{H}} T_e^{1/2} \exp\left(-\frac{51\,400}{T_e}\right) \,, \tag{37}$$

where the ratio

$$\frac{I(\mathrm{MgII})}{I(\mathrm{H}_\beta)} = \frac{I(2\,800\,\mathrm{MgII})}{I(\mathrm{H}_\beta)} \tag{38}$$

is taken from observations, while the universal abundance Mg/H is a known magnitude.

Table 7.6. Atomic parameters of Mg^{+r}

Ion	χ_r [eV]	$s_{\nu r,}$ [10^{-18}cm^2]	$\alpha_r(10^4)$, [10^{-13}cm^{3}s^{-1}]	η	$p_r(10^4)$, [10^{15}cm^3 deg^{-1}]
Mg^+	7.65	2.50	1.4	0.855	2.81
Mg^{++}	15.03	0.31	8.8	0.838	0.21
Mg^{+++}	80.14	5.2	35.0	0.734	25.7
Mg^{++++}	109.31	3.8	77.0	0.718	15.9
Mg^{+++++}	141.27	2.5	140.0	0.716	9.6

The described method of analysis of the ionization state of magnesium atoms is applied to PNs of widely varying excitation classes: NGC 6741 ($p = 9$), NGC 7027 ($p = 10$), IC 418 ($P = 1$), IC 4997 ($p = 8$), and M 1–2 ($p = 12$); the last two being superdense. The initial data, i.e. atomic parameters for Mg^{+r} ions, are collected in Table 7.6 (e.g. Barfield et al., 1972; Shevelko, 1973; Aldrovandi, Pequignot, 1973) together with the numerical values of the function $P_r(T_e)$ computed with the help of (27); $P_r(T_e)$ is shown also in graphical form in Fig. 7.8.

From (31) with the known coefficients a, b, c and d we obtain W, after which from (25) we get the ratios, N^{++}/N_0, N^{+++}/N_0, etc.

The initial data, T_e, T_* and n_e, for PNs under examination are collected in Table 7.7 along with the results of calculations, i.e. magnitudes of N^{+r}/N_0 and N_1/N_0; the former are obtained from the following relationships (for $r = 2$, 3 and 4):

$$\frac{N^{++}}{N_0} = W \frac{b}{n_e} \frac{N^+}{N_0} , \tag{39}$$

$$\frac{N^{+++}}{N_0} = W^2 \frac{bc}{n_e^2} \frac{N^+}{N_0} , \tag{40}$$

Table 7.7. Distribution of magnesium atoms by various states of ionization N^{+r} ($r = 1$, 2, 3 and 4) for five planetary nebulae

Planetary nebula	T_* [K]	T_e [K]	n_e [10^4cm^{-3}]	$\frac{N_1}{N_0}$	$\frac{N^+}{N_0}$	$\frac{N^{+2}}{N_0}$	$\frac{N^{+3}}{N_0}$	$\frac{N^{+4}}{N_0}$
NGC 6741	80 000	11 600	1.0	0.00012	0.054	20.908	0.031	0.00021
NGC 7027	170 000	12 400	8.0	0.00022	0.021	0.103	0.589	0.308
IC 418	35 000	9 700	1.4	0.00002	0.052	0.941	–	–
IC 4997	73 000	15 700	100	0.000005	0.013	0.787	0.200	0.00027
M 1–2	75 000	15 000	800	0.000002	0.030	0.844	0.124	0.00012

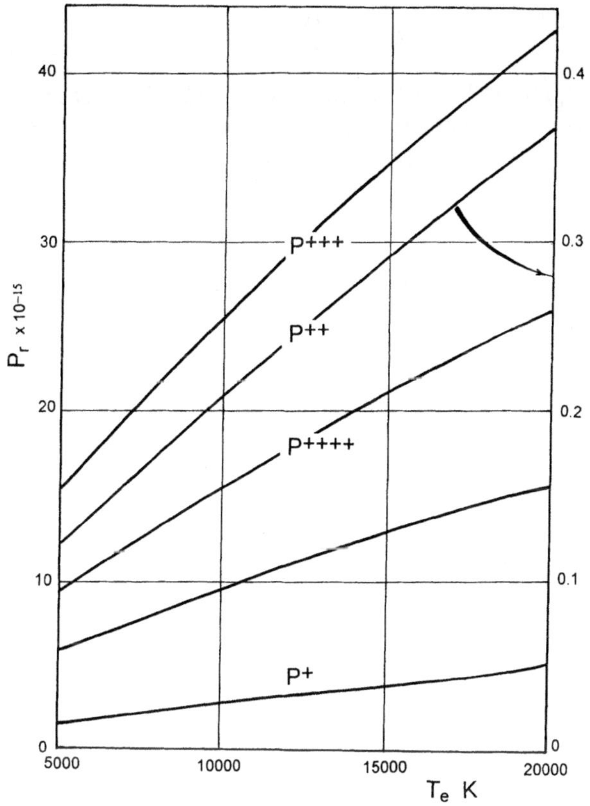

Fig. 7.8. Dependence of the function $P_r(T_e)$ on the electron temperature T_e. For $P_2(T_e)$ the vertical scale is to the right; in the remaining cases, it is to the left

$$\frac{N^{++++}}{N_0} = W^3 \frac{bcd}{n_e^3} \frac{N^+}{N_0}, \tag{41}$$

the latter, from

$$\frac{N_1}{N_0} = \frac{n_e}{aW} \frac{N^+}{N_0}, \tag{42}$$

where N^+/N_0 is given by (37). The values of $I(\mathrm{MgII})/I(\mathrm{H}_\beta)$ for these nebulae are taken from Table 7.7, and $\mathrm{Mg/H} = 2 \times 10^{-5}$ is adopted.

The values of W obtained from equations of type (31) for each nebula are presented in Table 7.8, along with the radius R_* of the nebula's central star calculated by means of the formula

$$R_* = 6.57 \times 10^8 \, W^{1/2} F^{1/3} \left(\frac{\tau}{n_e}\right)^{2/3}. \tag{43}$$

This relation is derived from (8.59), and the obvious relationship for the nebula's mass

$$\mathcal{M}_n = \frac{4\pi}{3} R_n^3 n_e m_H , \tag{44}$$

if we adopt $(T_e/10^4)^{-0.28} \approx 1$ for all nebulae. The distances r of the nebulae are given in the sixth column of Table 7.8, and the fluxes $F(H_\beta)$ in H_β were as follows (in units 10^{-11}ergs cm^{-2}s^{-1}):

	NGC 6741	NGC 7027	IC 418	IC 4997	M 1-2
$F(H_\beta)$	0.41	6.76	25.7	3.3	0.12

In the fourth column of Table 7.8 the radii of the nebula R_n, calculated from the relationship

$$R_n = 0.5\, R_* \, W^{-1/2}$$

are given. Finally, in the fifth column the calculated masses of the nebulae $\mathcal{M}_n$, obtained using (44), are given.

Quite a large value for W is obtained in the case of the star-like superdense nebula M 1-2, $W \approx 10^{-10}$. This yields an unlikely small radius for the nebula: $R_n = 3.4 \times 10^{14}$ cm $= 23$ a.u. However, the value of the nuclear radius obtained is $R_* = 0.10.\ R_\odot$ and the nebula's mass $\mathcal{M}_n \approx 10^{-4}\,\mathcal{M}_\odot$ seem quite reasonable.

Having N_1/N_0 (Table 7.7), one can obtain the optical thickness of the nebula t(MgII) in the line 2800 MgII:

$$t(\text{MgII}) = k_0 R_n N_1 = 2 \times 10^{-5} k_0 R_n n_e \frac{N_1}{N_0} , \tag{45}$$

where k_0 is the absorption coefficient for the ion Mg^+ in the line 2800 MgII. For this line $A_{21} = 2.67 \times 10^8$ s^{-1}, and then $k_0 = 3.53 \times 10^{-13}$ cm^2. The calculated values of t(MgII) obtained with the help of (45) are presented in the last column of Table 7.8.

As we can see, PNs are not always transparent in the line 2800 MgII, which, of course, will cause new problems, in particular, the necessity of taking into account the self-absorption when determining the true flux of emission in this line.

Table 7.8. Dilution coefficient W, radii of central star R_* and nebula R_n, nebula's mass $\mathcal{M}_n$ and optical depth t_*, obtained by means of the emission line 2800 MgII for a sample of planetary nebulae with known distances r_n

Planetary nebula	W	$R_*/R_\odot$	R_n [cm]	$\mathcal{M}_n/\mathcal{M}_\odot$	r_n [kpc]	t(MgII)
NGC 6741	7.75×10^{-14}	0.23	2.9×10^{16}	0.0008	1.25	0.25
NGC 7027	2.30×10^{-14}	0.06	1.3×10^{16}	0.0005	0.70	1.60
IC 418	2.50×10^{-12}	1.17	4.1×10^{16}	0.0034	0.41	0.10
IC 4997	2.94×10^{-11}	0.44	2.8×10^{15}	0.0001	1.20	0.02
M 1-2	1.05×10^{-10}	0.10	3.4×10^{14}	0.0001	1.90	0.48

These results indicate that in PN conditions less than 10% of the total number of magnesium atoms determine the emission doublet 2 800 MgII (N^+/N_0 in Table 7.7). The only *secret* is the excitation of these non-forbidden magnesium lines by electron inelastic collisions.

In PNs of both low (IC 418) and high (NGC 6 741, M 1-2) excitation classes, the largest number of magnesium ions are in the doubly ionized state, N^{++}/N_0, while in the case of NGC 7 027 this maximum corresponds to the third and fourth ionization states (Table 7.7). Hence, the possibility of observing of lines of [MgIV] and [MgV] in the spectra of this nebula might be expected. Indeed, two such lines were discovered in the far infrared – 4.5 μm [MgIV] and 5.6 μm [MgV] (Russell et al., 1977), another two lines in the ultraviolet – 2 783 [MgV] and 2 928 [MgV], and one line in the visible region – 3 488 [MgIV].

Thus we have briefly summarized the main aspects of the role of magnesium atoms in revealing the structure of PNs. However, another approach is also possible, and this is the number of interesting attempts to construct quantitative models in the case of any given object (e.g. for NGC 7 027, NGC 7 662, etc.).

7.9 Determination of Dimensions of Stellar Envelopes

At the not too high temperatures of the central star, i.e. at $T_* < 100\,000$ K and N_0/N^+, the dominating role in (31) is played by the third and fourth terms. Substituting into (31) the value of b from (33) and $P_2 = 0.21 \times 10^{15}$ cm^3 deg^{-1} from Table 7.6, we have

$$W = 4.76 \times 10^{-15} \frac{N_0}{N^+} \frac{n_e}{T_*} \left(\frac{T_e}{10^4}\right)^{-0.84} \exp\left(\frac{114\,260}{T_*}\right) \qquad (46)$$

This expression can be used for determining the parameter W for a star with a gaseous envelope, in the spectra of which the doublet 2 800 MgII is discovered. As for the ratio of N_0/N^+, it can be determined with the help of (37), according to the observed value of $I(2\,800\text{MgII})/I(\text{H}_\beta)$.

Having the W, we can obtain the radius of the gaseous envelope R_n from the following expression:

$$R_n = 0.5\, R_*\, W^{-1/2}\,, \qquad (47)$$

assuming that the radius of the star R_* can be estimated by an alternative method.

As a matter of fact, no general powerful methods exist for the determination of the linear dimensions of gaseous envelopes around the stars. The method just described and based on the exploitation of the observed intensity of the emission line 2 800 MgII, can be a rather efficient outcome of the situation.

7.10 The Problem of Calcium Lines

The emission lines K (3 933 Å) and H (3 963 Å) of singly ionized calcium, CaII, located in the visible region of spectra, originate as a result of the same transitions as the doublet 2 800 MgII. And although the universal abundance of calcium is an order, or more precisely, 18 times less compared with the magnesium abundance, the emission lines K and H CaII were expected to appear together with the *k* and *h* MgII. Meanwhile, in ground based conditions the doublet K and H CaII has never been observed.

One can think that the excitation of the *k* and *h* MgII doublet in PNs for some reason cannot be accompanied with the excitation of K and H CaII, both doublets being in emission. Note that in the usual stellar spectra the doublet 2 800 MgII *in absorption* and of photospheric origin is observed constantly along with the doublet 3 933 + 3 963 CaII also *in absorption*, and, which is important, both with the intensities corresponding to the universal abundance of these elements. Our aim is to remove this controversy (Gurzadyan, 1979).

In PNs both doublets K and H CaII and *k* and *h* MgII should be excited by the same mechanism – electron collisions. Moreover, this mechanism must work more effectively for Ca^+ ions in spite of fact that its energy of resonance level is noticeably lower, 3.16 eV, compared with 4.42 eV in the case of Mg^+ ions.

The reason for this discrepancy is as follows.

The ionization potentials of calcium in the second and third states of ionization are significantly lower than the ionization potentials of magnesium in the same ionization states, as follows from the following data (in eV):

	χ^+	χ^{++}	χ^{+++}
Mg	7.65	15.03	80.14
Ca	6.11	11.87	50.19

Clearly, in spite of this fact practically all calcium atoms must be accumulated in two, three and more times ionized states, which leads to a sharp reduction in the number of singly ionized calcium atoms. The distribution of the relative number of atoms by ionization states is seen in the following form (at $T_* = 50\,000$ K, $n_e = 10^3$ cm^{-3}, $W = 5 \times 10^{-13}$):

	n_1/n_0	n^+/n_0	n^{++}/n_0
Mg	2×10^{-4}	0.10	0.90
Ca	6×10^{-8}	0.0004	1

As we see, the fraction of Ca^+ ions is 250 times smaller than the fraction of Mg^+ ions. Hence, under the same conditions the intensity of CaII must be 250 times smaller compared with the intensity of MgII. In fact, however, the calcium is 20 times less abundant than magnesium. As a result, the K and H CaII lines must be four orders of magnitude weaker than the *k* and *h*

MgII lines. In such conditions, the detection of emission lines of singly ionized calcium in PNs really turns out to be a hard problem.

This conclusion is supported by observational data. For example, in the spectrum of seven PNs (NGC 7027, NGC 7662, NGC 2022 etc.) only the weak forbidden line of four-times ionized calcium, 5309 [CaV], and, rarely, 6087 [CaV] and 5615 [CaVII], has been discovered.

7.11 The Short Wavelength Radiation of Nuclei

Quite often, while solving certain problems of the physics of PNs, the short wavelength radiation of their nuclei is approximated by the Planckian law of a given effective temperature.

However, there are a number of problems which require knowledge of the true distribution of energy, especially in the far ultraviolet, i.e. at the ionization frequencies of many elements. Among these problems one can mention:

(a) Comparison of observational intensities of ultraviolet lines with the theory.
(b) Creation of the improved scale of Zanstra temperatures for the nuclei of PNs.
(c) Formation of ionization stratification in the nebula.
(d) Determination of the exact values of bolometric corrections.

In the Planckian approximation, the radiation capability of the nucleus in the ultraviolet is characterized by a single parameter, the effective temperature T_*. The actual radiation is determined at least by two parameters – the effective temperature T_* and the surface gravity g. The chemical composition, CC, of the upper layers of the photosphere is usually considered the third parameter. Thus, the numerous possible combinations between T_*, g and CC, define as many models of central stars and nebulae.

As an example, in Figs. 7.9 and 7.10 four calculated models for the central stars of PNs are given for two values of T_*: 40 000 K and 100 000 K and two values of $\log g$: 4.0 and 5.0 in the first case and 5.0 and 6.0 in the second (Bohm, 1968; Hummer, Mihalas, 1970b). In these curves, the boundaries of the jump-like variations of the non-transparency of the star's atmosphere, are indicated. The Planckian curves (B) are also shown for the same values of the temperatures.

As we see, the discrepancy from Planckian curves is rather significant, though it is different at various wavelengths.

In general, the model fluxes in the far ultraviolet turn out to be systematically smaller than the Planckian ones. This is the main conclusion following from model calculations.

Consider a nebula that is optically thick at the frequencies of the Lyman continuum. The condition of equality of the number of photons which ionize

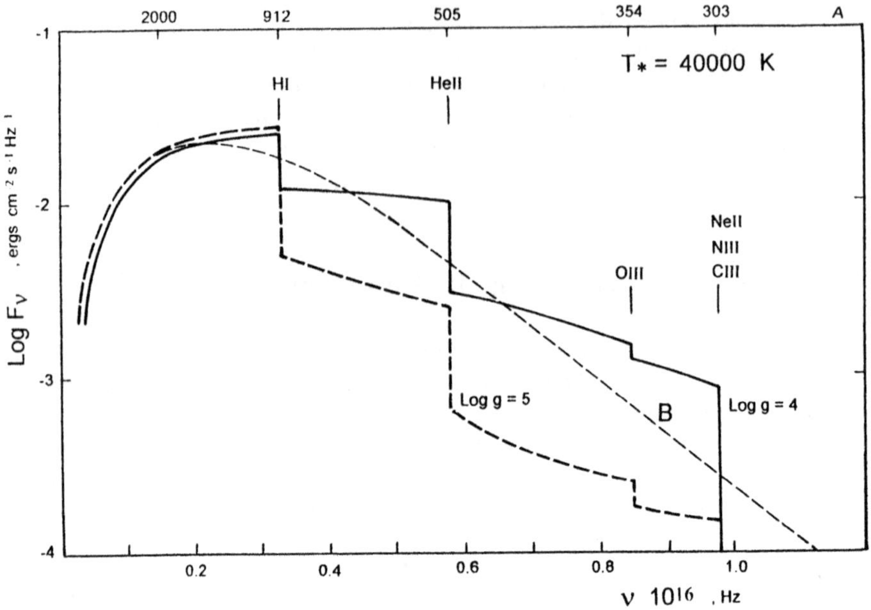

Fig. 7.9. Calculated model of the short wavelength spectrum for the central star of a planetary nebula at the effective temperature $T_* = 40\,000$ K and for two values of $\log g = 4.0$ and 5.0

the hydrogen being emitted by the nucleus per unit time, and the number of recombinations at all levels, up to $i = 1$, can be written in the form

$$\int_{\nu_0}^{\infty} \frac{L_\nu(T_*)}{h\nu}\,\mathrm{d}\nu = \int N^+ N_\mathrm{e} \sum_1^{\infty} C_i(T_\mathrm{e})\,\mathrm{d}V\,, \tag{48}$$

where $L_\nu(T_*)$ is the model luminosity of a nucleus with an effective temperature T_*, and the integration is carried out over the whole volume of the nebula.

The luminosity of a nebula in the hydrogen H_β line also depends on the number of recombinations, though up to the level $i = 3$. Therefore we have

$$\frac{L(\mathrm{H}_\beta)}{h\nu_\beta} = \int N^+ N_\mathrm{e} \sum_3^{\infty} C(T_\mathrm{e})\mathrm{d}V. \tag{49}$$

From (48) and (49) we have for a homogeneous nebula

$$\int_{\nu_0}^{\infty} \frac{L_\nu(T_*)}{h\nu}\,\mathrm{d}\nu = \frac{L(\mathrm{H}_\beta)}{h\nu_\beta}\,\frac{\sum_1 C_i(T_\mathrm{e})}{\sum_3 C_i(T_\mathrm{e})}\,. \tag{50}$$

In the same manner we can write for the case of Planckian radiation of the central star of the Zanstra temperature T_z

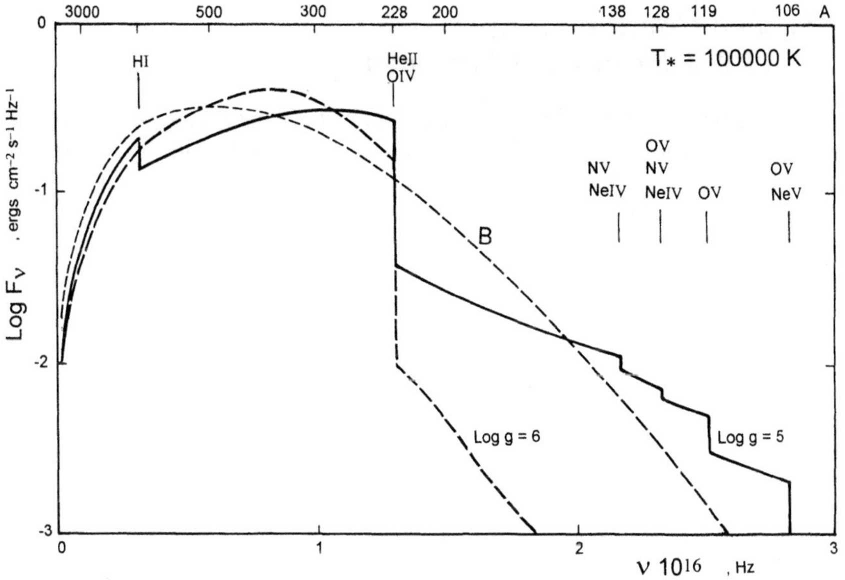

Fig. 7.10. The same as in Fig. 7.9 but for $T_* = 100\,000$ K and $\log g = 5.0$ and 6.0

$$\int_{\nu_0}^{\infty} \frac{B_\nu(T_z)}{h\nu} \mathrm{d}\nu = \frac{L(\mathrm{H}_\beta)}{h\nu_\beta} \frac{\sum_1 C_i(T_e)}{\sum_3 C_i(T_e)} . \tag{51}$$

In Table 7.9, the values of T_* and T_z, obtained in the described way, are presented for nuclei of a sample of PNs (Capriotti, Kovach, 1968). As we can see, the model atmosphere temperatures are somewhat higher than the Planckian ones.

Table 7.9. Temperatures of planetary nebula central stars obtained using the Zanstra method for two cases: the model atmosphere T_* and Planckian radiation T_z

Planetary nebula	Planck T_z	Model atm. T_*	Planetary nebula	Planck T_z	Model atm. T_*
NGC 2 392	68 000	84 000	NGC 6 826	69 000	85 000
NGC 3 242	93 000	109 000	NGC 6 853	132 000	148 000
NGC 3 587	105 000	123 000	NGC 7 009	81 000	98 000
NGC 6 543	66 000	82 000	NGC 7 662	100 000	118 000
NGC 6 572	61 000	77 000	IC 351	91 000	108 000

7.12 The Continuous Spectra of Nuclei in the Ultraviolet

Claiming the exceptional importance and informativeness of spectral observations of the nuclei of PNs in the far and near ultraviolet, one should mention that direct photometry of the same nuclei could also be rather useful for the solution of a number of problems, in particular:

(a) To obtain information on the interstellar extinction in the direction of a given nebula.
(b) To separate the continuous emission of the nebula in the ultraviolet region.
(c) To measure the fluxes in the most powerful emission lines generated in the nebula, in particular, 1 550 CIV and 1 640 HeII.
(d) To collect information on the state of the atmosphere of the central star.

The solution of these problems with the help of photometers of even moderate spectral resolution has an obvious advantage: the possibility of covering too faint objects during space observations that are inaccessible for spectral investigations. As an example we recall the space observatory *ANS* (The Netherlands).

Historically *ANS* was the first to carry out photometric measurements with the transmission band – 150 Å, not too narrow and not too wide – in five channels, centred on the wavelengths 1 500, 1 800, 2 200, 2 500 and 3 300 Å. (Van Duinen et al., 1975). Over 30 PNs of different excitation classes have been observed, both the nebula and its nucleus. By comparison of observed continuous spectra with various Planckian curves, the searched for nuclear temperatures T_* were determined; results for ten PNs are given in Table 7.10 along with temperatures obtained by the HeII/H_β method (Chap. 6). As we can see, for nearly all PNs the temperatures are smaller than T_*[4 686

Table 7.10. Temperatures of the nuclei of planetary nebulae obtained by wide photometric observations and by the 4 686 HeII/H_β method

Planetary nebula	Excit. class, p	T_*, K *ANS*	T_*, K HeII/H_β	τ_c
A 33	11	75 000	224 000	0.058
A 43	12	140 000	250 000	0.89
A 78	12	140 000	562 000	0.42
NGC 246	12^+	100 000	367 000	0.14
NGC 1 360	12	100 000	263 000	0.20
NGC 1 535	8	45 000	115 000	0.002
NGC 2 392	11	36 000	188 000	–
NGC 3 587	8	95 000	117 000	1.52
NGC 6 210	4	38 000	76 700	–
NGC 7 293	7	95 000	87 100	–

HeII/H_β]; hence for these nebulae $\tau_c < 1$. If so, then with the help of (5.19), we can determine the value of τ_c for these nebulae; the last column of Table 7.10.

Analogous observations of the nuclei of seven PNs, mainly of low excitation, have been carried out with the help of the ultraviolet spectrometer S 2/68 (*TD–1*) in the wavelength region 1 400–2 740 Å and with a spectral resolution of 40 Å. These nebulae are NGC 246, 1 360, 1 535, 6 543, 6 826, as well as IC 418 and IC 4 593 (Lutz, Carnochan, 1979). The results of these early space observations besides being of historical content are still of interest.

7.13 Nuclei Temperatures According to Their Ultraviolet Spectra

All Zanstra's methods for determining temperatures of the nuclei of PNs have a common defect: they all require knowledge of the crucial parameter – the optical depth τ_c of the nebula at the frequencies of the Lyman continuum of hydrogen. The determination of τ_c itself requires knowledge of the ionization state within the nebula and, hence, of the precise nucleus temperature T_*.

To exit from this closed circle, one needs direct and independent ways to determine T_*. This problem can be solved uniquely by direct spectral recordings of the nucleus in the region of 1 000–2 000 Å. By comparison of the obtained distribution of energy in the continuous spectrum of the nucleus with the model calculations or, at least with the Planckian law, one can obtain the desired temperature T_*.

This method seems to be the one already well applied in the optical region. The problem is, however, that at very high temperatures the Planckian curves in the optical region for various T_* hardly differ from each other.

The situation changes sharply when we move to the far ultraviolet; here the sensitivity of the gradient of the continuous spectra on the star's temperature is quite strong.

In Fig. 7.11, two groups of energy distribution curves computed by Planck's formula at different T_*, above 30 000 K, are given. In both cases, the dependence of the declinations of the curves on the temperature in the optical region, 3 000–5 000 Å (curves are normalized at $\lambda = 5\,000$ Å), is extremely weak. At the same time, these curves sharply deviate in the far ultraviolet, 1 000–2 000 Å.

As an illustration, the *IUE* data of the nuclei of NGC 2 371 and NGC 7 009 are plotted. For NGC 2 371 the fluxes are in frequency scale (Pottasch et al., 1981), for NGC 7 009 in wavelength scale (Perinotto, Benvenuti, 1981). As we see, in both cases observational points are near the curves with $T_* =$ 100 000 K in the region shorter than 1 300 Å, even above 10^5 K. A similar result is obtained also for NGC 7 662, Fig. 7.12 (Harrington et al., 1982).

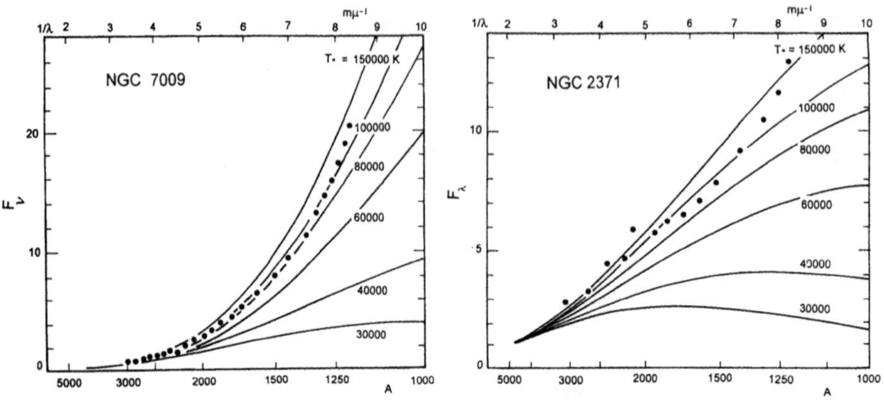

Fig. 7.11. Energy distribution curves in the spectrum of black-body radiation in the wavelength interval 1 000–5 000 Å and at temperatures from 30 000 K up to 150 000 K. Dots denote *IUE* observations of the nuclei of the planetary nebulae NGC 2 371 and NGC 7 009, corrected for interstellar extinction ($c = 1.18$ and $c = 0.09$, respectively). The curves are normalized at 5 000 Å. The intensities are in the frequency scale (NGC 2 371) and in the wavelength scale (NGC 7 009)

Thus for these three nebulae we have:

	$T_*(ANS)$	$T_*(\mathrm{HeII}/\mathrm{H}_\beta)$	$T_*[\mathrm{OIII}]/[\mathrm{OII}]$
NGC 2 371	100 000 K	250 000 K	123 000 K
NGC 7 009	100 000	107 000	60 200
NGC 7 662	100 000	160 000	84 000

Recall that the HeII method gives the upper limit, while the [OIII]/ [OII] method gives the lower limit of T_*.

As we can see, the application of the method just described for the determination of nuclei temperatures with the help of their short wavelength spectral recordings yields quite high values – higher than 100 000 K. To apply this method one needs the short wavelength recordings of the central star, which is too hard for faint objects. Therefore the indirect methods of determining nuclei temperatures should be preserved along with the direct ones.

7.14 Nuclei with Superhigh Temperatures

In Sect. 5.8, we had concluded that PNs with extremely high nuclei temperatures – 200 000 K and higher, up to 500 000 K – exist. Is it possible to support this conclusion on the basis of spectral observations of the nuclei in the far ultraviolet?

At temperatures above 100 000 K the distribution of energy in the optical region is insensitive to temperature variations; here at the Rayleigh–Jeans spectrum the ratio of intensities at two points, λ_1 and λ_2, does not depend

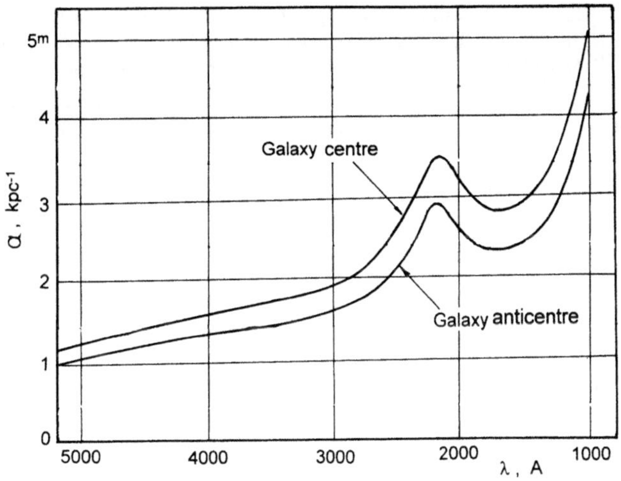

Fig. 7.12. *IUE* recording of the spectrum of the central star of the planetary nebula NGC 7662 in the region 1 100–3 100 Å corrected for interstellar extinction ($c = 0.23$). The black-body curves are for effective temperatures 100 000 K, 150 000 K and $T = \infty$

on the temperature, while in the region shorter than 2 000 Å up to 1 000 Å the ratio of the intensities at two points of continuous spectra can be easily determined.

The probability of discovering the nuclei with superhigh temperatures must be higher in the case of a very large PN (in terms of its absolute dimensions), e.g. from Abell's list, which can hardly be well illuminated by a low temperature star.

Therefore, if we can receive the recording of the continuous spectra of such a nebula for the nucleus of such a nebula in the region of the far ultraviolet, then there is a chance of revealing among them nuclei with superhigh temperatures.

The *IUE* observations of 32 PNs with radii larger than 0.2 pc (Kaler, Feibelman, 1985), performed in the region 1 200–2 000 Å, reveal the Planckian temperatures T_* of nuclei obtained by the ratio of intensities $\lambda 1\,300/\lambda 1\,750$ and $\lambda 1\,500/\lambda 1\,800$ Å. For at least 10 PNs from 32, the nuclear temperatures are greater than 200 000 K; Table 7.11.

Note the remarkable accordance in the temperatures obtained by various methods. Three of these PNs are indeed of giant dimensions – over 1 pc ($\sim 3 \times 10^{18}$ cm). The detected strong emission line 4 686 HeII again supports the concept of their hot nuclei.

Table 7.11. Planetary nebulae with superhigh nuclei temperatures obtained from *IUE* observations, T_* (*IUE*). For comparison, the highest, T_* (HeII/HI), and the lowest, T_* ([OIII]/[OII] temperatures are also given. R_n is the nebula's radius

Planetary nebula	R_n [pc]	$T_*(IUE)$ [K]	T_* (HeII/HI) [K]	T_*([OIII]/[O II] [K]
NGC 246	0.22	430 000	367 000	243 000
NGC 6 058	0.26	>320 000	209 000	110 000
A 20	0.46	>280 000	457 000	360 000
A 31	1.20	270 000	–	–
A 33	1.04	>220 000	221 000	126 000
A 34	1.02	>330 000	–	–
A 36	0.70	270 000	316 000	–
A 51	0.44	>260 000	243 000	160 000
K 1-16	0.33	410 000	343 000	320 000
K 1-27	–	270 000	–	–

7.15 X-Rays from Planetary Nebulae

Planetary nebulae were observed in X-rays by the space observatory *Einstein* (Appararo, Tarafder, 1988; Brown et al, 1994) and later by *EXOSAT* (Appararo, Tarafder, 1989), and *ROSAT* (Kreysing et al., 1992). For many others the upper limits of their X-ray fluxes have been obtained, including IC 4 997, NGC 3 242, NGC 6 905, etc. (Leahy et al., 1994). The *Einstein* observatory also detected ultrasoft X-ray sources in Magellanic Clouds, in particular, in two PNs – P67 and P22 (Brown et al., 1994). The list of the nine PNs mentioned is given in Table 7.12. For NGC 246, the X-ray point-like source coincides with the central star of this nebula.

The X-ray emission in these nuclei can be a result of the following two processes: (a) thermal emission with effective temperature of nuclei; (b) coronal emission. As we shall see below, in the case of the observed PNs the first mechanism seems to be the dominant.

In the case of the thermal X-ray radiation the following expression can be derived for the observed X-ray flux F_X as a function of "X-ray" temperature T_X:

$$F_X = 1.286 \times 10^{-7}\, 10^{-0.4 m_V}\, T_X \left(e^{26\,160/T_X} - 1\right) (x^3 + 3x^2 + 6x + 6)^{-1}\,, \quad (52)$$

where $x = 1\,740\,800/T_X$ corresponds to the energy of X-ray photons, 0.15 keV, and m_V is the visual magnitude of nucleus. The values of T_X obtained with the help of this formula are given in the third column of Table 7.12. They are close to the values of T_*(HeII) obtained by the HeII method. With known distances D_n of PNs (fourth column), the X-ray luminosity, L_X, and the bolometric luminosity, L_{bol}, were obtained (fifth and sixth columns).

Table 7.12. X-ray radiation of planetary nebulae. D_n is the nebula's distance in pc. The observed X-ray flux, F_X, is in units of 10^{-13} ergs cm^{-2} s^{-1}

Planetary nebula	F_X	T_X [K]	D_n [pc]	L_X [ergs s^{-1}]	L_{bol} [ergs s^{-1}]	$\frac{L_X}{L_{bol}}$
NGC 246	31.4	86 000	370	4.2×10^{34}	2.3×10^{36}	1.8×10^{-2}
NGC 1 360	38.0	92 500	260	1.1×10^{35}	6.0×10^{35}	2.0×10^{-1}
NGC 1 535	3.2	79 500	1 700	8.0×10^{35}	1.9×10^{37}	4.2×10^{-2}
NGC 3 587	2.7	68 000	570	6.3×10^{32}	1.4×10^{35}	4.5×10^{-3}
NGC 4 361	3.9	116 000	940	1.1×10^{36}	5.5×10^{36}	0.2×10^{-1}
NGC 6 853	7.5*	92 000	400	1.4×10^{34}	3.8×10^{35}	3.7×10^{-2}
NGC 7 293	2.7*	81 000	116	1.1×10^{30}	1.3×10^{35}	1.0×10^{-5}
A 33	1.3*	93 000	580	4.9×10^{30}	1.9×10^{35}	3.8×10^{-5}
A 36	1.5	93 000	312	1.3×10^{35}	1.9×10^{36}	8.7×10^{-2}

*Observations by *Einstein*, interval 3.5–62 Å. The rest observations are by *EXOSAT*, interval 6–250 Å.

The PN nuclei should possess a corona which itself can be a source of X-ray emission via thermal bremsstrahlung. Therefore, the observed X-ray fluxes, F_X, and X-ray luminosities, L_X, in Table 7.12 should be represented as the sum of two components: X-ray Planckian radiation of the star's photosphere at the effective temperature T_{eff} and X-ray emission of the corona. In all cases, the possibility of the two-temperature X-ray emission from NGC 7 293 was discussed as well (Leahy et al., 1994).

Extremely interesting are the *ROSAT* data on the discovery of X-ray emission from a number of *nebulae* themselves and not only from the nucleus (Kreysing et al., 1992): the PNs NGC 6 543, A 12, LoTr 5, BD+30°3639 as well as NGC 4 361 and NGC 6 853, Table 7.12. In all cases, with only the exception of BD+30°3639, *extended* sources of X-ray emission have been detected, with sizes comparable with the visual extension of the nebulae. For example, the sizes, e.g. of LoTr 5, appeared to be huge – 10′ in diameter. The X-ray luminosities of these objects are scattered within enormous limits – from 10^{30} up to 10^{35} ergs s^{-1}. Their X-ray spectrum turned out to be in agreement with theoretical models, i.e. thermal bremsstrahlung at temperatures of 10^5–10^6 K.

Thus, the PN are X-ray sources, point-like and extended. The first is connected with the nebula's central stars, the second, with the nebula itself. If in the case of the point-like component, the emission mechanism is clear – black-body radiation – its nature is unclear for the nebula. The following mechanisms are possible:

(a) Scattering of X-rays emitted by the nebula's nucleus on the dust particles of the nebula. In this case, the X-ray emission must be essentially weak compared with that of the nucleus.

(b) Thermal bremsstrahlung at temperatures of 10^6 K within the nebula. In this case, the problems of the origin of the high temperature plasma itself, and its surviving along with a cool nebular plasma of temperature of 10^4 K, arise.

(c) Interaction of the stellar wind blowing out from the central star with the velocity of the order of 1 000 km s^{-1} with a slowly expanding ($\sim$ 30 km s^{-1}) nebula, with shock waves appearing. Observations confirm the existence of stellar wind in the majority of nuclei ($\sim$ 60%) with a very high velocity (Kwok et al., 1978; Patriarchi, Perinotto, 1991), which makes this mechanism most probable. The interacting winds model of PNs aims to reveal the structure of the hot bubble, i.e. of those regions with temperatures 5×10^5–10^6 K (Zhekov, Perinotto, 1996). Then PNs are expected to be extended sources of X-ray (0.05–2.5 keV) and EUV (70–700 Å) emission.

(d) The nonthermal bremsstrahlung of so-called *fast electrons*, with energies of the order of 3 MeV, which may appear in the outer regions of the nucleus atmosphere and in the nebula itself (see Sect. 18.24).

Obviously, the X-rays are powerful tools for deciphering key properties of the nebulae and their nuclei.

7.16 Correction for Interstellar Extinction of Emission in Ultraviolet Lines

The ultraviolet radiation and, particularly, the emission lines undergo strong absorption in the interstellar medium. The curve of this absorption (extinction) for wavelengths from 5 000 Å up to 1 000 Å with a maximum at ~2 200 Å is given in Fig. 7.12; a is the amount of the extinction in stellar magnitudes, m, at distance of 1 kpc.

In the wavelength interval from 3 000 Å up to 1 000 Å, there are few groups of emission lines in the spectrum of PNs. The correction procedure itself is usually realized with the help of the so-called *extinction function* $f(\lambda)$. Then the extinction at a particular wavelength will be (Seaton, 1979)

$$10^{-c_\beta f(\lambda)}\,, \tag{53}$$

where c_β is logarithmic extinction at $\lambda = 4\,861$ Å, and is connected with the colour excess E(B–V) by the relationship

$$c_\beta = 1.47\,E(\mathrm{B}-\mathrm{V})\,. \tag{54}$$

The intensities of emission lines in PNs are usually expressed in units of the intensity of the H_β line of hydrogen. Then for the relative intensity of some line λ corrected for interstellar extinction we can write

$$\left[\frac{F(\lambda)}{F(\mathrm{H}_\beta)}\right]_{\mathrm{cor}} = \left[\frac{F(\lambda)}{F(\mathrm{H}_\beta)}\right]_{\mathrm{obs}} 10^{c_\beta f(\lambda)}\,, \tag{55}$$

where $f(\mathrm{H}_\beta) = 0$ for the line H_β.

Table 7.13. Numerical magnitudes of the extinction function $f(\lambda)$ for a sample of emission lines

Line [λ, Å]	$f(\lambda)$	Line [λ, Å]	$f(\lambda)$
4861 H_β	0.00	1640 HeII	1.14
3968 CaII	0.20	1550 CIV	1.18
3727 [OII]	0.22	1486 NIV	1.23
3426 [NeV]	0.34	1400 SiIV	1.31
3204 HeII	0.42	1356 CI, OI	1.39
3133 OIII	0.45	1340 SiIV	1.41
2800 MgII	0.63	1335 CII	1.42
2734 HeII	0.70	1305 OI	1.46
2423 [NeV]	1.12	1240 NV	1.64
2326 CII,[OIII]	1.35	1218 OV	1.75
1909 CIII]	1.23	1216 L_α	1.78
1892 SiIII	1.21	1206 SiIII	1.84
1815 SiII	1.02	1190 SiII	1.89
1666 OIII	1.12	1026 L_β	2.86

The numerical values of the function $f(\lambda)$ for most interesting emission lines are represented in Table 7.13. As for c_β, this parameter includes the integral effect of interstellar extinction, i.e. with the consideration of the distance of the given planetary nebula from us, and should be determined individually for each PN by various methods.

7.17 Determination of Interstellar Extinction

By comparison of the observed ratio of the intensities of hydrogen and helium emission lines with their theoretical values we can estimate the interstellar differential absorption between two wavelengths, λ_1 and λ_2. This is the essence of the method of determining interstellar differential extinction with the help of the emission line of gaseous nebulae.

Strictly speaking, this method is a kind of Balmer decrement method as described above. Now, by its application in relation to the Balmer decrement of the lines of ionized helium, the possibility of receiving quantitative information about the interstellar extinction in the *ultraviolet* region appears.

Let F_λ/F_β be the observational ratio of intensities (fluxes) in the emission line λ to the hydrogen emission line H_β, and let J_λ/J_β be the theoretical value of the same ratio derived from recombination theory. Then the difference in interstellar extinction will be

$$A_\lambda - A_\beta = 2.5\left[\log\frac{J_\lambda}{J_\beta} - \log\frac{F_\lambda}{F_\beta}\right] . \tag{56}$$

This relationship can also be applied for the radio region with the only difference that the interstellar medium is completely transparent for radio frequencies (continuous spectra in the cm range), i.e. $A_\lambda = A(\text{radio}) = 0$, and the theoretical ratio $J_\lambda(\text{radio})/J_\beta$ is known from recombination theory, if we bear in mind that the radio emission is generated only for free–free processes.

Then we have from (56) for the value of the interstellar extinction A_β in the hydrogen H_β line

$$A_\beta = 2.5 \left[\log \frac{F_\lambda(\text{radio})}{F_\beta} - \log \frac{J_\lambda(\text{radio})}{J_\beta}\right], \tag{57}$$

where $F_\lambda(\text{radio})/F_\beta$ is the observed ratio of the flux at radio frequencies to the flux of H_β line.

On the other hand, A_β is connected with the logarithmic extinction c_β and colour index $E(B-V)$ in the following manner:

$$A_\beta = 2.5 c_\beta\,, \tag{58}$$

$$c_\beta = \log(J_\beta/F_\beta)\,, \tag{59}$$

$$c_\beta = 1.47 E(B-V)\,. \tag{60}$$

Therefore, we have for the observational value of c_β

$$c_\beta = 0.4 \left[\log \frac{F_\lambda(\text{radio})}{F_\beta} - \log \frac{J_\lambda(\text{radio})}{J_\beta}\right]. \tag{61}$$

In the case of an optically thin model for the nebula, we have for the theoretical ratio of the radio flux $J_\lambda(\text{radio})$ in the cm range (1–5 cm) to the H_β emission J_β (Brocklehurst, 1971)

$$J_\lambda(\text{radio})/J_\beta \approx 2.0 \times 10^{-8}\,.$$

For example, for the PN NGC 7027, we have $F_\beta = 6.25\times10^{-11}$ ergs cm^{-2} s^{-1} and $F_\lambda(\text{radio}) = (6-7.5)\times 10^{-26}$ W m^{-2} Hz^{-1}. From these data we obtain from (61), on average, $c_\beta = 1.43$ or, from (58), $A_\beta = 3.58$.

With known A_β we can obtain the values of A_λ for each emission line from the expression

$$A_\lambda = 3.58 + 2.5 \left[\log \frac{J_\lambda}{J_\beta} - \log \frac{F_\lambda}{F_\beta}\right]. \tag{62}$$

For the line H_α, for example, we have from the recombination theory (optically thin nebula) $J_\alpha/J_\beta = 2.85$. We have from observations for NGC 7027 $F_\alpha = 4.85 \times 10^{-10}$ ergs cm^{-2} s^{-1}, and, hence, $F_\alpha/F_\beta = 7{,}76$. From these data we obtain from (62) $A_\alpha = 2.49$.

As an example, with this method the values of A_λ for NGC 7027 were obtained in the wavelength range from 21 665 Å up to 1 640 Å using the lines

Table 7.14. Interstellar extinction A_λ for the planetary nebula NGC 7 027 obtained with the help of emission lines from the infrared ($\sim 2\mu$m) up to the ultraviolet (1 600 Å)

Line	Å	A_λ	$A_\lambda(X)$	Line	Å	A_λ	$A_\lambda(X)$
Radio	1–5 cm	0.00	0.00	HI Ba β	4 861	3.58	3.58
HI Br γ	21 655	0.28	0.33	HeII Pa α	4 686	3.72	3.71
HI Ps γ	10 938	1.08	1.10	HI Ba γ	4 340	4.02	3.98
HeII Br α	10 123	1.28	1.26	HeII Pa β	3 202	5.07	5.03
HI Ps δ	10 049	1.28	1.26	HeII Pa γ	2 734	6.10	6.03
HI Ba α	6 563	2.47	2.46	HeII Ba α	1 640	7.53	7.78

of HI and HeII of the Brackett, Paschen and Balmer series; the results are presented in the third column of Table 7.14 (Seaton, 1979).

How well do the values obtained for A_λ for NGC 7 027 fit the values of A_λ derived from the averaged curve of interstellar extinction in the long wavelength and ultraviolet regions (Fig. 3.1, Table 7.13)? In latter case, the magnitude of A_λ must be obtained from the relationship $A_\lambda = A_\lambda(X) = X_\lambda E(B-V)$, where $E(B-V) = c_\beta/1.47 = 0.972$ in accord with (31) for NGC 7 027. The magnitudes of $A_\lambda = A_\lambda(X)$ obtained in such a way are presented in the last column of Table 12.7. As we can see, the values A_λ, obtained by both methods, are in perfect agreement.

8. Pseudoresonance Lines

8.1 Pseudoresonance Absorption Lines

Assume a situation in which the continuous radiation spectrum of the central star-nucleus of a PN undergoes selective absorption within the nebula. As a result absorption lines in the spectrum of the *nucleus* should be observed. Moreover, if one has strong evidence that this line originates neither in the photosphere of the central star, nor in the interstellar medium, then any information available by direct observations of such an absorption line should be related to the nebula. Thus this is a way of obtaining valuable data on the absorption features of nebulae. To evaluate its consequences one has to recall that the only source of our knowledge of PNs has been the emission lines.

In PNs these absorption lines cannot belong to subordinate series of atoms and ions because of their extremely low abundance in excitation levels. In the nebula the degree of excitation, n_i/n_1, for an ordinary, i.e. not metastable, level is extremely small and in the first approximation is given by the expression

$$\frac{n_i}{n_1} \approx W \frac{1}{e^{h\nu_i/kT_*} - 1}, \tag{1}$$

where W is the dilution coefficient and T_* is the temperature of the central star. In PNs $W \sim 10^{-13}$, and, hence, we have $n_i/n_1 \sim 10^{-14}$–10^{-15} for ordinary excitation levels of hydrogen or for any other element when $T_* \sim$ 50 000 K.

These absorption lines cannot belong to resonance series also, first, because in the majority of cases the lines of this series are in the far ultraviolet region – shorter than 1 000 Å – and, second, it would be hard to distinguish these lines from the same ones of photospheric origin.

The situation is different, however, when one considers the spectral lines originating at the *permitted pseudoresonance transitions above the metastable levels.*

The abundance of atoms and ions in metastable levels is large and can even be compared with the number of atoms (ions) in the ground state. Denoting by n_i^* the number of atoms in some metastable level i, instead of (1) we have approximately (see Sect. 2.5)

$$\frac{n_i^*}{n_1} \approx W \frac{A_{i1}}{A_{i1}^*} \frac{1}{e^{h\nu_i/kT_*} - 1}, \tag{2}$$

where A_{i1} and A^*_{i1} are the Einstein coefficients of spontaneous downward transitions from ordinary and metastable levels, respectively. The ratio A_{i1}/A^*_{i1} is usually of the order of 10^{10}–10^{14}.

Comparing (2) with (1) we see that the relative number of atoms in metastable levels must be 10^{10}–10^{14} larger then that of the atoms in the usual excitation levels. Therefore, at "metastable level – ordinary level" permitted transitions the absorption lines can have an almost equal intensity to the resonance lines. For such permitted lines the metastable levels serve as somehow elevated – in the energy sense – ground states. Let us call *pseudoresonance* lines those arising at upward transitions from the metastable to the ordinary subordinate levels (Gurzadyan, 1981).

The formation of pseudoresonance lines, in principle, is impossible in stellar photospheres because of the necessary conditions for the accumulation of atoms in metastable levels. The formation of these lines is impossible in the interstellar medium also because of the too small number of ions possessing metastable levels. The pseudoresonance lines become excited only in planetary and diffuse nebulae, and can be discovered only in the spectra of their central stars. Finally, the observable intensities of pseudoresonance lines – equivalent widths or residual intensities – are absolutely free from the effect of interstellar absorption.

The list of possible pseudoresonance lines, far from being complete, belonging to oxygen, nitrogen, silicon, sulphur, iron, aluminum, phosphorus and nickel, is as follows (atom or ion and λ, Å):

OI	1413	NII	1307	AlII	1858	NiII	2278
	1727		3140		1862		2296
	1893	SII	1102	CII	1010		2316
	2561		1540	CIII	1176	FeII	2756
OII	1073		1811	SiII	1251	FeIII	1026
	1260	SIII	1077		1671		1066
OIII	1002		1373	SiIII	1108		1078
	2497	AlII	1719		1296		1142
NI	1244		1721		1303	PII	1250
NII	1065		1725	NiII	2165		1486
	1300		1856		2216		

The pseudoresonance lines cannot be formed or observed in the following cases:

(a) When the pseudoresonance lines fall in the region of wavelengths inaccessible for observations even in space conditions, i.e. shorter than 1 000 Å. To this group in particular belong all pseudoresonance lines of Ne^{++}, Ne^{+++}, N^{++}, S^{++}, Al^{++}, O^{+++}, Fe^{++}, etc.

(b) When atoms and ions do not have metastable levels and, hence, can never in principle excite pseudoresonance lines. To this category of ions belong Mg^{+} , Si^{+}, Si^{++} etc.

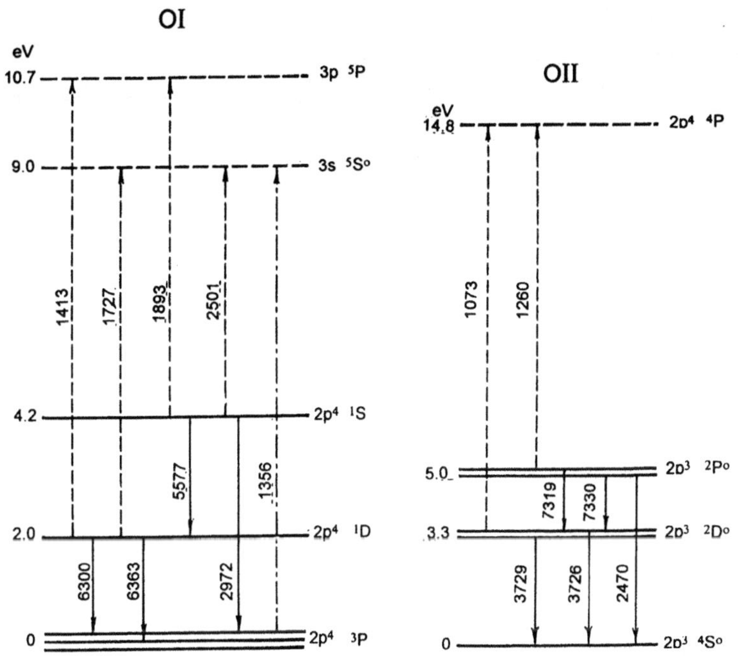

Fig. 8.1. Energy levels for OI and OII with the indicated classes of transitions – forbidden (*solid lines*), pseudoresonance (*broken lines*) and resonance (*dot-dashed lines*). Horizontal solid lines denote metastable levels; horizontal broken lines, ordinaryexcitation levels. Numbers to the left are the energy in eV

(c) When the neutral atoms of the given type cannot exist under PN conditions. These include, for example, easily ionized atoms, for which the second, third and higher levels of ionization are at the ground state.

The pseudoresonance lines discussed above – denoted by brackets, e.g. 1 893 <OI> or 2 165<NeII> – fall in the near and far ultraviolet, 1 000–3 000 Å. Therefore, their discovery should be the subject of space astrophysics. There are, however, cases when the pseudoresonance lines are in the optical region. For example, singly ionized iron possessing a large number of metastable levels can excite a whole series of pseudoresonance lines in the visible band – a triplet, 5 169, 5 018, 4 923 <FeII>, a doublet 4 583 + 4 548 <FeII>, and the lines 5 316 <FeII>, 3 824 <FeII> etc.

As for the strength of the pseudoresonance lines, in particular their equivalent widths, detailed calculations with a number of known atomic parameters are required. However, one can mention the strongest lines, e.g. 1 413<OI> and 1 727<OI>, 1 300<SiIII>, 1 540<SII>, 1 720 <AlII> etc.

In Figs. 8.1 and 8.2 the energy levels for OI, OII, NII and SII along with all three classes of transitions – forbidden (solid lines), pseudoresonance (broken lines) and resonance (dot-dashed lines), are shown. Traditionally in

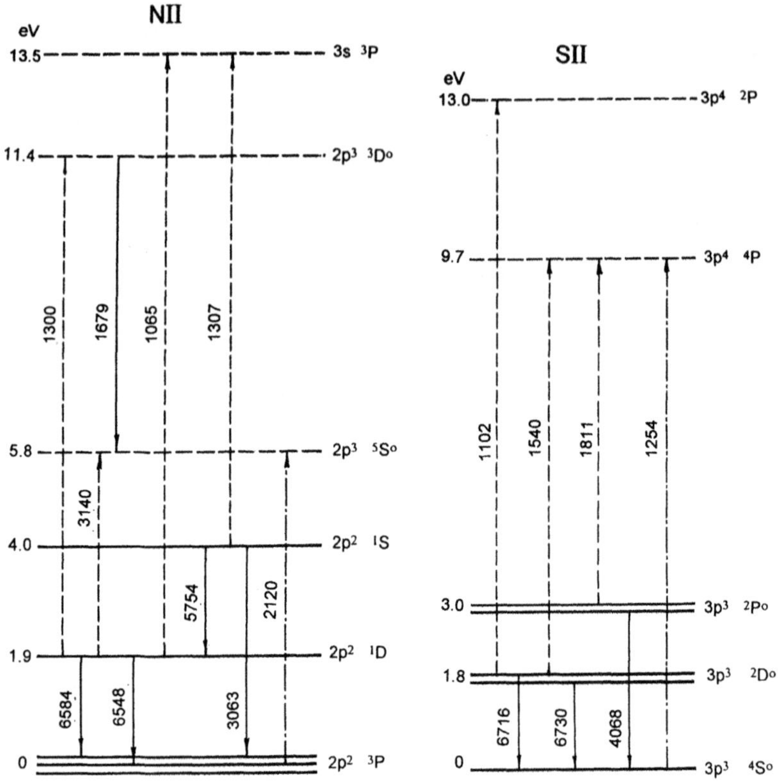

Fig. 8.2. The same as in Fig. 8.1, but for NII and SII

the study of planetary and gaseous nebulae, two categories of spectral lines – fluorescence and forbidden – are dealt with, both in *emission*. Now we have a third category, *the pseudoresonance absorption lines*, observed in the spectrum of PN nuclei.

8.2 Transition Probabilities of Pseudoresonance Lines

The atomic parameters, e.g. the oscillator strengths f and Einstein coefficients of spontaneous transitions A_{ij} for pseudoresonance lines both listed and not listed above, are unknown. Therefore, before we speak about their exploitation for a certain problem of the physics of gaseous nebulae, it is necessary to carry out corresponding quantum mechanical computations for the determination of those parameters.

Such a study has been realized for the pseudoresonance lines of OI, OII, OIII, NI, NII, SII, SIII, SiII, SiIII etc. (Egikyan, 1984, 1985). The results of computations, i.e. oscillator strengths f and probability coefficients A_{ij} for each pseudoresonance transition, are given in Table 8.1.

Table 8.1. Oscillator strengths f and probability coefficients A_{ij} for a sample of pseudoresonance lines

Ion	λ, Å	Transition	f	A_{ij}, s^{-1}
OI	995.5	${}^1D-{}^1P^0$	0.0214	1.60(+8)
	1 152.2	${}^1D-{}^1D^0$	0.0532	2.73(+8)
	1 217.7	${}^1S-{}^1p^0$	0.0536	2.62(+8)
OII	1 075.0	${}^2D^0-{}^4p$	7.55(−7)	3.84(+3)
	1 259.9	${}^2p^0-{}^4p$	5.30(−7)	1.59(+3)
OIII	1 002.5	${}^1D-{}^3D^0$	6.25(−6)	3.75(+4)
	1 008.1	${}^1S-{}^3p^0$	2.47(−6)	1.16(+4)
	1 300.8	${}^1S-{}^3D^0$	5.97(−7)	1.50(+3)
NI	1 143.7	${}^2p^0-{}^2S$	1.96(−2)	1.11(+8)
	1 243.3	${}^2D^0-{}^2D$	3.65(−2)	1.73(+8)
	1 492.6	${}^2D^0-{}^2p$	3.42(−2)	1.15(+8)
NII	1 065.0	${}^1D-{}^3p^0$	4.87(−7)	2.66(+3)
	1 300.0	${}^1D-{}^3D^0$	2.66(−6)	7.92(+3)
SII	1 096.6	${}^2D^0-{}^2p$	6.45(−1)	1.37(+10)
	1 102.4	${}^2D^0-{}^2p$	7.95(−1)	1.69(+10)
	1 234.2	${}^2p^0-{}^2p$	5.95(−1)	1.08(+10)
	1 102.0	${}^2D^0-{}^2p$	1.47(−1)	3.11(+9)
	1 226.7	${}^2p^0-{}^2p$	4.79(−1)	8.71(+9)
	1 233.4	${}^2p^0-{}^2p$	2.50(−1)	4.53(+9)
	1 227.4	${}^2p^0-{}^2p$	1.49(−1)	2.71(+9)
SIII	1 077.1	${}^1D-{}^1D^0$	1.38(+0)	2.65(+10)
	1 374.0	${}^1D-{}^3D^0$	2.24(−4)	2.01(+6)
SiII	1 248.4	${}^4p-{}^4S^0$	3.92(−1)	1.79(+9)
	1 251.2	${}^4p-{}^4S^0$	5.88(−1)	2.67(+9)
SiIII	1 108.4	${}^3p^0-{}^3D$	1.05(+0)	5.72(+9)
	1 109.9	${}^3p^0-{}^3D$	2.61(−1)	1.42(+9)
	1 110.0	${}^3p^0-{}^3D$	7.83(−1)	4.28(+9)
	1 113.2	${}^3p^0-{}^3D$	8.75(−1)	4.76(+9)
AlII	1 710.4	${}^3p^0-{}^3D$	1.02(+0)	2.29(+9)
	1 721.2	${}^3p^0-{}^3D$	7.65(−1)	1.72(+9)
	1 725.0	${}^3p^0-{}^3D$	8.65(−1)	1.92(+9)
	1 855.9	${}^3p^0-{}^3S$	1.46(−1)	2.81(+8)
	1 858.0	${}^3p^0-{}^3S$	1.46(−1)	2.81(+8)
	1 862.3	${}^3p^0-{}^3S$	1.46(−1)	2.79(+8)
PII	1 249.8	${}^1D-{}^1P^0$	3.36(−1)	1.31(+9)
	1 485.5	${}^1S-{}^1P^0$	5.23(−2)	1.23(+8)

In two cases, OII and OIII, the probability coefficients for pseudoresonance lines turn out to be too small – 10^3–10^4 s^{-1}. In other words, those ions which excite the strongest forbidden lines in the conditions of nebulae, are useless for the excitation of pseudoresonance lines. There is an analogous situation with the ion NII.

In contrast to these cases, the sulphur ions SII and SIII look quite promising; the probability coefficients of their pseudoresonance transitions are large – 10^9–10^{10} s^{-1}. The possibility is not excluded that pseudoresonance lines of SII and SIII can be as informative as the OII and OIII forbidden lines.

After the sulphur ions, the pseudoresonance lines of silicon, SiII and SiIII, with probability coefficients 10^9 s^{-1}, are of special importance. Then come the pseudoresonance lines of singly ionized aluminium and phosphor, AlII and PII.

The excitation probability of pseudoresonance lines is the same and sufficiently high – of the order of 10^8 s^{-1} – for neutral oxygen and nitrogen. However, in nebular conditions the abundance of neutral oxygen and nitrogen is small; therefore, it seems more hopeful to detect their pseudoresonance lines in diffuse nebulae, as well as in the low excitation stellar envelopes.

8.3 Possible Combinations Between Pseudoresonance and Forbidden Lines

The same metastable level can serve as an initial state for the excitation of both the pseudoresonance and forbidden groups of lines. The difference is in the directions of the transitions and the means of their realization – induced transitions *upward* under the action of radiation in the former case, and spontaneous transitions *downward* to the ground state, in the latter. Correspondingly one observes an absorption line – a pseudoresonance line – and an emission line – a forbidden line. The peculiarity of the situation is that in both cases the same number of atoms (ions) "act" in the given metastable level. The identity between the forbidden and pseudoresonance lines is very helpful for the solution of problems concerning the physics of gaseous nebulae. In this connection the question of the search for the most effective combinations between forbidden and pseudoresonance lines for the given metastable level arises.

In Table 8.2 the possible combinations between the mentioned categories of pseudoresonance and forbidden lines are given for OI, SII, SIII and AlII, with high enough probabilities of pseudoresonance transitions, of the order of 10^8–10^{10} s^{-1}. The ions SII and SIII can generate the strongest pseudoresonance lines along with forbidden lines of moderate strength. The next are the pseudoresonance lines of OI. Lines of other elements (ions) can be added as well.

Table 8.2. Possible combinations between pseudoresonance lines and forbidden lines for one and the same metastable level

Ion	Initial metast. level	Forbidden line			Pseudoresonance line		
		Transition	λ [Å]	A_{ki} [s^{-1}]	Transition	λ [Å]	A_{ki} [s^{-1}]
OI	^1D	$^1D-^3P$	6 300	5.1(−3)	$^1D-^1P_0$	999	1.60(+8)
			6 363	1.7(−3)		1 152	2.73(+8)
	^1S	$^1S-^3D$	5 577	1.34	$^1S-^1P_0$	1 218	2.63(+8)
SII	$^2D_0$	$^2D-^4S^0$	6 716	0.5(−3)	$^2D^0-^2P$	1 102	1.69(+10)
						1 102	0.31(+10)
						1 097	1.37(+10)
	$^2P60$	$^2P^0-^4S^0$	4 068	0.34	$^2P^0-^2P$	1 234	1.08(+10)
						1 227	0.27(+10)
						1 233	0.45(+10)
						1 227	0.87(+10)
SIII	^1D	$^1D-^1S$	6 312	2.5	$^1D-^1D^0$	1 077	2.65(+10)
AlII	$^3P^0$	$^3P^0-^1S$	2 667	4.7	$^3P^0-^3D$	1 725	2.26(+9)
						1 721	2.29(+9)
						1 719	2.29(+9)
					$^3P^0-^3S$	1 862	2.79(+8)
						1 858	2.81(+8)
						1 856	2.81(+8)

8.4 The Optical Depth of a Nebula in Pseudoresonance Lines

The possibility of detecting pseudoresonance absorption line depends, in fact, on the optical depth of the nebula t_0 in that line. The latter itself depends on the coefficient of selective absorption s_0 i.e. on the oscillator strength or coefficient of spontaneous transition (see Sect. 8.2), as well as on the abundance of atoms (ions) n_i in the metastable level i. Thus the problem comes down to the study of the ionization equilibrium of the nebula.

Thus, for the determination of the optical depth of the nebula t_0 in the given pseudoresonance line it is necessary to have: (a) a model of the ion – for determining atomic parameters; (b) a model of the nebula – for determining populations of levels.

This problem was quantitatively analysed for a sample of pseudoresonance lines of SiIII, when the lower levels are metastable, and the absorption lines are formed within the nebula. These pseudoresonance lines are (see Fig. 8.3) 997, 1 108, 1 113, 1 296 and 1 303 Å SiIII. The computations have been performed for a model of the nebula and its nucleus: $T_* = 100\,000$ K, radius

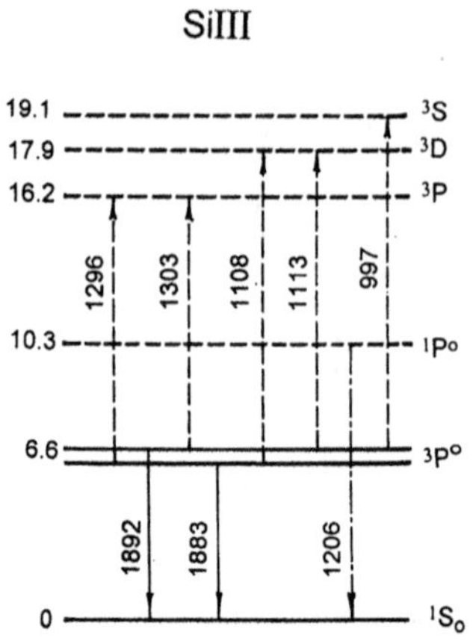

Fig. 8.3. Energy levels for SiIII: five pseudoresonance lines (*broken lines*), two forbidden lines (*solid lines*) and the resonance line 1 206 SiIII (*dot-dashed line*)

of the central star $R_* = 10^{10}$ cm, radius of the nebula $R_n = 1.2 \times 10^{17}$ cm, the linear thickness of the nebula $\Delta H = 10^{17}$ cm, the concentration of hydrogen atoms $n_{\mathrm{H}} = 10^4\ \mathrm{cm}^{-3}$, and the universal abundance of silicon Si/H $= 5 \times 10^{-5}$. As a result, the populations of ions n_i(SiIII) up to the 8th level, and for 20 points along the nebula's radius were obtained (Egikyan, 1986). Combined with the absorption coefficient s_0 in the lines, the optical depths t_0 of the nebula in the five pseudoresonance lines mentioned above were obtained; the results are given in Table 8.3. For comparison, the optical depth of the usual resonance line 1 206 SiIII is also given.

An absorption line can appear if $t_0 > 1$. This condition is fulfilled for all five pseudoresonance lines in Table 8.3. The strongest lines among them are 1 303 <SiIII> and 1 113 <SiIII>. In the last column of Table 8.3, the residual intensities, r_{ν_0}, for these lines computed according to the following

Table 8.3. Optical depth t_0 and residual intensity r_{ν_0} for five pseudoresonance lines of SiIII for a model of a planetary nebula (see in text). The line 1 206 SiIII is a resonance

Line λ, Å	Transition	A_{ki}, s^{-1}	t_0	r_{ν_0}
1 296 SiIII	$^3P^0 - ^3P$	9.90(+7)	0.406	0.77
1 108 SiIII	$^3P^0 - ^3D$	1.25(+8)	0.535	0.71
1 303 SiIII	$^3P^0 - ^3P$	1.72(+9)	2.42	0.35
1 113 SiIII	$^3P^0 - ^3D$	1.43(+9)	2.09	0.39
997 SiIII	$^3P^0 - ^3S$	3.04(+9)	0.64	0.67
1 206 SiIII	$^1S - ^1P^0_0$	1.21(+9)	366.0	0.0036

relationship are given:

$$r_{\nu_0} = 1 \Big/ \left(1 + \frac{3}{4} t_0\right) .$$

Note the high value of t_0 for the resonance line 1 206 SiIII (the last line in Table 8.3).

Thus, on the example of SiIII, the possibility of detecting a pseudoresonance absorption line in PN conditions is shown.

8.5 The Pseudoresonance Line 10 830 HeI in the Spectra of Gaseous Nebulae

The infrared emission line 10 830 HeI is one of the strongest in the spectra of PNs and gaseous nebulae; e.g. in the case of NGC 7027 it is ten times stronger than the H_β line.

The line 10 830 HeI arises at transitions from the ordinary level $2p\,^3P^0$ of neutral helium to the metastable level $2s\,^3S$ (Fig. 8.4). Owing to a low energy of excitation, 1.14 eV, the dominating transitions are from the $2s\,^3S$ state to the level $2p\,^3P^0$ by means of electron collisions, with subsequent spontaneous jumps $2p\,^3P^0 \rightarrow 2s\,^3S$ and emission in the line of 10 830 Å. Owing to the character of excitation, this line should be considered a pseudoresonance *emission* line.

The line 10 830 HeII has been detected in emission in the spectra of a number of PNs and diffuse nebulae (Scrimger, 1984; Aller, 1985). Its observed intensities in units of the neutral helium line, 5 876 HeI, for a sample of PNs and diffuse nebulae are given in Fig. 8.4. The ratio of intensities of these lines

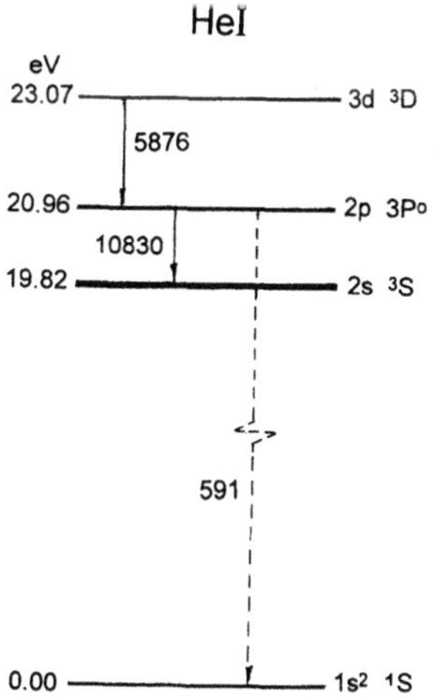

Fig. 8.4. Energy levels of neutral helium, HeI. The subordinate level $2p\,^3P^0$ is only 1.14 eV above the metastable level $2s\,^3S$

depends on the electron temperature T_e and the electron concentration n_e (LeVan, Rudy, 1983):

$$\frac{I(10\,830)}{I(5\,876)} = 1.6T_e^{1/3}\frac{1+0.37/n_eT_e^{0.6}+7.0T_e^{1.15}}{1+0.37/n_eT_e^{0.6}+0.16T_e^{3.15}}, \quad (3)$$

where T_e is in units of 10^4 K, and n_e in units of 10^4 cm^{-3}.

However, the ratio $I(10\,830)/I(5\,876)$ obtained with the help of (3) at the known electron temperature of a given nebula T_e, in the majority of cases turns out to be smaller than the observed ratio. In other words, accordance with the observations may be reached at an electron temperature $T_e(I)$ smaller than T_e, typically 1.5–2.5 times smaller than the latter.

The conclusion is clear: the line 10 830 HeI is excited in the outer regions of the nebula where the helium is neutral and the electron temperature is lower.

Thus, even a single emission line, 10 830 HeI, turns out to be suitable for the solution of the problem of determining electron temperatures in the outer regions of PNs – in the zone of neutral helium.

8.6 Discovery of Pseudoresonance Lines

The calculated intensities of pseudoresonance absorption lines, as was shown above (Sect. 8.4), are quite small, and, therefore, their discovery by direct observations seems to be a hard task. The main hope is to connect with the pseudoresonance line 1 300 <SiIII>, and the blending of six lines of SiIII within the interval 1 298 Å–1 304 Å. The absorption line 1 300 <SiIII> should be searched for, according to the definition, in the continuous spectra of the central star of the nebula.

Table 8.4. Planetary nebulae with the revealed pseudoresonance absorption line 1 300 <SiIII> in the spectra of their nuclei: with a high confidence level (the first three objects) and low confidence (the next three objects). $W(1\,300)$ and t_0 are the observed equivalent widths and optical depths of 1 300 <SiIII>

Planetary nebula	Excit. class, p	T_* [K]	R_0 [10^{17}cm]	$W(1\,300)$ [Å]	t_0
NGC 2610	12	100 000	5.9	0.5	0.4
NGC 3587	8	112 000	8.5	0.9	0.45
NGC 6891	4	70 000	2.0	0.4	0.4
NGC 6210	4	60 000	1.8	0.8:	3.6:
IC 3568	3	60 000	3.7	0.6:	2.6:
IC 4776	2	–	1.5	1.1:	1.9:

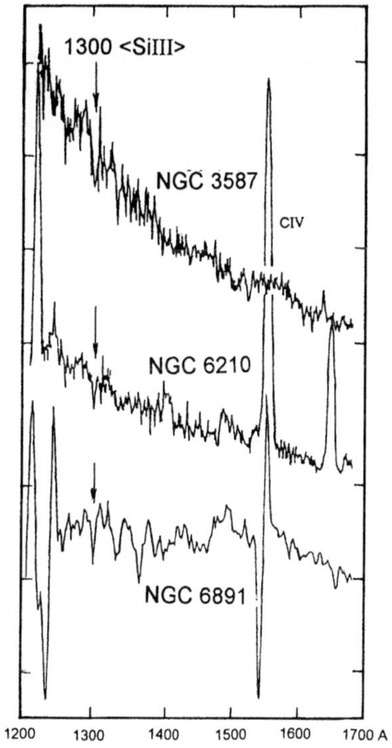

Fig. 8.5. Details of *IUE* recordings of the nuclei of NGC 3 587, NGC 2 610 and NGC 6 891; the pseudoresonance absorption line 1 300 <SiIII> is indicated

Though the region of the spectrum close to 1 300 Å is too noisy, nevertheless, the line 1 300 <SiIII> in absorption was discovered on a sample of *IUE* recordings (Table 8.4).

In Fig. 8.5 the details of *IUE* recordings of the nuclei of three PNs, NGC 2 610, NGC 3 587 and NGC 6 891 (Feibelman et al., 1988), are given together with an indication of the line 1 300 <SiIII> in absorption. As we can see, it is too faint – with the equivalent width being about 1 Å or less (fifth column in Table 8.4). Clearly, the following factors have to be taken into account: (a) the influence of interstellar selective absorption; (b) the blending by nearby absorption lines of various elements and ions; (c) the possibility of formation of this absorption line in the photosphere of the central star.

The analysis indicates (Gurzadyan, Egikyan, 1990), that these factors, at least in the case of the three PNs mentioned above are not crucial, and, thus, the existence of a pseudoresonance absorption line 1 300 <SiIII> in the spectra of NGC 2 610, NGC 3 587, and NGC 6 891 must be certain. For the remaining three objects, NGC 6 210, IC 3 568 and IC 4 776, the existence of this line in absorption seems to be highly probable.

The values of W(1 300<SiIII>) obtained from observations can also be used for the determination of the optical depth of the nebula in the centre of the line 1 300 <SiIII> (Table 8.4). As we can see, t_0 is of the order of unity in accord with theoretical predictions.

The obtained values of t_0 enable one to obtain the total number of absorption ions $N_2(Si^{++}) = t_0/k_0$, at the level $3s\,3p\,{}^3P^0$, where k_0, the absorption coefficient in the line centre of 1 300 <SiIII>, is given by

$$k_0 = 1.5 \times 10^{-2} f\,\lambda/v = 6.68 \times 10^{-15} (\Delta\lambda)^{-1}\ \mathrm{cm}^2\,, \tag{4}$$

where v is the mean expansion velocity and $f = 0.79$ is the oscillator strength for the multiplet. The result is: $k_0 \sim 5 \times 10^{-14}$ cm^2, $N_2 \sim 10^{13}$ cm^{-2} and $n_2 = N_2/R_n \sim 10^{-5}$ cm^{-3} for the concentration of Si^{++} ions in the nebula within a radius R_n.

9. Continuous Radiation of Nebulae. Two-Photon Emission

9.1 Statement of the Problem

The entire history of the two-photon emission phenomenon can be considered as an example of how astrophysical observations can support the solution of the basic problems of theoretical physics.

We live in a world of one-photon processes. Everywhere, in atmospheres and within all types of star, in the interstellar medium, in planetary atmospheres, etc., all kinds of atomic transitions from one energy level to another are accompanied by the escape of a single photon.

However, in the late 1920s, Goppert-Mayer (1929), guided by the ideas of newly developed quantum mechanics, theoretically predicted that atomic transitions from one energetic level to another can be realized not only by the escape of one photon, as usual, but also of two photons. In this case, the frequency of each photon can vary depending on the transitions, so that the sum of the energies of both photons must remain constant and equal to the difference of energies between the initial and final states.

At the beginning of the 1940s, it became obvious that two-photon emission of hydrogen arising at the transition from the nearest level $2S$ down to the ground state $1S$ can be of special interest (Fig. 9.1). Following the quantitative theory by Breit and Teller (1940), in the early 1950s, Kipper (1950) and Spitzer and Greenstein (1951), independently, considered this problem in an astrophysical context.

The predicted spectrum of two-photon emission in hydrogen, as we will see below, starts at $\lambda = 1\,216$ Å and extends up to $\lambda \to \infty$. The emission has a maximum near $\lambda = 1\,400$ Å and decreases relatively quickly at long wavelengths. Thus, at $\lambda = 1\,400$ Å, for example, it comprises nearly 5% and at $\lambda = 6\,000$ Å only 1% of the maximum emission at $\lambda \sim 1\,400$ Å. The main part – over 90% of two-photon emission – is in the short wavelength region of spectra, shorter than $3\,000$ Å; only the "tail" of the spectrum extends to the optical band.

Incidentally, this the only case when a spontaneous transition from one discrete state ($2S$) to another ($1S$) is not accompanied by escape of one photon, as usual, but gives a continuous spectrum of infinite width, from $\lambda = 1\,216$ Å up to infinity.

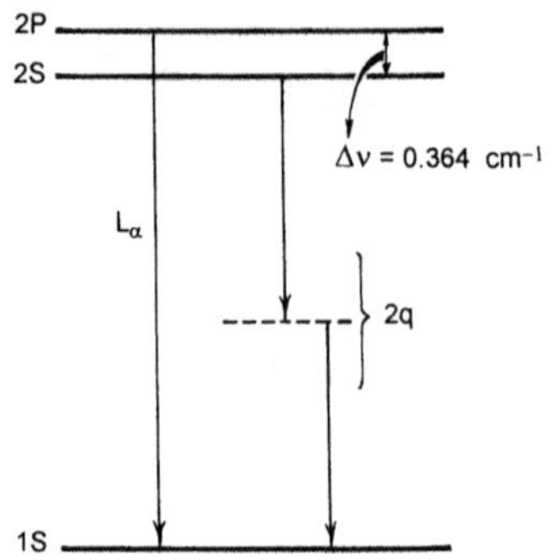

Fig. 9.1. The problem of two-photon emission ($2q$) of the hydrogen atom

To induce two-photon emission the hydrogen atom has to "live" in a metastable state ($2S$) for a sufficiently long time, i.e. about 0.1 second. The electron collisions or L_α photons are two possible disturbances. To avoid these one has to have a medium with low enough density both of matter and radiation. This factor – the "deactivation effect" – has a role for electron concentrations from 10^3 to 10^4 cm^{-3}. At high concentrations the spontaneous transition $2S \rightarrow 1S$ occurs only via emission of a single photon, the L_α line (1 216 Å), and not of two photons.

As we can see, the limiting electron concentrations are extremely low – millions of times smaller than in laboratory vacuum cameras. This should be considered the reason why this fundamental prediction of quantum mechanics – two-photon emission – has not been observed in laboratory conditions.

Electron concentrations of the order of 10^3–10^4 cm^{-3} and lower are typical for PNs. Here the density of radiation is quite low as well. It is natural, therefore, to hope for the discovery of two-photon emission of hydrogen and, hence, to confirm the two-photon phenomenon, in PNs.

Since the overwhelming part of two-photon emission lies, as was outlined above, in the region shorter than 3 000 Å, the corresponding observations have to be performed only in space conditions. Thus the idea of space observations of the continuous spectra of PNs in the ultraviolet with the aim of discovering two-photon emission are intriguing. The first such successful attempt used the PN IC 2 149 short wavelength spectrogram, obtained with the help of the space observatory *Orion-2* in 1973.

9.2 The Theory of Two-Photon Emission of Hydrogen

The nearest level of the hydrogen atom of $n = 2$ is represented by two sublevels, $2P$ and $2S$ (Fig. 9.1). Atoms arrive in the $n = 2$ state from above, by downwards cascade transitions due to the capture of an electron by a proton. The level $2P$ is not metastable, and when an atom reaches the level $2P$ the transitions must be continued up to the ground state, emitting an L_α line. But a certain fraction of the atoms, say X, in the process of recombination will reach the $2S$ level – favourable for two-photon emission. The quantity X, generally speaking, is determined by physical conditions in the nebula, but in a first approximation at low electron densities it is equal to the ratio of statistical weights of the $2S$ and $2P$ levels. This ratio is of the order of 0.3. Consequently, nearly one third of the atoms participating in the downwards cascade transitions reach the ground state $1S$ by means of two-photon emission.

In the theory of luminescence of gaseous nebulae we saw that one L_α photon is produced by each L_c photon absorbed in the nebula. Now we see that this is not exactly so: among the total number N_c of L_c photons on average $2/3\ N_c$ photons are emitted in the line L_α, the remaining $1/3\ N_c$ escape via two-photon emission. Strictly speaking, this does not imply a "decay" of the L_α photon into two photons (see Sect. 9.4), though the total energy in two-photon emission equals the total energy of the "lost" L_α photons.

Denote by $y\nu_\alpha$ and $(1-y)\nu_\alpha$ the frequencies of photons emitted via two-photon transitions in hydrogen, where ν_α is the frequency of an L_α photon, and by $A(y)\,\mathrm{d}y$ the probability that a photon is emitted in the frequency interval $y\,\nu_\alpha$ to$(y+\mathrm{d}y)\nu_\alpha$ per unit time. The function $A(y)$ has the form

$$A(y)\frac{9\alpha^6\nu_0}{2^{10}}\,\psi(y)\,, \tag{1}$$

where ν_0 is the ionization frequency of hydrogen, and $\alpha = 2\pi e^2/hc$ is the fine-structure constant. The energy radiated per unit frequency is proportional to $h\nu A(y)$ or, in accord with (1) is proportional to $y\psi(y)$. The numerical values of $\psi(y)$ computed by quantum-mechanical methods are given in Table 9.1, along with the relative intensity $y\psi(y)$. Since $\psi(1-y) = \psi(y)$, the values of $\psi(y)$ at $y > 0.5$ are omitted. Note that the sum of energies (frequencies) of both photons is equal to the energy (frequency) of the L_α photon, i.e. $y\nu_\alpha + (1-y)\nu_\alpha = \nu_\alpha$.

The Einstein coefficient for the $2S \to 1S$ transition with the escape of two photons is given by the following obvious relationship:

$$A_{2S,\,1S} = \frac{1}{2}\int_0^1 A(y)\,\mathrm{d}y = \frac{9\alpha^6\nu_0}{2^{11}}\int_0^1 \psi(y)\,\mathrm{d}y\,, \tag{2}$$

where the factor 1/2 implies the emission of two photons.

The numerical integration of (2) yields

$$A_{2S,\,1S} = 8.227\ \mathrm{s}^{-1}\,. \tag{3}$$

As we can see, the probability of transitions with two-photon emission is 6–7 orders smaller than the probability of the usual transitions for hydrogen atoms.

Table 9.1. Numerical values of the functions $\psi(y)$ and $y\psi(y)$ determining the spectra of two-photon emission of hydrogen

y	λ, Å	$\psi(y)$	$y\psi(y)$	y	λ Å	$\psi(y)$	$y\psi(y)$
0.00	∞	0	0	0.55	2210	4.889	2.688
0.05	24313	1.725	0.086	0.60	2026	4.824	2.894
0.10	12157	2.783	0.278	0.65	1870	4.711	3.062
0.15	8105	3.481	0.522	0.70	1731	4.546	3.182
0.20	6078	3.961	0.792	0.75	1621	4.306	3.229
0.25	4862	4.306	1.077	0.80	1519	3.961	3.168
0.30	4052	4.546	1.363	0.85	1430	3.481	2.958
0.35	3473	4.711	1.649	0.90	1351	2.783	2.505
0.40	3039	4.824	1.929	0.95	1280	1.725	1.639
0.45	2702	4.889	2.200	1.00	1216	0	0
0.50	2431	4.907	2.454				

The radiative capability of a hydrogen atom with two-photon emission yields $j_\nu \sim yA(y) \sim y\psi(y)$, in the frequency scale, and $j_\lambda \sim y^3\psi(y)$, in the wavelength scale.

The volume emission coefficient of $2q$ emission $\varepsilon_\nu(2q)$ depends also on n_{2s} – on the number of hydrogen atoms per unit volume in the state $2S$. Then we have for the volume emission coefficient

$$\varepsilon_\nu(2q) = \frac{1}{4\pi} n_{2S} A(y)\, h\nu = n_{2S} \frac{h}{4\pi} y\, A(y) \tag{4}$$

per unit frequency interval. We still have to determine the value of n_{2s}.

The number of recombinations at all levels above $n = 2$ is

$$n_e n^+ \sum_2^\infty C_i(T_e)\,.$$

Among all these transitions, according to the definition, only the fraction X falls to the level $2S$. Therefore we will have for the rate of transitions to the level $2S$ per unit volume

$$X\, n_e\, n^+ \sum_2^\infty C_i(T_e)\,. \tag{5}$$

The transitions $2S \to 1S$ occur only via the emission of two photons, so that the rate of such photons $N(2q)$ escaping per unit volume will be

$$N(2q) = 2X\, n_e\, n^+ \sum_2^\infty C_i(T_e)\,. \tag{6}$$

This number is compatible by an order of magnitude with the number of photons emitted at recombination. Hence, we have the first conclusion:

two-photon transitions should have an essential role for the formation of the continuous spectra of nebulae.

Moreover, the number of spontaneous transitions $2S \to 1S$ with the escape of two photons equals $n_2S\, A_{2S,1S}$, i.e.

$$n_{2S}\, A_{2S,\, 1S} = X\, n_{\rm e}\, n^+ \sum_2^\infty C_i(T_{\rm e}) \, . \tag{7}$$

Substituting from here the value of n_{2S} into (3), we obtain for the volume coefficient of $2q$ emission

$$\varepsilon_\nu(2q) = X\, n_{\rm e}\, n^+ \frac{h}{4\pi} \frac{y\, A(y)}{A_{2S,\, 1S}} \sum_2^\infty C_i(T_{\rm e}) \, , \tag{8}$$

or

$$\varepsilon_\nu(2q) \sim y\psi(y) \sum_2^\infty C_i(T_{\rm e}) \, .$$

From here, we get the second and third conclusions: the spectrum of $2q$ emission is determined by the form of the function $y\psi(y)$ in the frequency scale or $y^3\psi(y)$ in the wavelength scale.

The final expression for the determination of a specific emission coefficient for the two-photon radiation of hydrogen, i.e. for unit electron concentration, $n_{\rm e} = 1$ cm^{-3}, has the form

$$\begin{aligned} \varepsilon_\nu(2q) &= \frac{9\alpha^6 \nu_0}{2^{10}} \frac{X}{A_{2S\, 1S}} \frac{h}{4\pi} y\, \psi(y) \sum_2^\infty C_i(T_{\rm e}) \\ &= 2.80 \times 10^{-28} y\psi(y) \sum_2^\infty C_i(T_{\rm e}) \text{ ergs cm}^{-3}\ \text{s}^{-1} \, . \end{aligned} \tag{9}$$

Table 9.2. The specific coefficient of two-photon emission of hydrogen, $\varepsilon_\lambda(2q)$, as a function of wavelength for three electron temperatures $T_{\rm e}$: 10 000, 15 000 and 20 000 K, and in units of 10^{-36} ergs cm^{-3} s^{-1} Å^{-1}

λ	$T_{\rm e}$, K			λ	$T_{\rm e}$, K		
[Å]	10 000	15 000	20 000	[Å]	10 000	15 000	20 000
∞	0	0	0	2 210	4.843	4.342	2.924
12 157	0.0166	0.0126	0.0100	2 026	6.219	4.678	3.757
8 105	0.0702	0.0527	0.0403	1 870	7.710	5.802	4.662
6 078	0.1892	0.1421	0.1145	1 738	9.300	6.995	5.622
4 862	0.4009	0.3020	0.2424	1 621	10.827	8.147	6.538
4 052	0.731	0.550	0.442	1 520	12.095	9.093	7.300
3 473	1.204	0.905	0.726	1 430	12.619	9.485	7.623
3 039	1.845	1.388	1.115	1 351	12.102	9.091	7.299
2 702	2.649	1.993	1.599	1 280	8.825	6.628	5.328
2 431	3.664	2.753	2.210	1 216	0	0	0

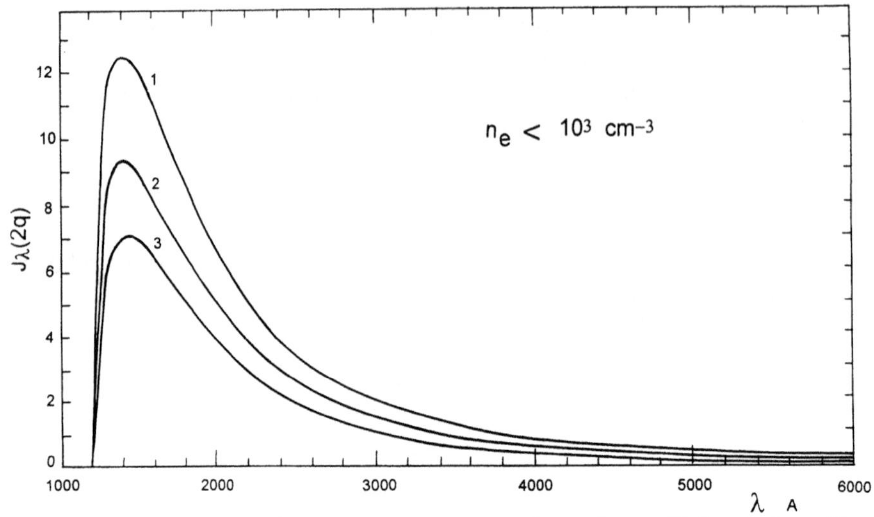

Fig. 9.2. The theoretical continuous spectrum of two-photon ($2q$) emission of hydrogen for three values of the electron temperature T_e: 10 000, 15 000 and 20 000 K and low electron concentrations

Thus, the elementary act of escape of two photons by a hydrogen atom, as a purely quantum mechanical phenomenon described by the function $\psi(y)$ is, obviously, independent of electron temperature. However, the volume emission determining the relative number of emitting atoms does depend, through the recombination coefficient $C_i(T_e)$, on the electron temperature. As a result, the intensity of two-photon emission turns out to be dependent on the electron temperature of the medium.

In Table 9.2, the numerical values of the specific coefficient for the $2q$ emission of hydrogen, $\varepsilon_\lambda(2q)$, computed with the help of (9), are given for three values of T_e: 10 000, 15 000 and 20 000 K. The same results are represented in graphical form in Fig. 9.2. As we can see, $2q$ emission begins from $\lambda = 1\,216$ Å and reaches a maximum at a wavelength $\sim 1\,400$ Å or at a frequency $\sim 2.14 \times 10^{15}$ s^{-1}, extending up to $\lambda \to \infty$. The location of the maximum is the same for all values of T_e. The low level of $2q$ emission in the optical region, longer than 3 000 Å, is clearly seen.

The last conclusion implies the difficulty of separating $2q$ emission from the background of the usually continuous PN spectra, and confirms that efforts should be concentrated on observations in the near and far ultraviolet, at 1 000–3 000 Å.

9.3 The Role of Collisions

Atoms can leave the state $2S$ through excitation to the higher nearby $2P$ level not only spontaneously, but also via collisions with electrons or protons.

Evidently this process leads to a reduction in the intensity of $2q$ emission. Denoting by $W_{2S,1S}$ the probability of collisional $2S \to 2P$ transitions, we can write for the fraction X of atoms in the $2S$ state which undergo two-photon transitions to the ground state

$$X = \gamma_{2S} \frac{A_{2S,\,1S}}{A_{2S,\,1S} + n_e W_{2S,\,1S}} = \frac{\gamma_{2S}}{1 + n_e \dfrac{W_{2S,\,1S}}{A_{2S,\,1S}}},$$

where γ_{2S} is the fraction of atoms reaching the level $2S$ from their total number reaching, as a result of recombination, the level $n = 2$, i.e. including both sublevels $2S + 2P$. Therefore in the "pure" case, i.e. when collisions are neglected, we have $\gamma_{2S} = 0.32$. However, taking account of collisions (deactivation factor), we readily have

$$X = \frac{0.32}{1 + n_e \dfrac{W_{2S,\,1S}}{A_{2S,\,1S}}}. \tag{10}$$

It appears that in deactivation processes the protons play a fundamental role: $W_{2S,2P} \approx 5 \times 10^{-4}$ cm s^{-1} at $T_e = 10^4$ K. Then instead of (10) we can write

$$X = \frac{0.32}{1 + 0.65 \times 10^{-4} n_e}, \tag{11}$$

and we finally obtain for the specific coefficient of two-photon emission, instead of (9).

$$\varepsilon_\nu(2q) = 0.84 \times 10^{-28} \frac{1}{1 + 0.65 \times 10^{-4} n_e} y\psi(y) \sum_2^\infty C_i(T_e) \text{ ergs cm}^{-3}\text{s}^{-1}. \tag{12}$$

As we can see, the intensity of $2q$ emission decreases with the increase of electron concentration: at $n_e \sim 10^4$ cm^{-3} nearly two times, and, for $n_e \sim 10^5$ cm^{-3}, almost ten-fold compared with its theoretically possible maximum.

Thus, $2q$ emission will form a considerable fraction of the total intensity for low density nebulae, with $n_e < 10^3$ cm^{-3}; for dense nebulae, with $n_e > 10^4$ cm^{-3}, it will be negligible.

9.4 The Decay of L_α Photons

The total number of L_α photons within the nebula is very large. The question arises whether the decay of L_α photons in nebular conditions into two photons is possible. This problem concerns the possibility of alternative mechanisms of continuous emission in nebulae.

At least three decay mechanisms for L_α photons are known. None seems to play an essential role under nebular conditions. Nevertheless, they are of particular interest from a heuristic point of view.

The first mechanism of L_α decay was pointed out by Spitzer and Greenstein (1951), and is as follows. A neutral hydrogen atom moves from the state $1S_{1/2}$ to one of the $2P_{1/2,2/3}$ states by absorbing an L_α photon. Because of the high density of L_α radiation within the nebula, the populations of $2P_{1/2}$ and $2P_{3/2}$ states might be not negligible compared with the ground state. Therefore, there is a certain probability that an atom in one of the $2P_{1/2,3/2}$ states, via collisions with free electrons, moves higher into the $2S_{1/2}$ state and then to the ground state via two-photon emission. The probability of the decay of the L_α photon by this mechanism – the cycle $1S \to 2P \to 2S \to 1S$ – was derived by Yada (1954)

$$P_1 = \frac{3.765 \times 10^{-13} n_e}{1 + 8.58 \times 10^{-5} n_e} . \tag{13}$$

The quantity p_1 is of the order of 10^{-9}–10^{-10} for planetary nebulae, 10^{-11}–10^{-12} for diffuse nebulae and 10^{-17} for the interstellar medium.

The actual probability of decay for the L_α photon during its stay within the nebula depends on the total number, $N(t_\alpha)$, of scatterings within the nebula at an optical depth t_α in the L_α line and is given by the relation (see Chap. 11)

$$N(t_\alpha) = 2t_\alpha \sqrt{\pi \ln t_\alpha} . \tag{14}$$

This gives $N(t_\alpha) \sim 10^6$. The total probability of decay of an L_α photon during its stay within the nebula by this mechanism will be $p_1 N(t_\alpha) = 10^{-3}$–$10^{-4}$.)

Thus, by the first mechanism, only a negligible number of L_α photons – 0.1–0.01% – can decay within the nebula.

The essence of the second mechanism for L_α photons is as follows (Kipper, Tiyit, 1958). The transition $2P \to 1S$ with the emission of one L_α photon has a probability for hydrogen of $w_1 = 6.24 \times 10^8\ \mathrm{s}^{-1}$. This transition can also occur with the emission of two photons, however, with much smaller transition probability, $w_2 = 1.938 \times 10^{-3}\ \mathrm{s}^{-1}$. The probability that for each single-photon transition of a hydrogen atom one two-photon transition will occur is

$$P_2 = \frac{W_2}{W_1} = 3 \times 10^{-14} . \tag{15}$$

This is just the probability of the decay of an L_α photon by the following mechanism: the cycle $1S \to 2P \to 1S$. Note that this does not depend on the physical conditions of the medium.

One can still mention another, third, mechanism for the decay of L_α photons in nebular conditions (Gurzadyan, 1961). This is connected with the fact that the energy difference between the $2P$ and $2S$ levels in hydrogen is small and is comparable with the kinetic energy of thermal motions of atoms. In frequency units, this difference yields $1.092 \times 10^{10}\ \mathrm{s}^{-1}(0.364\ \mathrm{cm}^{-1})$, which is equivalent to a difference of $1.32\ \mathrm{km\ s}^{-1}$ in the velocities of two hydrogen atoms. Therefore, if the atom absorbing an L_α photon has a negative velocity of 1.32 km s^{-1} with respect to the atom emitting the L_α photon, the absorbing atom can occur in the metastable level $2S$, with a subsequent transition to the ground state, $1S$, by two-photon emission.

The efficiency of this mechanism – the cycle $1S \to 2S \to 1S$ – obviously depends on the probability of the transition $1S \to 2S$ in hydrogen. The experimentally determined lifetime of the state $2S$ is of the order of 2.4×10^{-3} s (Fite et al., 1959). Hence, we have for the Einstein coefficient for a spontaneous, single-quantum, transition $A_{2S,1S} \approx 0.4 \times 10^3$ s^{-1}, approximately 50 times larger than we have for a two-photon transition A_{2q}.

The velocity difference 1.32 km s^{-1} corresponds to a kinetic temperature of the medium of about $\sim$ 100 K. Therefore, the present mechanism for L_α photon decay can be essential when the electron temperature of the medium exceeds 100 K.

For the probability p_3 of L_α photon decay by means of the cycle $1S \to 2S \to 1S$ we have

$$\begin{aligned} P_3 &= \frac{N_{1S,\,2S}}{N_{1S,\,2P}} \frac{A_{2q}}{A_{2q} + A_{2S,\,1S}} \\ &= \frac{g_{2S} A_{2S,\,1S}}{g_{2P} A_{2P,\,1S}} \frac{A_{2q}}{A_{2q} + A_{2S,\,1S}} \approx \frac{g_{2S}}{g_{2P}} \frac{A_{2q}}{A_{2P,\,1S}}, \end{aligned} \tag{16}$$

since $A_{2S,1S} \gg A_{2q}$. Substituting $g_{2S}/g_{2P} = 1/3$, $A_{2q} = 8.227$ s^{-1}, $A_{2P,1S} = 6.24 \times 10^8$ s^{-1}, we find that

$$P_3 = 0.44 \times 10^{-8}\,. \tag{17}$$

This value is at least one or two orders of magnitude larger than with the first mechanism, and five orders of magnitude larger than with the second one.

A stricter formulation, taking into account the influence of electron and proton collisions on the level $2S$ (deactivation factor), gives

$$P_3 = \frac{0.44 \times 10^{-8}}{1 + 0.6 \times 10^{-4}\, n_e}\,. \tag{18}$$

The total probability of decay by this mechanism will be $p_3 N(t_\alpha)$, where $N(t_\alpha)$ is given by (14), and then we have for the total number of scatterings required for the decay 10^8. In PNs the possible number of scatterings of an L_α photon can be about 10^6, so that only 1% of L_α photons decay into two quanta.

9.5 Theoretical Continuous Spectra of Nebulae in the Ultraviolet

In the conditions of gaseous nebulae, the generation of the continuous spectrum in the ultraviolet is inevitable by the following processes:

(a) Recombination of free electrons with the ions of hydrogen H^+.
(b) Recombination of free electrons with singly, He^+, and doubly, H^{++}, ionized helium.
(c) Free–free transitions of electrons (thermal bremsstrahlung).
(d) Two-photon emission of hydrogen.

The radiative capacity of the medium for each of these mechanisms is determined by a volume (specific) coefficient of emission, ε_ν, per electron and per ion in the frequency scale, ergs cm^{-3} s^{-1} Hz^{-1}. Energy, emitted isotropically per unit volume per unit time and in the frequency interval $d\nu$ of continuous spectra is equal to $\varepsilon_\nu \, n_e \, n^+ \, d\nu$, where n^+ is the concentration of ions of hydrogen (protons) or helium. In parametric form, ε_ν depends mainly on the electron temperature of the medium T_e, and in some cases also on n_e, the electron concentration.

Theoreticians have spent much time deriving in parametric form the function ε_ν for the mentioned processes, taking into account the peculiarities of atomic configurations of hydrogen and helium. At present our knowledge in this field can be considered sufficiently detailed, at least for astrophysical purposes.

The volume emission coefficient $\varepsilon_n(\nu)$ at the frequency ν of the continuous spectrum caused by recombination of a free electron at the level n of a hydrogen ion H^+ (proton) or ionized helium He^{++} is given by the following expression:

$$\varepsilon_n(\nu) = 2n^2\, h(h\nu)^3 \alpha_n \left(\frac{2}{\pi}\right)^{1/2} \frac{e^{I_n/kT_e}}{c^2(mkT_e)^{3/2}} e^{-\frac{h\nu}{kT_e}} , \tag{19}$$

where $I_n = Z^2\kappa^+/n^2$; κ^+ is the potential of ionization for hydrogen: $Z = 1$ for hydrogen, $Z = 2$ for doubly ionized helium; α_n is the effective cross section of photoionization; m is the electron's mass; k is Boltzmann's constant; h is Planck's constant.

The volume emission coefficient $\varepsilon_n(\nu)$ for continuous emission at the frequency ν caused by thermal bremsstrahlung of free electrons (free–free transitions) is given by the expression

$$\varepsilon_b(\nu) = \frac{2^5 Z^2 e^4 h}{3m^2c^3} \left(\frac{\pi\kappa^+}{3kT_e}\right)^{1/2} g_G e^{-\frac{h\nu}{kT_e}} , \tag{20}$$

where g_G is the Gaunt factor (e.g. Seaton, 1960).

The complete volume emission coefficient $\varepsilon(\nu)$ can be represented as the sum of all acts of recombination at all levels n of H^+ (H^{++}) and of thermal bremsstrahlung

$$\varepsilon(\nu) = \sum_n^\infty \varepsilon_n(\nu) + \varepsilon_b(\nu) \, . \tag{21}$$

Table 9.3. Volume emission coefficients of continuous radiation for hydrogen, ε_λ(HI), and for ionized helium, ε_λ(HeII), in units of 10^{-36} ergs cm^{-3} s^{-1} Å^{-1}, as a function of wavelengths λ, for the optical, 3 000–10 000 Å, and ultraviolet, 1 000–3 000 Å, regions and for three values of the electron temperature T_e: 10 000, 15 000 and 20 000 K

λ	ε_λ(HI)			ε_λ(HeII)		
[Å]	10 000	15 000	20 000	10 000	15 000	20 000
10 000	0.175	0.160	0.149	0.864	0.720	0.616
8 204$^+$	0.192	0.194	0.189	0.941	0.869	0.808
2 202$^-$	0.514	0.371	0.304	0.614	1.238	1.044
7 000	0.531	0.421	0.362	1.655	1.704	1.238
5 696$^+$	0.511	0.468	0.435	1.577	1.549	1.485
5 696$^-$	0.511	0.468	0.435	3.959	2.858	2.331
4 500	0.427	0.487	0.504	3.305	2.994	2.697
4 000	0.365	0.571	0.523	2.827	2.889	2.810
3 646$^+$	0.312	0.455	0.531	2.414	2.764	2.852
3 646$^-$	5.599	3.362	2.405	13.638	8.930	6.819
3 122	4.062	3.021	2.415	9.763	7.931	6.776
2 600	2.393	2.404	2.251	5.678	6.247	6.238
2 300	1.474	1.888	1.996	3.498	4.891	5.534
2 000	0.750	1.327	1.642	1.800	3.457	4.550
1 700	0.292	0.785	1.205	0.690	2.038	3.341
1 400	0.0692	0.341	0.710	0.1640	0.8864	1.995
1 216	0.0191	0.1600	0.428	0.0453	0.4222	1.1571
1 000	0.0022	0.0423	0.180	0.0510	0.1098	0.4980

The numerical values of ε_λ(HI) and ε_λ (HII) for H$^+$ and He^{++} in the wavelength scale and in units of ergs cm^{-3} s^{-1} Å^{-1}, obtained with the help of these formulae both for optical, 3 000–10 000 Å, and ultraviolet, 1 000–3 000 Å, bands are given in Table 9.3.

More complicated is the situation with the volume emission coefficient for singly ionized helium, He$^+$ (Robbins, 1968):

$$\varepsilon_{nl}(\nu) = \left(\frac{2}{\pi}\right)^{1/2} \frac{e^{\kappa^+_{nl}/kT_e}}{c^2(mkT_e)^{3/2}} \frac{\omega(nl)}{2} \alpha_{nl} h(h\nu)^3 e^{-\frac{h\nu}{kT_e}} , \qquad (22)$$

where α_{nl} is the cross section of photoionization of neutral helium, κ^+_{nl} is the ionization potential of helium from the level nl and ω(nl) is the statistical weight of the level nl. The complete emission coefficient ε_λ(HeI) for singly ionized helium we obtain by summing (22), analogously to (21), over all states n, l. The numerical values of ε_λ(HeI), again in the wavelength scale, are given in Table 9.4.

The volume emission coefficients for two-photon emission of hydrogen, $\varepsilon_\lambda(2q)$ are given in Table 9.2, without taking into account the factor of deac-

Table 9.4. Volume emission coefficients of continuous radiation for neutral helium, ε_λ(HeI), in units of 10^{-36} ergs cm^{-3} s^{-1} Å^{-1} as a function of wavelength λ and for three values of the electron temperature t_e: 10 000, 15 000 and 20 000 K

λ	ε_λ(HeI)			λ	ε_λ(HeI)		
[Å]	10 000	15 000	20 000	[Å]	10 000	15 000	20 000
10 000	0.176	0.161	0.150	4 000	0.387	0.491	0.535
8 268$^+$	0.191	0.192	0.187	3 680$^+$	0.320	0.453	0.524
8 268$^-$	0.224	0.210	0.199	3 680$^-$	1.257	0.968	0.855
8 197$^+$	0.225	0.212	0.201	3 422$^+$	1.106	0.934	0.863
8 197$^-$	0.263	0.232	0.215	3 422$^-$	5.888	3.553	2.550
8 195$^+$	0.263	0.232	0.215	3 122$^+$	4.888	3.373	2.590
8 195$^-$	0.380	0.297	0.256	3 122$^-$	5.375	3.644	2.772
7 849$^+$	0.393	0.313	0.272	2 600$^+$	3.170	2.930	2.624
7 849$^-$	0.516	0.382	0.317	2 600$^-$	5.270	4.085	3.365
7 440$^+$	0.542	0.411	0.344	2 300	3.249	3.198	2.982
7 440$^-$	0.552	0.416	0.347	2 000	1.672	2.257	2.452
6 636$^+$	0.590	0.473	0.468	1 700	0.641	1.330	1.794
6 636$^-$	0.619	0.488	0.416	1 400	0.146	0.579	1.055
5 696	0.625	0.546	0.490	1 216	0.0420	0.2760	0.5075
4 500	0.490	0.537	0.542	1 000	00474	0.0720	0.2685

tivation. Two-photon emission of ionized helium is insignificant and can be neglected.

Thus, the integral volume emission coefficient ε_λ can be represented as a sum of all four kinds of emission:

$$\varepsilon_\lambda = \varepsilon_\lambda(\mathrm{HI}) + \varepsilon_\lambda(\mathrm{HeII})\,\frac{n(\mathrm{He}^{++})}{n(\mathrm{H})} + \varepsilon_\lambda(\mathrm{HeI})\,\frac{n(\mathrm{He}^{+})}{n(\mathrm{H})} + \varepsilon_\lambda(2q)\,. \tag{23}$$

To apply this expression for real nebulae, it is necessary to know also the fraction of singly and doubly ionized helium, $n(\mathrm{He}^+)/n(\mathrm{H})$ and $n(\mathrm{He}^{++})/n(\mathrm{H})$, which depends on the temperature of the central star T_* and the dilution factor W. However, bearing in mind that the contribution of He^+ in (23) is not so large and that in average $n(\mathrm{He})/n(\mathrm{H}) \approx 0.1$, we can carry out with sufficient accuracy the corresponding computations based on (23).

The results of computations for a typical PN, i.e. with $T_e = 10\,000$ K and He/H = 0.1, are presented in in Fig. 9.3. Here the curves of all functions are shown: ε_λ(HI), ε_λ(HeI), ε_λ(HeII) and $\varepsilon_\lambda(2q)$ (by replacing ε by J), along with the resulting curves for volume emission coefficients in two cases, i.e. with helium's contribution in the form of doubly ionized helium (the upper curve) and of singly ionized helium (the second curve from the top). All curves concern the near and far ultraviolet, starting from the Balmer jump (3 646 Å) up to 1 000 Å.

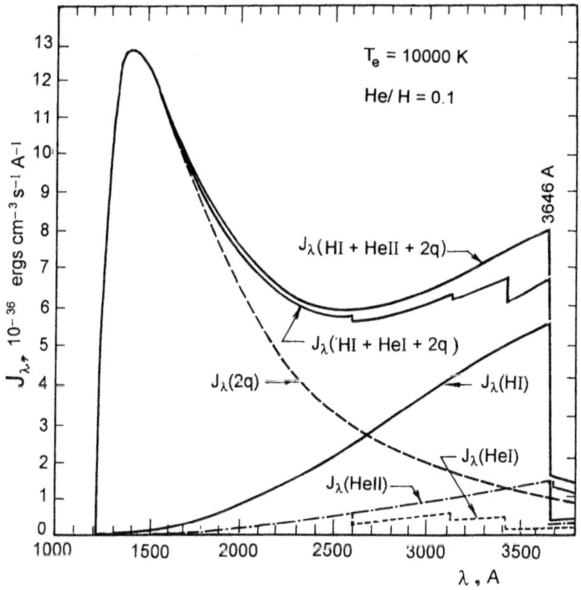

Fig. 9.3. Theoretical continuous spectra of gaseous nebulae in the near and far ultraviolet, 1 000–3 700 Å, for the following processes: Balmer continuum of hydrogen ε_λ(HI); continuum of singly ionized helium ε_λ(HeI); continuum of doubly ionized helium ε_λ(HeII); two-photon emission of hydrogen $\varepsilon_\lambda(2q)$. The upper two curves are the summed emissions. The ordinate is the volume emission coefficient in units of 10^{-36} ergs cm^{-3} s^{-1} $Å^{-1}$

The peculiarity of the continuous spectra of PNs in the ultraviolet is the almost complete absence of emission in the wavelength interval 912–1 216 Å. Then, at 1 216 Å, a sharp, jump-like increase of emission occurs owing to two-photon emission of hydrogen. Reaching a maximum at $\sim$ 1 400 Å in the wavelength scale (or at $\sim$ 1 500 Å in the frequency scale), the spectrum falls and increases quickly up to $\sim$ 2 000 Å. In the interval from 2 000 Å up to the Balmer jump (3 646 Å) the level of continuum emission is nearly constant.

9.6 Theoretical Continuous Spectra of Nebulae in the Visual Region

The theoretical continuous spectrum of gaseous nebulae in the optical range, from 3 000 Å up to 10 000 Å, is shown by a thick line in Fig. 9.4. This spectrum, drawn in accordance with the data of Tables 9.2, 9.3 and 9.4, represents the sum of three components: the hydrogen continuum ε_λ (HI), the continuum of neutral helium ε_λ (HeII) at He/H = 0.1, and the two-photon emission of hydrogen $\varepsilon_\lambda(2q)$. The helium continuum is characterized by a large number of weak jumps (Table 9.4).

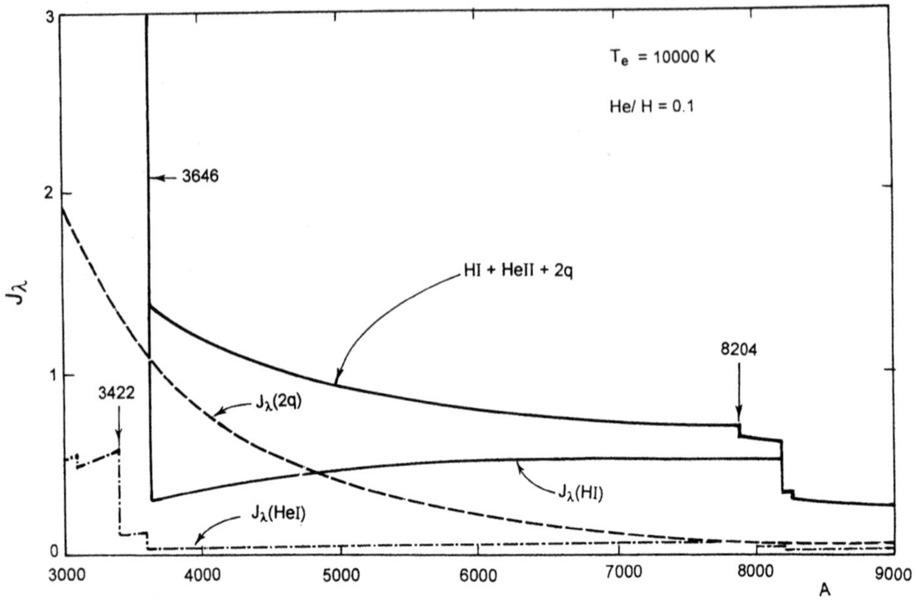

Fig. 9.4. Theoretical continuous spectra of gaseous nebulae in the optical region, 3 000–9 000 Å, for the following processes: continuum of hydrogen (HI); continuum of neutral helium (HeI); two-photon emission of hydrogen ($2q$)

The peculiar properties of the summed continuous spectrum in Fig. 9.4 are, first, the almost constant level within a broad interval – from 3 700 Å up to 8 000 Å – and, second, the presence of quite a strong jump at 3 646 Å – at the boundary of the Balmer continuum, the so called *Balmer jump.* In contrast with the Balmer jump of the stellar continuous spectrum, when the level of the continuous spectrum, from long wavelengths to short ones, undergoes a sharp *fall* at 3 646 Å – *a negative* jump – the sign of the Balmer jump in the case of the continuous spectrum of gaseous nebulae is *positive.* In this case the level of the continuous spectra *increases* sharply, from long wavelengths to short ones. Numerically, the magnitude of the Balmer jump D is determined by the expression

$$D = \log \frac{J_{3\,646-}}{J_{3\,646+}} \,. \tag{24}$$

The predicted value of D depends on the electron temperature T_e and decreases with the increase of T_e in the following manner:

T_e, K	6 000	10 000	15 000	20 000
$D(a)$	0.95	0.70	0.51	0.41
$D(b)$	1.70	1.27	0.84	0.63

Here, two values of D are adopted: (a) with the $2q$ emission accounted for, i.e. when the continuous spectrum is presented as a sum of three components:

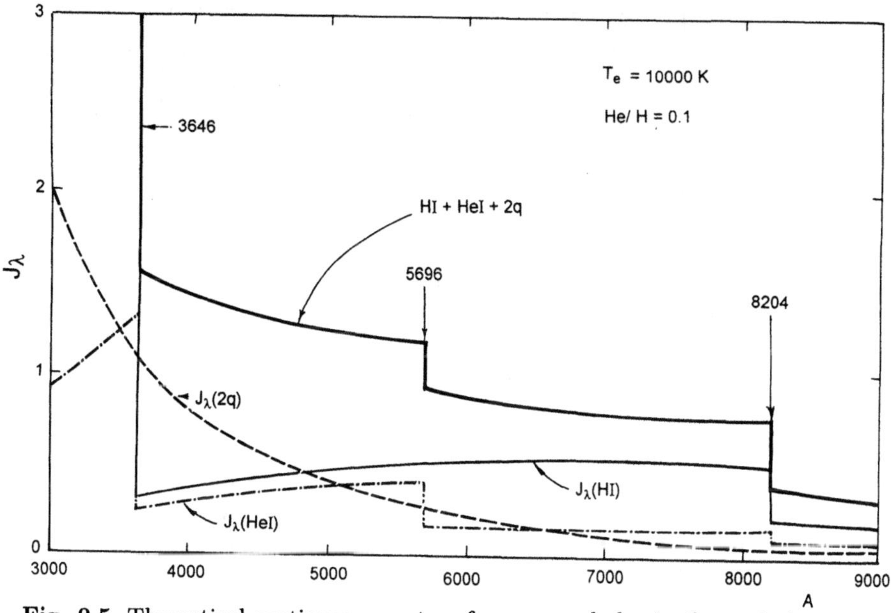

Fig. 9.5. Theoretical continuous spectra of gaseous nebulae in the optical region, 3 000–9 000 Å, for the following processes: continuum of hydrogen (HI); continuum of ionized helium (HeII); two-photon emission of hydrogen ($2q$)

HI + 0.1 HeI + $2q$; (b) without $2q$ emission, i.e. as a sum of two components: HI + 0.1 HeI. One can see that the $2q$ emission has an essential role for the Balmer jump at lower values of T_e.

The spectrum in Fig. 9.4 is related to low excitation nebulae where helium is predominantly neutral. The recombination of free electrons takes place with singly ionized helium He^+, so that $He^+ + e = He^0 + h\nu$.

The situation is different in the case of strongly excited nebulae in which helium is predominantly in the singly ionized state: recombination takes place with doubly ionized helium He^{++}, i.e. $He^{++} + e = He^+ + h\nu$. The summary continuous spectrum is shown in Fig. 9.5 by a solid line in accordance with the data of Tables 9.3 and 9.4 at He/H = 0.1.

Qualitatively the spectrum in Fig. 9.5 almost does not differ from the one in Fig. 9.4, if one neglects the weak jump at 5 696 Å ($D = 0.089$), due to ionized helium. The magnitude of the Paschen jump at 8 202 Å is nearly the same in both cases: $D(8\,202) = 0.30$–0.32. Figures 9.4 and 9.5 represent the details of the Balmer jump; the whole picture is given in Fig. 9.3.

The following conclusions can be drawn:

(a) The contribution of neutral helium in the continuous spectra of gaseous nebulae in the visible region is negligible – of the order of 5%.

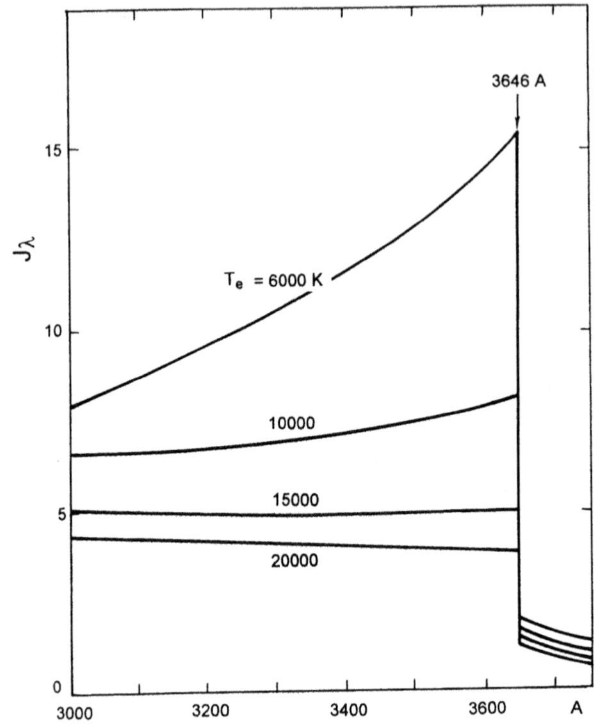

Fig. 9.6. Main features of the continuous spectrum after the Balmer jump, in the region of $\lambda < 3\,650$ Å, at four values of the electron temperature T_e. The slope of the spectrum is positive at $T_e < 15\,000$ K and negative at $T_e > 15\,000$ K

(b) The contribution of ionized helium is comparable with the hydrogen contribution only in the short wavelength region, shorter than 5 500 Å. At $\lambda > 6\,000$ Å the HeII contribution is negligible.
(c) The contribution of $2q$ emission is not large, smaller than 10% at $\lambda > 6\,000$ Å. In the region $\lambda \sim 5\,000$ Å this contribution is comparable with the recombination radiation of hydrogen. Only the 1/8 of the Balmer jump $J(3646^-)$ is due to two-photon emission.
(d) The main features of the continuous spectra at $\lambda > 3\,650$ Å are less sensitive to the electron temperature, so that this region cannot be used for determining T_e.
(e) In principle, the Balmer decrement, D, can serve as an indicator for T_e. However, the difficulties connected with the estimation of the true value of D from the observed spectrum make it less reliable.
(f) The declination of the mean level of the continuous spectra after the Balmer jump, i.e. at $\lambda < 3\,650$ Å, is noticeably sensitive to the electron temperature, as seen in Fig. 9.6. This declination can even change its sign with the variation of T_e; it is positive at $T_e < 15\,000$ K, and is negative

when $T_e > 15\,000$ K. The continuous spectrum in the wavelength scale is almost constant at $T_e > 15\,000$ K.

9.7 Comparison with Observations

A continuous spectrum in the optical region, longer than 3 300 Å, has been discovered in the spectrum of many PNs. The level of the continuous spectra is approximately constant at 3 600–4 800 Å, which qualitatively supports the theory. Qualitative comparisons have been carried out between observational magnitudes of the Balmer jump D_{obs} and predicted values of D_{cal} with the consideration of the $2q$ emission of hydrogen. For some early studied PNs the observed values of D_{obs} were in good accordance with the D_{cal} (Seaton, 1955), being interpreted as confirmation of $2q$ emission in PNs.

However, in view of the manipulation with logarithms of the initial magnitudes – the intensities of continuous spectra – which results in a sharp fall of the sensitivity of the method itself, one has to be cautious in drawing final conclusions on the presence or absence of $2q$ emission in PNs.

Reliable solution of the problem should be reached, obviously, by means of detailed observations of PNs carried out in the ultraviolet, at wavelengths shorter than 3 000 Å.

9.8 *Orion-2*: Two-Photon Emission of the Planetary Nebula IC 2 149

IC 2 149 was the first PN for which an ultraviolet spectrogram was obtained in 1973 by means of the space observatory *Orion-2* (Gurzadyan, 1975). This spectrogram, extending up to 2 400 Å, revealed signs of two-photon emission.

The brightness of the central star (nucleus) of this nebula ($\sim 10^{\text{m}}$) is approximately equal to that of the nebula itself. Therefore the spectrogram obtained could be represented as two superimposed spectra – the continuous spectrum of the nucleus and of the nebula. Separating them from each other, we obtained, in particular, the observed spectrum of the nucleus. After correction for interstellar extinction, it was in good agreement with a model of an O7 class star with an effective temperature $T_* = 50\,000$–$66\,000$ K: Table 5.3 and (O'Dell, 1962).

In Fig. 9.7, the *Orion-2* spectrum of IC 2 149 is presented along with that of the nucleus at $T_* = 50\,000$ K (solid line). The latter fits the observations only at wavelengths shorter than 3 800 Å up to 2 800 Å, while at shorter wavelengths a remarkable departure from the model distribution is notable: the observations yield greater flux than the model.

The marked *excess emission* – the shaded area in Fig. 9.7 – can be due to $2q$ emission generated by hydrogen atoms in the nebula itself. This conclusion can be checked by two methods.

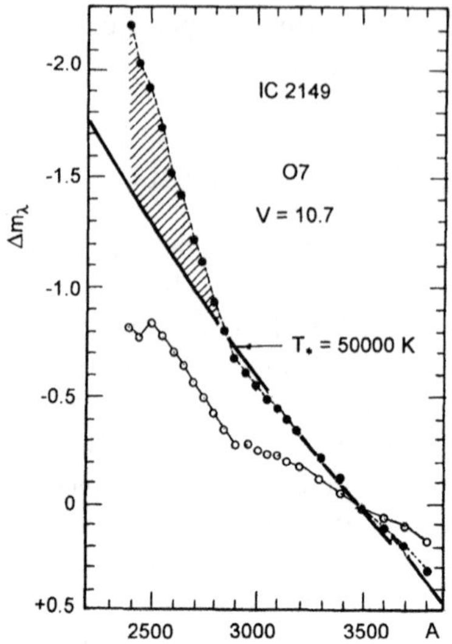

Fig. 9.7. The observed distribution of energy in the ultraviolet spectrum of the nucleus of the planetary nebula IC 2 149. Circles denote observations by *Orion-2*; the dots are corrected for interstellar extinction. The solid line indicates the nucleus of $T_* = 50\,000$ K; the flux excess (*shaded area*) is the suggested two-photon emission of the nebula

The first is the comparison of the observed distribution of energy in the spectrum of excess emission with the theoretical distribution.

Using the volume emission coefficient data for HI, HeII and $2q$ emission from (21) and (9), the ultraviolet continuous spectrum is recalculated (in the wavelength scale) for IC 2 149, at $T_e = 12\,400$ K, $n_e = 3.2 \times 10^3\,\text{cm}^{-3}$ and in wavelength region 2 400–3 800 Å. See Fig. 9.8, where the curves denote the various components of the continuous spectrum, the solid line the sum of these components, and the points observations from Fig. 9.7. The satisfactory accordance of observations and the theory can be interpreted as confirmation of $2q$ emission in this nebula.

9.9 The Analogue of Zanstra's Formula for Two-Photon Emission

The second method to check the conclusion drawn above is based on the estimation of the total amount of additional emission and its comparison with the theoretical value for two-photon radiation. It can be realized by means of a kind of Zanstra formula for two-photon emission (Gurzadyan, 1976).

First, one obtains the observed value of the ratio $Q(\lambda_1\lambda_2)/N(\lambda_1\lambda_2)$, where $Q(\lambda_1\lambda_2)$ is the additional emission (number of photons) between λ_1 and λ_2.

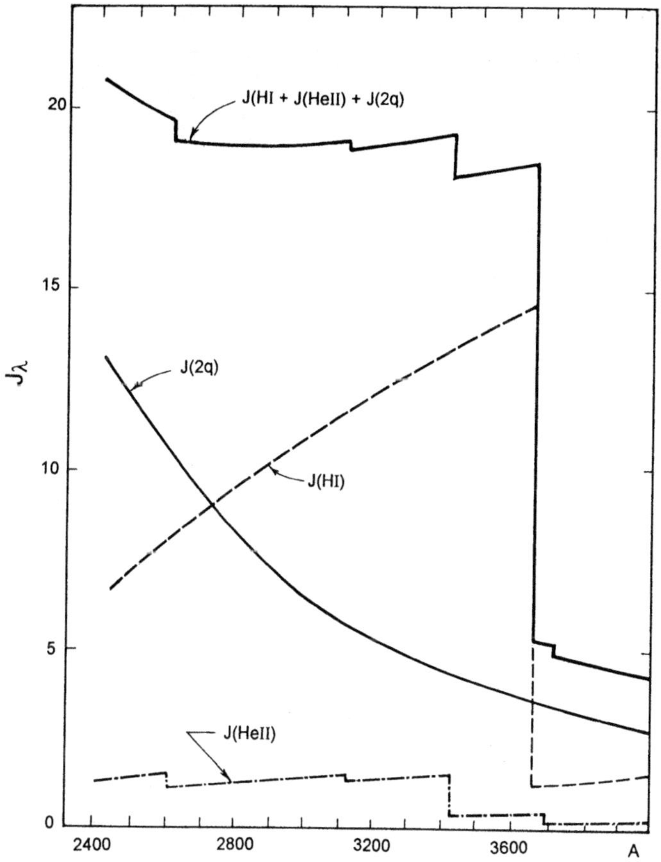

Fig. 9.8. Representation of the sum of the components of the continuous spectrum in the ultraviolet for the planetary nebula IC 2 149. The thick solid line is the sum of the hydrogen continuum (HI), ionized helium continuum (HeII) and two-photon emission ($2q$). Data are from the observations of Fig. 9.7

The function $N(\lambda_1\,\lambda_2)$ is the number of photons emitted by the central star in the same wavelength region. Taking the region $\lambda_1 - \lambda_2$ broad enough and close to the short wavelength region of spectrum, we can increase the fraction of $2q$ photon emission, thus increasing the reliability of its detection.

One also can obtain the ratio $Q(2q)/Q(\lambda_1\,\lambda_2)$ from the theoretical spectrum of two-photon emission, where $Q(2q)$ is the total number of $2q$-photons from $\lambda = 1\,216$ Å to $\lambda \to \infty$.

Finally, denoting by N_c the emission rate in the Lyman continuum ($\lambda <$ 912 Å) of the nucleus, we can derive the following relationship for the ratio $Q(2q)/N_\mathrm{c}$ and the temperature of the nucleus T_*, an analogue of Zanstra's formula for two-photon emission:

$$\frac{Q(2q)}{N_c} = \frac{Q(\lambda_1\,\lambda_2)}{N(\lambda_1\,\lambda_2)}\,\frac{Q(2q)}{Q(\lambda_1\,\lambda_2)}\,\frac{\displaystyle\int_{x_1}^{x_2}\frac{x^2\mathrm{d}x}{e^x-1}}{\displaystyle\int_{x_0}^{\infty}\frac{x^2\mathrm{d}x}{e^x-1}}\,, \tag{26}$$

where $x = hc/\lambda kT_*$.

In our case we have $\lambda_1 = 2\,400$ Å, $\lambda_2 = 2\,850$ Å, and $Q(\lambda_1\,\lambda_2)/N(\lambda_1\,\lambda_2) = 0.50$, estimated from Figs. 9.7 and 9.8, and $Q(2q)/Q(\lambda_1\,\lambda_2) = 14$ from the theoretical spectrum of $2q$ emission (Fig. 9.2). From here we derive from (26) $Q(2q)/N_c$ for two probable values of the nucleus temperature:

T_*, K	$Q(2q)/N_c$
55 000	0.67
66 000	0.45

According to the theory, approximately 1/3 of the hydrogen atoms formed by recombination at all levels, starting from the second, find themselves in the $2S$ state, inducing $2q$ transitions. Therefore, in the "pure" case, i.e. when the collisions are neglected, we have for the theoretical ratio $Q(2q)/N_c = 2 \times 0.33 = 0.66$. Assuming, of course, $t_c > 1$ (deactivation factor),

$$\frac{Q(2q)}{N_c} = \frac{0.66}{1 + 0.63 \times 10^{-4} n_e}\,, \tag{27}$$

we obtain, for IC 2149 $Q(2q)/N_c = 0.55$, which is in good agreement with observations at $T_* = 60\,000$ K.

Thus, two independent methods, both the spectral distribution (Fig. 9.8) and the amount of radiation emitted, support the existence of hydrogen two-photon emission in IC 2149.

9.10 Two-Photon Emission of the Planetary Nebulae NGC 7009 and NGC 7662

The PN NGC 7009 gives persuasive observational confirmation of the two-photon emission directly on the spectral recording. Owing to its large visible size ($10'' \times 23''$), the separate recordings, i.e. for only the nucleus and for only the pure nebula, were obtained by *IUE* in the region of 1 200–3 100 Å. In Fig. 9.9, the spectral recording of only the nebula, without the nucleus, is shown. Note that together with strong and faint emission lines, the well expressed continuous spectrum with a maximum near 1 400 Å is clearly distinguished. Undoubtedly the latter belongs to the nebula itself. The problem is to reveal its nature.

It is not difficult to be convinced that this continuous spectrum cannot be caused by hydrogen or ionized helium. This readily follows from the comparison of the observed spectrum with theoretical ones of hydrogen and ionized

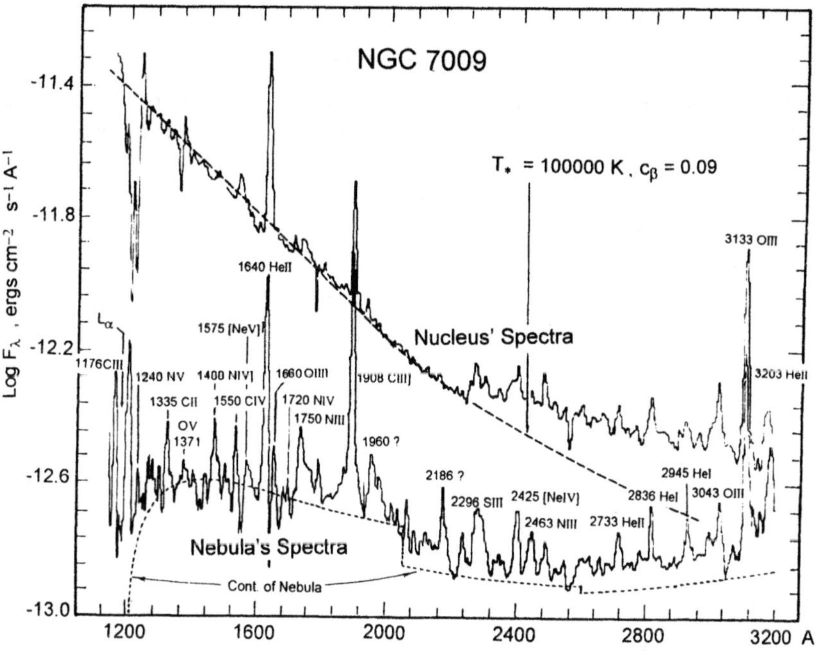

Fig. 9.9. *IUE* recording of the spectrum of the planetary nebula NGC 7 009. *Above:* The observed spectrum of the central star of the nebula and the black-body curve at $T_* = 100\,000$ K (*broken line*). *Below:* The observed spectrum of the nebula: continuum and emission lines; $2q$ emission (*broken line*), the sum of three components, hydrogen (HI), ionized helium (HeII), and two-photon emission ($2q$) (*solid line*). The continuum in the region shorter than 2 000 Å is almost completely due to two-photon transitions

helium, shown in Fig. 9.9 in accordance with to Tables 9.2 and 9.3. The curve of $2q$ emission and that of the sum HI + HeII = $2q$, are shown. As we can see, the region shorter than 2 000 Å and up to 1 216 Å belongs completely to two-photon emission. The summary curve is in complete accordance with observations at all wavelengths, from 1 200 Å up to 3 200 Å.

This conclusion has been confirmed by repeated *IUE* observations of NGC 7 009.

The picture is even more impressive with quantitative comparison, i.e. while applying (26), the analogue of Zanstra's formula for $2q$ emission. Directly from Fig. 9.9 we obtain for the numerical value of the ratio $Q(\lambda_1\,\lambda_2)/N(\lambda_1\,\lambda) = 0.125$, where $Q(\lambda_1\,\lambda_2)$ and $N(\lambda_1\,\lambda_2)$ are the numbers of photons emitted by the nebula and its nucleus, respectively, at wavelengths from $\lambda_1 = 1\,320$ Å up to $\lambda_2 = 2\,240$ Å.

First, we estimate from Fig. 9.2 for the ratio, $Q(2q)/Q(\lambda_1 - \lambda_2) \approx 3.0$. Then we obtain from (26) the numerical value of $Q(2q)/N_c$ at a given T_*:

T_*, K	100 000	50 000	40 000
$Q(2q)/N_c$	0.055	0.37	0.53

The predicted ratio $Q(2q)/N_c$ is equal to 0.40 at $n_e = 10^4$ cm^{-3} and when the electron collisions are neglected (Sect. 9.9), which yields, if we interpolate within the above data, $T_* = 48\,000$ K.

However, the true temperature of the nucleus of this nebula is higher, of the order of 100 000 K (see Fig. 7.11).

The reason for this discrepancy is clear: (26), as well as all Zanstra's formulae, are valid for $\tau_\kappa > 1$, i.e. when all L_c photons emitted by the nucleus are absorbed. Since quite often $\tau_c < 1$, the more exact formula reads

$$\frac{Q(2q)}{N_c} = \frac{1}{1-e^{-\tau_c}} \frac{Q(\lambda_1\,\lambda_2)}{N(\lambda_1\,\lambda_2)} \frac{Q(2q)}{Q(\lambda_1\,\lambda_2)} \frac{\displaystyle\int_{x_1}^{x_2} \frac{x^2 \mathrm{d}x}{e^x - 1}}{\displaystyle\int_{x_0}^{\infty} \frac{x^2 \mathrm{d}x}{e^x - 1}} . \tag{28}$$

For $T_* = 100\,000$ K, and $Q(2q)/N_c = 0.40$, we find from (28) that $\tau_c = 0.15$ for the nebula NGC 7 009.

What is the situation with the two-photon phenomenon in other PNs? First of all, $2q$ emission cannot be generated in dense nebulae, with electron

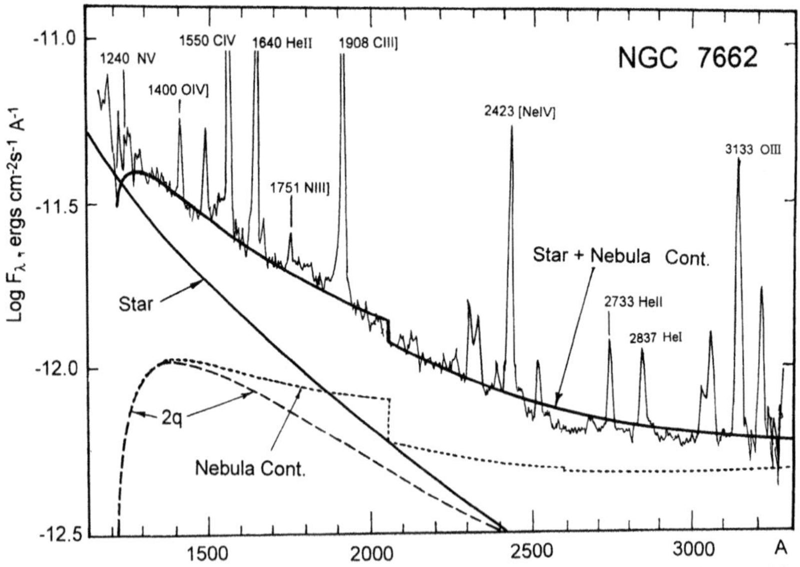

Fig. 9.10. *IUE* recording of the spectrum of NGC 7 662. The curves denote: black-body radiation of the central star of the nebula at $T_* = 100\,000$ K (*solid line*); the sum of three components of the continuum (*dotted line*), hydrogen (HI), ionized helium (HeII) and $2q$ emission (*broken line*). The region shorter than 2 000 Å is almost all due to two-photon transitions

concentrations of 10^4 cm^{-3} and higher. For example, for NGC 7027, we have $n_e = 4 \times 10^5$ cm^{-3}, and in accordance with the deactivation factor (11), the intensity of $2q$ emission must be diminished 25 times, so that $2q$ emission in this nebula cannot be observed.

Moreover, the role of interstellar extinction can also be crucial; e.g. in the direction of NGC 7027, $A(2\,200) - A(\mathrm{H}_\beta) = 6.3$ is obtained. This implies over a 300-fold weakening at ~2 200 Å compared with the emission line H_β, while one can show that the absolute weakening must be over 5 000 times! Indeed, *IUE* spectral recordings reveal no trace of any continuous emission in the ultraviolet.

In contrast to NGC 7027 with $c = 1.37$, the interstellar extinction is quite small for another PN, NGC 7662: $c = 0.23$ is close to that of NGC 7009: $c = 0.17$. The electron concentration of NGC 7662 is also not large – 3×10^3 cm^{-3} (Table 6.5). Hence, one could expect the presence of $2q$ emission in the continuous spectrum of this nebula. This is the case, as is seen in Fig. 9.10, of the *IUE* recording of NGC 7662 given for the wavelength interval 1 200–3 200 Å with the curve of the black-body spectrum for the nucleus at $T_* = 100\,000$ K (broken line) and the calculated continuous spectrum of the nebula as the sum HI + HEII+$2q$ (solid line), where, as in the case of NGC 7009, the region shorter than 2 000 Å belongs completely to $2q$ emission.

Thus, the short wavelength space observations confirm the phenomenon of the generation of two-photon emission of hydrogen in gaseous nebulae.

10. Infrared Radiation of Planetary Nebulae

10.1 The Structure of the Infrared Spectrum

The emission lines of gaseous nebula with wavelengths from 1 μm up to 100 μm are attributed to the infrared band. The corresponding composition and typical structure of the PN spectra is shown in Fig. 10.1.

It is convenient to divide this range into the following three intervals:

Near infrared	1–5	μm
Mid infrared	5–30	μm
Far infrared	30–100	μm

The strongest in the near infrared is the well known line of neutral helium 10 830 Å $= 1.083$ μm HeI; usually it is as intense as the H_β line of hydrogen. With the line 1.083 μm HeI is blended the third line of the Paschen series of hydrogen, 1.09 P_γ, though it is fainter by an order of magnitude. Then comes the weak line 2.17 μm Br_γ of the Brackett series, and the rather faint line 3.74 μm Pf_γ of the Pfund series.

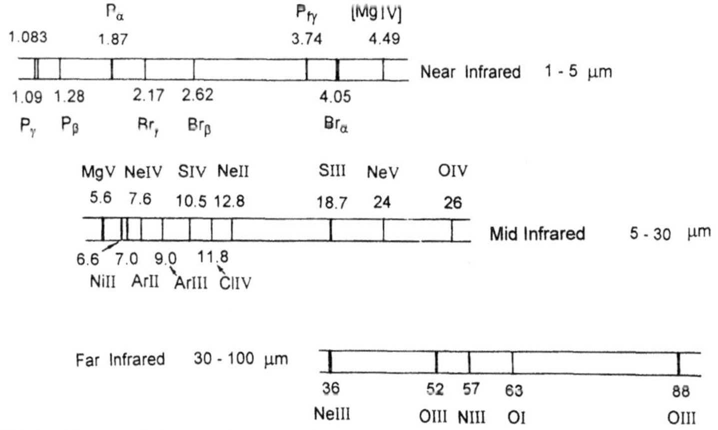

Fig. 10.1. Composition and general structure of the infrared spectrum of gaseous nebulae in the region of wavelengths 1–100 μm. Conditional limits of different parts: near infrared – 1–5 μm; mid-infrared – 5–30 μm; far infrared – 30–100 μm. All lines in the mid- and far infrared are forbidden

The middle infrared is quite rich, 5–30 μm, band. Here the strongest structure consists of two lines: 10.52 [SIV] and 15.56 [NeIII], of nearly the same intensities compared with the strength of the H_β line. In certain nebulae these forbidden lines are even stronger – 2–4 times – than the H_β line (e.g. NGC 7009, NGC 6153, NGC 7354). These lines are detected for practically for all PNs. The cases when the infrared spectrum of the nebula is given only by a single pair of lines, 10.5 [SIV] and 15.6 [NeIII], are common.

Next are the lines 9.0 [ArIII] and 18.7 [SIII]. The faintest in the mid infrared are the lines 12.8 [NeIII] and 14.3 [NeV], by an order weaker than the mentioned lines, 10.5 [SIV] and 15.6 [NeV].

The most exotic infrared line of five-fold ionized neon 7.65 [NeVI] has been also discovered (e.g. in the high excitation class PN NGC 6302 ($p = 1$)).

The far infrared, 30–100 μm, is represented by two forbidden lines, the doublet 52 [OIII] and 88 [OIII], arising at the transitions between sublevels of the ion O^{++}. The line 52 [OIII] is weaker in absolute intensity than the line 5007 [OIII] by three (NGC 3242, NGC 6543, NGC 6826) up to ten times (NGC 2440) (Pottasch et al., 1986). The critical electron concentration for these lines is relatively small: $\sim 6 \times 10^3$ cm^{-3} for 52 [OIII] and $\sim 7 \times 10^2$ cm^{-3} for 88 [OIII], and therefore in many PNs these lines would be strongly suppressed. This is the reason for the weakness and even the absence of these lines in the spectra of dense PNs (NGC 7027).

Both the infrared forbidden lines, 52 [OIII] and 88 [OIII], were observed in 1981 in the spectra of PNs of various excitation classes (NGC 6543, NGC 6720 etc.) as well as in diffuse nebulae (M 17, W 43, W 49 etc.) (Watson et al., 1981). The infrared line of doubly ionized nitrogen, 57.3 [NIII], also exotic, has been detected in the spectra of diffuse nebulae (M 17, W 43).

10.2 Infrared Emission Lines of Nebulae

The presence of four categories of infrared emission lines should be expected in the spectrum of planetary and diffuse nebulae: (a) recombination lines of hydrogen and helium; (b) forbidden lines of different ions; (c) molecular bands of hydrogen and other elements; (d) Molecular emission bands of dust component of nebula – heated crystals and dust particles.

Let us illustrate the situation with the example of the PN NGC 7027, which has been studied in great detail. This nebula has been observed in the ranges 0.9–2.7 μm, 2–4 μm, 2–14 μm, 8–13 μm. Observations have indicated the following:

(a) Emission lines of hydrogen of the following series:

Paschen	$(i \to 3)$:	1.87 P_α, 1.28 P_β, 1.09 P_γ,
Brackett	$(i \to 4)$:	4.05 Br_α, 2.62 Br_β, 2.17 Br_γ up to 1.53 Br_{19},
Pfund	$(i \to 5)$:	3.74 Pf_α .

(b) Rather faint emission lines of molecular hydrogen (H_2): 2.42 μm, 2.41 μm, 2.12 μm.
(c) Faint emission lines of neutral helium : 1.083 HeI, 2.06 HeI, as well as one or two lines of ionized helium: 1.01 HeII(4–5).
(d) Forbidden lines 1.04 [NI], 1.03 [SII], 0.95 [SIII], 0.91 [SIII].

Intensities $F(\lambda)$ of the strongest among these lines corrected for interstellar extinction are given in Table 10.1.

Over a hundred emission lines in the region 0.68–1.05 μm have also detected in the spectrum of NGC 7027 (Pequignot, Baluteau, 1988). Among the identified ones were the recombination lines of hydrogen HI in the Paschen series (up to the line $n = 40$), of ionized helium HeII in the series $(5, n)$ and $(6, n)$ (up to line $n = 43$) and neutral helium HeI of few series $(3l, nl)$. In all cases the observed intensities fit the predictions of photoionization theory, perfectly.

Estimation of theoretical intensities of infrared forbidden lines do not represent any difficulty if the numerical values of the atomic parameters are known. In fact, there are only two: the Einstein coefficients of spontaneous transitions A_{ij} and the "collision strength" $\Omega(ij)$. Corresponding calculations for a large number of infrared forbidden lines have been performed by Petrosian (1970), as well as by Simpson (1975). Results for most interesting ions and configurations p and p^5 as well as p^2 and p^4 (doublets) are given in Table 10.2.

Historically, the first infrared line of three times ionized sulphur 10.5 [SIV] was simultaneously detected in the early 1970s in the spectra of three PNs – NGC 7027, NGC 6572 and NGC 7009.

The second infrared line, 9.0 [ArIII], was detected in NGC 7027 and NGC 6572, with intensities two orders smaller than the calculated ones.

The third line, 12.8 [NeII], was observed in five PNs, NGC 7027, 7662, 6572, IC 418 and BD+30°3639. Along with [SIV], [ArIII] and [NeII] other infrared lines were also observed in the spectrum of NGC 7027, e.g. at 3.09, 3.3, 11.3, 11.4 μm, etc. Obviously, these first airborne observations were continued with tremendous intensity by space missions.

Table 10.1. Observed intensities, $F(\lambda)/F(H_\beta)$, of the most intense infrared emission lines in the spectrum of the planetary nebula NGC 7027. $F(H_\beta) = 21.9 \times 10^{-10}$ ergs $cm^{-2}s^{-1}$

Line [λ, μm]	$F(\lambda)/F(H_\beta)$	Line [λ, μm]	$F(\lambda)/F(H_\beta)$
1.083 HeI	0.854	2.17 Br_γ (7–4)	0.038
0.95 [SIII]	0.492	3.74 Pf_γ (8–5)	0.007
1.28 P_β (5–3)	0.112	0.4686 H_β	1
1.09 P_γ (6–3)	0.069		

Table 10.2. Calculated intensities $F(\lambda)$ of a number of infrared forbidden lines of some ions in units of F(10.5 [SIV]) = 100

Line, λ, μm	$F(\lambda)$	Line, λ, μm	$F(\lambda)$
Configuration p		Configuration p^5	
10.53 [SIV]	100	12.79 [NeII]	40
25.87 [OiV]	10	3.21 [CaICV]	0.7
34.8 [SiII]	2	6.98 [ArII]	0.3
Configuration p^2		Configuration p^4	
88.16 [OIII]	3.9	147 [OI]	0.06
51.69 [OIII]	14.9	63.07 [OI]	0.7
24.15 [NeV]	1.0	35.20 [MeIII]	0.9
14.32 [NeV]	0.7	15.35 [NeIII]	10
33.65 [SIII]	17	21.84 [ArIII]	0.2
18.68 [SIII]	21	9.00 [ArIII]	3

Do the *infrared emission lines* have a crucial role by virtue of their informativeness and significance for the physics of PNs ? It can seem strange but the answer to this question seems to be negative. Infrared emission lines can provide us mainly with information on three parameters: the electron temperature T_e (not always), electron concentration n_e and relative abundance of a given ion, N_i/N_p, while we have already learned how to successfully determine these parameters with the help of forbidden lines in the visual, and now also in ultraviolet region.

The infrared observations of the *continuous* spectra of nebulae nevertheless have led to a discovery of exceptional importance, i.e. of the *dust particles as permanent components of PNs.*

In spite of this remarkable fact, the observations in *infrared lines* can hardly have fundamental consequences for the physics of gaseous nebulae.

10.3 *IRAS*: Infrared Spectra of Planetary Nebulae

Observations of all infrared lines presenting astrophysical interest turn out to be possible only if the infrared telescope is located outside the Earth's atmosphere.

The first intense infrared observations of PNs in space conditions were connected with the launching in 1983 on a polar orbit of a special astrophysical satellite *IRAS* – the *InfraRed Astronomical Satellite*, with the infrared telescope on board equipped with a cryogenic system. The observations have been carried out in four bands centred at 12, 25, 60 and 100 μm.

The first list of *IRAS* observations already contains the data on the radiation fluxes of 64 PNs in the following lines (Pottasch et al., 1986):

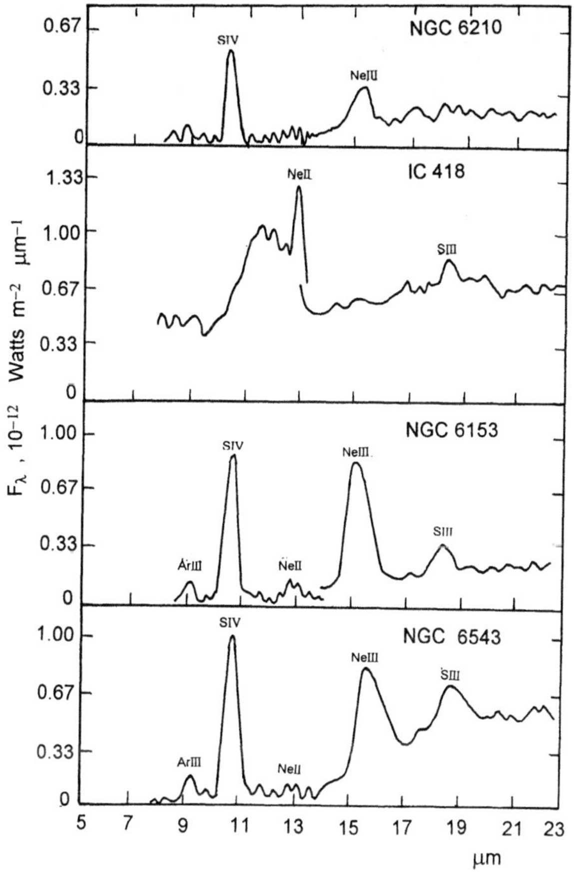

Fig. 10.2. *IRAS* recordings of the infrared spectra for four planetary nebulae in the wavelength region of 7–23 μm. The strongest lines are indicated

7.65 [NeVI]	10.52 [SIV]	14.3 [NeV]	18.7 [SIII]
9.0 [ArIII]	12.8 [NeII]	15.56 [NcIII]	

IRAS recordings of infrared spectra in the region of 7–23 μm for four PNs are shown in Fig. 10.2. Certain features are common to all these spectra; they contain both components – the continuous spectrum and emission lines. The behaviour of the continuous spectrum is not the same, however. In some cases it falls quickly starting from ~ 14 μm and even disappears at $\sim 10\mu$m (NGC 6 210, NGC 6 153) at the level of flux sensitivity of $\sim 3 \times 10^{-14}$ W m^{-2} μm^{-1}. For other objects, on the other hand, the level of the continuous spectra varies weakly (IC 418). The composition of lines seems more stable: the forbidden lines of ArIII, Siii, SIV, NeII, NeIII are typically fixed.

In some nebulae, (NGC 6 153 and NGC 6 543), the line 15.5 [NeIII] is the strongest, the first component of the doublet [NeIII]. The second component

Table 10.3. Results of *IRAS* observations of the infrared emission line 15.5 μm [NeIII] in the spectrum of four planetary nebulae

Planetary nebula	p	$F(H_\beta)$ $[10^{-14} Ws^{-1}]$	$\frac{F(15.5)}{f(H_\beta)}$	$\frac{F(3686)}{F(15.5)}$	$T_e(155)$ [K]	$T_e([OIII])$ [K]
NGC 6153	6	15	4.7	0.20	7200	8400
NGC 6210	4	10	1.3	0.60	9000	9570
NGC 6543	5	28	2.5	0.20	7200	8230
IC 418	1	50	0	–	–	8420

of this doublet, 35.2 [NeIII], in accordance with calculations, must be weaker by an order of magnitude, i.e. below the *IRAS* working sensitivity. Note that in many PNs the line 15.5 [NeIII] is stronger than H_β (Table 10.3).

The next line, by its strength, is 10.52 [SIV] (p). The line of moderate strength 18.7 [SIII], the first component of the doublet [SIII] (p^2), is visible in almost all nebulae. The second component, 33.6 [SIII], nearly of the same strength, again was out of the working range of the telescope. The remaining lines, 8.99 [ArIII] and 12.82 [NeII], are rather weak and at the limit of discovery.

Besides these lines, a rather powerful emission band was fixed at 11.3 μm in the spectrum of IC 418 and Cn 1–1, both of low excitation classes. Clearly, the excitation of this band is connected with dust particles.

The line 15.5 [NeIII] is more sensitive with respect to the electron temperature T_e than to the electron concentration n_e. For example, the ratio of intensities $F(3869$ Å $[NeIII])/F(15.5$ μm $[NeIII])$ depends weakly on n_e, and therefore can be used as an indicator for the determination of T_e. This dependence is given in Fig. 10.3 (see Pottasch, 1984). Using the data of ground based observations of the line 3 869 [NeIII] (Pottasch et al., 1986), the values of $T_e = T_e(15.5)$ were estimated by known values of the ratio of $F(3868))/F(15.5)$ (fifth column of Table 10.3) for a sample of four low excitation (p) PNs. The results are given in the sixth column of Table 10.3. In the last column the values of $T_e = T_e([OIII])$, obtained using the "[OIII] method" i.e. with the help of the ratio $N_1 + N_2/4363$ (Table 6.1), are given. In all three cases the accordance between both T_e is quite good; hence, the electron temperatures obtained are related to one and the same region.

To summarize, we may conclude that space observations in the infrared emission line of PNs in the range 7–2/3 μm do not lead to extraordinary conclusions nor to the necessity of revising existing viewpoints on the composition of PNs.

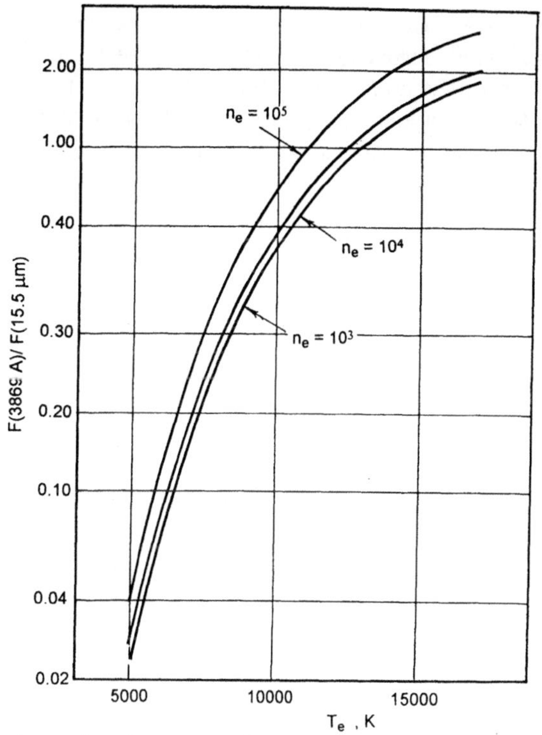

Fig. 10.3. The theoretical dependence between the electron temperature of nebulae T_e and the ratio of the intensities of the lines 3 896 Å [NeIII] and 15.5 μm [NeIII], for three values of the electron concentration n_e: 10^5, 10^4 and 10^3 cm^{-3}

10.4 The Nature of Infrared Emission of Planetary Nebulae

From the energy point of view, the observed infrared emission of PNs in the region of 1–30 μm occurs almost completely in continuous emission; the contribution of infrared emission lines is small – of the order of 5–10%. The absolute flux of infrared emission is much larger compared with the continuum of all atoms and ions as well as thermal bremsstrahlung.

From here, the first important conclusion can be drawn: *infrared continuous emission of PNs cannot be of atomic origin.*

The spectrum of infrared emission as a rule becomes stronger towards the long wavelength region. Usually the maximum of the spectrum occurs at 10–30 μm, which corresponds to a black-body temperature of 300–100 K, i.e. two orders smaller than the nebula's electron temperatures.

Hence, the second important conclusion is: *infrared emission of PNs is not determined by the high electron temperature of nebulae.*

Obviously at such low temperatures, 100–300 K, gas in the nebula cannot radiate. Hence, the observed infrared emission should be related to the thermal emission of the dust particles within the nebula. The heating can be performed in any way – either by absorption of powerful ultraviolet radiation of the central star, or the no less powerful L_α radiation of the nebula itself, or by collisions with the high energy particles, such as electrons, proton beams, etc. At the same time, heating up to a temperature of 100–300 K, clearly, does not result in the evaporation of crystals themselves.

Thus, the third important conclusion is: *the dust particles or the crystals of solid matter are, together with the hot gas – plasma – the permanent component of PNs.*

The idea of the existence of dust particles in PNs goes back to the 1960s, when quite unexpectedly powerful infrared emission was discovered in NGC 7027 (Gillet et al., 1967), and the impossibility of its generation by free–free emission was demonstrated (Delmar et al., 1967).

The continuous spectrum in the region $\lambda > 2.9$ μm is much higher (by two orders of flux magnitude) than the extrapolated level of radio emission (free–free continuum). This additional emission is attributed to the thermal emission of dust particles.

According to direct measurements, the absolute flux of infrared emission is of the order and even larger than the L_α flux of nebulae. This makes it highly probable that the assumption on the heating of dust particles in nebulae via the absorption of L_α photons is correct. Meanwhile, in optically thick nebulae 75% of L_c energy transforms to L_α energy (Chap. 11); hence, the flux of L_α radiation in nebulae really must be very high. The possibility of heating of dust particles by L_α radiation was noticed just after the discovery of infrared emission in NGC 7027 in 1967.

Hence, the fourth important conclusion is: *the heating of dust particles in nebulae is realized via the hydrogen ionizing energy emitted by the central star of the nebula.*

There are no theoretical reasons excluding the existence of dust particles in strongly heated gaseous media. The collisions of dust particles with thermal ions or electrons at the kinetic temperatures $\sim$ 10 000 K is not connected with any destruction effects either for the crystals (silicates, carbonates etc.) or dust particles (see Sect. 10.6).

However, the conclusions that the infrared emission is issued by the heated dust particles and that the heating of dust particles is realized by L_α photons have to be checked quantitatively.

10.5 Parameters of Dust Particles

Consider the question of the determination of the parameters of dust particles in the nebula by the data from infrared observations.

Let N be the number of L_α photons appearing in an ionized zone per unit time and per unit volume. Obviously this rate must be equal to that of the recombinations of free electrons with protons: $n_1 = n_e n^+ \alpha_r = n_\alpha^2 \alpha_r$, where α_r is effective coefficient of recombination on levels $n > 2$. The total luminosity in L_α photons of the nebula with a volume V will be

$$L_\alpha = n_e^2 \alpha_r \, h\nu_\alpha \, V \,, \tag{1}$$

where $h\nu_\alpha$ is the energy of an L_α photon.

Let q be the fraction of L_α energy spent on heating dust particles within the nebula. According to the conjecture made above, the energy qL_α absorbed by dust particles is reemitted in the infrared range. Therefore we can write for the infrared luminosity L_{IR}

$$L_{IR} = q n_e^2 \, \alpha_r \, h\nu_\alpha \, V \,. \tag{2}$$

Assuming the same mean mass for all PNs,

$$n_e \, V = \text{const} \,, \tag{3}$$

we have from (2)

$$L_{IR} \sim n_e \,. \tag{4}$$

This is a remarkable relationship indicating a linear correlation between the infrared luminosity and the electron concentration in the nebula, if we adopt the "dust" nature of infrared radiation and the heating of dust particles by L_α radiation.

The results of homogeneous infrared observations by Cohen and Barlow (1974) for a sample of PNs and in the wavelength interval 2.2–22 μm, which are still of importance, are given in Table 10.4. As an immediate result one has T_{IR}, the effective temperature obtained by the Planckian fit of the observed distribution of energy, and L_{IR}, the absolute infrared luminosity obtained according to the expression

$$L_{IR} = 4\pi \, D^2 \, F_{IR} \,, \tag{5}$$

where D is the distance of the nebula and F_{IR} is the observed IR flux.

The infrared energy emitted by a particle of radius a by the black-body temperature will be $4\pi a^2 \sigma \, T_{IR}^4 \, Q_i$, where σ is the Stefan–Boltzmann constant and Q_i is the efficiency of emission by particle in infrared. The total number of particles is $n_\alpha V$, where n_d is the particle concentration. Therefore we have for the total infrared luminosity of the nebula

$$L_{IR} = 4\pi \alpha^2 \sigma \, T^4 Q_i n_d \, V \,. \tag{6}$$

Equating this expression with (2), we obtain

$$T = n_e^{1/2} \left(\frac{q\alpha_r h\nu_\alpha}{4\pi n_d a^2 Q_i \sigma} \right)^{1/4} \sim n_e^{1/2} \,, \tag{7}$$

i.e. the "infrared" temperature of the nebula T_{IR} must vary, depending on the electron concentration, relatively slowly, according to $\sim n_e^{1/2}$.

Table 10.4. Infrared effective temperatures T_{IR} and infrared luminosity L_{IR} for a sample of planetary nebulae

Planetary nebula	T_{IR} [K]	n_e [$10^3 cm^{-3}$]	$\frac{L_{IR}}{L_\odot}$	$\frac{L_{IR}}{L_\odot}$
NGC 2440	185	3.27	105	0.4
NGC 6210	175	2.43	1000	3.8
NGC 6302	185	5.53	520	2.5
NGC 6439	–	1.38	520	2.8
NGC 6543	–	7.60	130	2.7
NGC 6572	174	8.42	4900	4.5
NGC 6741	205	12	700	0.5
NGC 6751	170	0.7	480	5.5
NGC 6790	245	37.3	14600	3.2
NGC 6803	190	4.16	1100	2.0
NGC 6807	185	13	3800	2.3
NGC 6826	–	2.3	123	0.2
NGC 6881	200	11.2	1700	1.2
NGC 6884	210	6.62	410	0.5
NGC 6886	165	9.85	1400	1.1
NGC 6891	–	1.66	40	0.3
NGC 6905	–	0.53	22	0.5
NGC 7009	350	4.73	210	0.6
NGC 7026	200	6.15	710	2.0
NGC 7027	210	25.5	11400	3.6
NGC 7662	–	3.06	23	0.08
IC 418	250	14.1	950	0.5
IC 2149	170	6.36	390	0.7
IC 2165	190	7.68	500	0.5
IC 3568	–	4.57	24	0.1
IC 4593	155	2.2	155	1.7
IC 4634	155	3.14	300	1.7
IC 4997	175	18.2	15000	6.5
IC 5117	205	30.4	11400	3.0
IC 5217	–	7.0	250	0.3
A 30	155	0.035	255	7.8
A 78	135	0.024	300	4.6
BD+30°3639	210	15.2	25500	13
Cn 3–1	155	8.56	2600	2.3
Hu 1–2	–	5.88	130	0.2
Hu 2–1	200	12.9	3100	1.9

The values of $T_{\rm IR}$ and $L_{\rm IR}/L_\odot$, obtained in this manner are given in the second and third columns of Table 10.4. In the fourth column, the values of $n_{\rm e}$ are collected from different sources. First, note that neither the law $L_{\rm IR} \sim n_{\rm e}$, nor $L_{\rm IR} \sim n_{\rm e}^{5/2}$ fit the data, and, second, the dependence $T_{\rm e} \sim n_{\rm e}^{1/2}$ again is far from satisfactory.

We have for the complete mass of dust particles $\mathcal{M}_\odot$

$$\mathcal{M}_d = n_d V \rho_d \frac{4\pi}{3} a^3 . \tag{8}$$

Comparing this with (6) we obtain

$$\mathcal{M}_d = \rho_d \frac{a}{3Q_i} \frac{L_{\rm IR}}{\sigma T_{\rm IR}^4} , \tag{9}$$

where r_d is the density of the dust particle.

Using the numerical values of constants and bearing in mind that $Q_i = 2\pi a/\lambda_m$, where λ_m is the wavelength of the maximum on the IR spectral curve of the nebula, we obtain at $\lambda_m \approx 20$ μm and $r_d \approx 3$ g cm^{-3}

$$\frac{\mathcal{M}_d}{\mathcal{M}_\odot} = 9.21\, T_{\rm IR}^{-4} \frac{L_{\rm IR}}{L_\odot} . \tag{10}$$

The scatter in values of $T_{\rm IR}$ of PN included in Table 10.4 is not large, on average $T_{\rm IR} \sim 200$ K, while the scatter in the magnitudes of $L_{\rm IR}/L_\odot$ is large. Therefore, we have from (10)

$$\frac{\mathcal{M}_d}{\mathcal{M}_\odot} = 5.8 \times 10^{-9} \frac{L_{\rm IR}}{L_\odot} . \tag{11}$$

The limiting values of the ratio $L_{\rm IR}/L_\odot$ vary from 20 (NGC 7662, NGC 6826) to 2.5×10^4 (BD+30°3639). The limiting values of $\mathcal{M}_d$ should be in the range $(10^{-7}$–$10^{-4})\mathcal{M}_\odot$, respectively.

Adopting for the mass of purely gaseous nebula $\mathcal{M}_g \approx 0.1\mathcal{M}_\odot$, we then have

$$\frac{\mathcal{M}_d}{\mathcal{M}_g} \sim 10^{-6} - 10^{-3} .$$

For the interstellar medium the ratio $\mathcal{M}_d/\mathcal{M}_g$ is of the order 10^{-3}-10^{-2}. This discrepancy is real, and in all cases PNs cannot be considered as suppliers of dust to interstellar medium.

10.6 Heating of Dust Particles in Nebulae

In the last column of Table 10.4, the ratios of infrared luminosity, $L_{\rm IR}$, to the Lyman alpha luminosity, $\rm L_\alpha$, are given. The latter, $\rm L_\alpha$, is obtained from the observed radio emission S_ν on the frequency ν, generated by thermal electrons at free-free transitions and on the assumption that total number of ionized photons $\rm L_c$, emitted by the central star, is equal to the total number of $\rm L_\alpha$ photons, emitted by the nebula (Chap. 11). The following relationship for the $\rm L_\alpha$ luminosity of nebula was derived by Rubin in 1968:

$$\mathrm{L}_{\mathrm{L}_\alpha} = \mathrm{L_c} h\nu_\alpha = 7.77 \times 10^{54} \nu_\alpha^{0.1} S_\nu T_{\rm e}^{-0.45} D^2 \ \mathrm{ergs\,s^{-1}} \,. \tag{12}$$

From (5) and (11) we have

$$\frac{L_{\rm IR}}{\mathrm{L}_{\mathrm{L}_\alpha}} = 4.68 \times 10^{-56} T_{\rm e}^{0.45} \frac{F_{\rm IR}}{S_\nu} \,, \tag{13}$$

i.e. the ratio $L_{\rm IR}/\mathrm{L}_{\mathrm{L}_\alpha}$ is independent of the nebula's distance and therefore acquires the status of an independent parameter.

In accordance with to the conjecture of the heating of dust particles by $\rm L_\alpha$ radiation, one must expect the condition: $L_{\rm IR}/\mathrm{L}_\alpha < 1$ to be fulfilled. Meanwhile, judging by the data in Table 10.4, for the majority of nebulae (~60%), $\mathrm{L_{IR}}/\mathrm{L}_{\mathrm{L}_\alpha} > 1$, in certain cases even being about 10.

Clearly the $\rm L_\alpha$ radiation cannot be the only source of heating of dust particles in PNs.

A way out of this situation may be the assumption that the heating of dust particles is realized under the action of direct $\rm L_c$ radiation of the central star, part of which is the $\rm L_\alpha$ radiation of the nebula. Note that only those $\rm L_c$ photons are transformed into $\rm L_\alpha$-photons which are located near the limiting frequency $\nu_0 n$ (912 Å), while at high frequencies this efficiency falls sharply, according to $\sim \nu^{-3}$. Thus, the efficiency of absorption of a photon by hydrogen falls to half at $\lambda \sim 725$Å, being only 10% at $\lambda \sim 425$Å and 1% at $\lambda \sim 200$ Å. One can therefore assume that the wavelength region shorter than 500 Å remains practically unused by hydrogen and, hence, cannot be transformed into $\rm L_\alpha$-photons. In other words, of the total ultraviolet energy E_u, emitted by the central star, only a small fraction $E_{\rm c}$ is used by the nebula and transformed into $\rm L_\alpha$ photons. This unused resource of energy $E_{\rm c}(E_u/E_{\rm c} - 1)$ can lead to the direct heating of dust, without transforming into $\rm L_\alpha$ photons.

Obviously, this resource of ultraviolet energy, $E_u/E_{\rm c}$, is equal

$$\frac{E_u}{E_{\rm c}} = \frac{1}{1 - e^{-\tau_{\rm c}}} \, \frac{\displaystyle\int_{x_0}^{\infty} \frac{x^3 \mathrm{d}x}{e^x - 1}}{\displaystyle x_0^3 \int_{x_0}^{\infty} \frac{\mathrm{d}x}{e^x - 1}} \,, \tag{14}$$

where $x_0 = h\nu_0/kT_* = 159\,000/T_*$, and the term $(1 - e^{-\tau_c})$ takes into account that the nebula can be optically thin ($\tau_c < 1$) at frequencies of hydrogen ionization.

Clearly the "resource factor", E_u/E_c, can be identified with L_{IR}/L_α. As the calculations indicate, even at $T_* \sim 30\,000$ K and $\tau_c \gg 1$, this factor is more than unity, of the order of 10 at $\tau_c \sim 0.2$. This resource is quite large, of the order of 10 and larger for PN with high temperature nuclei, $T_* \sim 60\,000$ K and higher.

The question, to what extent this resource of ultraviolet energy is used and absorbed by the dust particles needs separate examination. There are reasons to assume that the efficiency of absorption of ultraviolet photons by dust particles should be higher than that of the L_α photons.

Thus, our conjecture made above and its consequences on the heating of dust particles by L_α radiation can be generalized in the form of *a fifth main conclusion:* the heating of dust particles in nebulae is realized mainly by radiation of the central star in the far ultraviolet, shorter than 500 Å.

10.7 The Morphology of the Dust Component of a Nebula

The comparison of infrared maps of individual nebulae with their optical images leads to the following conclusions by their morphological character:

(a) In some cases (NGC 7 027, BD+30°3639) the infrared images of different wavelength are in mutual correlation and even correlate with the images in hydrogen lines (Bentley et al., 1984).
(b) The phenomenon of bipolarity of the nebula in certain cases is evident in both bands – in dust emission and in the emission of gas.
(c) For some objects (NGC 6 302) the infrared morphology is in marked discrepancy with that in optical band (Lester, Dinerstein, 1984).

Thus, in general, one cannot exclude the possibility of "morphological surprises" on the basis of infrared observations. On this occasion, the detailed consideration of the influence of the genuine magnetic field of nebulae on the distribution of dust particles is of particular interest (see Chap. 17).

10.8 Infrared Lines of [OIII]

The thin structure of the ground state ^3P of the ion O^{++} includes three sublevels, $^3P_0$, $^3P_1$ and P_2^3: transitions between them lead to the forbidden infrared lines, 52 μm [OIII] and 88 μm [OIII] (Fig. 10.4). These lines are associated with the known forbidden lines N_1 (5 007 Å) and N_2 (4 959 Å)

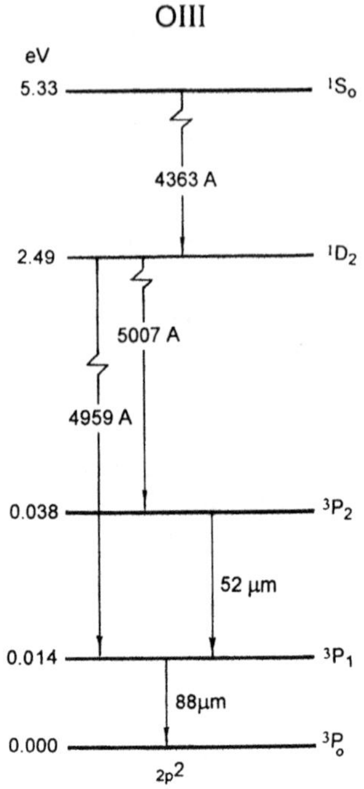

Fig. 10.4. Energy levels of [OIII]

appearing at the transitions from the upper level $^1D_2$ to $^3P_2$ and $^3P_1$, respectively.

The intensities of the lines 52 [OIII] and 88 [OIII] do not depend on the electron temperature T_e but do depend on the electron concentration n_e. However, the intensities of the lines N_1 and N_2 both depend on T_e and n_e.

All these lines are excited by electron collisions, and the ratio of intensities, e.g. of the line N_1 to the intensity of the line 52 [OIII], must depend on both T_e and n_e. In this connection the following question arises: can one gain by

Table 10.5. Observed fluxes $F(\lambda)$, in units of 10^{-10}ergs cm^{-2} s^{-1}, in forbidden lines of [OIII], 52 μm, 88 μm and 5 007 Å, and values of n_e (in 10^3 cm^{-3}) and T_e, obtained with the help of these lines for four planetary nebulae

Planetary	$F\lambda$			n_e, cm^{-3}		T_e, K	
nebula	52	88	5 007	52/88	[OII]	5 007/52	[OIII]
NGC 2 440	1.72	0.48	19.4	1.6	1.6	13 000	14 300
NGC 3 242	4.41	1.63	19.2	1.0	2.0	9 800	11 500
NGC 6 543	8.79	1.10	27.0	10	4	5 800	8 000
NGC 6 826	3.66	0.96	9.2	1.6	1.6	7 700	8 800

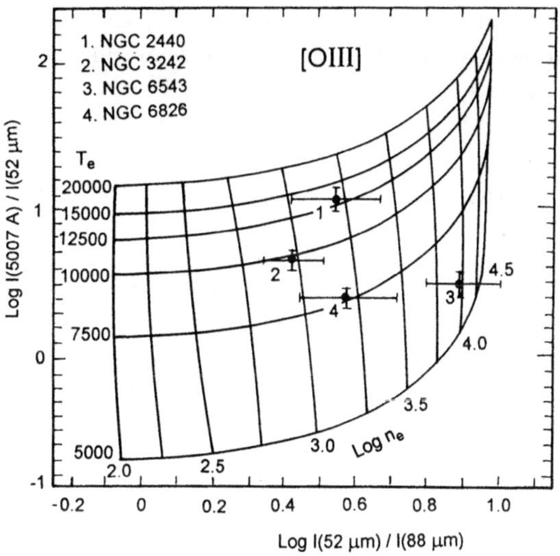

Fig. 10.5. Dependence between the intensities of lines of [OIII], in the infrared, 52 and 88 μm, and the optical, 5 007 Å, as calculated for various values of n_e and T_e. The locations of four planetary nebulae are shown: 1 – NGC 2 440, 2 – NGC 3 242, 3 – NGC 6 543, 4 – NGC 6 826

combined observations in the optical lines of N_1 and N_2, and in the infrared ones of 52 [OIII] and 88 [OIII] for the physics of PNs?

Ground based observations of the lines 52 [OIII] and 88 [OIII] are impossible, but can be performed, say, in airborne conditions. In Table 10.5 the results of measurements by a 91 cm infrared airborne telescope (Dinerstein et al., 1985) of the 52 [OIII] and 88 [OIII] line fluxes for a sample of PNs are presented, along with the fluxes in line 5007 [OIII]. Using the known atomic parameters for these lines, the computed curves of dependence between the ratios of the fluxes F(52)/F(88) and (F(5 007)/F(52) were constructed for various values of n_e and T_e (Fig. 10.5).

Using the observed values of the ratios $F(52)/F(88)$ and $F(5\,007)/F(52)$, we can obtain n_e and T_e. The results are presented in Table 10.5 along with the values of n_e and T_e obtained earlier by means of traditional methods, i.e. based on the forbidden doublet 3 726 [OII]/3 729 [OII] (for n_e) and $N_1 + N_2/4363$ [OIII] (for T_e).

From these results the following conclusions can be drawn:

(a) Electron concentrations obtained using the infrared lines [OIII] almost coincide with those derived from the optical lines of [OIII].

(b) Electron temperatures obtained with the help of the infrared lines [OIII] are systematically lower than the temperatures obtained by the optical lines of [OIII]. The discrepancy between these "infrared" and "optical" temperatures is too large – 1 000–2 000 K – and has a systematic character.

The second conclusion is of special importance, since once more it confirms that nebulae are non-isothermal systems and seem to possess a separate zone of infrared emission.

10.9 The Temperature of Dust in Planetary Nebulae

Obviously, one of the remarkable features of PNs, the identity of electron temperatures for all PNs, should be considered. It varies within quite narrow limits with respect to the mean value, 10 000 K, for PN with enormous scattering in the main physical parameters, including the diversity of temperatures of their central stars.

What is the situation with the dust temperature in the nebula? The results of infrared observations carried out in the range of 3–108 μm (Moseley, 1980) combined with the observations mentioned above (Table 10.4), enable one to construct the infrared spectra in the wavelength range from 2.2 μm up to 108 μm. These spectra obtained, with two or three exceptions, have common features for all PNs: a clearly expressed maximum near 30 μm corresponding to a black-body temperature of $\sim$ 100 K, i.e. what was discovered earlier for NGC 7 027 (Fig. 14.11).

Thus, the mean temperature of the dust is also a kind of a universal parameter for all PNs, analogous to the electron temperature of the gas.

The sample of PNs in Table 10.4 includes objects of various ages, very young, such as IC 4 997, 150 years old, and rather old, and as A 30, 20 000 years old; in both cases the infrared temperatures are practically the same. One can conclude that the dust particles do not undergo essential evolution of *physical* and *optical* properties during the evolution of the nebula itself. If so, then the dust particles in the nebula must be of primordial origin.

10.10 The Survival of Dust Particles in Nebulae

Dust particles cannot be formed in a purely gaseous medium of the nebula by a kind of atomic collision, even if the gas density is rather large – for this the nebula is too hot.

Infrared observations confirm the presence of dust–gas envelopes surrounding many cool giants and supergiants and, hence, the very process of formation of dust particles within the dense cool atmospheres, in principle, is possible.

So, dust particles do exist primordially in the outer parts of stellar atmospheres, far away from stars, i.e. in kinds of surrounding and presumably expanding envelopes. One can think that at a definite stage of its expansion this envelope can evolve to what we call a *planetary nebula* – an ionized medium with a temperature of 10 000 K, under the action of short wavelength radiation.

In this connection the crucial problem is that of the survival of dust particles under the influence of intense central radiation.

The elementary act which can destroy the dust particle is the tearing off of electrons from its surface under the action of ultraviolet radiation, i.e. the "photon alienation" of electrons. As a result, the positively charged dust particle starts to merge with either free electrons or positive ions, finally becoming either more positive or negative.

In conditions of stationarity the following three processes can occur: (a) the photon alienation of an electron from the dust particle under the central ultraviolet radiation; (b) the impact or capture of an electron by dust particles; (c) the impact or capture of a proton by dust particles.

The solution of the corresponding differential equations describing these processes reveals the evolution of dust particle properties with time. The preliminary conclusions are as follows:

(a) In central parts of the nebula the photon tearing processes dominates, and the dust particles have to be charged positively.
(b) In the outer parts of the nebula the impacts of electrons with dust particles dominate, and particles have to be charged negatively.
(c) The charge of dust particles should be very high: $Z \approx +400$ in the central parts of the nebula and $Z \approx -400$ in the outer ones.

In the latter case the electrostatic force on the particle's surface should: $Ze^2/a \approx 0.005Z \approx 2$ eV, which coincides with the tearing off energy of atoms from the particles' surface (≈ 2 eV), as well as with the photon mean energy within the nebula.

Thus, the destruction of dust particles seems to be possible. The question is how efficient this process can be, i.e. how fast a particle can vanish. Rough estimation yields for the duration of the dust particle's life $t_{\text{in}} \approx 10^{15}a/N_p$ year in central regions of nebula and $t_{\text{out}} \approx 3 \times 10^{12}a/N_p$ in outer ones. Adopting $a \sim 3 \times 10^{-5}$ cm and $N_p \approx 10^{-3}$ cm^{-3} for their concentration, we obtain $T_{\text{in}} \approx 10^7$ year and $t_{\text{out}} \approx 10^5$ year. This result seems to be paradoxical, so far as it predicts that particles live in central parts of a nebula longer than in its periphery. However, in both cases the lifetime of the dust particles exceeds the lifetime of the PN itself, i.e. 100 000 years.

10.11 Infrared Lines and the Temperatures of Nuclei

The presence of emission lines of FeVII, NV, NeV or NeVI in the spectrum of a PN is an obvious indication of a very high temperature of its nucleus – higher than 100 000 K. Ions of higher degrees of ionization, however, rarely give emission lines in optical and even ultraviolet regions convenient for registration.

In certain cases the presence of highly ionized ions can be established by emission lines emitted by these ions also in the infrared region. Such an example was mentioned above (Sect. 3) and was related to the line 7.65 [NeVI] – five-times ionized neon; it is detected in the spectrum of high excitation nebulae (NGC 6302) ($p = 10$). Ionization potentials of NeV and NeVI are equal to 126 eV and 158 eV, which correspond to nuclear radiation in the region shorter than 100 Å and 80 Å.

There are two more infrared forbidden lines, 1.96 [SiVI] and 2.48 [SiVII], belonging to five and six times ionized silicon, both simultaneously resonances, with the ionization potentials of 167 eV and 205 eV. These lines enable one to penetrate too far – up in to 74 Å and 60 Å in the nucleus spectra. These lines were also observed in PN spectra (NGC 6302 and NGC 6537) (Ashley, Hyland, 1988).

The ratio of intensities of the lines 1.96 [SiVI] and 2.48 [SiVII] depends on the nucleus temperature T_*. The corresponding relationship may be derived by applying the condition of detailed ionization equilibrium. As a result, $T_* \approx 430\,000$ K for NGC 6302 and $T_* \approx 240\,000$ K for NGC 6572 have been obtained.

Thus, the infrared lines of [SiVI] and [SiVII] can be used successfully as powerful indicators of the PN nuclei of superhigh temperatures. From this point of view the importance of these kind of infrared observations is obvious.

10.12 Evolution of Dust Particles in Planetary Nebulae

Do the main parameters of dust particles – the effective temperature T_d, sizes a, their total mass in nebula $\mathcal{M}_d$, their total number N_p, the ratio of mass of both dust and gas components, $\mathcal{M}_d/\mathcal{M}_g$ etc. – remain constant during the lifetime of a PN? The importance of this question is unquestionable for our understanding of the evolution of PN themselves.

This problem was considered by Lenzuni and collaborators (1989) for a sample of 233 PNs observed by *IRAS*. Their methods were basically those described in Sect. 5. Note that the expansion velocity does not depend on the linear sizes of nebulae, therefore one can consider the dependence of parameters only on the nebula radius R_n.

The main results are as follows:

(a) The sizes of the dust particle decreases with the increase of the nebula's radius R_n according to:

$$a \propto R_n^{-1.86}\,, \qquad (15)$$

as is shown in Fig. 10.6(a). The points denote the *IRAS* data, i.e. the fluxes in bands 25, 60 and 100 μm; circles denote data only in bands 25 and 60 μm.

The limiting values of dust particle sizes are: from 4×10^{-5} cm at $R_n = 10^{16}$ cm up to 3×10^{-8} cm at $R_n = 10^{18}$ cm. This means that with the aging of nebulae, the dust particles become smaller while bigger particles dominate in young, new-born nebulae.

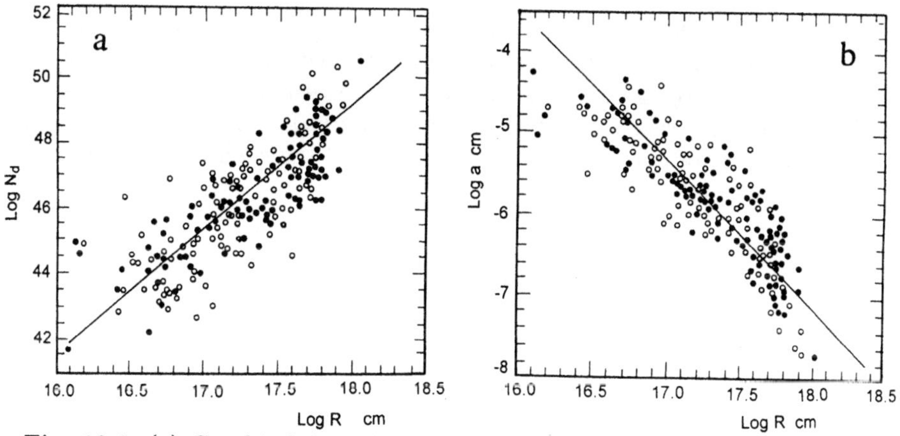

Fig. 10.6. (**a**) Graphical dependence between the sizes of dust particles a and the radius of the nebula R_n according to *IRAS* data for 233 PNs. Dots denote observations in the bands of 25, 60 and 100 μm; circles, those only in the bands of 25 and 60 μm. (**b**) The same as in (a) but for the dependence between the total number of dust particles N_d and R_n

(b) The total number of dust particles N_d increases rapidly in accordance with the law

$$N_d \propto R_n^{+4.19}, \tag{16}$$

as is shown in Fig. 10.6(b). The limiting values of $N_d \sim 10^{42}$ at the birth of a nebulae up to $\sim 10^{50}$ particles at the end of its life – the scatter yields eight (!) orders of magnitude. The crucial role of decaying processes is obvious.

(c) The total mass of dust in the nebula $\mathcal{M}_d$ decreases according to the law

$$\mathcal{M}_d \propto R_n^{-1.00}, \tag{17}$$

however, with strong scattering of observational points. On average, $\mathcal{M}_d \sim 3 \times 10^{-4}\,\mathcal{M}_\odot$ at $R_n \sim 2.5 \times 10^{17}$ cm. The limiting magnitudes are from $\sim 10^{-2}\,\mathcal{M}_\odot$ at the moment of formation of the nebula, and up to $10^{-5}\,\mathcal{M}_\odot$ at the end of its life. The sweeping-up of dust particles from the nebula is obvious.

(c) The effective temperature of dust particles T_d statistically is less sensitive, in accordance with the conclusion drawn in Sect. 10.5, to the nebula's size

$$T_d \propto R_n^{-0.4}. \tag{18}$$

The limiting values of T_d differ slightly as well: only three times, from 50 K up to 150 K.

(d) The ratio of the mass of both dust and gas components, $\mathcal{M}_d/\mathcal{M}_g$ varies in wide limits – from 0.2 in young nebulae up to 10^{-4} in old ones. On average $\mathcal{M}_d/\mathcal{M}_g \sim 0.001$. This is at least two orders larger than the values mentioned in Sect. 10.5.

(e) The optical depth of dust particles at the frequencies of the Lyman continuum t_α was obtained as well: they were within limits from 0.005 up to 1–2, on average 0.1, however, without noticeable signs of evolution.

(f) The limiting values of infrared luminosities of PNs $L_{IR}/L_\odot$, bolometric luminosity of nucleus $L_{bol}/L_\odot$ in the case of the mentioned selection of 233 PNs are as follows:

$L_{IR}/L_\odot$:	min	5	NGC 6853, IC 1295, A 20
	max	7 800	NGC 7027
$L_{bol}/L_\odot$:	min	300	NGC 5315, NGC 1501, Hu 1–2
	max	14 400	NGC 7027
L_{IR}/L_{bol}:	min	0.2	NGC 6853, H 2–41
	max	70	M 2–9

Of special importance from the point of view of the theory of heating of dust particles is the ratio L_{IR}/L_{bol}. For a sample of 36 PNs (Table 10.4), this ratio is larger than unity for 60% of objects. For the sample of 233 PN this ratio turns out to be larger than unity in 80% (!) of cases. For six nebulae this ratio is even larger than 10. Thus, now we have another impressive indication that the L_α radiation of nebulae cannot be the only source of heating of dust particles. A decisive role here should be played by heating by the direct radiation of the central star in the far ultraviolet, at wavelengths shorter than 500 Å, as was outlined above, in Sect. 10.6.

Moreover, judging by the fact that in certain cases the ratio L_{IR}/L_{bol} is larger than 0.8, i.e. quite near to unity, one can conclude that in those PNs (M 2–9, M 1–26, He 2–108 etc.) the heating of dust particles under the action of the stellar wind or the fast electrons from the central star should have a dominant role.

The discrepancy in the values of the main parameters of dust particles in both samples, with 36 and 233 objects, is of a systematic nature; in the latter ($N = 233$) the values of T_d are on 1.5–2 times smaller than we have in the first case, while just T_d plays crucial role in computations of the remaining parameters. Nevertheless, it is difficult to doubt the correctness of the main conclusion: all or almost all the parameters of the dust particles undergo evolutionary variations.

Thus, the dust particles in PNs continuously undergo processes of decay, erosion and sweeping out under the actions, presumably of corpuscular beams of nuclei, stellar wind and hard ultraviolet radiation.

The reasons for variations and the details of evolutionary parameters of dust particles in PNs, as well as the role of numerous other effects, such as "particle–particle" collisions, the final stages of the evaporation of dust particles, are a matter for special study.

11. Radiative Equilibrium: Classical Theory

11.1 Statement of the Problem

Planetary nebulae are transparent for radiation in forbidden lines and in lines of subordinate series of atoms and ions. They are transparent even in all resonance lines of atoms and ions – 1 550 CIV, 1 240 NV, 1 400 SiIV, 1 909 CIII, 2 800 MgII etc. – with the exception of hydrogen and helium. Under these conditions the problem of the radiation field at the frequencies of emission lines can be solved quite easily. The density of radiation ε_λ, e.g. in the lines of the Balmer series of hydrogen, is proportional to the square of the electron concentration in it: $\varepsilon_\lambda \sim n_e^2$. Therefore, the distribution of density of the radiation in these lines within the nebula will be proportional to the distribution of the electron concentration itself. Then, the intensity of radiation emerging from a nebula in any line of this series can be determined simply by summing the radiation from each unit volume along the line of sight. In the same way one can find the intensity of radiation emerging from a nebula in forbidden lines, with the only difference that in this case the volume emission coefficient is proportional to the product of the electron and atom (ion) concentrations of $n(A)$, exciting the given line: $\varepsilon_\lambda \sim n_e\, n(A)$. Nothing else is required to solve the radiative transfer problem at the frequencies of these categories of lines in the visible part of the spectrum.

The situation is completely different for the radiation in the principal series, say, in the Lyman series of hydrogen, i.e. in the region of wavelengths shorter than 912 Å. For the radiation in the continuous spectrum, i.e. in the continuum of this series (L_c radiation) the optical thickness τ_c of the nebula is of the order of unity. Therefore, processes of absorption and reemission of L_c photons also play an essential role in the nebulae. In particular, these processes produce photons in the lines of the Lyman series – L_α, L_β etc., the absorption coefficients of which are tens of thousands of times larger than that beyond the series limit. All Lyman photons produced in this way are finally converted into L_α photons, along with the creation of photons of subordinate series, which escape from the nebula. Thus, only L_c and L_α photons are "imprisoned" in the nebula. Therefore, the problem of radiative transfer or, in other words, of radiative equilibrium in nebulae will arise only for L_c and L_α radiation. In contrast to L_c radiation, for which the nebula is partially transparent, L_α radiation sees the nebula as quite opaque, during all stages of its evolution. Moreover, these radiation fields interact with each other. The determination of the radiation field in these conditions, generally speaking, is a rather complicated problem.

The problem of finding the radiation fields of L_c and L_α radiations in PNs was formulated and solved by Ambartsumian in 1933. He developed a method of *separation* of L_c and L_α fields, which greatly simplifies the solution of the problem. Although the results obtained by this method were subsequently refined and generalized by other authors, especially by Chandrasekhar in 1935 and Sobolev in 1947 (see Sobolev, 1986), the fundamental idea, the separation of the fields, was preserved.

In view of the difficulty of the problem in its general formulation – for a spherical nebula with a variable dilution coefficient – a simpler geometrical model of the nebula is usually used. In the following, we assume that the nebula is bounded by two concentric spheres of radii r_1 and r_2, with a central star which excites the nebula. The linear thickness is assumed to be small in comparison with the linear distance from the centre, i.e. $r_2 - r_1 \ll r_1$. In this case the nebula is approximated with "plane parallel" layers with a constant dilution coefficient. Though considerable work was performed on the study of radiative equilibrium in PNs, taking into account a number of possible effects (see, e.g., Unno, 1952; Busbridge, 1960; Hummer, 1962; Wing, 1962), those complicated solutions add little new concerning the behaviour of L_c and L_α fields within the nebula. Therefore, below we shall confine ourselves with the statement of basic results of the so-called classical theory.

11.2 The L_c Radiation Field

Denote by πS_* the number of photons in the Lyman continuum, i.e. in L_c radiation emitted by the nucleus per second from 1 cm^2 of its surface. Obviously the number of L_c photons falling onto 1 cm^2 of the inner surface of the nebula will be $\pi S = \pi S_*(R_*/r_1)^2 = 4W\,\pi S_*$, where R_* is the radius of the central star and W is the dilution coefficient.

Let κ be the mean absorption coefficient per neutral hydrogen atom in the Lyman continuum[1], and let τ be the corresponding optical depth, measured from the inner boundary of the nebula up to a given layer, i.e.

$$\tau = \int_{r_1}^{r} \kappa n_1 \mathrm{d}r \,, \tag{1}$$

[1] Strictly speaking, the mean absorption coefficient κ should be represented in the form

$$\kappa = \kappa_* \phi(\tau, \tau_*) \,,$$

where κ_* is the absorption coefficient in the continuum at the series limit ($\lambda =$ 912 Å) and $\phi(\tau, T_*)$ is a function depending on τ and temperature of the star T_*. Usually $\phi(\tau, T_*) \approx 1.1$, i.e. κ only slightly exceeds κ_* because of the rapid decrease of the absorption coefficient with the increase of frequency ($\kappa \sim \nu^{-3}$).

where r is the linear distance from the centre of the nebula to the layer and n_1 is the concentration of neutral hydrogen atoms.

The absorption of L_c photons results in the ionization of hydrogen. Subsequently some of these photons will be reemitted in recombinations into the ground state. Denote by p the fraction of all recombinations which go directly into the ground state among all recombinations at all levels, from $n = 1$ up to $n = \infty$. Numerically p depends, very weakly, on the electron temperature of the nebula and is determined by (4.3). Obviously p is the probability of reemission of the absorbed L_c photons, while the probability of emission of L_α photons will be $1 - p$.

Of all L_c photons falling onto the inner boundary of the nebula at the rate πS per cm^2, $\pi S e^{-\tau}$ photons reach the optical depth τ. Among these, $n_1 \kappa \pi S e^{-\tau}$ photons will be absorbed per unit volume, where n_1 is the concentration of hydrogen atoms. Diffuse L_c radiation, arising from recombinations into the ground state, is also absorbed by hydrogen atoms. The absorption rate of diffuse L_c photons per unit volume will be

$$n_1 \kappa \int I(\tau, \theta)\, d\omega ,$$

where $I(\tau, \theta)\, h\nu_c$ is the intensity of diffuse radiation under the angle θ to the optical depth τ and the integration is performed over the complete solid angle.

Thus, the total number of L_c photons absorbed per unit volume per second at an optical depth τ is

$$n_1 \kappa \left[\pi S e^{-\tau} + \int I(\tau, \theta)\, d\omega \right] . \tag{2}$$

From this number only the p part will be reemitted again in the frequencies of L_c radiation.

The number of L_c photons emitted within the nebula per unit volume per second will be

$$4\pi n_1 \kappa C(\tau) . \tag{3}$$

We are now able to write the condition of radiative equilibrium by equating (2) and (3):

$$C(\tau) = p \int I(\tau, \theta) \frac{d\omega}{4\pi} + p \frac{S}{4} e^{-\tau} . \tag{4}$$

The quantities $C(\tau)$ and $I(\tau, \theta)$ are also related via the equation of radiative transfer, which for plane-parallel layers has the form

$$\cos\theta \frac{I(\tau, \theta)}{d\tau} = C(\tau) \quad I(\tau, \theta) . \tag{5}$$

The joint solution of (4) and (5), together with the boundary conditions, gives the quantities $C(\tau)$ and $I(\tau, \theta)$. While writing down the boundary conditions one must keep in mind that the inner surface of the nebula receives not only L_c radiation from the central star but also L_c radiation from the opposite side of the nebula. Consequently, the first boundary condition can be written in the form

$$I(0,\, \theta) = I(0,\, \pi - \theta)\,, \tag{6}$$

i.e. the diffuse radiation entering the nebula at any point of the inner boundary at an angle θ to the normal is nothing other than the radiation leaving the nebula at an angle $\pi - \theta$ on the opposite side (Fig. 11.1).

The second boundary condition is written in the form

$$I(\tau_0,\, \theta) = 0 \quad \text{for} \quad \theta > \frac{\pi}{2}\,. \tag{7}$$

It expresses the obvious fact that no radiation returns from the outer boundary of the nebula.

Equations (4) and (5) can be solved by means of any of the approximation methods developed in the theory of radiative transfer. Ambartsumian, for example, has used the Schwarzschild–Schuster approximation, Chandrasekhar–Eddington approximation. However, Sobolev gave a more accurate solution of these equations based on a derivation of an integral equation. The integral equation for $C(\tau)$ has the form

$$C(\tau) = \frac{P}{2} \int_0^{\tau_0} [E_1(\tau - \tau') + E_1(\tau + \tau')]\; C(\tau')\, \mathrm{d}\tau' + p\, \frac{S}{4}\, e^{-\tau}\,, \tag{8}$$

where

$$E_1(\tau) = \int_\tau^\infty e^{-\tau}\, \frac{\mathrm{d}\tau}{\tau}\,.$$

Since our goal is to reveal the role of diffuse radiation in the transfer processes of L_c radiation, we can adopt $\tau_0 = \infty$ in (8). The latter then will be written in an approximate form:

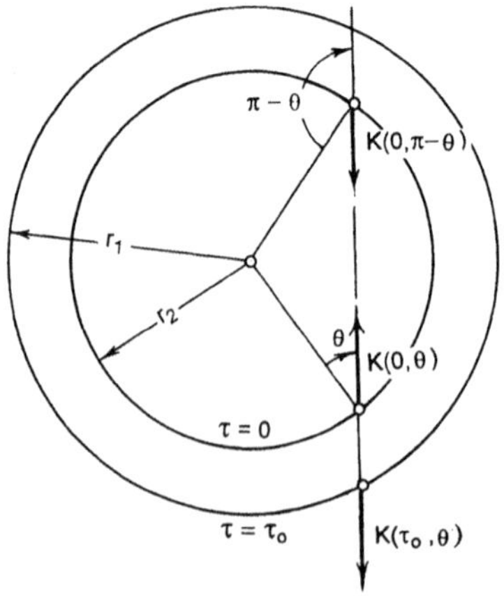

Fig. 11.1. Geometrical scheme for planetary nebula radiation field analysis

$$C(\tau) = \frac{P}{2} \int_{-\infty}^{\infty} E_1(|\tau - \tau'|)\, C(\tau')\, d\tau' . \tag{9}$$

Let us look for the solution of (9) in the form:

$$C(\tau) = A\, e^{-k\tau} , \tag{10}$$

where A is an arbitrary constant. Substituting this into (9), we find for k the equation

$$\frac{p}{2k} \ln \frac{1+k}{1-k} = 1 . \tag{11}$$

The value of the constant A has to be determined from the condition that (8) is satisfied exactly, considering the function (10) as only its approximate solution. Then we obtain

$$A = \frac{kpS}{4(1-p)} . \tag{12}$$

Since the electron temperatures of nebulae are of the order of 10 000 K, one can adopt $p \approx 0.4$ on average. For this value of p we find from (11) that $k = 0.958 \approx 1$. Hence, we can write for the required function

$$C(\tau) = S \frac{p}{4(1-p)}\, e^{-\tau} . \tag{13}$$

Substituting this expression into (4), we see that the number of diffuse L_c photons is approximately equal to the number of L_c photons arriving directly from the star (the second term on the right side). Then, we can conclude that the role of the diffuse component of L_c radiation in a nebula is not large.

Averaging (5) over the outer hemisphere we obtain

$$\frac{1}{2} \frac{dI(\tau)}{d\tau} = C(\tau) - I(\tau) , \tag{14}$$

where $I(\tau)$ is the outward component of L_c radiation. Substituting the value of $C(\tau)$ from (13) into (14) and integrating, we find, at the outer boundary of the nebula ($\tau = \tau_0$),

$$I(\tau_0) = \frac{pS}{2(1-p)}\, e^{-\tau_0} \approx 0.4\, S\, e^{-\tau_0} . \tag{15}$$

For the flux of diffuse L_c photons we have

$$H_c(\tau_0) = \int I(\tau_0) \cos\theta \, d\omega = \pi\, I(\tau_0) = 0.4\pi\, S\, e^{-\tau_0} . \tag{16}$$

The flux of direct L_c radiation near the outer boundary is equal to $\pi S e^{-\tau_0}$, while the total flux (diffuse + direct) is $1.4\pi S\, e^{-\tau_0}$.

The number of L_c photons falling onto the inner surface of the nebula is πS. The number of L_c photons escaping from the nebula is $1.4\pi S e^{-\tau_0}$, so that $(1 - 1.4e^{-\tau_0})pS$ photons are converted into photons of other frequencies. Since only one L_α photon is formed from each L_c photon, the number of L_α photons at the outer boundary of the nebula should be $(1 - 1.4e^{-\tau_0})\pi S$. At $\tau_0 \gg 1$, this yields just πS, i.e. the result obtained earlier, in Chap. 2, from purely qualitative considerations.

11.3 The L_α Radiation Field

Denote by k the absorption coefficient per hydrogen atom in the L_α line and define the corresponding optical depth t into the nebula by

$$t = \int_{r_1}^{r} n_1 k \, dr \,. \tag{17}$$

From (17) and (1) we have

$$\frac{t}{\tau} = \frac{k}{\kappa} = \frac{1}{q} \,. \tag{18}$$

The quantity k depends on the mean velocity of the hydrogen atoms. For the thermal velocities occurring in nebulae, the quantity q is of the order 10^{-4}–10^{-5}. Since we assume that the total optical thickness τ_0 of the nebula for L_c radiation is of the order of unity or larger, we find that the optical thickness t_0 of the nebula in the L_α line must be extremely large, of the order 10^4–10^5. In strongly evolved PNs, when $\tau_0 \sim 0.001$, we have $t_0 \sim 10$–100. Therefore the PN should be opaque for the radiation in the L_α line.

On the one hand, we saw that one L_α photon is formed from each L_c photon absorbed in the nebula. On the other hand, because of high opacity, the L_α photons are unable to escape, and the accumulation of large number of those photons is inevitable. In other words, the density of L_α radiation will become larger than of L_c radiation, hence, the priority of the role of the L_α radiation field in the dynamical problems of PNs.

Denote by $K(t, \theta)$ the intensity of L_α radiation directed at an angle θ to the normal and at an optical depth t, and by $4\pi n_1 k G(t)$, the number of L_α photons emitted per unit volume per second at the same time. Then we can write the equation of transfer for L_α radiation as

$$\cos\theta \, \frac{dK(t, \theta)}{dt} = G(t) - K(t, \theta) \,. \tag{19}$$

To derive the equation of radiative equilibrium one must keep in mind that not only does the conversion of L_c photons to L_α photons occur, but also scattering of the latter. As regards the L_α radiation coming directly from the central star, it can be neglected in comparison with the L_c radiation.

As we have seen above, the unit volume of the nebula emits L_c photons at a rate of $4\pi n_1 \kappa C(\tau)$. The latter should be equal to the recombination rate directly into the ground state per unit volume. Obviously, the number of recombinations to the states other than the lowest will be

$$\frac{1-p}{p} 4\pi n_1 \kappa C(\tau) \,. \tag{20}$$

The number of diffuse L_α photons absorbed per unit volume per second will be equal to $n_1 k \int K(t, \theta) \, d\omega$. Therefore the total number of L_α photons absorbed per unit volume per second at an optical depth t is given by

$$n_1 k \int K(t,\,\theta)\,\mathrm{d}\omega + 4\pi\,\frac{1-p}{p}\,n_1\kappa\,C(\tau)\,. \tag{21}$$

The number of corresponding L_α photons equals $4\pi\,n_1 kG(t)$. Consequently we have for the condition of radiative equilibrium

$$G(t) = \int K(t,\,\theta)\,\frac{\mathrm{d}\omega}{4\pi} + q\,\frac{1-p}{p}\,C(\tau)\,, \tag{22}$$

where the value of function $C(t)$ is taken from (13).

The solution of (19) and (22) gives the values for the functions $G(t)$ and $K(t,\,\theta)$. In the Eddington approximation we obtain

$$\overline{K(t)} = a - 3bt - \frac{3}{4}\,\frac{S}{q}\,e^{-tq}\,, \tag{23}$$

$$F_\alpha(t) = b - \frac{S}{4}\,e^{-tq}\,, \tag{24}$$

where a and b are arbitrary constants and $\overline{K(t)}h\nu_\alpha$ and $4\pi F_\alpha(t)h\nu_{+}\alpha$ are, respectively, the mean intensity and flux of the L_α radiation, i.e.

$$\overline{K(t)} = \int K(t,\,\theta)\,\frac{\mathrm{d}\omega}{4\pi}\,; \tag{25}$$

$$F_\alpha(t) = \int K(t,\,\theta)\,\cos\theta\,\frac{\mathrm{d}\omega}{4\pi}\,. \tag{26}$$

In determining the constants of integration one must take into account the kinematic state of the nebula. In the particular case when it is motionless (stationary nebular model) or expands with a velocity not exceeding the mean thermal velocity of the atoms, we have the following boundary conditions, analogous to (6) and (7):

$$F_\alpha(0) = 0\,, \qquad 2F_\alpha(t_0) = \overline{K(t_0)}\,. \tag{27}$$

From (27) and (24) we find approximately, for $\tau_0 = qt_0 \gg 1$, that

$$a = \frac{3}{4}\,St_0\,, \qquad b = \frac{1}{4}\,S\,.$$

Substituting these values into (23) and (24) we find for the functions $F_\alpha(t)$ and $\overline{K(t)}$ that

$$\overline{K(t)} = \frac{3}{4}\,S\left[(t_0 - t) - \frac{1}{q}\,e^{-tq}\right]\,, \tag{28}$$

$$F_\alpha(t) = \frac{1}{4}\,S(1 - e^{-tq})\,. \tag{29}$$

From these formulae we are able to calculate the intensity $\overline{K(t)}h\nu_\alpha$ and the flux $4\pi F_\alpha(t)h\nu_\alpha$ of the L_α radiation at any optical depth in the nebula.

For the density of L_α photons we have

$$\rho_\alpha(t) = \frac{1}{c}\int K(t)\,d\omega = \frac{4\pi}{c}\,\overline{K(t)}\,. \tag{30}$$

We can use (30) and (28) at $t = 0$ to estimate the number of L_α photons per cm^3 at the inner boundary of the nebula

$$\rho_\alpha(0) = \frac{\pi S}{c}\,3t_0\,. \tag{31}$$

But $\pi S/c$ is just the number of L_c photons per cm^3 at $t = 0$. Consequently the density of L_α radiation at the inner boundary of the nebula is $3t_0$ times larger, i.e. tens of thousands of times larger than the density of L_c radiation. In its turn the density of L_α radiation at star temperatures of 50 000 K is roughly 5×10^4 times larger than the density of L_α radiation arriving from the star. Therefore, the density of diffuse L_α radiation at the inner boundary of the nebula is 10^9 times larger than the direct L_α radiation from the nucleus. This density greatly exceeds the radiation density in all other lines of the hydrogen spectrum. Nevertheless, if W is of the order of 10^{-13}, then the density of diffuse L_α radiation is still 10^4 times smaller than the density on the surface of the star.

Consider the situation when the nebula expands with a velocity significantly greater than the mean thermal velocity of atoms. Then again one has a stationary model, with the only difference being in boundary conditions: L_α photons leaving the inner boundary of the nebula will not be absorbed on the opposite side because of the Doppler effect. These conditions are

$$2F_\alpha(0) = -\overline{K}(0)\,, \qquad 2F_\alpha(0) = \overline{K}(t_0)\,. \tag{32}$$

Then for the constants a and b we have

$$a = \frac{3}{4}\frac{S}{q}\,, \qquad b = \frac{1}{4}\frac{S}{qt_0}\,.$$

As a result we obtain

$$\overline{K(t)} = \frac{3}{4}\frac{S}{q}\left(\frac{t_0 - t}{t_0} - e^{-tq}\right)\,, \tag{33}$$

$$F_\alpha(t) = -\frac{S}{4}\left(e^{-tq} - \frac{1}{qt_0}\right)\,. \tag{34}$$

For the flux of L_α photons at the inner boundary, $t = 0$, which is expanding with high velocity, we have from (34)

$$F_\alpha(t) = -\pi S\left(1 - \frac{1}{qt_0}\right) = -\pi S\left(1 - \frac{1}{\tau_0}\right)\,, \tag{35}$$

and at the outer boundary, assuming $q\,t_0 = \tau_0 \gg 1$,

$$F_\alpha(t_0) = \frac{\pi S}{qt_0} = \frac{\pi S}{\tau_0}\,. \tag{36}$$

For a motionless nebula we had, respectively, $F_\alpha(0) = 0$ and $F_\alpha(t_0) = \pi S$. Comparing these results with Es.(35) and (36), we see, first, that the L_α flux is negative at the inner boundary of the rapidly expanding nebula, i.e. the flux is directed inward. Besides this, the majority of L_α photons escape from the nebula through its outer boundary, and their fraction increases as the optical thickness τ_0 of the nebula increases. The rest of the L_α photons leave the nebula through the inner boundary.

Obviously, somewhere within a rapidly expanding nebula, the flux of L_α radiation must vanish. The optical depth of the layer τ_{cr} where the L_α flux changes its sign is determined by (34) at the condition $F_\alpha(\tau_{cr}) = 0$:

$$\tau_{cr} = \ln \tau_0 . \tag{37}$$

Thus, the character of the L_α radiation field within a nebula depends strongly on its kinematics.

11.4 The Radiation Field in an Expanding Nebula

The results obtained in the preceding section are valid for stationary nebulae, i.e. either motionless or expanding with high velocity, while in reality PNs are gaseous regions expanding with low velocities and with radiation fields different then those of motionless nebulae. Strictly speaking, the radiation field within a nebula expanding without a velocity gradient will be exactly the same as for the motionless one. However, even at constant expansion velocity an inevitable velocity gradient should appear between the layers of the expanding nebula, provided the thickness of the nebula is not too small. Indeed, the velocity gradient dv/ds within an expanding region will be

$$\frac{dv}{ds} = \frac{dv}{dr} \cos^2\theta + \frac{v}{r} \sin^2\theta , \tag{38}$$

where dv/dr is the velocity gradient along the radius and θ is the angle between a given direction and the radius vector.

If the expansion velocity is constant and equal to v, then

$$\frac{dv}{dt} = \frac{2}{3} \frac{v}{r} . \tag{39}$$

Even if at some moment the expansion occurs without velocity gradients, the radiation pressure inevitably will set up differences in expansion velocities of various layers of the nebula.

Clearly, for L_c radiation the gradients of the expansion velocity are of no importance; the field is the same as for a stationary nebula, while for L_α radiation it has a crucial effect. Therefore, in the following we confine ourselves to investigating the transfer problem for L_α radiation in a nebula expanding with a velocity gradient. The correct solution of this problem was first given by Sobolev (1947).

The role of expansion gradients on the radiation field can be seen from the following argument. When the medium is stationary, i.e. when the relative velocity of atoms is zero (with the subtraction of thermal velocity), then a photon emitted by an atom at point A will be absorbed by an atom at B, located a distance s from A. However, if the relative velocity of these points is non-zero, the frequency of the photon emitted at A will be shifted owing to Doppler effect at B. If the shift is large enough the atoms at B will be unable to absorb the photon, which can then escape from the medium. In other words, the photon can escape from interior of the nebula, owing to the Doppler effect, and boundary regions.

Thus, the role of the velocity gradient is in the reduction of the radiation density in the nebula.

Denote by β the fraction of photons leaving the medium owing to the Doppler shift caused by the velocity gradient. Obviously, β will depend on the kinematic and physical parameters of the medium: the velocity gradient $\mathrm{d}v/\mathrm{d}s$, the volume absorption coefficient k, and the mean thermal velocity u of the atoms. This dependence can be derived from the following considerations. The radiation emitted from the point A, and arriving at the point B at a distance s, has to be weakened by a factor $e^{-\mathrm{ks}}$ and shifted in frequency by

$$\nu - \nu' = \frac{\nu}{c}\frac{\mathrm{d}v}{\mathrm{d}s}\, s\,. \tag{40}$$

Consequently, only the following fraction of the radiation emerging from A will be absorbed in the medium:

$$\int_0^\infty e^{-\mathrm{ks}}\left(1 - \frac{\nu - \nu'}{2\Delta\nu}\right)\mathrm{kds} = 1 - \frac{1}{2u}\frac{1}{k}\frac{\mathrm{d}v}{\mathrm{d}s}\,, \tag{41}$$

where $\Delta\nu = 2(u/c)\nu$. The upper limit on the integral in (41) has only symbolic meaning, indicating that the product ks, i.e. the optical thickness, becomes substantially larger than unity.

Equation (41) defines the fraction of the radiation absorbed in the medium. Obviously, the fraction of radiation escaping from the medium will be

$$\beta = 1 - \left(1 - \frac{1}{2uk}\frac{\mathrm{d}v}{\mathrm{d}s}\right) = \frac{1}{2uk}\frac{\mathrm{d}v}{\mathrm{d}s}\,. \tag{42}$$

Now the problem is to introduce β into the equation of radiative transfer. The solution of radiative equilibrium will obviously give the radiation field in a nebula expanding with a velocity gradient.

Substituting $k = n_1\kappa/q$ into (41), we find for β, in the case of L_α radiation:

$$\beta = \frac{q}{2un_1\kappa}\frac{\mathrm{d}u}{\mathrm{d}s} = \frac{q}{2u}\frac{\mathrm{d}v}{\mathrm{d}\tau}\,, \tag{43}$$

where $d\tau = \kappa ds$. Recall that τ is the optical depth of the nebula for L_c radiation of the layer under consideration.

Assuming that β is constant and positive within the nebula, the solution of the transfer equation together with the condition of radiative equilibrium will give for the function $G(t)$

$$G(t) = 3\frac{q}{\beta}\left\{\psi(\tau) - \left(1 - \frac{2}{3}\sqrt{\beta}\right)\left[\psi(0)e^{-t\sqrt{\beta}} + \psi(\tau_0)e^{-(t_0-t)\sqrt{\beta}}\right]\right\}, \quad (44)$$

where

$$\psi(t) = \frac{3}{4}Se^{-k\tau}, \qquad k \sim 1. \quad (45)$$

The L_α flux can be expressed in the form

$$F_\alpha(t) = 2\pi\left[\int_0^t E_{i_2}(t-t')\,G(t')\,dt' - \int_t^{t_0} E_{i_2}(t'-t)\,G(t')\,dt'\right], \quad (46)$$

where

$$E_{i_2}(x) = \int_1^\infty e^{-xz}\,\frac{dz}{z^2}.$$

For internal regions of the nebula (44) and (46) give

$$F_\alpha(t) = -4\pi\,\frac{q^2}{\beta}\,\frac{d\psi}{d\tau}, \quad (47)$$

while at the boundaries the flux is

$$F_\alpha(0) = -2\pi\psi(0)\,\frac{q}{\sqrt{\beta}}, \quad (48)$$

and

$$F_\alpha(t_0) = +2\pi\psi(0)\,\frac{q}{\sqrt{\beta}}. \quad (49)$$

Comparing (48) and (49) with (29), on the one hand, and (30) and (28) with (48) and (49) on the other, we see that the the gradient in the nebula essentially reduces the flux and the density of L_α radiation. Suppose, for example, that the velocities $dv/d\tau$ and u are of the same order. Then (48) gives for the L_α flux at the inner boundary a value some hundred times smaller than follows from (35). For the L_α density deep in the nebula one obtains from (47) a value some ten thousand times smaller than that found from (30) and (33).

Despite these differences, the L_α field in a nebula expanding with a velocity gradient has the same character as in nebula expanding with constant velocity. In particular, in both cases the L_α flux is directed inwards at the inner boundary and outwards at the outer boundary. The optical depth τ_{cr} of the layer where the L_α flux vanishes is determined by a relation analogous to (37):

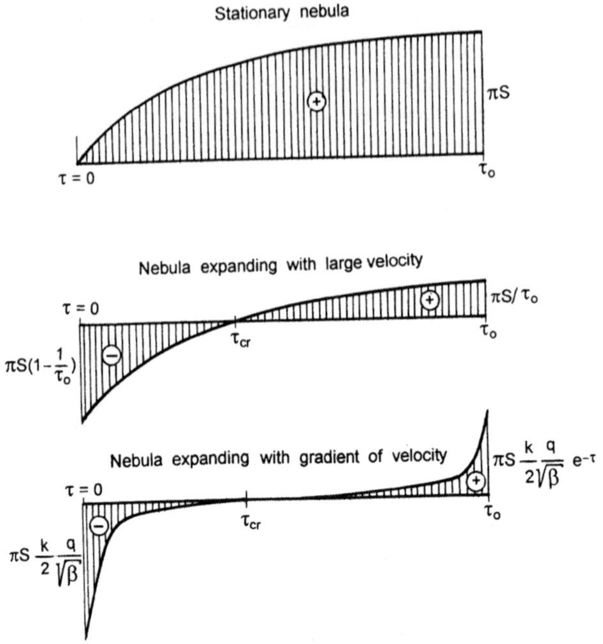

Fig. 11.2. Distribution of L_α radiation throughout the nebula in three different kinematic conditions. *Top:* Stationary nebula. *Centre:* Nebula expanding with high velocity. *Bottom:* Nebula expanding with a velocity gradient. The point $\tau = 0$ is on the inner boundary of the nebula; $\tau = \tau_0$ is on the outer boundary. The point τ_{cr} corresponds to the zero value of the flux. In the case of a motionless nebula (*top*), the L_α flux is zero on the inner boundary of the nebula

$$e^{-\tau_{cr}} = \left(1 - \frac{2}{3}\sqrt{\beta}\right)\left[e^{-\tau_{cr}\sqrt{\beta}/q} + e^{-k\tau_0}e^{-(\tau_0-\tau_{cr})\sqrt{\beta}/q}\right] . \tag{50}$$

The qualitative picture of the distribution of L_α flux throughout the nebula for all three cases considered above is shown in Fig. 11.2. These curves represent approximately the case when the optical thickness τ_0 for L_c radiation is of the order of three.

11.5 Diffusion of L_α Radiation

There is still another effect tending to reduce the density of L_α radiation in nebula. This is the so-called effect of redistribution of radiation by frequency or non-coherent scattering.

The point is that in the problem of resonance scattering of L_α photons (Sect. 11.2) it was assumed that the diffusion of L_α photons occurs without change of frequency. In other words, it was assumed that the absorption coefficient in the line is constant, i.e. the line has a rectangular profile.

In reality, the profile of the absorption coefficient in the nebula is due to radiation damping (natural width of the energy levels) and the Doppler effect arising from the thermal motions of the atoms. This profile takes the form of a Gaussian probability distribution, rather than a rectangular shape. Thus, for example, the atom absorbing a photon with the central frequency of the line will in general emit the photon in another direction and at a slightly displaced frequency. In this way a fraction of L_α photons will shift from the centre of the line to its wings, where the absorption coefficient is smaller and the probability of the photon escaping from the nebula is higher. This phenomenon is known as the *escape of photons on the line wings.*

The problem of diffusion of L_α radiation can be considered with either of two assumptions: the *actual* and *complete* redistribution of radiation by frequency. In the first an elementary volume absorbs photons of one frequency and emits them at other frequencies. Since the function characterizing this process is rather complicated, one adopts the assumption of complete redistribution in frequency. In this case, the probability of emission of a photon of one frequency is considered to be independent of the frequency of the absorbed photon. Although this assumption is not entirely justified, it simplifies the problem considerably.

The problem of diffusion of L_α radiation for the *actual* law of redistribution has been studied in detail (Unno, 1952; Sobolev, 1958). It appeared, in particular, that the density and flux did not differ greatly from those obtained with the assumption of *complete* redistribution. Therefore, the solution of the diffusion problem of L_α radiation in the case of *complete* redistribution can be sufficient for many problems.

The importance of redistribution in frequency for the diffusion of L_α radiation can be revealed more accurately if one can find out which of the following two effects is the dominant one in nebulae: the escape of photons from the nebula owing to the velocity gradient or to their escape in the line wings.

In the first case, of the effect of a velocity gradient, we have from (44) for the central parts of the nebula

$$G(t) = 3\frac{q}{\beta}\psi(\tau)\,, \qquad \tau = qt\,. \tag{51}$$

Here, obviously, $\beta/3q$ is the fraction of L_α photons escaping from a given point due to the velocity gradient.

In the second case we neglect the velocity gradient, and the absorption coefficient in the line is represented in the form

$$k_\nu = k_{\nu_0}\, e^{[(\nu-\nu_0)/\Delta\nu_D]^2}\,, \tag{52}$$

where $\Delta\nu_D$ is the Doppler half-width of the line. With the assumption of complete redistribution, the emission coefficient in the line is

$$\varepsilon_\nu = \varepsilon_{\nu_0} \exp\left[-\left(\frac{\nu - \nu_0}{\Delta\nu_D}\right)^2\right] . \tag{53}$$

The function $G(t)$ in this case is determined in the same way as for the diffusion of L_α radiation with a velocity gradient, i.e. by means of the approximate solution of the integral equation. In this manner one obtains for the central regions of the nebula (Sobolev, 1957)

$$G(t) = \frac{2q\psi(\tau)}{L(t) - L(t_0 - t)} , \tag{54}$$

where

$$L(t) = \frac{1}{\sqrt{\pi}} \int_{-\infty}^{+\infty} e^{-x^2} E_{i_2}(te^{-x^2}) \, dx . \tag{55}$$

For large values of t we have approximately for a three-dimensional medium:

$$L(t) = \frac{1}{2t\sqrt{\pi} \ln t} . \tag{56}$$

The quantity $1/2[L(t)+L(t_0-t)]$ in (54) obviously represents the fraction of photons escaping in the line wings. Therefore, comparing (51) with (54), we see that when the inequality

$$\frac{\beta}{2} \ll \frac{1}{2} \left[L(t) + L(t_0 - t)\right] \tag{57}$$

is satisfied, the relative role of the velocity gradient will be small. Note that if this inequality is satisfied in the central layers of nebula, it will also be satisfied in its outer layers.

Assuming $t = t_0/2 = \tau_0/2q$ in (57) and using (56), we obtain

$$\frac{1}{u} \frac{dv}{d\tau} \ll \frac{1}{\tau_0} , \tag{58}$$

where the value of β is taken from (43). On the other hand, we have $d\tau = n_1 \kappa dr$, and $\tau_0 = n_1 \kappa \Delta r$, where Δr is the thickness of the nebula. Therefore, we find from (58) that

$$\frac{1}{u} \frac{dv}{dr} \ll \frac{1}{r} , \tag{59}$$

or approximately

$$\frac{1}{u} \ll \frac{r}{\Delta r} . \tag{60}$$

The expansion velocity v of the nebula is of the order of the mean thermal velocity u, while its thickness amounts to a few tenths of its radius. In other words, the quantities v/u and $r/\Delta r$ in PNs are of the same order, and therefore the inequality (60) is not satisfied for these objects. In this case both effects under discussion – the escape of photons due the velocity gradient and the escape of photons in the line wings – have approximately equal roles.

11.6 The Relaxation Time

Thus, a PN is an optically thick medium, practically opaque for the L_α line of hydrogen, and in addition has an extremely high density of L_α radiation. These facts lead to certain questions related with the physics and dynamics of PNs. The first concerns the determination of the so-called *relaxation time*, i.e. of the time scale during which the radiative equilibrium will be established in the PN. Obviously, this parameter will determine the role of L_α radiation, in particular, of the pressure caused by this radiation in the dynamics of the PN.

The relaxation time can be defined as a mean time scale T_1 of the stay of an L_α photon in the nebula. To determine T_1 it is necessary to know the total number of L_α photons in a nebula as well as the number of L_α photons leaving the nebula per second. Obviously, the ratio of the first and the second quantities will just give the sought for relaxation time T_1.

For the total number of L_α photons in a nebula Q_α we can write on the basis of (30)

$$Q_\alpha = 4\pi r_1^2 \int_0^{t_0} \rho_\alpha(t)\, \frac{\mathrm{d}t}{n_1 k} = 4\pi r_1^2 \frac{4\pi}{c} \int_0^{t_0} K(t)\, \frac{\mathrm{d}t}{n_1 k}\,.$$

Substituting the value of $K(t)$ from (22), (25) and (44), derived for the effect of the velocity gradient, and performing the integration, we find approximately, for $\tau_0 \gg 1$ and $\mathrm{d}v/\mathrm{d}t \sim u$,

$$Q_\alpha = \frac{4\pi r_1^2}{c n_1 k} \frac{3}{\beta} \pi S\,. \tag{61}$$

As for the number of L_α photons Q_{tr} leaving the nebula, it obviously equals to the number of L_c photons falling on the inner boundary of the nebula from the central star. The latter number is just equal to

$$Q_{tr} = 4\pi r_1^2 \pi S\,. \tag{62}$$

Therefore, with the effect of a velocity gradient, we have for the mean time scale T_1 of stay of photons in the nebula

$$T_1 = \frac{Q_\alpha}{Q_{tr}} = \frac{3}{n_1 k c} \frac{1}{\beta}\,. \tag{63}$$

It is interesting that when one takes into account only the effect of a velocity gradient, the mean lifetime of an L_α photon diffusing within the nebula does not depend on its optical thickness t_0.

Adopting in (63) $n_1 = 1\ \mathrm{cm}^{-3}$, $k = 10^{-12}\ \mathrm{cm}^2$, $\beta \sim q \sim 10^{-4}$–$10^{-5}$, we find that the mean time scale of the stay of L_α photons in the nebula is of the order of several months.

Knowing T_1, we can determine the number of scatterings N_1 experienced by L_α photons before they emerge from the nebula. Obviously, to determine this one must divide the quantity T_1 over the average interval of time between two successive scatterings. The mean distance between two successive scatterings is equal to $1/n_1k$ and the mean interval is then $1/n_1kc$. Therefore, dividing T_1 by $1/n_1kc$, we find for the mean number of scatterings

$$N_1 \approx \frac{1}{\beta}, \tag{64}$$

i.e. each photon undergoes on average several tens or hundreds of thousands of scatterings before escaping from the nebula.

Analogous formulae can be derived also for the case when the density of L_α radiation is determined by the effect of redistribution in frequency. From (54) and (56) one can write for the number $N_2(t)$ of scatterings, which incidentally depends on the optical thickness of the nebula,

$$N_2(t) = 2t\sqrt{\pi \ln t}\,. \tag{65}$$

For the time of diffusion of an L_α photon we have

$$T_2 = \frac{2t\sqrt{\pi \ln t}}{n_1kc}\,. \tag{66}$$

With $t \sim 10^4$–10^5 and the same values of the remaining quantities we find for T_2 a value of the order of one month to a year.

Finally, for a motionless nebula without the effect of redistribution of radiation in frequency, we have

$$T_3 = \frac{3}{2}\frac{t^2}{n_1kc} \tag{67}$$

and

$$N_3(t) = \frac{3}{2}t^2\,, \tag{68}$$

which gives $T_3 \approx 10^3$ years and $N_3 \approx 10^9$ for $t = 10^4$–10^5.

Thus, even for motionless nebulae the time of capture of L_α photons in the nebula, i.e. the relaxation time, is significantly smaller than the lifetime of the nebula. Therefore, one can assume that the observed nebulae are in a state of radiative equilibrium.

11.7 The Degree of Excitation of the Resonance Level of Hydrogen

The high density of L_α radiation will lead to an increase in the number of $1 \to 2$ transitions and, hence, to a greatly increased population n_2 of atoms in the second – resonance – state. The density of diffuse L_α radiation, as we have seen above, exceeds the density of direct radiation of the star in this line over 10^9 times on average for a motionless model, and 10^7 times for a velocity gradient model. Consequently for the degree of excitation of the resonance state we have (for the second model):

$$\frac{n_2}{n_1} \approx 10^7 \frac{W}{e^{h\nu/kT_*} - 1} . \tag{69}$$

For $T_* = 50\,000$ K and $W = 10^{-13}$ we obtain

$$\frac{n_2}{n_1} \approx 10^{-7} .$$

Thus, the degree of excitation of the second level is over a million times larger than in normal conditions, when the population of the levels is determined by photoionization and subsequent recombination (see Chap. 2). Even at $T_* \sim 100\,000$ K, which is clearly characteristic for the majority of nuclei of PNs, the magnitude just obtained, $n_2/n_1 \sim 10^{-7}$, remains unchanged. The question is, whether such a high degree of excitation can lead to an opaque medium for radiation in the lines of the Balmer series.

Denoting by t_i the optical depth of the nebula in a line of the Balmer series, we have

$$t_i n_2 s_i \Delta r = 10^{-7} n_1 s_i \Delta r ,$$

where s_i is the absorption coefficient in those lines. For Doppler broadening corresponding to an electron temperature of 10 000 K, we have $s_i \sim 10^{-12}$–10^{-13} cm^2 for the first lines of the series. If we take $n_1 \sim 1$ cm^{-3} and the linear thickness of the nebula $\Delta r = r_2 - r_1 \approx 10^{17}$ cm, we find that $t_i \approx 0.01$ or less. Hence, one can conclude that the real PN under the effects of velocity gradient and redistribution of radiation are practically transparent for the lines of the Balmer series of hydrogen.

For a motionless model of a nebula, however, the result is somewhat different: $n_2/n_1 \sim 10^{-5}$ and $T_i \sim 1$.

Note that among real PNs there are objects opaque in the Balmer lines, i.e. when t_i is not negligibly small and can even be of the order of unity. These are dense and superdense nebulae for which $n_e \sim 10^6$–10^7 cm^{-3}, i.e. two to three orders higher than the concentration in ordinary PNs. In such nebulae the observed Balmer decrement reveals the signs of self-absorption, so that the decrement will differ from the theoretical one. However, the Balmer decrement can be distorted also under the influence of interstellar extinction. The main difficulty is that qualitatively the Balmer decrement distorted by one of mechanisms can hardly be distinguished from the other. This problem will be examined in detail in Chap. 16.

12. Distances and Dimensions of Planetary Nebulae

12.1 Formulation of the Problem

One of the most difficult problems associated with PNs is that of their distances and dimensions. No other PN problem seems to have attracted so much effort with so little outcome than this "bone in the throat", as Aller stated (1991) characterizing the crisis concerning this problem. In fact, at present we have no single method enabling one to determine the distances of PNs with an accuracy comparable to that of determining the distances, say, of ordinary stars and galaxies.

The reason for this situation in determining the distances of PNs lies in the fact that not one of the basic physical or geometrical parameters is constant for all PNs.

One can mention at least eight or so parameters which to a certain degree can be used as starting points for the determination of the distances of PNs:

(a) The absolute stellar magnitude M_n of the nebula (plus that of the nucleus).
(b) The absolute stellar magnitude M_* of the nucleus only.
(c) The difference δ between the apparent stellar magnitude m_* of the nucleus and the integrated magnitude m_n of the nebula: $\delta = m_* - m_n$.
(d) The linear diameter of the nebula D.
(e) The electron concentration n_e of the nebula.
(f) The mass $\mathcal{M}_n$ of the nebula.
(g) The absolute flux of radiation in an emission line, for example, H_β of hydrogen.
(h) The absolute flux at radio wavelengths.

The values of some of these parameters (M_n, M_*, $\mathcal{M}$) are different in various nebulae; others (R_n, δ, n_e), as well as the former ones, vary over time for one and the same nebula. Some parameters, e.g. $\mathcal{M}$ and H_β, are interrelated, so that in fact the question is about one parameter. Meanwhile, existing methods for determining the distances of PNs are based on the assumption of the constancy of certain parameters for all objects and at all stages of evolution.

Even assuming that certain parameters are preserved constant for some limited number of nebulae grouped by definite signs, the problem of determining the zero point of distance still remains. In other words, the problem is to define the "candle" to fix the distance scale.

Note that the classical methods of astrometry and stellar astronomy – trigonometric parallaxes and proper motions – are unacceptable because of the large distances of PNs. The same is true for the method of spectral parallaxes, because of the extreme inhomogeneity of PNs both in their luminosities and spectral compositions.

The importance of the estimation of the distances and dimensions of PNs is exceptional. The distances are necessary, in particular, for consideration of a number of important problems, such as the total number of PNs, their space density and birth rate in the Galaxy, properties of the spatial distribution of PNs and of the subsystems containing these objects.

Knowledge of the linear dimensions of PNs will enable the determination of their key physical parameters: electron concentration, total mass, lifetime of the nebula and so forth, and also the dispersion of these quantities. All this will eventually lead to the understanding of the nature of PN and their role in the evolution of stars and stellar systems.

The distances of PNs in other galaxies are known with higher accuracy than the distances of PNs in our Galaxy; here the accuracy is determined by that of the measured distances of those galaxies. However, in this way one discovers and determines M_n of the *giant* PNs only. The fact of this selection alone is evidence of the non-constancy of M_n for all PN.

The global solution of the problem of determining the distances of PNs in our Galaxy can be reached by: (a) the application of exotic methods at least for a limited number of PNs; (b) the assumption on the constancy of one of these parameters of the nebula or of its nucleus being the least sensitive to the distance.

In case (a) the results obtained can be used, at the least, for determining the *zero point* of the distance scale. Case (b) implies a *statistical* method of determining distances.

12.2 Basic Methods

The methods for determining the distances of individual PNs are numerous. The most interesting ones can be represented under the following conditional names:

(a) Expansion parallaxes.
(b) Spectral distances.
(c) The Balmer decrement.
(d) Interstellar extinction.
(e) Astrophysical methods.

(f) Surface gravity.
(g) Helium lines.
(h) The 21 cm HI absorption line.
(i) Pseudoresonance lines.

Among the statistical methods for determining the distances of PNs are those of:

(a) Constant mass (Shklovsky, 1956)
(b) Non-constant mass (Pottasch, 1984)
(c) Radio flux (Milne, Aller, 1975).
(d) Mass–radius correlation (Maciel, Pottasch, 1980).
(e) Radio temperature–radius correlation (Zhang, Kwok, 1993).
(f) The combined statistical method (Acker, 1978; Daub, 1982).
(g) Standard distances (Gathier, 1987).

The reliability and accuracy of these methods vary greatly. On the other hand, the diversity of the methods is not a guarantee for the solution of the problem, first, owing to the increase of scatter among the results obtained, and second, because of ambiguity in the choice of the most probable result. For example, for the PN He 2-131 there are 14 determinations of distance by various methods, with the range from 0.6 to 4.5 pc (Terzian, 1993). Another example is the number of PNs among the same 40 objects within 1 kpc: it yields 1 according to estimations by Zhang and Kwok (1993), 11 according to Gathier (1987) and 20 according to Daub (1982). Again the discrepancy is significant.

Clearly, of particular importance is the search for new methods of determining the distances of PNs. In the future, most probably, one will abandon the existing tradition of using a single *general* method for all PNs. Instead, the development of *individual* methods applicable only for definite categories of PNs may be more efficient. Thus, for example, the return to trigonometric parallaxes for individual nebulae via CCD techniques (Pier et al., 1992) can ensure accuracy up to 0.001 arc sec over a period of three years.

Below, these methods are examined separately.

12.3 The Expansion Parallax Method

Because of difficulties in the study of sharply expressed details on photographic images of PNs, the measurements of their angular expansions are realized only in exceptionally favourable conditions. Much easier is the measurement of the velocity of radial expansion – by splitting of the emission lines. With the presence of both these quantities the distance D of the nebula is determined readily:

$$d = 21.1 \frac{V_r}{\varphi} \text{ pc}, \quad (1)$$

Table 12.1. Distances D of a sample of planetary nebulae obtained by the expansion parallax method. $\Delta\varphi$ is in arcsec per 100 years, V_r is the expansion velocity in km s^{-1}, D_Z is Zhang's distance

Planetary nebula	$\Delta\varphi$	V_r	D [pc]	D_Z [pc]	Planetary nebula	$\Delta\varphi$	V_r	D [pc]	D_Z [pc]
NGC 246	1.4	38	570	990	NGC 6 572	0.81	16	430	1 020
NGC 2 392	0.72	18	530	1 170	NGC 6 720	0.9	30	700	1 130
NGC 3 242	0.83	30	760	940	NGC 7 009	0.75	21	590	1 090
NGC 3 587	2.00	41	430	1 500	NGC 7 662	0.8	26	680	1 170

where φ is the angular expansion rate in angular seconds per 100 years, V_r is the velocity of radial expansion in km s^{-1} and the distance D is in parsecs.

Despite the simplicity of the method, its application is strongly limited by the difficulties of measuring angular expansions of nebulae (Gomez et al., 1993). The results for a sample of PNs are presented in Table 12.1 (Pottasch, 1987). For comparison, in the last column, their distances taken from Zhang's catalogue (1995) are also given. As we can see, the expansion distances are systematically smaller. The accordance is better with Daub's (1982) distances.

The discrepancies in expansion parallaxes are more clearly seen when one uses high resolution radio images. For example, using VLA maps of NGC 7 027 separated by $\sim$ 3 year period, a distance of 880 $\pm$150 pc is derived (Masson, 1989), in good agreement with Zhang's distance of 640 pc. The same technique applied for the PNs NGC 6 210, NGC 6 572, NGC 2 392, NGC 3 242 as well as BD+30°3639 and NGC 6 572 at two epochs separated by seven years, for the last two objects gives distances 2 and 4 times larger the previous estimations.

The prospects for the expansion parallax method, especially when it involves radio maps, seem to be good (Hajian et al., 1993, Hajian et al., 1995a, Hajian et al., 1995b).

12.4 Spectral Distances

Among the nuclei of PNs we occasionally observe binary systems with one of the companions clearly being a cool star of A, G or even K type. Assuming that the latter are ordinary stars of the main sequence and, hence, their absolute magnitudes M_v are known, we can obtain the distance D of the nebula, using the visual magnitude m_v of that star.

Despite the obvious logic of the method, its application is not easy. First, the reliability of the identification of the cool component with a main sequence star is unknown. There is ambiguity also in the attribution of m_v to one of the

Table 12.2. Spectral distances D for a sample of planetary nebulae with binary nuclei. D_Z is Zhang's distance

Planetary nebula	Spectral type of nucleus	m_v	M_v	D [pc]	D_Z [pc]
NGC 246	K0 V	$14^m.3$	$5^m.9$	470	990
NGC 1514	A0 III	9.42	-0.2	400	1040
NGC 2346	A2 V	11.12	1.4	640	2070
NGC 3132	A0 V	10.06	0.7	670	1500
A 35	G8 III–V	9.63	1.9	360	740

stars of the binary system. For the same reason the true value of the colour index $E(B-V)$ cannot be established and, hence, neither can reliable account be taken of the interstellar extinction. As a result, the distances obtained by this method can differ widely from those obtained by other methods. For example, spectral distances for 5 PNs are given in Table 12.2 (Pottasch, 1987), together with Zhang's distances D_z (last column). As we can see, in all cases the spectral distances are 2–3 times smaller than Zhang's distances.

From the comparison of the optical (ground-based) observations with model spectra for a late-type star in a binary nucleus of LoTr 5 a gravity $\log g = 2.7$ and $T_{\mathrm{eff}} = 5\,250$ K were derived in full agreement with a spectral type G5 III for cool component of this nuclei (Jasniewicz et al., 1996). This corresponds to a distance $\sim$ 400 pc, again two times smaller than Zhang's' distance, i.e. 890 pc.

In the future the inverse problem may also make sense, i.e. by using a reliable distance to find out the absolute magnitude M_v of the cool star and, hence, to reveal the true nature of conditions in the nuclei of PNs.

12.5 The Balmer Decrement Method

The relative intensities of hydrogen emission lines and the Balmer decrement must be practically the same for all PNs, without exception. However, owing to selective interstellar absorption it becomes distorted, i.e. steeper. Therefore the observed decrement will differ from the theoretical one; the stronger the interstellar absorption, the farther is the nebula from the Earth. The steepness of the decrement is large for the first lines of the series and falls for the next lines, so that the largest divergence should be expected for the ratio H_α/H_β. The theoretical value of H_α/H_β is equal to 2.85 for optically thick nebulae (in the Lyman continuum) and $\sim$ 1.90 for optically thin ones (Table 3.8). The observed values of H_α/H_β are often of the order at 5–8, sometimes 10 at higher (see Table 3.10).

Thus, the degree of divergence of the observed ratio H_α/H_β from its theoretical value of 2.85 is an indicator of the distance D of a PN from the Earth. Then we can obtain D by the difference of both values of H_α/H_β, if the law of interstellar absorption and its value a_λ per kpc, i.e. the specific absorption coefficient, are also known.

Denote by a_α and a_β the specific absorption coefficients for lines H_α and H_β. Then we have

$$\left(\frac{H_\alpha}{H_\beta}\right)_{obs} = \left(\frac{H_\alpha}{H_\beta}\right)_{thr} e^{-(a_\alpha - a_\beta)\,D} .$$

From here we obtain for the PN distance D, in kpc, from the Earth

$$D = \frac{2.30}{a_\alpha - a_\beta} \log \frac{(H_\alpha - H_\beta)_{obs}}{(H_\alpha - H_\beta)_{thr}} , \tag{2}$$

where the numerical values of $a_\alpha(\lambda = 6\,563$ Å) and $a_\beta(l = 4\,861$ Å) may be taken from standard curves of the type shown in Fig. 7.13. In the direction, say, of the Galactic plane, we have $a_\alpha = 0^m.67$ kpc^{-1} and $a_\beta = 1^m.30$ kpc^{-1}, and then we get, also taking $(H_\alpha/H_\beta)_{thr} = 2.85$

$$D = 3.65 \; [\log (H_\alpha/H_\beta)_{obs} - 0.455] \;\; \text{kpc} . \tag{3}$$

If, for example, $(H_\alpha/H_\beta)_{obs} = 4.0$, we find for the distance of the object $D = 540$ pc.

In this case again, the method is simple and logical. However, its practical application again meets certain difficulties. The main one is connected with the knowledge – in the case of individual applications of this method – of the values of a_α and a_β in the given direction; the latter may sometimes vary strongly, depending on the Galactic latitude and longitude.

There is one more difficulty, i.e. the distortion of the ratio H_α/H_β as a result of self-absorption into the nebula. Therefore, first, one has to distinguish the contributions of both types of absorption, interstellar and nebular.

The statistical Balmer decrement method has found wide application based on the use of some mean magnitudes of a_α and a_β (Stasinska et al, 1993).

12.6 The Radio Flux Method: Individual and Statistical

The total radio power P_ν emitted in all directions over a 1 Hz band at the frequency ν by an optically thin HII region of angular radius θ at a distance D from the Earth can be written as

$$P_\nu = \frac{4\pi}{3} (D\theta)^3 \, 4\pi\varepsilon(\nu) , \tag{4}$$

where $\varepsilon(\nu)$ is the free–free emission, both by hydrogen and helium atoms, per unit volume per second.

But P_ν is also equal to

$$P_\nu = 4\pi D^2 S_\nu , \tag{5}$$

where S_ν is the flux density measured at the Earth.

The interstellar medium is practically transparent for the radio frequency $\nu \sim 5$ GHz. Using the known expressions for $\varepsilon(\nu)$ (e.g. Oster, 1961), we can derive from (4) and (5) the following relationship for the distance D (in pc) of the nebula from the Earth (Milne, Aller, 1975):

$$D = \left[\frac{3.23 \times 10^{19} \mathcal{M}^2 \ln(9\,980\, t^{3/2}}{t^{1/2}\, \theta\, S_{5\ \mathrm{GHz}}} \psi \right]^{1/5} , \tag{6}$$

where $t = T_e/10^4$ K, $\mathcal{M}$ is the nebula's mass in solar units, and ψ is a factor of the order of unity which takes into account the influence of helium atoms.

Equation (6) if applied to a single PN is considered the *individual radio flux* determination method of the distance.

However, if we adopt on average for all PNs He/H =0.11, $\mathrm{He}^+/\mathrm{He} = 0.5$, $\mathcal{M}/\mathcal{M}_\odot = 0.16$ and $T_e = 10^4$ K, we will find from (6) for D (in pc) that

$$D = 6\,180\, \theta^{-3/5}\, S_{5\ \mathrm{GHz}}^{-1/5} , \tag{7}$$

where θ is expressed in arc seconds, and $S_{5\ \mathrm{GHz}}$ is in flux units ($\mathrm{W\,m^{-2}s^{-1}}$). In this case we have the *statistical radio flux method* for determining the PN distances. In fact (7) is the radio frequency version of the constant mass method.

The distances and sizes of 165 PNs have been determined (Milne, Aller, 1975) by means of the radio flux method.

12.7 The Interstellar Extinction Method

Consider a group of stars in the direction and neighbourhood of the given PN known both by their distances and magnitudes of interstellar absorption, i.e. by the values of $E(B-V)$. Then we can obtain the dependence of absorption on the distance D, i.e. $D \sim E(B-V)$. With the help of "absorption–distance" graphics, then, we can find out the distance of the PN from the known value of $E(B-V)$ (Martin, 1994; Saurer, 1995).

This method can be applied only to PNs located very near the Galactic plane. The difficulties in finding proper stars in the vicinity of PN limit the wide application of this method. Nevertheless it has been applied both for individual PNs, and in a statistical version (Pottasch, 1984).

12.8 The Astrophysical Method

The electron concentration in PNs can be determined both with the help of forbidden lines, when it is not necessary to know the distance of the nebula from the Earth, and with the help of recombination emission lines, when the distance of the nebula should be known. Then, we can obtain the distance D from the known electron concentration n_e, and this is the essence of the astrophysical method of determining the nebula's distance.

The final expression for the distance D has the following form:

$$d = 2.12 \times 10^{41} \frac{F(H_\beta)}{n_e^2 \theta^3 \varepsilon} \text{ cm} = 6.9 \times 10^{22} \frac{F(H_\beta)}{n_e^2 \theta^3 \varepsilon} \text{ pc}, \quad (8)$$

where θ is the radius of the nebula in arc seconds, $F(H_\beta)$ is expressed in units of ergs $cm^{-2}s^{-1}$ and ε is the so-called filling factor, i.e. the fraction of radiating gas in the nebular volume (Botti, Stanghellini, 1994).

Equation (8) has a weak point: D depends on the square of n_e, and on the cube of θ. The magnitude of n_e is determined with an accuracy of 2, while θ is determined quite ambiguously, especially for small angular size and non-spherical nebulae. As a result, one cannot expect reliable results from this method. In Table 12.3 the values of D determined by (8) are given (fifth column) for a sample of PNs (assuming $\varepsilon = 0.80$ for NGC 7 293 and $\varepsilon = 1$ for the remaining nebulae) together with Zhang's D_Z and Daub's D_D distances. Note the large discrepancies between all three determinations.

Table 12.3. Distances D for a sample of planetary nebulae obtained by the astrophysical method. D_Z and D_D are Zhang's and Daub's distances

Planetary nebula	θ	n_e [$10^3 cm^{-3}$]	$\log F(H_\beta)$	D [pc]	D_Z [pc]	D_D [pc]
NGC 246	118″	0.09	−10.20	325	990	460
NGC 6 210	8.1	5.0	−10.08	430	1 730	1 190
NGC 6 572	6.2	5.0	−9.81	1 790	1 020	474
NGC 6 720	35	0.63	−10.06	350	1 130	790
NGC 6 818	9.1	2.70	−10.52	380	1 630	1 290
NGC 7 009	13.4	2.2	−9.78	980	1 090	759
NGC 7 293	330	0.10	−9.35	100	420	150

12.9 The Surface Gravity Method

This method should be attributed to astrophysical methods of determining nebulae distances. Its basic idea is simple, i.e. to fit the high resolution profiles of absorption lines in the spectrum of a nebula's central stars with the model atmosphere computed profiles. Then one can estimate the parameters of the central star, including the effective temperature and the surface gravity. Combined with the surface gravity g of the accepted mass of the central star, its radius and, hence, the absolute luminosity can be obtained. Finally, the obtained luminosity together with the visual magnitude of the central star will give the distance of the star (nebula) from the Earth (Mendez et al., 1983, 1988, 1992).

This approach has become possible as a result of extensive computational work on non-LTE model atmospheres and an enormous number of predicted profiles of absorption lines. On the other hand, the high resolution lines profiles can well be obtained from observations. Although the application of this method is limited to PNs with visible central stars and possessing absorption spectra, the method itself seems promising. Indeed, in Table 12.4 the distances D_g determined by the surface gravity method for six PNs, along with results D_Z by Zhang (1995) and D_C by Cahn et al.(1992), are presented for comparison. Clearly, the distances D_g are in surprisingly in good accordance with the other estimations, D_Z and D_C.

Table 12.4. Distances D_g obtained by the surface gravity method for a sample of planetary nebulae. D_Z and D_C are the distances by Zhang (1995) and Cahn et al. (1992)

Planetary nebula	$\log g$ [cm s^{-2}]	m_v	D_g [pc]	D_Z [pc]	D_C [pc]
NGC 1360	5.25	11.35	550	980	850
NGC 1535	4.5	11.92	1 500	2 140	2 250
NGC 4361	5.0	13.33	1 400	1 250	900
NGC 7293	7.5	13.43	125	420	160
A 7	7.0	15.43	500	700	820
A 36	5.2	11.71	400	1 020	380

12.10 The Helium Lines Method

The sizes of monochromatic images of some PNs in the line 4 686 HeII are noticeably smaller than in the hydrogen lines. This means that the optical depth in the frequencies of singly ionized helium is close to unity on the boundary of the He^{++} zone, i.e. on the edge of the nebula's image in the line 4 686 HeII. This may be used to derive a relationship connecting the distance of the nebula D with the observed angular size of the nebula in HeII.

The final expression for determining the distance D of a nebula by this method has the following form (Gurzadyan, 1969):

$$D = 3.61 \times 10^{-8} \left(\frac{R_*}{n_e}\right)^{2/3} \frac{[f(T_*)]^{1/3}}{\varphi''}(\text{HeII}) \text{ pc}, \tag{9}$$

where

$$f(T_*) = \frac{1}{2} \left(\frac{T_e}{T_*}\right)^{1/2} \frac{(2\pi\mu k T_*)^{2/3}}{h^3} \exp\left(-\frac{\chi^{++}}{kT_*}\right) \tag{10}$$

In these formulae φ''(HeII) is the angular size (diameter) of the nebula in arc seconds in the line 4686 HeII, χ^{++} is the ionization potential for He^{++}, T_* and R_* are the temperature and the radius of the nebula's nucleus, T_e and n_e are the electron temperature and concentration, respectively.

The main idea of the method is to use only the observational parameter φ''(HeII) directly from the nebula's spectral image in the line 4686 HeII.

One can act also in another manner: (9) can be used for determining not the distance of the PN but its nucleus radius R_*. In any case there exist a number of methods for determining the nebulae distances, while the radii of nuclei are usually known very badly.

From (9) we can write for the nucleus radius

$$R_* = 2.70 \times 10^{-17} n_e \left[\varphi''(\text{HeII})\right]^{3/2} \left[f(T_*)\right]^{-1/2} D^{3/2} \text{ cm}, \tag{11}$$

where D is expressed in cm and φ'' in arc seconds.

As an illustration, (11) has been used for determining the nucleus radii R_* for 12 PNs for which $\varphi(\text{HeII}) < \varphi(\text{HI})$; the results are presented in Table 12.5. In the computations Daub's distances D_D were used, as well as the values of n_e and T_* from different sources.

Table 12.5. Radii R_* of planetary nebula nuclei and dilution factors W obtained by the HeII method

Planetary nebula	φ''(HeII)	T_* [K]	n_e [$10^3 cm^{-3}$]	D_d [pc]	R_* [$R_\odot$]	R_n [pc]	W
NGC 2392	6.4	100 000	5.6	1 230	0.30	0.139	6.0×10^{-16}
NGC 2440	2.6	122 000	2.8	1 010	0.021	0.132	3.3×10^{-16}
NGC 3242	7.4	78 000	14.1	730	0.074	0.074	1.3×10^{-16}
NGC 6572	2.0	58 000	20.0	474	0.17	0.017	1.3×10^{-14}
NGC 6720	19	63 000	0.63	790	0.26	0.164	3.2×10^{-16}
NGC 6818	7.2	75 000	3.3	1 290	0.40	0.069	4.3×10^{-15}
NGC 6886	1.7	69 000	5.6	1 790	0.17	0.026	5.3×10^{-15}
NGC 7009	9	100 000	12.5	759	0.54	0.055	1.2×10^{-14}
NGC 7027	7	150 000	63.0	178	0.11	0.007	3.2×10^{-14}
NGC 7662	6.5	100 000	3.2	837	0.10	0.061	3.5×10^{-15}
IC 2165	2.2	74 000	2.8	1 370	0.07	0.030	7.2×10^{-16}
IC 5217	2	69 000	15.8	1 620	0.85	0.028	1.2×10^{-13}

As we can see, the nuclear radii are systematically smaller than the solar radius. The smallest radius $R_*/R_\odot = 0.02$ corresponds almost to a white dwarf (NGC 2 440).

In combination with Daub's radii for the nebulae R_n (seventh column) the dilution factor W for these nebulae was obtained too; the results are given in the last column.

12.11 The 21 cm HI Absorption Line Method

This method, in fact, is similar to that of interstellar extinction, since is based on the properties of the 21 cm HI interstellar absorption line. This is along the line of traditional methods for determining the distances of objects with unknown luminosities in certain wavelengths (Gathier et al., 1986)

We already know that PNs reveal radio continuous emission (Chap. 14). The interstellar neutral hydrogen of concentration $n_{\rm H}$ along the Earth–nebula line will cause an absorption line at 21 cm. The intensity of this absorption line is proportional to the column density of neutral hydrogen $N(\rm HI)$, as well as to the distance of the nebula D, i.e. $N(\rm HI) = n_{\rm H}D$. Then we can obtain the distance of the nebula: $D = N(HI)/n_{\rm H}$.

If the 21 cm line is identified with confidence, using the radial velocity of the nebula, then the numerical magnitude of $N(\rm HI)$ can be obtained also with confidence. The main difficulty is connected with $n_{\rm H}$; it can be different in various directions in the Galaxy and even along the line of sight to the nebula. Despite this uncertainty, the 21 cm HI line method in some cases gives results in good accordance with the distance determined by other methods. See below the obtained distances $D(21$ cm) for six PNs compared with Zhang's distances $D_{\rm Z}$ (in pc):

Planetary nebula	D(21 cm)	$D_{\rm Z}$	Planetary nebula	D(21 cm)	$D_{\rm Z}$
NGC 2 440	2 000	1 570	NGC 7 026	3 500	2 030
NGC 6 537	2 500	1 620	NGC 7 027	1 000	640
NGC 6 572	450	1 020	NGC 7 354	1 800	1 300

12.12 The Pseudoresonance Line Method

This method of determining distances to individual PNs is based on the use of pseudoresonance absorption lines (Chap. 8).

Recall that the pseudoresonance absorption line is excited within the nebula at transitions from some lower metastable level i to some subordinate level J, and will appear with residual intensity J_{ij} in the background of the continuous spectrum of the central star. Then we will have approximately for the optical depth $t_{\rm o}$ of the nebula in this pseudoresonance line

$$t_o = \ln J_o / J_{ij} ,$$

where J_o is the level of the continuous spectrum near the line in question. Both the quantities J_o and J_{ij}, can be easily determined from the observed spectrum of the nebula's central star.

However, from the level i we will have also transitions $i \to 1$ corresponding to forbidden emission line of intensity I_f; the latter is expressed in units of intensity I_β of the hydrogen line H_β. Hence, the ratio I_f/I_β is also a known quantity. Then we can find the distance d of the nebula with the following relationship (Gurzadyan, Egikyan, 1988):

$$D = 5.71 \times 10^7 \frac{A^*_{i1} h\nu^*_{i1}}{\varphi''} \frac{T_e}{n_e^2} \frac{I_\beta}{I_o} \frac{t_o}{s_{ij}} \text{ pc} , \tag{12}$$

where ν^*_{i1} and A^*_{i1} are the frequency and Einstein probability coefficient of the forbidden transition $i \to 1$, s_{ij} is the absorption coefficient in the given pseudoresonance line, φ'' is the angular size of the nebula.

Note that both lines forbidden in emission and permitted in absorption must appear in a definite combination, i.e. with a common starting level.

Among the recommended combinations (see Chap. 8) is the forbidden doublet of twice ionized silicon 1 883+1 892 <SiIII> and the pseudoresonance line 1 300 <SiIII>.

The pseudoresonance line method has still not been applied to determine the distances of PNs.

12.13 The Constant Mass Method

Above, in Chap. 6 an expression (55) was derived for the ionized mass $\mathcal{M}$ of the nebula, its luminosity L and volume V:

$$\mathcal{M} \sim (LV)^{1/2} . \tag{13}$$

For the luminosity of the nebula in some emission line with the observed flux on the Earth F_λ, we have $L = 4\pi\, D^2\, F_\lambda$, where D is the distance of the nebula from the Earth. For the linear radius of the nebula we have $R_n = \varphi\, D$, where φ is the angular size (radius) of nebula, and $V = (4\pi/3)\, R_n^3 = (4\pi/3)(\varphi\, D)^3$. Then we obtain

$$D = C \frac{\mathcal{M}^{2/5}}{F_\lambda^{1/5}\, \varphi^{3/5}} . \tag{14}$$

This relationship, based on the assumption $\mathcal{M}$ = const for all PNs, has been recommended by Minkowski and Aller (1954) and Shklovsky (1956) for the statistical determination of distances of PNs. The idea behind it lies in the weak dependence between the distance and the ionized mass: $D \sim \mathcal{M}^{0.4}$. The zero-point scaling, i.e. the numerical value of the constant C in (14), is obtained by applying this expression for PNs with known masses $\mathcal{M}$ and distances D, obtained in another way.

The constant mass method has been used extensively (O'Dell, 1962; see also Seaton, 1968; Cudworth, 1974; Milne and Aller. 1975, etc.).

Originally the (14) was applied in relation to PNs with the known integral stellar magnitude m_n, using $F_\lambda \sim 10^{-0.4m(n)}$ in (14). Soon after, O'Dell (1962) recommended the use of the observed fluxes $F(\mathrm{H}_\beta)$ in the hydrogen emission line H_β measured in the spectra of a number of PNs. As a result the following working formula for the determination of PN distances by known values of $F(\mathrm{H}_\beta)$ and φ appears:

$$D = 22.8 \frac{\mathcal{M}^{2/5}}{F(\mathrm{H}_\beta)^{1/5} \varphi^{3/5}} , \tag{15}$$

where D is expressed in kpc, $F(\mathrm{H}_\beta)$ in flux units, 10^{-11} ergs $\mathrm{cm}^{-2}\mathrm{s}^{-1}$ the mass $\mathcal{M}$ in solar masses and the angular size (radius) in arc seconds.

For density limited nebulae, i.e. for those optically thin to ionizing radiation, this method can give a reasonable estimate of the distance. The main problem, therefore, is to decide whether the nebula is optically thin or not. Moreover, there is evidence that a remarkable fraction of PNs are optically thick, thus limiting the application of the constant mass method.

We mention that as a rule this method "prefers" larger distances to PNs, while the extinction methods indicate shorter ones.

12.14 The Non-constant Mass Method

Soon after the formulation of the constant mass method it was realized that the values of $\mathcal{M}$ can vary by up to two orders of magnitude. Obviously, the most reasonable way out of the situation might be the accumulation of data on the masses of individual PN obtained with individual methods. From this point of view the dependence in Fig. 12.1 between $\mathcal{M}$ and the electron concentration n_e is of particular interest (Maciel, Pottasch, 1980). The existence of a definite tendency of a nebula's ionized mass $\mathcal{M}$ to decrease with the increase of the electron concentration n_e is obvious. From here one can write for $n_e > 10^3\,\mathrm{cm}^{-3}$

$$\frac{\mathcal{M}}{\mathcal{M}_\odot} = 10^2/n_e , \tag{16}$$

where n_e is in cm^{-3}.

Rewrite (10) in the following form, taking for the filling factor $\varepsilon = 1$:

$$4\pi\, D^2 F(\mathrm{H}_\beta) = 4\pi\, \varepsilon(\mathrm{H}_\beta)\, n_e^2\, V , \tag{17}$$

where V is the volume of the ionized part of the nebula.

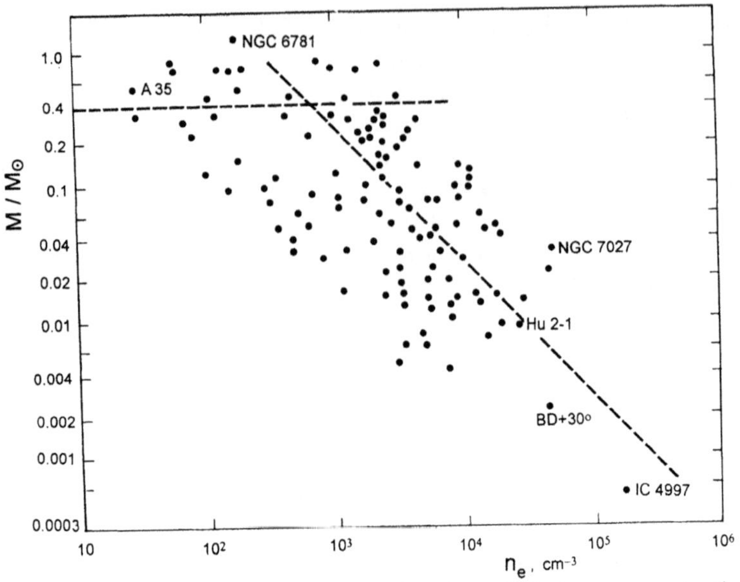

Fig. 12.1. Ionized nebular mass $\mathcal{M}$ as a function of electron concentration n_e. Dots indicate the nearby planetary nebula; the broken horizontal line, optically thin ones; and the broken sloping line, optically thick ones

We have for the mass of the ionized part

$$\mathcal{M} = 1.4\,V\,n_e\,m_H\,, \tag{18}$$

where m_H is the mass of the hydrogen atom, and the coefficient 1.4 arises owing to the fact that $\sim$ 10% of the total number of atoms in the nebula or $\sim$ 40% of its mass is composed of helium atoms.

Combining (16), (17) and (18) and using the numerical values of the constants, we have for the distance D

$$D = 2.92 \times 10^{16}\, F(\mathrm{H}_\beta)^{-1/2}\,\mathrm{cm} \tag{19}$$

or

$$D = 3\,F(\mathrm{H}_\beta)^{-1/2}\ \mathrm{kpc}\,, \tag{20}$$

where the observed flux $F(\mathrm{H}_\beta)$ is expressed in units of 10^{-11} ergs cm^{-2} s^{-1}.

Equation (20) gives the statistical "non-constant mass" distance of each PN with a known observational value of the flux $F(\mathrm{H}_\beta)$. It is interesting that, in contrast with the initial relationship, (14) or (15), it is now not necessary to know not only the mass of the nebula $\mathcal{M}$ but even its angular size φ.

In fact, the only problem is proper consideration of the interstellar extinction.

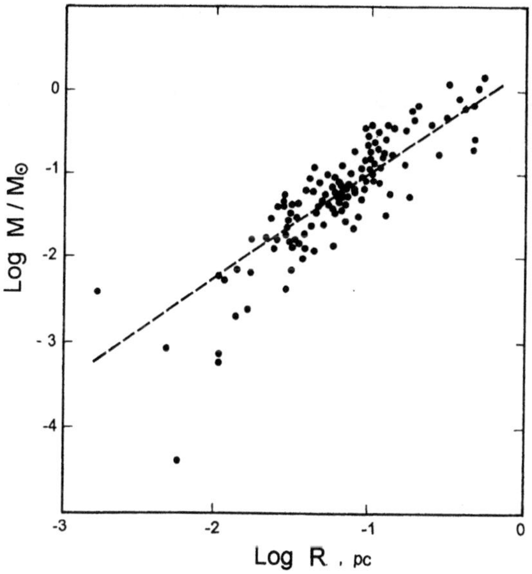

Fig. 12.2. The relationship between the ionized mass and the nebular radius. Dots denote planetary nebulae with well-determined individual distances

12.15 The Mass–Radius Correlation Method

The constant mass conjecture neglects one more obvious consideration: with the expansion of a nebula, i.e. with the increase of its linear sizes the amount of its ionized mass should also be increased. Thus, the idea of the existence of a "mass–radius" type correlation has arisen (Pottasch, 1980; Maciel, Pottasch, 1980).

The mass–radius relationship also also its theoretical explanation in the framework of the model of nebular expansion coupled with the evolution of the central stars (Kwok, 1983; Schmidt-Voigt, Köppen, 1987). The results of these investigations have been used for the derivation of a correlation between the ionized mass and the nebular radius. Shown graphically in Fig. 12.2, it can also be given in the following empirical form (Zhang, 1995):

$$\log \mathcal{M}_i = 0.473 + 1.414 \ \log R\,. \tag{21}$$

On the other hand, for a nebula which is optically thin at 5 GHz (6 cm) radio frequency, the ionized mass $\mathcal{M}_i$ can be written as (Gathier, 1987)

$$\mathcal{M}_i = 5.170 \times 10^{-5} \, (S_{5\ \mathrm{GHz}})^{1/2} \, \theta_r^{3/2} \, D_i^{5/2}\,, \tag{22}$$

where $S_{5\ \mathrm{GHz}}$ is the radio continuum flux density (in mJy) at 6 cm, θ_r is the angular radius of radio image in angular seconds, and D_i is the distance in kpc.

Taking (21) into account, we have for the correlation between the nebula's distance D_i

$$\log D_i = 1.458 - 0.161 \log \theta_r - 0.419 \log S_{5\ \mathrm{GHz}} \tag{23}$$

in the case of the "mass–radius" distance scale (in kpc).

The calibration of the mass-radius distance scale is realized for 132 PNs with well determined individual distances (Zhang, 1995).

12.16 The Radio Temperature–Radius Correlation

From the known radio continuum flux density $S_{5\ \mathrm{GHz}}$ at 5 GHz, the radio continuum surface brightness temperature T_b at 5 GHz is obtained via the relationship

$$T_b = \frac{c^2}{2\pi k \nu^2} \frac{S_{5\ \mathrm{GHz}}}{\theta^2}\,. \tag{24}$$

This formula applied to 23 nearby PNs with individually determined distances D leads to a noticeable correlation between D and T_b (Van de Steene, Zijlstra, 1994). This correlation is confirmed by the data of 134 PNs, again with individually determined radii (Zhang, 1995). The following linear relationship between $\log T_b$ and $\log R$ is derived:

$$\log R = -0.56 - 0.32 \log T_b \tag{25}$$

or, rewritten for a nebula's distance D_b (in kpc)

$$\log D_b = 1.748 - 0.31 \log T_b - \log \theta_r \tag{26}$$

and is called the "radio temperature–radius" distance scale.

12.17 The Combined Statistical Method

This method has been recommended by Acker (1978) and, in fact, combines two different statistical distance scale methods. It compares these scales by dividing the PNs into two groups in accordance with two principles: constant nebular luminosity – for ionization limited PNs (optically thick nebulae) – and constant nebular ionized mass – for density limited PNs (optically thin nebulae). The definition of the boundary between these two groups, and the procedure of separating optically thick nebulae, which are the majority, remains not quite certain. Hence, by comparing several statistical distance scales for density limited and ionization limited PNs, with a number of individual distance determinations distances the final *combined distance scale* is defined.

Despite the above mentioned uncertainties, the distances derived in this way are expected to be on average more accurate than distances based, say, on the constant mass assumption.

12.18 Standard Distances

Below in Table 12.6 we present a sample of 30 PNs, the distances of which have been determined by means of the methods described above – *spectroscopic, expansion, reddening and 21 cm HI absorption* – with uncertainties less than 40% (Gathier, 1987). These data are presumed to be the most accurate PN distances.

Table 12.6. Distances of a sample of 30 planetary nebulae

Planetary nebula	D, pc	Planetary nebula	D, pc	Planetary nebula	D, pc
NGC 246	500	NGC 5 189	1 730	NGC 6 884	1 800
NGC 1 514	500	NGC 5 315	2 620	NGC 6 886	1 700
NGC 2 346	800	NGC 6 369	2 000	NGC 7 009	580
NGC 2 440	2 190	NGC 6 537	2 400	NGC 7 026	2 180
NGC 2 452	3 570	NGC 6 565	900	NGC 7 027	1 100
NGC 2 792	1 910	NGC 6 567	1 680	NGC 7 354	1 500
NGC 3 132	600	NGC 6 572	410	NGC 7 662	980
NGC 3 211	1 910	NGC 6 578	2 000	IC 1 747	2 500
NGC 3 242	500	NGC 6 720	650	He 2–131	600
NGC 3 918	2 240	NGC 6 803	3 000	K 648*	10 000

*The planetary nebula K 648 is a member of the globular cluster M 15 with an adopted distance of 10 ± 1 kpc.

One can add, perhaps, one more object, NGC 7 293, in fact the nearest PN, the distance of which has been determined by at least seven different authors, in good agreement with each other (see Acker et al., 1992): 150, 180, 160, 120, 150, 180 and 200 pc, the mean distance being $D = 163 \pm 24$ pc.

Individual distances of 50 nearby PNs are determined using a variety of methods, but excluding statistical methods or distance scales (Pottasch, 1996). These distances are used to determine local PNs space density and formation rate. A local formation rate 3×10^{-3} PN kpc^{-3} yr^{-1} is found. For these 50 PNs, which have a distance less than 1 kpc from the Sun, the scale height is $\sim$ 160 pc.

12.19 Catalogues of Distances of Planetary Nebulae

Progress in the problem of determining the distances of PNs is obvious, clearly expressed by the appearance of a number of catalogues of distances and dimensions of PNs. Thus, the first catalogue compiled by Cahn and Kaler (1971) included distances for 610 PNs based on the constant–mass conjecture with $\mathcal{M}/\mathcal{M}_\odot = 0.18$ for all PNs.

The next was Acker's catalogue for 330 PNs (Acker, 1978), in which, besides the individual methods, a combined methods were also used, namely, constant mass for optically thin nebulae and constant luminosity for optically thick ones.

An analogue of the Cahn–Kaler catalogue is the catalogue by Milne and Aller (1975, 1982), which includes 397 objects, predominantly of the southern sky. Here, the constant mass value ($\mathcal{M}/\mathcal{M}_\odot = 0.16$) was adopted, along with the use of angular sizes φ'', the radio brightness S of PNs at a frequency of 14.7 MHz (2.04 cm), for which the interstellar medium is transparent even in the direction of the Galactic centre.

The first catalogue based on the concept of non-constant mass was Daub's catalogue for 299 PNs (Daub, 1982). The starting point in this catalogue was the above mentioned relationship (16) or Fig. 12.2; the results of the Milne–Aller radio observations were used as well. This catalogue includes, besides the distances, also the angular sizes, linear radii and radio fluxes at a frequency of 5 GHz.

The most complete catalogue of distances and dimensions of 663 PNs, was compiled by Maciel in 1984, again using the non-constant mass method (Maciel, 1984). Here, not only the H_β fluxes but also the H_α ones were involved. Good agreement exists between the last two catalogues, Daub's and Maciel's the largest discrepancy in the distances being not larger than two-fold. However, the number of objects in Maciel's catalogue is much larger.

Of special importance is the catalogue of expansion velocities for 288 PNs by Weinberger (1989). Here a set of PN distances is presented along with their adopted linear radii and expansion velocities.

An essential step was made by Cahn, Kaler and Stanghellini (1992) in compiling the statistical distances for 620 Galactic PNs derived from H_β fluxes. The statistical distance scale based only on the empirical correlation between the radio continuum brightness temperature and the nebula's radius for 400 Galactic PNs given represented by Van de Steene and Zijlstra (1994).

The fairly complete statistical distance scale proposed by Zhang (1995) is the average of distances obtained by two methods, based on the mass–radius correlation (D_i) and radio temperature–radius correlation (D_b) for 647 Galactic PNs. The last two distance scales, by Zhang and by Van de Steene and Zijlstra, are in good agreement with each other. This fact should be considered an event of exceptional importance characterizing progress in this area. The PN distance problem has never been so close to its solution.

13. Double-Envelope Nebulae

13.1 Observational Aspects

One often talks about PNs as objects consisting of a single gaseous envelope surrounding the central star – the nucleus. However, the nuclei are sometimes surrounded simultaneously by two concentric envelopes. In such cases we speak of *double-envelope planetary nebulae.* The outer (second) envelope is as a rule considerably less bright; typically it is so weak that it can be discovered only with great difficulty mainly by CCD technique. Examples of double-envelope PNs are given in Chap. 1, and Plates II, IV–VI, VIII and IX.

At the time when the double- or two-envelope PNs first attracted attention in 1953 (Gurzadyan, 1953), the number of known objects was less than twenty. Forty years later (1993) over a hundred of these objects are known. This indicates that double-envelope PNs are not an exotic or chance phenomenon but rather are associated with the basic physical processes occurring in a PN and its nucleus. First, let us see to what extent the observational picture – the presence of two envelopes around the nucleus – is a real physical event.

It is not difficult to be convinced that the double-envelope phenomenon cannot be a result of a stratification effect (Chaps. 1 and 11) because:

(a) The stratification effect should lead to not only two but many envelopes, since there are many lines that can produce intense monochromatic images of the nebula.
(b) Some monochromatic images of double-envelope PNs also show two concentric circles. This means that the emission of both envelopes has identical spectral composition.

Thus, the presence of two envelopes around certain nuclei is real, and does not depend on the observational conditions.

13.2 Morphological Peculiarities

In double-envelope PNs the edges of both envelopes tend to be sharp and are characterized by definite apparent diameters of the inner, d_1, and outer, d_2, shells. The ratio d_2/d_1 does not depend on the distance to the nebula and can be obtained without any difficulty as it is one of the basic parameters of two-envelope nebulae. The ratio d_2/d_1 varies in broad limits, on average being between 1.5–2 (Sect. 13.7). This means that the linear depth of the second (outer) envelope, i.e. $\Delta r = (D_2 - d_1)/2$, is of the same order as the size of the nebula itself.

The analysis of images of two-envelope nebulae, obtained in more or less homogeneous conditions, enables one to divide them into two main types, I and II, depending on their morphological peculiarities (Chu et al., 1987).

Type I contains the objects the outer envelope of which seems to be isolated from the inner envelope, i.e. not having any contact with the inner one. Example of such objects are NGC 2438, NGC 2022 (Plate VIII). To *type II* are attributed those two-envelope nebulae with the outer envelope not isolated from the inner one. The two envelopes, however, can be clearly distinguished. An example of type II objects is NGC 1535 (Plate VIII).

The subclasses *a*, *b*, *c* and *p* are distinguished and characterize the degree of homogeneity, filamentary structure, etc., of the outer envelope in the following manner:

Subclass *a*: there is complete homogeneity of brightness of the outer envelope (NGC 1535, IC 3568).

Subclass *b*: the outer envelope is visible everywhere; however, the brightness is not homogeneous (NGC 3242, NGC 7662).

Subclass *c*: the outer envelope is seen partly, i.e. in one direction, and is absent in others (NGC 7354, IC 289).

Subclass *p*: the outer envelope is irregular in its brightness with a strongly developed filamentary structure (NGC 1514, NGC 4361).

13.3 The Origin of the Second Envelope

There are a few proposed mechanisms explaining the origin of two-envelope PNs. According to the first one proposed as early as 1954, the outer envelope originates as a result of dynamical separation of the main envelope (Gurzadyan, 1954). This mechanism is based on the possibility of splitting off the main envelope under the action of L_α radiation pressure on some stage of evolution of the nebula. This dynamical splitting mechanism of the origin of two-envelope nebulae has been supported also by later considerations (e.g. Jewitt et al., 1986).

Alternative proposed mechanisms include:

(a) Multiple ejection of gaseous matter from the central star (Tuchman, Barket, 1980).
(b) Long duration stellar winds (Kwok et al., 1978) or even superwinds (Renzini, 1981).

(c) Interacting stellar winds (Kahn, 1989; Kahn, Breitschwerdt, 1990; Mellema, 1994).

Moreover, note that the second (outer) envelope is in fact the hydrogen ionization zone of interstellar gas. At a definite moment of expansion the nebula should become transparent for radiation beyond the Lyman limit, and therefore a fraction of the ultraviolet radiation will penetrate the medium surrounding the nebula, and in principle, can create a second envelope around the first. This assumption, however, cannot explain many properties of two-envelope nebulae. Moreover, no strong concentration in the Galactic disk region of this type of nebula takes place; two-envelope nebulae are found even at high Galactic latitudes – up to $b \sim 60°$.

13.4 The Similarity of Both Envelopes

The mutual similarity of both outer and inner envelopes is an important feature of two envelope nebulae. They are similar in several ways:

(a) In the general forms of the outer and inner envelopes (NGC 6 826, NGC 2 022, NGC 2 438, He 2–149).
(b) In the *parallel* outer edges or boundaries of both envelopes (NGC 7 009, NGC 7 293, NGC 3 132, IC 1 295).
(c) In the *bipolarity* of forms of both envelopes (NGC 6 720, NGC 3 242, NGC 7 293, NGC 7 009).
(d) In the degree of *inhomogeneity* of both envelopes (NGC 7 354, NGC 2 392).

In many cases the following are true. (i) Both the outer and inner envelopes are elongated to the same degree. This is of special importance because it indicates one and the same ratio of expansion velocities along both axes – large and small – for both envelopes. (ii) The orientation of the long axes in both the outer and inner envelopes is the same. (iii) The outer limbs of both envelopes are strongly parallel even in detail, including the variation of local curvature and the sizes of *fragmentations*.

The similarity of both envelopes in nebulae with *thin* outer envelopes (NGC 3 587, NGC 6 804), is of particular interest. For example, for IC 1 295 the inner envelope is of oval shape, while the outer envelope has the form of a very thin ring, strongly parallel to the inner oval (Plates VI and IV). Here, the thin outer envelopes seem to preserve even the the slightest peculiarities on the edge of the inner ones.

As we will see below (Sec. 13.17) these peculiarities can hardly be explained by the models of repeated outbursts, while they can be easily explained in the framework of the dynamic separation of envelopes.

13.5 Are Repeated Low Velocity Outbursts Possible?

The confirmation of the impossibility of the deceleration of gaseous envelopes ejected from the central stars with initial velocities of 1 000 km s^{-1} up to a few tens of km s^{-1} is one of the main results of the theory of PNs (Chap. 16). Just this has stimulated the idea of the "slow expansion–inflation" scenario of outer atmospheres of red giants with velocities of tens of km s^{-1} (Chap. 18)

The process of slow expansion of the outer atmosphere of red giants at low temperatures can occur only once in the star's life, since a "naked" red giant of high effective surface temperature will produce higher velocities of ejected mass, while no high velocity expansions, i.e. of the order of 1 000 km s^{-1}, have been detected for any of two-envelope PNs.

The observations therefore support also the *kinematic* identity of both envelopes, i.e. correlations between their velocity fields along with their shapes and elongations.

13.6 Physical Conditions in the Outer Envelope

From the spectra of the second (outer) envelope, e.g. of NGC 6 369, NGC 7 026, Ap 2–1 (Jewitt et al., 1986), the ratios of the intensities of the emission lines 6 716 [SII] and 6 731 [SII] have been derived. From these the electron concentrations obtained for the outer envelopes of NGC 6 369 and Ap 2–1 yield $n_e(\mathrm{II}) = 708\ \mathrm{cm}^{-3}$ and $426\ \mathrm{cm}^{-3}$, correspondingly, while the electron concentrations $n_e(\mathrm{I})$ in the main (inner) envelopes for these nebulae are 2.5×10^3 and $1.12 \times 10^3\ \mathrm{cm}^{-3}$, respectively, so that the ratio $n_e(\mathrm{II})/n_e(\mathrm{I})$ is equal to 0.28 and 0.38, respectively.

Note that at $d_2/d_1 = 1.6$, typical for the majority of two-envelope nebulae and $\mathcal{M}_2/\mathcal{M}_1 = 1$, we have for the ratio $n_e(\mathrm{II})/n_e(\mathrm{I}) = 0.32$ and $n_e(\mathrm{II})/n_e(\mathrm{I}) = 0.14$ if $d_2/d_1 = 2$. The mean electron concentration in the outer envelopes for five nebulae is of the order of $140\ \mathrm{cm}^{-3}$, which corresponds to 1/10 of the electron concentration of the main envelope. At $d_2/d_1 = 2$ this gives for the mass ratio $\mathcal{M}_2/\mathcal{M}_1 = 1$.

As for the electron temperatures in the outer envelopes, the data indicate that they are slightly, $\sim$ 10%, lower than that of the main envelope of $\sim$ 10 000 K.

13.7 The Mass of the Outer Envelope

The mass of the outer envelope, more precisely, the ratio of the two masses, $\mathcal{M}_2/\mathcal{M}_1$, of the outer $\mathcal{M}_2$ and the inner one $\mathcal{M}_1$, is among the key parameters of two-envelope nebulae. Knowledge of the ratio $\mathcal{M}_2/\mathcal{M}_1$ may throw light on the nature of the origin of two-envelope nebulae.

The determination of the ratio $\mathcal{M}_2/\mathcal{M}_1$ from observations is not difficult. The optical depth τ_c of both envelopes at the frequencies of the Lyman continuum of hydrogen is smaller than unity and, hence, all the mass of both envelopes will be in an ionized state. In this case, the relationship $\mathcal{M} \sim (LV)^{1/2}$, derived above, in Chap. 6, can be used for the determination of the nebula's mass from its luminosity L and volume V. Writing this expression for both envelopes we have

$$\frac{\mathcal{M}_2}{\mathcal{M}_1} = \left[\frac{\mathrm{L}_2}{\mathrm{L}_1}\frac{V_2 - V_1}{V_1}\right]^{1/2} = \left(\frac{\mathrm{L}_2}{\mathrm{L}_1}\right)^{1/2}\left[\left(\frac{d_2}{d_1}\right)^2 - 1\right]^{1/2} . \qquad (1)$$

The main difficulty is associated with the determination of the ratio of the luminosities L_2/L_1 of the two envelopes. However, even in this case the ratio L_2/L_1 cannot be estimated with enough accuracy because of the impossibility of accounting for the effect of projection of the outer envelope Start Stop onto the inner one. Also the question of the determination of the ratio of d_2/d_1 in the case of non-spherical and non-symmetrical nebulae is not completely clear either. Some compensation to these uncertainties is due to the weakness of the dependence of both $\mathcal{M}_2/\mathcal{M}_1$ on L_2/L_1 and on d_2/d_1.

The observational values of L_2/L_1 and d_2/d_1, collected from various sources for 24 two-envelope PNs (Chu et al., 1987; Jewitt et al., 1986; Daub, 1982; Maciel, 1984), along with the values of $\mathcal{M}_2/\mathcal{M}_1$ obtained with the help of (1), are given in Table 13.1.

As we can see, for the overwhelming majority of objects, about 70%, the ratio $\mathcal{M}_2/\mathcal{M}_1$ is less than unity. This percentage, however, might increase

Table 13.1. The mass ratio $\mathcal{M}_2/\mathcal{M}_1$, obtained by known values of the luminosity and diameter ratios, L_2/L_1 and d_2/d_1, for a sample of two-envelope planetary nebulae

Planetary nebula	d_2/d_1	$\mathrm{L}_2/\mathrm{L}_1$	$\mathcal{M}_2/\mathcal{M}_1$	Planetary nebula	d_2/d_1	$\mathrm{L}_2/\mathrm{L}_1$	$\mathcal{M}_2/\mathcal{M}_1$
NGC 1514	1.4	0.2	0.59	NGC 6852	3.7	0.03	1.20
NGC 1535	2.14	0.15	1.16	NGC 6891	2.0	0.18	1.10
NGC 2022	1.70	0.25	1.00	NGC 6894	1.5	0.2	0.69
NGC 2392	2.50	0.14	1.41	NGC 7009	1.7	0.20	0.89
NGC 2610	1.45	0.25	0.72	NGC 7354	1.5	0.2	0.69
NGC 3242	1.76	0.2	0.95	NGC 7662	2.0	0.25	1.30
NGC 3587	1.16	0.025	0.09	He 1–5	1.41	0.3	0.74
NGC 6058	1.6	0.15	0.69	IC 289	1.35	0.25	0.60
NGC 6369	1.84	0.07	0.59	IC 1295	1.50	0.3	0.85
NGC 6720	2.0	0.005	0.02	IC 3568	2.73	0.16	1.75
NGC 6804	1.6	0.2	0.79	IC 4593	2.1	0.003	0.16
NGC 6826	2.2	0.45	2.10	M 2–2	2.8	0.4	2.9

owing to selection effects connected with the difficulties of the discovery of outer envelopes and, hence, of the variation of the ratio $\mathcal{M}_2/\mathcal{M}_1$. As a result, the two-envelope nebulae with $\mathcal{M}_2/\mathcal{M}_1 < 1$ should be the absolute majority, of the order of 90% and higher.

13.8 The Ratio of Diameters of Both Envelopes

The edges of both envelopes for the overwhelming majority of two-envelope nebulae are well outlined. This fact enables one to estimate the visual sizes of both the inner (d_1) and outer (d_2) envelopes and then the ratio $R = d_2/d_1$, one of the most important and independent parameters describing the two-envelope nebulae. Numerically R varies within wide limits. Its upper limits should clearly be $R \sim 3$; values up to 10 and higher are attributed to the halo (see the next section).

Are there any regularities in the values of R from one nebula to another? Consider the data for 72 two-envelope PNs given in Table 13.2 (Gurzadyan, 1962; Chu et al., 1987; Jewitt et al., 1986). The existence of a number of two-envelope nebulae with $R \sim 1.5$ is noticeable. Other properties of this distribution can be followed in Fig. 13.1, (a) being drawn from the data of Table 13.2 (crosses) for values $R = 1$–3.

It is interesting that the distribution of the number of two-envelope nebulae by R is well fitted by a Gaussian within the interval of $R = 1$–2.5:

$$\frac{N}{N_m} = \exp\left[\frac{(R - R_0)^2}{2\sigma^2}\right], \qquad (2)$$

where N_m is the number of two-envelope nebulae at the maximum $R = R_m$ of the normal distribution and σ is the dispersion in the magnitudes of R. The curve in Fig. 13.1(a) corresponds to the following parameters of the distribution: $R_0 = 1.6$, $2\sigma^2 = 0.230$, $N_m = 14$.

Two-envelope nebulae with $R > 3$ do really exist, though their total number is less than 10%. Those are PNs with a giant halo and are not governed by any distribution law.

Table 13.2. Distribution of the number N of two-envelope planetary nebulae as a function of the ratio of the diameters of both envelopes, $R = d_2/d_1$

R	N	R	N	R	N
1.0–1.2	3	2.0–2.2	7	3.0–4.0	3
1.2–1.4	10	2.2–2.4	3	4–5	1
1.4–1.6	14	2.4–2.6	2	5–6	2
1.6–1.8	8	2.6–2.8	2	6–14	3
1.8–1.2	10	2.8–3.0	1	14–30	3

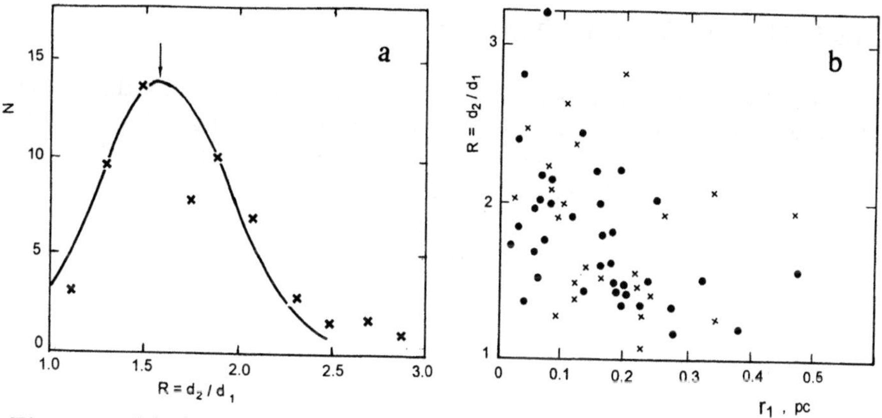

Fig. 13.1. (**a**) Distribution of the number of two-envelope planetary nebulae as a function of the ratio of the diameters of both envelopes, $R = d_2/d_1$. The solid curve is a Gaussian with parameters $R = R_0 = 1.6$, $N_0 = 14$, $2\sigma^2 = 0.23$. (**b**) Distribution of the ratio $R = d_2/d_1$ for two-envelope planetary nebulae versus the radius of the inner envelope r_1. Dots denote the data by Chu et al. (1989), crosses those by Jewitt et al. (1986)

The splitting of the population of two-envelope nebulae into two groups with various magnitudes of R and various statistical behaviours, can be interpreted as an indication of the existence of at least *two different mechanisms* of their formation. If the random initial conditions can determine the Gaussian law for the group A, nothing can be claimed for group B because of insufficient data.

The dependence of R on the linear radius of the inner envelope r_1 is shown in Fig. 13.1(b). Though no definite relationship can be claimed here, nevertheless some tendency of R to decrease with the increase of r_1 is noticeable; at least there are no cases of large R (larger than 1.5–2) at large r_1 (larger than 0.3–0.4 pc).

The reason for such behaviour between R and r_1 can be understood. If the masses of PNs on average are the same, then on the expansion of the nebula, i.e. at large r_1, the surface brightness of the nebula decreases, and it will become more ring-like. Therefore, the distribution in Fig. 13.1(b) should be interpreted as a result of observational selection, and giant haloes cold presumably be discovered also for the largest PNs.

Judging from Fig. 13.1(b), the outer envelope has been observed already in nebulae with a radius of the inner envelope of about 0.02–0.03 pc or 4 000–6 000 a.u. (NGC 6 369, 7 662, M 3–28 etc). For them $R \sim 2$, and, hence, the linear thickness of the outer envelope will be of the same order. It should be concluded, therefore, that the formation or origin of the outer envelope starts at still smaller sizes of nebulae, probably of a few thousand a.e.

13.9 Haloes Around Nebulae

At $R > 3$ we speak about *planetary nebulae with a halo.* One can think of at least two mechanisms responsible for the formation of typical *two-envelope nebulae* (type A) and of *nebulae with a halo* (type B). PNs with halo are listed in Table 13.3; in its last column the ratio, $d_{\rm H}/d_1$, of halo's diameter $d_{\rm H}$ to the diameter of inner envelope d_1 is given (mainly from Chu et al., 1987). If in typical two-envelope PN the ratio d_2/d_1 is of the order of 1.5–2, in those with a halo the ratio $d_{\rm H}/d_1$ is much greater – 5, 10, even up to 30.

With the application of CCD technique two envelopes of the barred - shaped core and a jet-like feature of NGC 40 have been revealed (Meaburn et al., 1996). A faint diffuse inner halo is of diameter $\sim 90''$, whereas the outer halo has prominent filaments embedded in patchy emission within a region of diameter $\approx 4'$.

Haloes can surround any nebula, independent of the number of usual envelopes – one, two or three. We suppose, in particular, that a halo should exist around the three-envelope nebula NGC 7293 (Plate II), at a much larger distance from the farthest, third envelope.

Presumably the halo is a new formation, and should be nothing other than a stellar wind, strongly modified, however, at the passage through the nebular matter – through the first, second or even the third envelopes. The interaction of the halo, a product of stellar wind, with the circumnebular medium can result in effects of Rayleigh–Taylor instability.

Table 13.3. Two-envelope planetary nebulae with a halo. d_1, d_2 and d_H are the diameters (in arcsec and only the largest size given) for the first envelope, for the second envelope and for the halo, respectively

Planetary nebula	d_1	d_2	$d_{\rm H}$	$d_{\rm H}/d_1$
NGC 7009	23	33	64	2.8
NGC 6804	37	63	110	3.0
M 3–28	9	–	32	3.5
NGC 2022	21	32	88	4.2
NGC 3132	30	81	128	4.3
NGC 6894	40	60	210	5.1
NGC 6891	12	23	68	5.7
NGC 6445	33	50	195	5.9
NGC 7662	20	33	134	6.7
NGC 6826	19	32	130	6.8
IC 4593	13	42	128	9.0
He 2-111	32	–	464	14.5
NGC 6543	14	21	315	22
NGC 3242	31	47	960	31

Thus, there are indications that the halo is neither a third envelope. nor an envelope in general. It may preserve its ideal spherical–symmetrical shape if the expansion occurs in a medium practically free of interstellar matter. Such a halo probably exists in NGC 2022 (Plate VIII). If the role of that matter is essential, its interaction with the halo can lead it to lose its stability, of Rayleigh–Taylor type distortions in the halo's shape; NGC 6543 (Plates V and XII).

13.10 The Envelope Separation Concept

The idea that the double-envelope PNs can originate as the result of two successive ejections of gaseous matter from the central star seems not only trivial but also attractive. However, observational data, as we will see below, do not confirm this version, although they do support the viewpoint of the appearance of an outer envelope in the definite stage of evolution of a single-envelope nebula owing to its dynamical splitting into two parts. If this is the case, then the dynamical separation of a nebula should be an inevitable stage through which all PNs must pass during their evolution.

We have at our disposal at least two types of observational fact directly related to the mechanism of the origin of two-envelope nebulae. One of them follows straight from Fig. 13.1*b*, the empirical dependence of R on r_1 for objects of group A with $R < 3$. As was mentioned above, no definite correlation between R and r_1 can be claimed, though marginally we can adopt

$$R = \text{const} \tag{3}$$

for r_1 at least up $r_1 \sim 0.6$–0.7 pc.

Another observational fact is the dependence of the linear depth of the outer envelope $\Delta r = r_2 - r_1$ on the radius of the inner envelope r_1. Very approximately it can be presented in the form

$$\Delta r = 0.7\, r_1 \,. \tag{4}$$

Consider now the predictions of each of the mechanisms of the origin of two-envelope nebulae concerning these relationships, $R \sim r_1$ and $\Delta r \sim r_1$.

We have three kinematic possibilities:

(a) The outbursts of both envelopes occur from the central star simultaneously but with different velocities, V_1 and V_2.
(b) The outbursts of both envelopes take place successively, with the time interval Δt and with velocities $V_1 \approx V_2$.
(c) The outer envelope appears as a result of dynamical separation of the main envelope into two parts.

Version (a), simultaneous outbursts with different velocities, results in the radii of the inner (r_1) and outer (r_2) envelopes

$$r_1 = V_1 t\,, \qquad r_2 = V_2 t\,, \tag{5}$$

where $V_2 > V_1$. Then we can derive

$$R = V_2/V_1 = \text{const}\,, \tag{6}$$

$$\Delta r(V_2/V_1 - 1)\, r_1 \sim r_1\,. \tag{7}$$

Comparing (6) with (3) and (7) with (4) we see that the agreement of version (a) with observations is with respect to both parameters R and Δr.

In version (b), two successive outbursts with time interval Δt, we have

$$r_1 = V_1(t - \Delta t)\,, \qquad r_2 = V_2 t\,. \tag{8}$$

from which we obtain, adopting $V_1 \approx V_2$,

$$R = (1 - \mathrm{d}t/t) \sim r_1^{-1}\,, \tag{9}$$

$$\Delta r = V_1 \Delta t = \text{const}\,. \tag{10}$$

Comparing these results with (3) and (5), we see that version (b) contradicts observations, with respect to two parameters, R and Δr. In version (c) the dynamical separation starts, say, at the moment $t = t_0$ starting from the moment of outburst $t = 0$, and the separated outer envelope departs with relative velocity ΔV. Then, we have at the moment t after the outburst and at t_0 after separation

$$r_1 = V_1 t, \qquad r_2 = r_1 + \Delta V(t - t_0)\,, \tag{11}$$

from which we obtain

$$R = 1 + (\Delta V/V_1)(1 - t_0/t)\,. \tag{12}$$

As we shall see later, the separation of the outer envelope starts very early, at sizes of the nebula $r_0 = V_1 t_0$, of the order of a few thousand a.u., i.e. an order of magnitude smaller than the normal sizes of nebulae. Therefore we have $t_0/t \approx 0.1$. Then we can write from (12)

$$R = 1 + \Delta V/V_1 = \text{const}\,. \tag{13}$$

We also have for $\Delta r = r_2 - r_1$

$$\Delta r = \Delta V t_0(r_1/r_0 - 1) \sim r_1\,. \tag{14}$$

Comparing (13) and (14) with (3) and (5), we see that the conjecture of dynamical separation agrees with observations with respect to both parameters R and Δr simultaneously.

Equating (4) and (14) and adopting $r_1/r_0 \gg 1$, we find that

$$\Delta V = 0.7V_1 , \tag{15}$$

i.e. the mean velocity δV of departure of the outer envelope should be slightly smaller than the expansion velocity V_1 of the nebula itself. Since typically $V_1 \sim$ 20–30 km s^{-1}, we have for the relative expansion velocity of the outer envelope $\Delta V \sim$ 15–30 km s^{-1}, i.e. velocities too small to be observed easily.

Thus, we see that both versions (a) and (c) fulfil the two observational facts considered above. What can this mean? The reason is trivial, since version (c) should coincide with (a), if the moment of separation t_0 is reduced up to the outburst moment at $t_0 \to 0$ when both velocities V_1 and V_2 are somehow created. Indeed, if we substitute $r_0 = V_1 t_0$ and $\Delta V = V_2/V_1$ and $t_0 \to 0$ in (12) and (14) we obtain

$$R = \frac{V_2}{V_1} , \qquad \Delta r = \left(\frac{V_2}{V_1} - 1 \right) r_1 ,$$

which coincides fully with (6) and (7).

Thus, double-envelope nebulae cannot be the result at least of successive outbursts from the nucleus. It is much harder to imagine the possibility of natural mechanisms of absolutely simultaneous ejections with quite different initial velocities. This scenario should seem even more artificial in view of the arguments discussed below, that the double-envelope property can be typical for any PN at a certain stage of its evolution.

13.11 Evolution of Envelopes

The main successive stages of evolution of the nebulae can be imagined in the following way. When the nebula is small it consists of a single envelope, appearing either round or oval in projection. These are the *planet-like* nebulae. Thus shape is preserved up to a certain moment and certain dimension for the given nebula. After this, the main envelope splits up, and the separated fraction of mass moves away with a low relative velocity, thus creating the second envelope around the first. The nebula enters the *double-envelope* phase.

The second envelope cannot always be observable. As both envelopes expand, their surface brightness decreases; however, the brightness of the second decreases more rapidly, since its expansion velocity is even higher. Therefore, the second envelopes can disappear from our telescopes even earlier. Now only the inner envelope can be seen, in a *ring-like* shape, since as far as we can observe no further emission of gaseous matter from the nucleus occurs.

Ring nebulae evolutionarily follow the double-envelope ones, which themselves are preceded by the planet-like nebulae. In other words, the evolution of PNs is as follows:

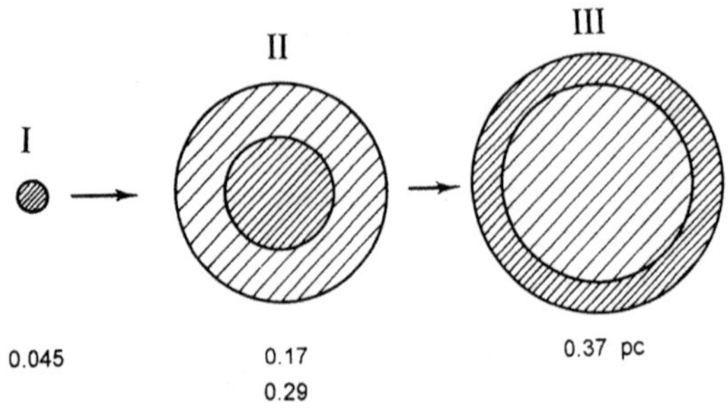

Fig. 13.2. Evolutionary sequence of planetary nebulae: I – planet-shaped, II – double-envelope, III – ring-like. The mean diameters (in pc) are given for each type

$$\textit{Planet-like} \rightarrow \textit{Two-Envelope} \rightarrow \textit{Ring-like}$$

Since during this evolution the nebula continuously expands, obviously on average the planet-like nebulae should have the smallest dimensions, while the ring nebulae should have the largest. The dimensions of the double-envelope nebulae should be somewhat intermediate.

The homogeneous CCD data by Jewitt et al. (1986) and Chu et al. (1987) enable one to perform a statistical analysis in order to check this evolutionary scheme. These images were used to classify the objects – planet-like, two-envelope and ring-like – while other sources were used for the PN sizes from catalogues (Daub, 1982; Maciel, 1984). Among the 205 PNs analysed the distribution by types and the average values of their radius R is as follows:

	n	R, pc
Planet-like	76	0.040
Two-envelope	60	0.174–0.290
Ring-like	69	0.370

Figure 13.2 illustrates these data. Thus, in spite of a number of uncertainties, the comparison of dimensions of the nebulae seems to confirm the proposed evolutionary sequence: planet-like, double-envelope and ring nebulae.

This analysis tends to support the idea that the appearance of a second envelope around the main one is a structural feature of any PN. *All planetary nebulae are born with one envelope but end their lives with two.*

13.12 The Frequency of Double Envelopes

A number of attempts have been made to estimate the fraction of double envelopes among all PNs. For example, the careful examination of a sample of 126 PNs revealed outer envelopes in 41 objects (Chu et al., 1987), i.e. in about 33%. Some authors taking into account additional considerations, recommend a higher value – 50%. The results of another homogeneous sample of 44 PNs revealed double envelopes in 29, i.e. in 66%.

If our evolutionary considerations described above are the case, than one can expect a further increase in the fraction of double-envelope PNs with higher accuracy observations. At the same time, because of the existence of single-envelope, i.e. planet-like nebulae, this frequency will never become 100% – it is impossible theoretically.

What can we gain from knowledge of the frequency of observation of double-envelope PNs? What types of problem of cosmogonical or physical character can be solved? Clearly, at least one: when can the process of separation of the nebula start or how long will the nebula preserve its two-envelope phase? Judging from the described estimations, the double-envelope phase lasts during more than half and not less than two-thirds of nebula's life.

13.13 Evolution of the Central Star

According to one of the early theories of the evolution of the central stars of PNs (Paczynski, 1971), the time dependent theoretical curves for the bolometric luminosity $L_{\rm bol}$, effective temperature T_*, and radius R_* can undergo loop-like variations at definite stages of their evolution. This can occur once with masses of $1.2\,\mathcal{M}_\odot$ and $0.8\,\mathcal{M}_\odot$ and twice in the case of $0.6\,\mathcal{M}_\odot$ (Fig 13.3). Later, the possibility of the reverse, i.e. loop-like variations in the same curves for $L_{\rm bol}$, T_* and R_*, again at the helium burning stage, was confirmed by more refined calculations by Schönberner (1979, 1981, 1983).

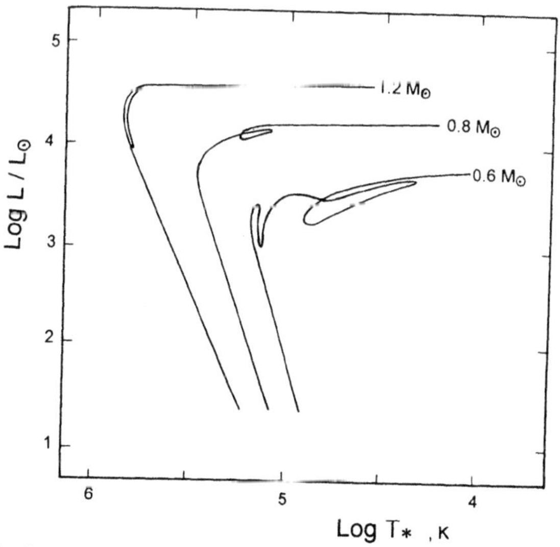

Fig. 13.3. Evolutionary tracks of the Hertzsprung–Russell diagram for models of nuclei of mass 0.6, 0.8 and 1.2 $\mathcal{M}_\odot$

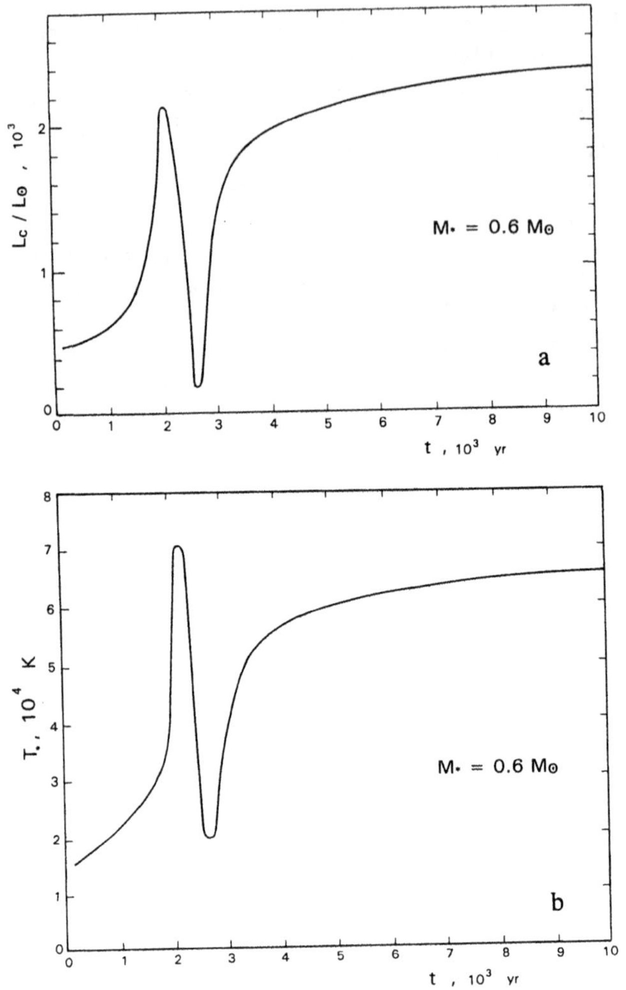

Fig. 13.4. Nucleus of a planetary nebula (model by Paczynski). (**a**) Variation of L_c luminosity with time t ($t = 0$ corresponds to the ejection and early expansion phase of the nebula). (**b**) Variation of the effective temperature T_* of the nebula's central star with time t

Paczynski's curves, shown in Fig. 13.4, relate two important parameters. One of them is the L_c luminosity of the star at frequencies of hydrogen ionization, i.e. at wavelengths shorter 912 Å. This curve is constructed on the basis of theoretical estimations of L_{bol} and the expression $L_c = p\,L_{bol}$, where p is the fraction of L_c luminosity within L_{bol}:

$$p = \frac{\displaystyle\int_{x_0}^{\infty} \frac{x^3\,\mathrm{d}x}{e^x - 1}}{\displaystyle\int_0^{\infty} \frac{x^3\,\mathrm{d}x}{e^x - 1}}, \tag{16}$$

$x_0 = h\,\nu_0\,k\,T_*$, where ν_0 is the frequency of hydrogen ionization. The curve of the dependence of p on T_* is shown (Gurzadyan, 1995).

As we can see in Fig. 13.4, during the first two thousand years after the birth of the nebula we have an increase of L_c luminosity up to a sharp maximum at $2 \times 10^3\,L_\odot$. Then the curve falls by up to $0.2 \times 10^3\,L_\odot$ at the beginning of the second thousand years. The consequent increase up to $2.5 \times 10^3\,L_\odot$ occurs at the end of ten thousand years. Qualitatively the same behaviour of the maxima and minima demonstrates the curve for the effective temperature T_*. Both curves in Fig. 13.4 play an important role in the theory of two and three-envelope PNs.

13.14 The Theory of Envelope Separation

The sequence of phases of formation of the second envelope within the concept of dynamical splitting of a single envelope due to L_α radiation pressure is as follows.

With small nebulae we have two structures: the inner, completely ionized, and the outer, completely neutral. The separation front between these zones moves outwards with the expansion of the nebula. As a result, the ionized zone will increase its size, while the neutral one will become narrower. Assume a spherical homogeneous nebula expanding with a constant velocity V_0. We have for its radius r_n and for the variation of the concentration of hydrogen atoms $n(t)$

$$r_n = V_0\,t\,, \tag{17}$$

$$n(t) = n_0 \left(\frac{r_0}{V_0\,t}\right)^3 = n_0 \left(\frac{r_0}{V_0}\right)^3 t^{-3}\,, \tag{18}$$

where n_0 is the concentration at the radius $r_0 = V_0\,t_0$.

The variation law of the radius of the ionization zone r with time, i.e. during the expansion can be derived from the ionization equilibrium condition. The total number of L_c photons emitted by the nebula's nucleus must be equal to the total number of recombinations of hydrogen ions with free electrons:

$$4\pi\,R_*^2 \int_{\nu_0}^{\infty} \frac{B_\nu\,(T_*)}{h\,\nu}\,\mathrm{d}\nu = \frac{4\pi}{3}\,r_1^3\,n_\mathrm{e}^2\,(t)\,\alpha\,(T_\mathrm{e})\,, \tag{19}$$

where R_* is the radius of the central star, $B_\nu\,(T_*)$ is the Planckian function of the effective temperature T_*, $n_\mathrm{e}\,(t)$ is electron concentration in the ionization zone and $\alpha\,(T_\mathrm{e})$ is the recombination coefficient of hydrogen ions with free electrons.

Assuming for the ionization zone $n_e = n^+ = n(t)$, we have from (19), using also (18),

$$r_1 = 0.57 \times 10^4 \, R_*^{2/3} \, n_0^{-2/3} \, T_* \left(\frac{V_0}{r_0}\right)^2 \left[\frac{J(T_*)}{\alpha(T_e)}\right]^{1/3} t^2 \,, \tag{20}$$

where

$$J(T_*) = \int_{x_0}^{\infty} \frac{x^2 \, dz}{e^x - 1} \,, \tag{21}$$

$$x_0 = h \nu_0 \,/\, k T_* \,. \tag{22}$$

The dependence of r_1 on $\alpha(T_e)$ is weak; therefore, taking $\alpha(10\,000 \text{ K}) = 4.13 \times 10^{-13} \text{ cm}^3 \text{ s}^{-1}$, we have from (20)

$$r_1 = 7.60 \times 10^7 \, T_* \left(\frac{R_*}{n_0}\right)^{2/3} \left(\frac{V_0}{r_0}\right)^{1/3} J^{1/3}(T_*) \, t^2 \text{ cm} \,. \tag{23}$$

As follows from (23) and (17), the radius of the ionization zone r_1 increases faster than the nebula's radius r_n. Obviously, the nebula will tend to become completely ionized when $r_n = r_1$ at some value of t_{cr} determined by the relationship

$$t_{cr} = 1.75 \times 10^{-4} \, R_*^{-2/3} \, n_0^{2/3} \, \frac{r_0^2}{V_0 \, T_*} \left[\frac{\alpha(T_e)}{J(T_*)}\right]^{1/3} . \tag{24}$$

The law of variation of the radius of the ionization zone r_1 by the time t, i.e. by the nebula's expansion, is given by (23). In the third column of

Table 13.4. Time variations of computed parameters of planetary nebulae of spherical form and mass $\mathcal{M}_n = 0.1 \mathcal{M}_\odot$, expanding with a velocity of $V_0 = 20 \text{ km s}^{-1}$: r_n is the radius of the outer boundary, r_1 is the radius of the ionized zone, $n(t)$ is the concentration of hydrogen atoms, τ_c is the full optical depth at the frequencies of L_c radiation, $\mathcal{M}(I)/\mathcal{M}_n$ is the relative mass of the neutral zone, T_* is the effective temperature of the central star. The time $t = 0$ corresponds to the moment of ejection of the nebula from the central star

t [years]	r_n [a.u.]	r_1 [a.u.]	$n(t)$ [cm^{-3}]	τ_c	$\frac{\mathcal{M}(I)}{\mathcal{M}_n}$	T_* [K]
0.5×10^3	2 100	370	92×10^4	150 000	0.99	17 000
1.0	4 200	1 500	11.5	29 000	0.95	20 000
1.5	6 300	3 340	3.4	9 500	0.85	26 000
2.0	8 400	5 950	2.0	3 100	0.64	41 000
2.25	9 450	7 540	1.0	1 800	0.49	74 000
2.50	10 500	9 290	0.73	830	0.31	30 000
2.75	11 550	11 250	0.55	15	0.07	23 000
2.825	11 870	11 870	0.51	-	-	27 000

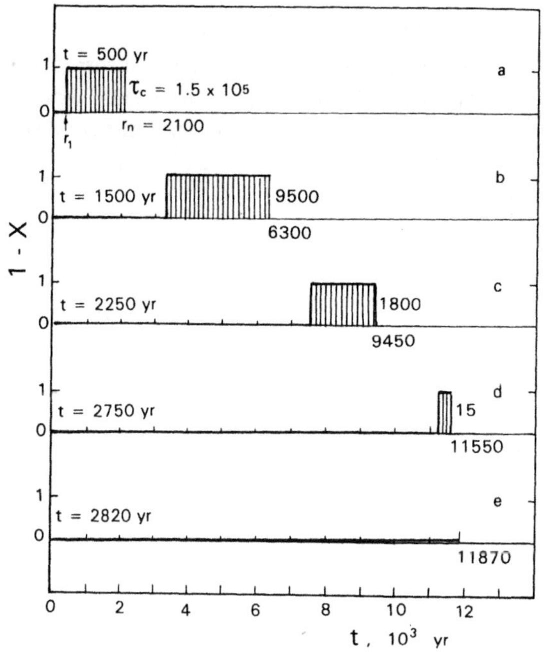

Fig. 13.5. The distribution of ionized and neutral (*lined*) zones within a spherical PN of outer radius r_n for five time periods of evolution and expansion, from $t =$ 500 yr ($r = 2\,100$ a.u.) up to $t = 2\,820$ yr ($r = 11\,870$ a.u.). r_1 is the radius of the ionized zone, $r_n - r_1$ is the size of the neutral zone, τ_c is the optical depth of the nebula, both of the ionized and the neutral zones, at the frequencies of the L_c radiation

Table 13.4 the values of r_1 obtained by means of (23) are given for a PN model with parameters $r_0 = 5\,000$ a.u. and $n_0 = 6.8 \times 10^4\,\mathrm{cm}^{-3}$. The latter quantities correspond to the nebula's mass $\mathcal{M}_n = 0.1\,\mathcal{M}_{\odot}$, $V_0 = 20\,\mathrm{km\,s}^{-1}$, $R_* = 10^{11}$ cm. Calculations were carried out for $T_* = 40\,000$ K and for various periods t of the nebula's expansion, taking $t = 0$ at the moment of departure of the nebula from the central star. In the second column of this table the values of the outer radii of the nebula r are given in accordance with (17).

As follows from these results, the radius of the ionized zone, r_1 increases much faster than the nebula's radius r_n, as is clearly seen in Fig. 13.5. The condition $r_1 = r_n$ is fulfilled at $t_{\mathrm{cr}} = 2\,825$ yr, in accordance with (24) at $n\,(t_{\mathrm{cr}}) = 0.51 \times 10^4\,\mathrm{cm}^{-3}$ (fourth column of Table 13.4).

In the fifth and sixth columns the values of the optical depth of the neutral zone, τ_1, at the frequencies of L_c radiation and the relative mass of the neutral zone $\mathcal{M}(I)/\mathcal{M}_n$ (shaded area in Fig. 13.5) are given, in accordance with the relationships

$$\tau_1 = n(t)\,(r_n - r_1)\,\kappa_c\,, \qquad (25)$$

$$\mathcal{M}(I) / \mathcal{M}_n = 1 - (r_1 / r_n)^3 , \tag{26}$$

where $\kappa_c = 0.63 \times 10^{-17}$ cm^2 is the continuous absorption coefficient per neutral hydrogen atom at the frequencies of the Lyman continuum, and the magnitude of $n(t)$ is given by (3). Note that the mass of the neutral zone $\mathcal{M}(I)$ is of the order of half of the nebula's mass, $\mathcal{M}(I)/\mathcal{M}_n \sim 0.5$, already at the phase of the nebula's expansion when $t \sim 2$–2.5×10^3 year. The latter circumstance will be of special importance for subsequent speculations.

13.15 The Structure of the Transition Zone

The description presented above was related to the macrostructure of the ionized zone. To complete that picture, one has to consider the problem of the small scale structure of the ionized zone and especially, of the transition zone. For simplicity, we shall limit our examination to a gaseous shell of inner radius r_0, linear depth Δr_0 and concentration of hydrogen atoms n.

The ionization formula for a hydrogen layer at a distance r from the nuclei has the following form:

$$\frac{x^2}{1-x} n = \frac{1}{4} \left(\frac{R_*}{r} \right)^2 f(T_*) \, e^{-\tau} , \tag{27}$$

where x is the degree of ionization:

$$n^+ = n_e = x\, n \qquad n_1 = (1-x)\, n , \tag{28}$$

where n_e, n^+ and n_1 are the concentrations of electrons, ions and neutral atoms, respectively, τ is the optical depth at the frequencies of L_c radiation, the linear interval is $r_1 - r_0$, and

$$f(T_*) = 2 \sqrt{\frac{T_e}{T_*}} \frac{(2\pi\, \mu\, k\, T_*)^{2/3}}{h^3} \, e^{-h\nu_0 / k T_*} . \tag{29}$$

Rewrite (27) in the form

$$\frac{x^2}{1-x} = K \frac{e^{-\tau}}{n r^2} , \tag{30}$$

where

$$K = \frac{1}{4} R_*^2 f(T_*) . \tag{31}$$

Differentiating (31) and taking into account $\mathrm{d}\tau = \kappa_c\, n_1\, \mathrm{d}r = (1-x)\, n\, \kappa\, \mathrm{d}r$, we obtain

$$\left[\frac{2}{x(1-x)} - \frac{1}{1-x} \right] \mathrm{d}x = -n\, \kappa\, \mathrm{d}r . \tag{32}$$

After integrating this, we obtain for the dependence of the degree of ionization x at a distance r, i.e. over the linear depth of the ionized zone,

$$2\ln\frac{x}{1-x}+\frac{1}{1-x}=-n\kappa r+C\,. \tag{33}$$

The constant of integration C is obtained from the following condition: at $r=r_0$, at the inner boundary we have $\tau=0$ and $x=x_0$, where x_0 is the solution of the equation

$$\frac{x_0^2}{1-x_0}=\frac{K}{n r_0^2}\,. \tag{34}$$

Then we have from (33)

$$\ln\left(\frac{x_0}{x}\,\frac{1-x}{1-x_0}\right)^2+\frac{1}{1-x_0}\left(1-\frac{1-x_0}{1-x}\right)=n\kappa\,\Delta r\,, \tag{35}$$

where $\Delta r=r-r_0$.

The dependence of the optical depth, τ, on the degree of ionization, x, can be obtained from the following relationship:

$$\tau=\ln\frac{K}{n r_1^2}-\ln\frac{x^2}{1-x}\,. \tag{36}$$

For the full optical depth τ_* of the transition layer, located between both the ionization and neutral zones, where $\Delta r_1=r_1-r_0$, we have

$$\tau_*=\ln\frac{K}{n r_1^2}-\ln\frac{x_*^2}{1-x_*}\,, \tag{37}$$

where x_* is the solution of the equation

$$\ln\left(\frac{x_0}{x_*}\,\frac{1-x_*}{1-x_0}\right)^2+\frac{1}{1-x_0}\left(1-\frac{1-x_0}{1-x_*}\right)=n\kappa\,\Delta r_1\,. \tag{38}$$

The dependence of $1-x$ on the linear distance within the envelope, described by these relationships, is given in Fig. 13.6 for a nebula with $T_*=40\,000$ K, $R_*=4\times10^{10}$ cm, $r_0=10^{17}$ cm and for two values of n: 10^4 cm^{-3} and 2×10^4 cm^{-3}.

There is nothing unexpected in the global structure of these curves. They confirm once again the well known fact that at the transition from the ionization zone to the neutral one, there is a mess. The calculations are of definite interest for the following two reasons. First, the linear thickness of the transition zone is much less, one hundred times, compared with the nebula's linear size. Second, the optical depth of the separation boundary, τ_*, corresponding to the degree of ionization $x_*\sim0.5$ and obtained from (37), is equal to $\tau_*=7.0$, and $\tau_*=7.3$ at $n=10^4\,\text{cm}^{-3}$ and $n=2\times10^4$ cm^{-3}, respectively.

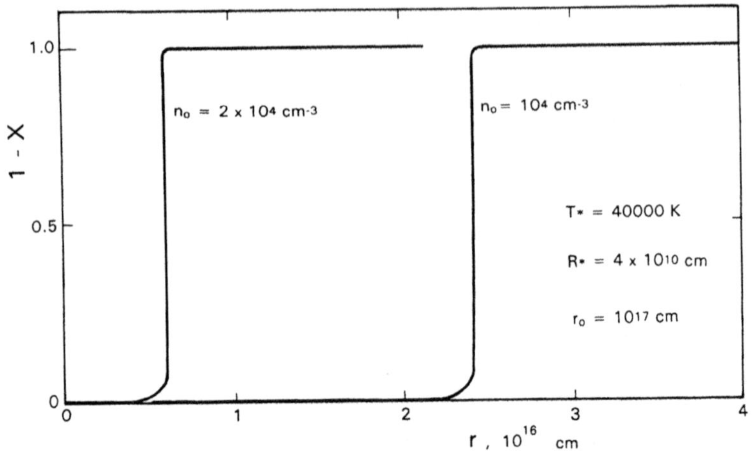

Fig. 13.6. Detailed structure of the transition zone where $1 - x \sim 0.5$ is the end of the ionized zone and the beginning of the neutral zone; $T_* = 40\,000$ K, $R_* = 4 \times 10^{10}$ cm, and two values of the concentration of hydrogen atoms n are used: $10^4\ \mathrm{cm}^{-3}$ and $4 \times 10^4\ \mathrm{cm}^{-3}$

13.16 Separation of the Second Envelope

One of the main results of the radiation equilibrium theory of real, i.e. expanding, PNs concerns the distribution of the L_α radiation field. The latter is directed outwards in the outer layers of the nebula, and inwards in the inner ones, and, hence, somewhere in the nebula this flux should be zero. In the simplest case the location of this zero-point, τ_0, is determined by the condition (11.37):

$$\tau_0 = \ln \tau_c\,, \tag{39}$$

where τ_c is the total optical depth of the nebula at the frequencies of the L_c radiation and is given by (25).

In the case when $1 - x$ is represented as depending on the linear depth of the nebula, corresponding to $t = 2\,250$ yr (scheme c, Fig. 13.5), we have $\tau_c = 1\,800$, which corresponds, according to (24), to $\tau_0 = 7.5$, i.e. with complete coincidence of the value of τ_* obtained above for the transition boundary between the ionized and neutral zones. This means that the L_α flux must be directed outside this boundary. Correspondingly, on this boundary a light pressure directed outwards is due to L_α radiation. As a result, dynamically this zone of neutral hydrogen will drastically differ from the conditions in the inner ionized zone.

To the left of the point τ_0, the L_α flux is directed inwards. It is not difficult to be convinced, however, that this flux cannot lead to any noticeable dynamical effect owing to the small number of neutral hydrogen atoms and, hence, the insignificance of the radiation pressure.

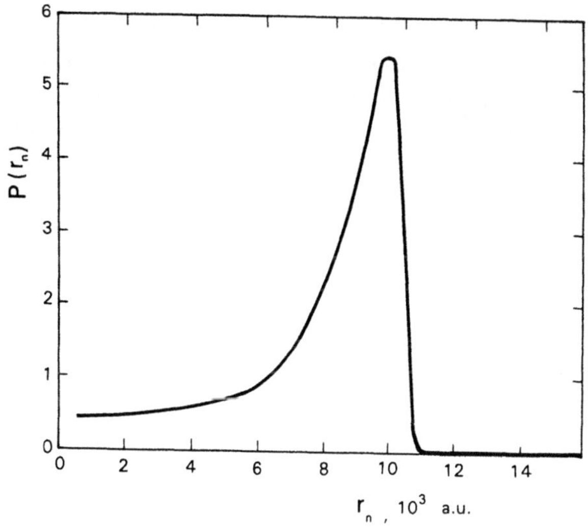

Fig. 13.7. Variation of the radiation pressure, $P(r_n)$, due to L_α radiation on the zone of neutral hydrogen as a function of the nebula's outer radius r_n . The radiation pressure increases with the increase of the nebula's radius r_n, reaches a maximum at $r \sim 10\,000$ a.u. and then vanishes completely

The situation in the neutral zone is quite different. Here all the hydrogen is in a neutral state, its relative number is ten thousand times larger than in the ionized zone and, hence, the outwards directed radiation pressure will be ten thousand times larger – this fact must be underlined. As a result, the neutral zone gains an additional velocity and will split, at least in principle, from the inner half of the nebula.

At $\tau_c \gg 1$, according to the theory, three-quarters part of the L_c radiation will be transformed into L_α radiation. In the completely opaque neutral zone in the L_α line all the momentum of this radiation, being $0.75\,L_c/c$, will pass to the mass of neutral zone $\mathcal{M}(I)$.

During the early period of the expansion of the nebula, the density of L_c radiation is small, and although $\mathcal{M}(I)$ is large, the specific momentum per unit mass, $L_c/\mathcal{M}(I)$, will be small. Later, the L_c luminosity increases and $\mathcal{M}(I)$ diminishes. Therefore, the ratio $L_c/\mathcal{M}(I)$ must increase sharply while approaching the peak of the L_c luminosity curve (Fig. 13.4); its subsequent behaviour is shown in Fig. 13.7. All this occurs when $\tau_c \gg 1$.

In the case when $\tau_c \sim 1$, the momentum of the radiation pressure will again be small, practically zero, because of the very small number of neutral atoms n_1. Fig. 13.7 exhibits the behaviour of the light pressure with the expansion of the nebula; it reaches a maximum at a definite stage of the expansion of the nebula. Figure 13.7 should be considered one of the most important results of the theory of the origin of double-envelope PNs within the framework of dynamic separation.

Here we arrive at the key point of this framework. Let us turn to Fig. 13.4, according to which in the early period of the expansion of the nebula ($t \sim$ 500–1 000 yr), the L_c-luminosity of its nucleus is small, in fact is a minimum. During this period the effective temperature of the nuclei is also very low, of the order of 15 000–20 000 K. Of course, in such conditions the radiation pressure will not have a major role. In this period the neutral zone must be much more extended than shown in schemes (a) and (b) in Fig. 13.5 for $T_* = 40\,000$ K. However, because of the quite low level of L_c radiation, the neutral zone cannot experience any pressure.

With the expansion of the nebula both the L_c luminosity and the temperature T increase, and as early as $t \sim 2\,000$ yr T_* reaches 40 000 K, and the power of the L_c luminosity will be at least 100 times larger than at $T \sim 20\,000$ K, and, hence, the light pressure will be 100 times larger.

The growth of both the L_c luminosity and the temperature T_* continues (see the last column of Table 13.4, where the nuclei temperatures T_* are given for different periods of expansion according to Fig. 13.4). At $t \sim 2\,000$ yr the temperature reaches $\sim 70\,000$ K or higher. This level is preserved, however, not for long, about 500 yr. Note that during this period, according to Fig. 13.4, the nuclei radius is $\sim 10^{11}$ cm, i.e. the quantity we had used during our calculations above.

The moment of the highest temperature, $\sim 70\,000$ K, corresponds to $t \sim 2\,250$ yr, i.e. to scheme c in Fig. 13.5, when $\tau_c = 1\,800$. The optical depth τ_0 at the point of reverse of the L_α radiation will be, according to (39), $\tau_0 = 7.5$. This value is in absolute coincidence (!) with that of τ_*, obtained above for the separation boundary between the ionized and neutral zones. This fact is also of special importance.

Thus, all the momentum of the light pressure onto the separation front of both zones will be completely passed to the mass of the neutral zone, which is nearly half (!) of the nebula's mass (penultimate column in Table 13.4). At that time the ratio $L_c/\mathcal{M}(I)$ reaches its maximum, and we can write for the equation of the motion of this mass, i.e. of the second envelope of a mass $\mathcal{M}_2$, adopting $\mathcal{M}_2 = \mathcal{M}(I)$, at $\tau_c \gg 1$,

$$\mathcal{M}(I)\,\frac{d\Delta V}{dt} = \frac{3}{4}\,\frac{L_c}{c}\,. \tag{40}$$

Hence, we have for the departure velocity of the second envelope ΔV with respect the initial nebula, which is now the first envelope,

$$\Delta V = \frac{3}{4c}\,\frac{L_c}{\mathcal{M}_2}\,\Delta t\,. \tag{41}$$

In Table 13.5, the values of ΔV obtained with the help of this relationship are given for two values of the mass of the second envelope, $\mathcal{M}_2 = 0.50\mathcal{M}_n$ and $\mathcal{M}_2 = 0.25\mathcal{M}_n$, where $\mathcal{M}_n = 0.1\mathcal{M}_\odot$ is the total mass of the nebula. Calculations are carried out for the period $t = 2.25 \times 10^3$ yr, i.e. when $r_n =$ 10 000 a.u. or $\tau_c = 1\,800$, the duration of the action of the light pressure

Table 13.5. Computed values of the relative velocity ΔV of departure of the outer (second) envelope from the inner (first) envelope at different magnitudes of effective temperature of the central star T_* and two values of the ratio of masses of the two envelopes, $\mathcal{M}_2/\mathcal{M}_1$

T_* [K]	p	L_c [$\times 10^{38}$ ergs s^{-1}]	ΔV km s^{-1} $\mathcal{M}_2/\mathcal{M}_1 = 1$	ΔV km s^{-1} $\mathcal{M}_2/\mathcal{M}_1 = 0.5$
50 000	0.57	0.25	1.0	2.0
60 000	0.69	0.64	2.4	4.8
70 000	0.77	1.3	5.0	10.0
80 000	0.83	2.4	9.2	18.4

$\Delta t = 500$ yr and for the temperature of the central star T_*. Judging from the results obtained, the relative departure velocity ΔV of the second envelope is of the order of 5–10 km s^{-1} in the first case ($0.5\mathcal{M}_n$), and 10–20 km s^{-1} in the second, ($0.25\mathcal{M}_n$).

The schedule of the radiation pressure at temperatures 50 000–80 000 K is preserved no longer than 500 yr; its action then falls sharply because of the falling of the L_c luminosity (the renewal of the effect is possible in about 500 yr, see the next section), therefore a further separation process should occur governed by the conservation of momentum at a constant velocity ΔV.

During 8 000 yr the nebula's radius will reach the sizes $r_n = 5 \times 10^{17}$ cm$\sim$ 33 000 a.u. at an expansion velocity of 20 km s^{-1}. The separation of the second envelope starts at $t \sim 2\,000$ yr. Over these 8 000 yr the growth of the radius will be $\Delta r = 1.9 \times 10^{17}$ cm$\sim$ 12 700 a.u. and 3.8×10^{17} cm$\sim$ 25 000 a.u. at separation velocities $\Delta V = 10$ km s^{-1} and 20 km s^{-1}, respectively. This gives for the radius of the second envelope $r_2 = r_n + \Delta r = 6.9 \times 10^{17}$ cm = 46 000 a.u. and 8.8×10^{17} cm = 58 700 a.u. Then we have for the ratio d_2/d_1 of the sizes of both envelopes, 1.4 and 1.8 respectively.

From direct observations for dozens of double-envelope nebulae one can derive the ratio of their diameters, d_2/d_1: this varies within the limits of 1.10–1.20 up to 2, occasionally 2,5. At $d_2/d_1 > 3$, we have a halo. Incidentally, the values of d_2/d_1 can be fitted by a Gaussian distribution with the mean $d_2/d_1 \sim 1.6$. The masses of both envelopes, $\mathcal{M}_2/\mathcal{M}_1$, vary within limits from 0.02–0.10 up to 1–1.5, occasionally 2. The average is $\mathcal{M}_2/\mathcal{M}_1 = 1$, i.e. the mass of the second (outer) envelope is nearly half of the nebula's total mass (Sect. 13.6) in agreement with the above theoretical results.

Despite the obvious achievements at the theory of the radiative equilibrium of PNs, we still have no clear view on the behaviour of L_α radiation. In particular, we know that the flux of L_α radiation onto the inner boundary of the neutral zone is directed outwards, since it is the flux from the opposite half of the nebula. However, the same boundary emits also diffuse L_α radiation directed inwards. The nature of this component is still unclear. If it can be of mirror-reflected type, then one has to expect a doubling of the total radiation pressure, $2\,L_\alpha/c$. In this case we will have values for d_2/d_1 sufficiently larger than those obtained above. Finally, for a ratio $\mathcal{M}_2/\mathcal{M}_1$ of the order of 0.1–0.2 we will have much larger values for d_2/d_1. And also, at nucleus temperatures T_* of the order of 100 000 K and higher, we must again have higher values for the ratio d_2/d_1.

In spite of the obvious necessity of more refined calculations, this brief survey seems to indicate that the concept of dynamic separation, in principle, can explain the observed values of the main parameters of double-envelope PN: the ratio of the diameters, d_2/d_1, and masses, $\mathcal{M}_2/\mathcal{M}_1$, of the envelopes.

In view of the fact that the outer envelope expands with a velocity higher than the velocity of expansion of the inner envelope, the concentration in the inner envelope must be as small as the concentration in the outer one. This should be also attributed to the consequences of the concept of dynamic separation.

13.17 Three-Envelope Planetary Nebulae: Observations

Planetary nebulae with *three* envelopes are known. Among them are NGC 7 009 and NGC 7 662, NGC 2 022, NGC 3 132, NGC 6 804, NGC 7 009, NGC 6 720. Of particular interest is the fragmentation of the third envelope discovered by Kwitter (1992) around the giant two-envelope nebula NGC 7 293 (Plate II). Also, in the case of LoTr 4 two clearly separated shells have been detected around the central star with radii of 7″.5 and 13″.5. Relative to there shells the central star appears to be off-centred by about 2″. A fragment of a *third* shell is found at a radius 22″.5 to south east of central star (Rouch et al., 1996). In a number of other cases the situation is much less clear. Most probably in those cases there are haloes with various relative diameters. Therefore, in Table 13.6 we have included the data for only PNs

Table 13.6. Three-envelope planetary nebulae: d_1, d_2, d_3 are the angular diameters of the envelopes (arcsec), r_1, r_2, r_3 are the linear radii of the envelopes, $\Delta r_3 = r_3 - r_2$ is the linear thickness of the third envelope

Planetary nebula	d_1	d_2	d_3	$\frac{d_2}{d_1}$	$\frac{d_3}{d_1}$	r_1 [pc]	r_2 [pc]	r_3 [pc]	Δr_3 [pc]
NGC 2 022	20″	32″	88″	1.6	4.4	0.12	0.19	0.51	0.32
NGC 3 132	30	66	128	2.2	4.3	0.08	0.16	0.32	0.16
NGC 6 720	76	150	270	2.0	3.6	0.13	0.25	0.46	0.21
NGC 6 804	34	56	100	1.6	3.2	0.10	0.15	0.31	0.16
NGC 7 009	18	30	64	1.7	3.6	0.03	0.06	0.12	0.06
NGC 7 293	840	1 500	2 500	1.8	3.0	0.30	0.54	1.50	0.96

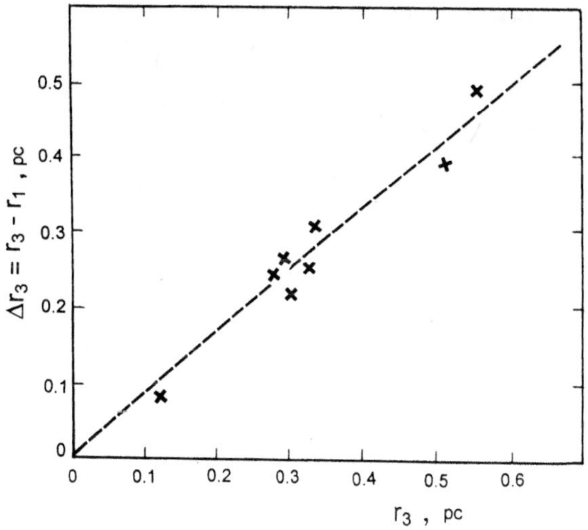

Fig. 13.8. Correlation between the linear thickness of the third envelope Δr_3 and the radius of the third envelope r_3 (Table 13.6)

with a confidently discovered third envelope: the angular diameters of all three envelopes, d_1, d_2, d_3 (arc sec), the ratios d_2/d_1 and d_3/d_1, the linear radii of three envelopes, r_1, r_2,r_3 (pc), as well as the linear thickness of the third envelope $\Delta r_3 = r_3 - r_2$.

In view of the above considerations concerning the two-envelope PNs, the three-envelope ones also hardly seem to be the result of multiple outbursts. The latter conclusion not supported, for example by the apparent correlation between the linear thickness of the third envelope, $\Delta r_3 = r_3 - r_2$, and the radius of the third envelope r_3, shown in Fig. 13.8. This correlation, $\Delta r_3 = 0.67 r_3$, will fail for the repeating outburst scenario, since this predicts the linear thickness of the final, envelope to remain constant and independent of its radius.

As we shall see in the next section, the origin of the third envelope in the framework of dynamical separation can be explained as an inevitable event during the evolution of the nebula's nucleus depending on its mass.

13.18 Three-Envelope Nebulae: Their Origin

As described above, the separation of the second envelope terminates nearly 2 500 yr after the birth of the nebula, and then it continues to depart from the initial mass of the nebula. Nearly at this time, the L_c luminosity, as noted above, rather quickly falls by 10 orders of magnitude (Figs. 13.9 and 13.10) while the effective temperature falls by three orders (Fig. 13.4). If the ionization boundary lies, say, near the outer edge of the second envelope, it must move inwards towards the nucleus; the latter will now appear within the

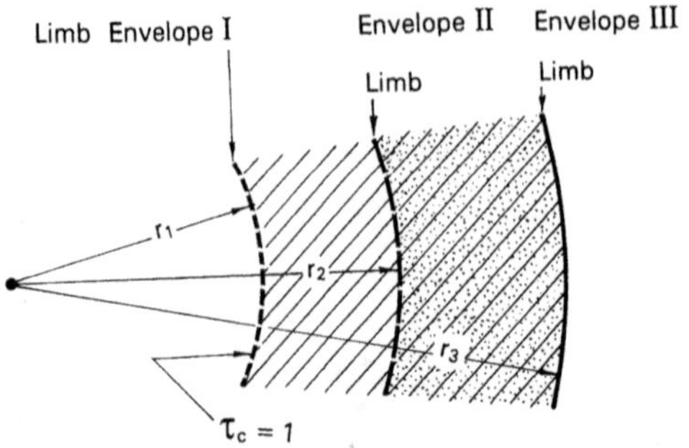

Fig. 13.9. The origin of three-envelope nebulae: r_1 is the radius of the outer boundary of the inner envelope, r_2 is the radius of the outer boundary of the second envelope, r_3 is the radius of the final (by origin) third envelope of very low temperature T_3. The radii r_2 and r_3 correspond to high temperatures T_2 and T_3 when the formation of three-envelope nebulae occurs. The dashed area indicates the neutral zone at the nucleus temperature T_3

limits of the inner (first) envelope, as shown by the broken line in Fig. 13.10. Here, all the regions on the right hand side, i.e. the outer (second) envelope completely and the inner envelope partly, now must find themselves outside the ionization zone. Therefore they cannot be visible, though physically they do exist (shaded area in Fig. 13.10).

The situation changes markedly. The nebula again becomes optically thick and again at the inner boundary of the neutral zone (the broken arc in Fig. 13.9) we have the condition $\tau_* \sim$ 7–8. Then the scenario described for a one-envelope nebula will repeat here, i.e. the single envelope will separate into two under the action of L_α radiation pressure, while the former second envelope, now the third one, also exists, although it is temporarily invisible. The nebula while expanding will become optically thin again, and the third envelope will shine again (in $t > 3\,000$ yr). We will observe a three-envelope nebula.

Consider the main points of a quantitative treatment of the described scenario of the origin of three-envelope nebulae.

Assume a double-envelope nebula with electron concentrations n_e and $\delta\, n_e$ in the first (inner) and second (outer) envelopes, respectively, where $\delta < 1$, and let r_1 and r_2 be their radii. Obviously, one can have such a value of temperature of the nuclei T_2 when the boundary of the ionization zone of the two-envelope system coincides with the edge of the second envelope, where $\tau_c = 1$. Then the condition of the equality of the total number of L_c photons emitted by the nucleus at the temperature T_2 with the total number of recombinations should be

$$4\pi R_*^2 \int_{\nu_0}^{\infty} \frac{B_\nu(T_2)}{h\nu}\,d\nu = \frac{4\pi}{3} r_1^3 \left[\left(\frac{r_2}{r_1}\right)^3 + \frac{1-\delta^2}{\delta^2}\right] n_e^2\,\alpha(T_e)\,. \qquad (42)$$

Substituting the values of constants, we obtain

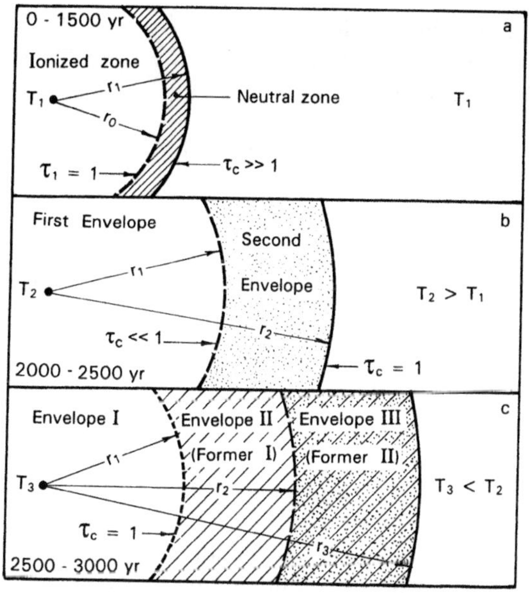

Fig. 13.10. The sequence of the origin of three-envelope nebulae. (**a**) The nucleus temperature T_1 is very low, $\sim$ 20 000 K, and the nebula consists of two zones, ionized and neutral (*lined*). (**b**) The process of the formation of double-envelope nebulae is finished; r_1 and r_2 are the radii of the first and second envelopes; the nucleus temperature $T_3 \sim 70\,000$ K is significantly higher than T_1. (**c**) The nucleus temperature T_3 is very low, $\sim$ 20 000 K, and the ionization front has moved up to the radius r_3. The outer envelope lies completely, and the inner envelope partially, in the neutral zone at a temperature T_3

$$r_1 \left[\left(\frac{r_2}{r_1} \right)^3 + \frac{1-\delta^2}{\delta^2} \right]^{1/3} = 7.66 \times 10^7 \, R_*^{2/3} \, n_e^{-2/3} \, T_2 \, J^{1/3}(T_2) \,, \tag{43}$$

where T_2 is known and is of the order of 70 000 K.

Assume, now, that a sudden decrease in the nucleus temperature takes place down to the value of T_3, so that the ionization boundary moves from r_2 to r_3 (Fig. 13.10). Then the condition of ionization equilibrium will have the form

$$4\pi \, R_*^2 \int_{\nu_0}^{\infty} \frac{B_\nu(T_3)}{h\nu} \, \mathrm{d}\nu = \frac{4\pi}{3} \, r_3^3 \, n_e^2 \, \alpha(T_e) \,, \tag{44}$$

which gives

$$r_3 = 7.66 \times 10^7 \, R_*^{2/3} \, n_e^{2/3} \, T_3 \, J^{2/3}(T_3) \,. \tag{45}$$

From (28) and (30) we obtain

$$\frac{r_1}{r_3} \left[\left(\frac{r_3}{r_1} \right)^3 + \frac{1-\delta^2}{\delta^2} \right]^{1/3} = \frac{T_2}{T_3} \left[\frac{J(T_2)}{J(T_3)} \right]^{1/3} \,. \tag{46}$$

Apply this relationship for the case when $r_1/r_2 = 2$, $\delta^2 = 0.1$ (the electron concentration of the outer envelope is three times smaller than that of the inner envelope), and taking also

$$T_2 = 70\,000 \text{ K},$$

we obtain from (46)

$$T_3\, J^{1/3}(T_3) = 1.47 \times 10^4\,, \tag{47}$$

where $J(T)$ is given by (21) and (22). From (47) we obtain

$$T_3 = 28\,000 \text{ K},$$

in complete coincidence with the temperatures ($T_3 = 20\,000$–$30\,000$ K) seen in Fig. 13.4. During about 1 000–1 500 yrs the outer envelope continues its invisible expansion by inertia with constant velocity up to the moment ($t \sim 4\,000$ yr) when the temperature again reaches 60 000–70 000 K. Then all three envelopes become visible.

In the case $t_c \ll 1$ on the outer boundary of the second envelope (Fig. 13.8), a relationship analogous to (46) can be derived by combining the ionization formula with the expression for the optical depth, $\tau_c = \kappa_c \int n_1\, dr$. Using also the equation for r_3 (45), we can write

$$T_3\, J^{1/3}(T_3) = 0.81 \times 10^4\, \frac{r_3}{r_2} \left(\frac{\tau_c}{\phi}\right)^{1/3} T_2^{1/3} \exp\left(-\frac{52\,400}{T_2}\right), \tag{48}$$

where

$$\phi = 1 + \delta^2 \left[\left(\frac{r_2}{r_3}\right)^3 - 1\right]. \tag{49}$$

At $r_3/r_2 = 0.5$, $\delta_2 = 0.1$ and $\tau_c = 0.01$ we obtain from (33), again, $T_3 \sim 27\,000$ K when $T_2 = 70\,000$ K.

Figure 13.10 illustrates the sequence of evolution of the above described scenario of formation, first, of a double-envelope and then, of a three-envelope nebula. Scheme (a) (top) concerns the period $t \sim 0$–$1\,500$ yr when the nucleus temperature T_1 is still low, $\sim$ 20 000–30 000 K, the nebula consists of two zones, ionized and neutral (shaded area), and the formation of the second envelope has not yet started. Scheme (b) (intermediate) concerns the period $t \sim 2\,000$–$2\,500$ yr when the nucleus temperature is the highest, $\sim$ 70 000–80 000 K, and when the separation of the second envelope is realized and on its outer boundary $\tau_c \sim 1$ or $\tau_c < 1$. Scheme (c) (bottom) is related to the period $t \sim 2\,500$–$3\,000$ yr when determined by the behaviour of the central star shown in Fig. 13.4; the temperature T_3 again becomes low, $\sim$ 20 000 K. The ionization zone moves inwards towards the nucleus down to r_3; as a result, both the outer envelopes completely and the inner envelope partly will be invisible.

Strictly speaking, the second envelope may continue to reradiate for some time after the modification of the ionizing front. The duration of reradiation t_r can be estimated from the relationship

$$t_r = \frac{1}{n_e\,\alpha(T_e)}\,, \tag{50}$$

which gives, for $n_e \sim 10^4$ cm^{-3} and $\alpha(T_e) \sim 4 \times 10^{-13}$ cm^3 s^{-1}, a value of t_r of the order of ten years, i.e. not very long.

One more scheme can be added in Fig. 13.10: on the increase of the nucleus temperature up to 70 000 K, the splitting of the inner envelope into two parts will be finished and the ionization zone will reach the outer regions of the final, third envelope.

As follows from Paczynski's model, the reverse in the evolution is expressed more strongly if the mass of the central star is $0.6\mathcal{M}_\odot$, while at $0.8\mathcal{M}_\odot$ and $1.2\mathcal{M}_\odot$ it is revealed less strongly. If this is the case, then the three-envelope phenomenon should be expected for nuclei with a mass $0.6\mathcal{M}_\odot$. Can one conclude that in the remaining cases PNs will have a nuclear mass of about $0.8\mathcal{M}_\odot$ of larger? Clearly, no, in view of the difficulties concerning, say, the discovery of the third envelopes.

As for the double envelopes, they can originate in PNs with any nuclear mass, since, first, the origin of the second envelope does not depend on the presence of the phenomenon of reversal, i.e. on strong variations of temperature and the L_c luminosity of the nucleus, and, second, the evolution of the nebula starts at any nuclear mass and at low temperatures, 15 000–20 000 K.

13.19 The Concept of Interacting Stellar Winds

A number of studies have been devoted to the gas dynamic problems of PNs with special reference to the concept of interacting stellar winds (Soker, Livia, 1989; Kahn, 1989; Kahn, Breitschwerdt, 1990; Breitschwerdt, Kahn, 1990; Marten, Schönberner, 1991; Frank, 1993, 1994; Marten; et al., 1993; Frank et al., 1993; Mellema, 1994; Frank, Mellema, 1994). The aim is to explain the diversity of features of the evolution and morphology of PNs, including the development of multi-shell structures. The main idea is the representation of the multiple shell PN as the result of the action of shock waves caused by the interaction of the so-called fast wind of the central star with a superwind of a progenitor AGB star (see Chap. 18).

This concept meets several difficulties. The numerical computations carried out in this framework assume a mass loss rate of $10^{-4}\mathcal{M}_\odot$ yr^{-1} during a time interval of 1 000 years, whereas the real mass loss rate obtained from direct observations for PN nuclei is of the order of $\mathcal{M} \sim 10^{-8} - 10^{-9}\mathcal{M}_\odot$ yr^{-1}, i.e. 4–5 orders of magnitude smaller (Perinotto, 1993).

In some cases the results of calculations of the dynamical (expansion) age of PNs within the same framework of interaction winds lead to contradictions.

For example, according to one of the models (Mellema, 1994), the expansion age of a nebula is less than its evolutionary age, i.e. nebulae are generally too small for their expansion age. Marten et al. (1993) derived the opposite result, i.e. that the expansion ages of old PNs are likely to be overestimated.

Not so clear is the discussion of the kinematic properties of the envelopes. Thus, certain models predict a constant relative velocity for the envelopes, whereas as a rule, the outer envelopes are expanding faster than the inner ones. Mellema (1994) presented remarkable data obtained by Chu and Jacoby, of the emission line [OIII] in IC 3 568 and NGC 2 022, where one can clearly see that the outer envelope expands with a velocity 1.5–2 times larger than that of the inner envelope. This fact led Chu to the conclusion that the interacting wind model cannot explain the existence of two-envelope nebulae.

These studies usually ignore an important fact: the possibility of loop-like variations of the luminosity–temperature evolutionary path of central stars of mass $0.6\mathcal{M}_\odot$ (Figs. 13.3 and 13.4), which was taken into account in the framework of dynamic separation.

It is not clear if the interacting winds can naturally explain the crucial observational phenomenon of *bipolarity* in both envelopes, with strong regularities in the orientation of "caps", in their shapes at the outer and inner envelopes, as well as in fragmentations on the edges of both envelopes (see Chap. 17).

At the same time, the strong stellar winds can be crucial for the formation of haloes, as well as for small radially directed cometary tails observed, for example, in NGC 7 293. Rather extended jet-like radial structures within the second (outer) envelope seen in overexposed images of NGC 7 293 and, especially, of NGC 6 853 (Kwitter, 1992), should also be connected with the activity of stellar winds. In this connection, of special importance are the detailed spectroscopic, kinematic, morphological, statistical etc. studies of so-called microstructures – small scale formations – in PNs. Such studies have been carried out, in particular, for NGC 6 543, NGC 6 826 and NGC 7 009 (Balick et al., 1994).

The central star of PN K 1-16, belonging to the class of pulsating pre-white dwarfs, has been observed with the *Hubble Space Telescope (HST)* telescope in the zone of the 1 550 CIV doublet, to ascertain the reality and examine the properties of the exceptionally fast wind, suggested to occur in this star. A close analysis making use of synthetic spectra supports the existence of a stellar wind with a terminal velocity 3 800 km s^{-1} (Patriarchi, Perinotto, 1996). It is one of the fastest stellar winds observed for a PN central star.

13.20 The Syndrome of the Nebula NGC 7 293

One of the interesting PNs, NGC 7 293 (Plate I), is one of the largest both in terms of its visible size – more than 20′ in diameter – as well as its absolute size – of the order of one parsec. This nebula is known as bipolar ring-like object. As for the outer, second envelope, only a little arc-like fragmentation has been discovered, by Minkowski as early as the 1960s. However, in an 1992 image (Kwitter et al., 1993) a fragment of the third envelope is visible rather far from the second envelope (Plate II). Thus, NGC 7 293 is a three envelope nebula of type B.

If the shape of the second envelope even partly repeats the basic form of the inner one, as we have seen above, it is serious evidence of the dynamic separation concept. The peculiarity of NGC 7 293 is that part of the third envelope is located exactly in the direction of the second envelope. From the viewpoint of the multiple outburst hypothesis, this means that with intervals of tens or hundreds of thousand of years, two successive outbursts from the central star (rotating!) occurred strongly *in the same direction!*, which is obviously hardly possible.

Since the dynamical separation concept is free of such difficulties, we conclude that similar objects (e.g. NGC 7 662) can be efficient tools enabling us to choose the mechanisms of shell formation.

13.21 The Mystery of the Nebula NGC 6 543

This remarkable object has been the centre of attention for a long time, because of its complex structure. The *Hubble Space Telescope* image of NGC 6 543, called the "Cat's Eye" (Plate V and XII), reveals a rich set of concentric shells, jets and unusual knots.

According to the model suggested by Harrington and Borkowski (1995), this nebula might contain a central binary system. A fast stellar wind of gas blowing off the central star created the elongated shell of dense, glowing gas. This structure is embedded inside two larger lobes of gas blown off the star at an earlier period. These lobes are "pinched" by a ring of a binary companion. It is suggested that this suspected companion star also might be responsible for a pair of high-speed jets of gas that lie at right angles to this equatorial ring. These jets would explain several features along the periphery of the gas lobes, particularly: the jets compress gas ahead of them, creating the "curlicue" features and bright arcs near the outer edge of the lobes.

However, a whole system of very thin and very long jets, spread outside of the two bright arcs symmetrically located relatively to the centre and directed strongly radially from the central star, call this scenario into question. These radial jets should be accounted for as a direct result of the dynamic interaction of the stellar wind with the nebular mass, and to suggest the same relation to the origin of such a completely different kind of formation as the system of the rings themselves would not be logical.

The following interpretation of the *Hubble Space Telescope* image of this extremely remarkable nebula seems more acceptable. Indeed, the central star might be a double-star system; however, both components may equally be protoplanets, i.e. cool giants, each of which may be an independent centre of the origin of a planetary nebula. Let us suppose that the separation of the outer layers of these star embryos takes place almost simultaneously, then we would have two different shells with all the consequences, namely, the evolutionary separation of each of these shells into two, the transition to the double-envelope phase, the formation of bipolar structure of one of these shells under the action of an intrinsic magnetic field, and in particular, to the formation of two "caps" – banana-like formations symmetrically located from both sides relative to the nebula's centre. etc.

Thus, the HST image of NGC 6543 seems to be a superposition of two PNs with one and the same geometrical but two physical centres and which were born almost at the same true. The outer diameter of the second (outer) envelope, i.e. the angular distance between bright "caps" is equal to $21''(d_2)$; the dimensions of the inner ring-like envelope are $10 \times 14''(d_1)$, i.e. one and a half times smaller than early accepted sizes for this nebula.

In addition, this nebula has a very extended halo with an exactly spherical form of diameter $d_{\mathrm{H}} = 315''$, which gives the ratio $d_{\mathrm{H}}/d_1 = 22$. The dynamical interaction of this halo with the surrounding interstellar medium leads to the loss of its stability, as a result of which many filamentary structures and formations of Rayleigh–Taylor instabilities and knots were formed as well.

On the cover of this book *the Hubble Space Telescope* colour image of NGC 6543 is shown.

14. Radio Emission of Planetary Nebulae

14.1 The Possibility of Radio Emission

A planetary nebula, is essentially a cloud of strongly ionized gas, and so the emission at radio frequencies should be inevitable. Continuous radio emission should be excited mainly by free–free transitions of electrons in the Coulomb field of charged particles. In this case, the total energy emitted at radio frequencies is proportional to $\int n_e^2 \, dV \sim n_e^2 V$, where V is the nebula's volume and n_e the electron concentration. The magnitudes of V and n_e for PNs are well known and, hence, the flux of radio emission detected on the Earth from PNs at a known distance can be easily computed. The expected radio flux is of the order of 10^{-26} W m^{-2} Hz^{-1}. This small magnitude is conditioned mainly by a relatively small emission region, the nebula's volume.

The first and basic conclusion, as we see, is unfavourable: *PNs should be weak radio sources.* This was the main reason that the first radio observations of PN were performed only in the 1960s. As for the observational region of wavelengths, no boundary exists from the short wavelengths, while the long wavelengths are limited at 10 m, determined by the transparency of the Earth's ionosphere. Radio observations of PNs are in fact carried out within the wavelength interval from 1–2 cm up to 1.5–2 m; at longer wavelengths the radio fluxes are too weak.

14.2 The Expected Radio Spectrum of Nebulae

The theoretical spectrum of gaseous nebulae at radio frequencies is characterized by certain properties. Since it is caused by free–free transitions of electrons, its intensity should be controlled by two parameters – by the electron temperature T_e, and the concentration of electrons (protons) n_e. The latter can be replaced by the optical depth of the nebula at radio frequencies $\tau_\nu \int k_\nu \, ds$, where k_ν is the volume absorption coefficient of the electron gas at the radio frequency ν. When the absorption is due to free–free transitions, we have

$$k_\nu = \xi \frac{n_e^2}{\nu^2} T_e^{-3/2} \text{ cm}^{-1} , \tag{1}$$

where ξ depends weakly on ν and T_e:

$$\xi(\nu, T_e) = 9.776 \times 10^{-15} \ln\left(49.503 \frac{T_e^{3/2}}{\nu}\right), \tag{2}$$

and ν is expressed in MHz = 10^6 Hz. In the interval of radio frequencies from 15 000 MHz (2 cm) to 200 MHz (1.50 m), ξ varies within 8–12$\times 10^{-14}$ at $T_e = 10\,000$ K. For the decimetre wavelengths $\xi \approx 10^{-13}$.

In view of (1), we can write

$$\tau_\nu = \frac{\xi}{\nu^2 T_e^{3/2}} \int n_e^2 \, ds = \frac{EM}{\nu^2 T_e^{3/2}} \xi(\nu, T_e), \tag{3}$$

where EM is the *emission measure*:

$$EM = \int_0^s n_e^2 \, ds \tag{4}$$

and is measured in units of cm^{-5} or cm^{-6} pc.

From the solution of the transfer equation at radio frequencies,

$$\frac{dI_\nu}{ds} = -k_\nu I_\nu + \varepsilon_\nu,$$

we find for the flux F_ν

$$F_\nu = \frac{\varepsilon_\nu}{k_\nu}\left(1 - e^{-\tau_\nu}\right). \tag{5}$$

Identifying F_ν and ε_ν/k_ν with the Planckian law, in the present case, in Rayleigh-Jeans form we have

$$F_\nu = \frac{2k\nu^2}{c^2} T \tag{6}$$

for the *brightness temperature* $T = T_{\rm eff}$ in the first case, and *the electron temperature* $T = T_e$ in the second, and we get

$$T_{\rm eff}(\tau_\nu) = T_e\left(1 - e^{-\tau_\nu}\right). \tag{7}$$

For nebulae that are optically thin in radio frequencies, i.e. when ($\tau_\nu \gg 1$), we have $T_{\rm eff} = T_e \tau_\nu$. From (6) and (7) we find for the flux of thermal radio emission

$$F_\nu = \frac{2k\nu^2}{c^2} T_{\rm eff}(\tau_\nu) = \frac{2k\nu^2}{c^2}\left(1 - e^{-\tau_\nu}\right) T_e, \tag{8}$$

where k is the Stefan–Boltzmann constant. Equation (8) gives the theoretical spectrum of thermal radio emission. For low frequencies, i.e for long wavelengths, τ_ν can be, as follows from (3), larger than unity. Then we have from (8), at $\tau_\nu \gg 1$,

$$F_\nu = \frac{2k\nu^2}{c^2} R T_e \sim \lambda^{-2}; \tag{9}$$

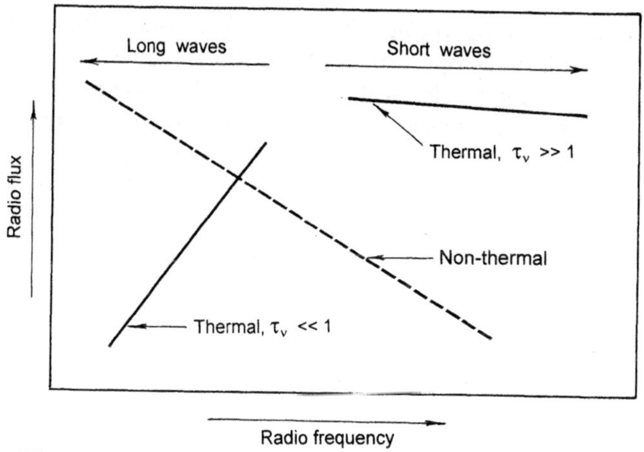

Fig. 14.1. Theoretical radio spectrum of planetary nebulae. *Solid lines:* thermal bremsstrahlung; *broken line*: non-thermal radio emission

i.e. the flux (intensity) of thermal radio emission should decrease in the region of long radio waves. At $\tau_\nu \ll 1$, i.e. for short wavelength radio emission, we get

$$F_\nu = \frac{2k\nu^2}{c^2}\, T_\mathrm{e}\, \tau_\nu \sim T_\mathrm{e}^{-1/2} = \mathrm{const}\,. \tag{10}$$

We see that the intensity of short wavelength radio emission does not depend on wavelength, while the dependence on electron temperature is weak. In fact, however, bearing in mind (2), i.e. the weak dependence of ξ on ν in the form $\xi \sim \nu^{-0.1}$, we have, instead of (10),

$$F_\nu \sim \nu^{-0.1}\, T_\mathrm{e}^{-1/2}\,, \tag{11}$$

which means the extremely slow decrease of radio flux as one moves to the short wavelength region.

In Fig. 14.1, the general structure of theoretical radio emission of thermal origin is shown (solid line) for the low frequency (long wavelength) region, as well as for the high frequency (short wavelength) one. Any deviation of observed radio spectrum from this picture should be interpreted as evidence for the non-thermal nature (broken line) of radio emission in a given object.

14.3 Radio Observations of Planetary Nebulae

NGC 6543, NGC 6572, NGC 6853, NGC 7293 and IC 418 were the first PNs observed in 1961 at radio frequencies by means of 25 m radio telescope (Lynds, 1961). Only approximate values of radio fluxes at frequencies of 1 420 and 3000 MHz (21.1 cm and 10 cm) were fixed at that time.

In 1964 the newly operating 90 m *NRAO* (National Radio Astronomy Observatory) radio telescope produced the first reliable data on the radio fluxes for 10 PN at frequencies of 750 and 1 410 MHz (Menon, Terzian, 1965). Almost at the same time, observations of PNs in the southern sky were begin. By means of the powerful 64 m radio telescope at Parkes (Australia), the radio fluxes were measured for 50 PNs at frequencies of 2 730 MHz ($\sim$ 11 cm) as well as of 620 and 1 420 MHz (Slee, Orchiston, 1965). Here, the problem of separating the studied objects from the background of cosmic radio sources arises.

An important event was the starting in 1966 of regular observations of PNs by the 300 m Arecibo radio telescope in Puerto Rico. This enabled observations to be initiated also at lower frequencies corresponding to the optically thick region of the nebula's radio spectrum. The first results (Terzian, 1966) already included measurements of radio fluxes for 130 PNs at frequencies of 430 MHz ($\sim$ 70 cm).

The situation was markedly changed with the appearance of radio interferometric methods of observations, especially after the coming into operation of the Very Large Array (VLA), observations with which with an extremely high angular resolution, up to $0''.4$ (in 5 GHz) were used particularly to map PNs at radio frequencies (Terzian et al., 1974; Kwok et al., 1981; Kwok, 1985; Masson, 1986; Terzian, 1989; Pottasch, Zijlstra, 1994,). The first observations performed by two 27 m radio telescopes working in a constant guidance regime, not only enabled the radio fluxes of 50 PNs to be measured but also solved two problems: the reliable separation of the investigated objects among other nearby radio sources of cosmic origin and the determination of their angular sizes (Thompson, Colvin, 1967). It appeared that for almost all PNs their angular sizes both in the radio and optical, are essentially the same.

The next series of radio interferometric observations (Cahn, Rubin, 1974) in the northern sky 69 PNs at frequencies of 2 700 MHz and 8 100 MHz (11 cm and 3.7 cm) radio fluxes up to 0.01 mJy to be approached; the flux unit = mJy = $10^{-26}\,\mathrm{ergs\,cm^{-2}\,s^{-1}\,Hz^{-1}}$ was introduced in honour of Carl Jansky, who discovered cosmic radio emission in 1931. Highly accurate observations by the Parkes radio telescope of 165 PNs at a frequency of 5 000 MHz (6 cm) served as initial data for the study of interstellar extinction and individual distances of these nebulae (Milne, Aller, 1975).

Various observations of PNs, along with the accumulation of data on radio fluxes at frequencies of, e.g. 400, 1 400 and 7 800 MHz (Thompson, Davies, 1970); 3 200, 6 600 and 10 600 MHz (Higgs, 1971); 400 MHz (Terzian, Dickey, 1973), etc., enabled millimetre wavelengths to be approached, i.e. up to frequencies 31 000 and 85 000 MHz or 9.68 mm and 3.53 mm (Johnson, 1973).

Measurements of radio fluxes for over 550 PNs over many decades reveal that the maximum radiation capability for almost all PNs is at a frequency of 5 000 MHz (6 cm). The highest flux of radio emission has been registered for NGC 7 027: 7.42 mJy, in the frequency interval from 5 000 MHz up to 16 000 MHz.

14.4 Observed Radio Spectra of Planetary Nebulae

The main goal of all types of radio observation of PNs was and remains the creation of their individual spectra at radio frequencies. The spectrum of continuous radio emission is the main source of information on the nature of the radio emission and, therefore, is an essential tool for determining of the physical parameters of the nebula.

Figure 14.2 exhibits the observed radio spectra for four PNs, NGC 7 027, NGC 6 853, NGC 6 572 and IC 418, the frequency interval from 195 MHz (1.54 *m*) up to 16 200 MHz (1.85 cm). The basic structure of these spectra is the same in all objects, i.e. there is an almost constant level of radio flux at high frequencies and a smooth fall at low frequencies. Clearly this structure is similar to that shown in Fig. 14.1 corresponding to the thermal mechanism of radio emission in PNs, i.e. with a constant level at $\tau_\nu \ll 1$, and smooth fall at $\tau_\nu \gg 1$. From here we draw the first conclusion: *the radio emission*

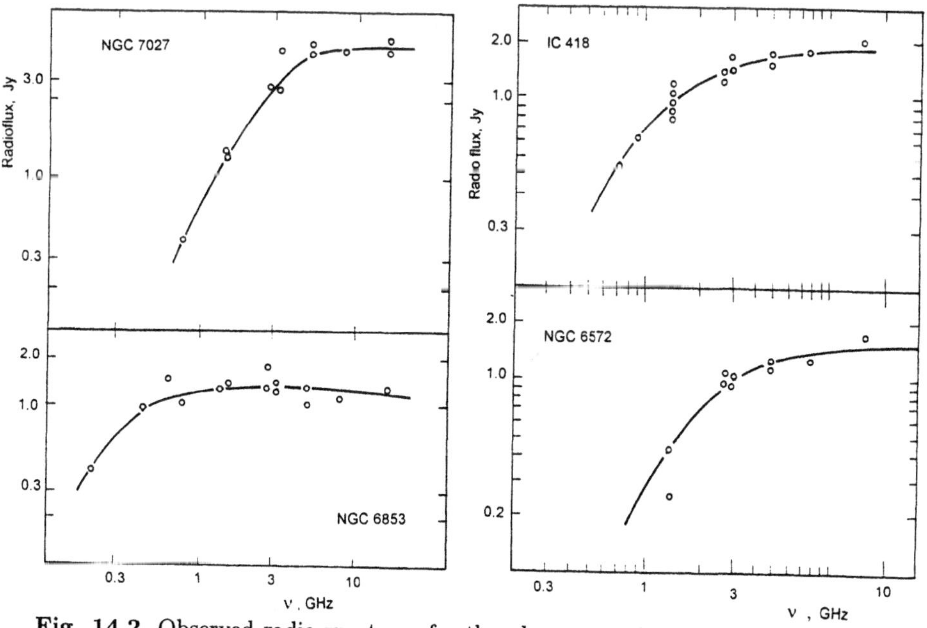

Fig. 14.2. Observed radio spectrum for the planetary nebulae NGC 7 027, NGC 6 853, NGC 6 572 and IC 418

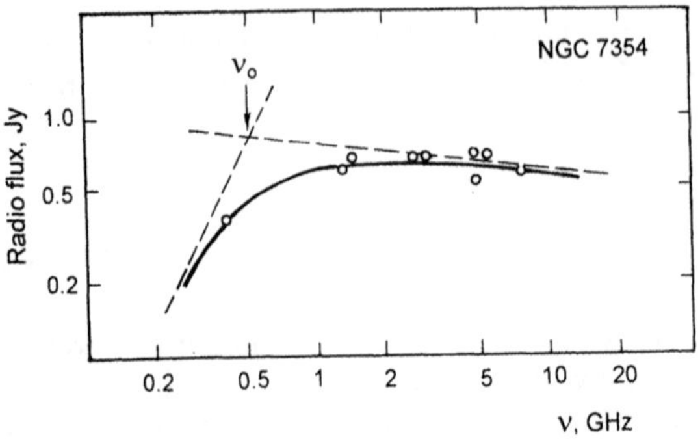

Fig. 14.3. Observed radio spectrum of the planetary nebula NGC 7354. The intersection point of both asymptotes (*broken lines*) corresponds to the critical frequency ν_0

in PN is of thermal nature and is induced by free–free transitions of thermal electrons – thermal bremsstrahlung.

Obviously, the conditions for the generation of radio emission cannot be uniform even within one and the same nebula; for example, the nebula can be transparent for short – decimetre and centimetre – wavelengths, and opaque for longer – metre–wavelengths. One can therefore introduce a *critical frequency* ν_0, separating one region of the spectrum from the other. This critical frequency ν_0 corresponds to an optical thickness $\tau(\nu_0)$ of the order of unity. In Fig. 14.3, the radio spectrum of the PN NGC 7354, we see an example of a graphical method for determining the critical frequency ν_0: this corresponds to the intersection point of the asymptotes of both the long and short wavelength parts of the spectrum.

The critical frequency ν_0 is different for various objects and can be used to estimate the emission measure EM:

$$EM = \frac{\nu_0^2 T_e^{3/2}}{\xi(\nu_0, T_e)}, \tag{12}$$

as follows from (3) when $\tau(\nu_0) = 1$. The values of ν_0 in MHz are taken directly from the observed radio spectra of the PN. The results for a sample of PNs are as follows (Terzian, 1978):

	ν_0, MHz	EM, $10^6 \mathrm{cm}^{-6}\mathrm{pc}$
NGC 6790	4500	92.5
NGC 7027	3500	70.5
NGC 6572	2900	30.0
NGC 6543	1800	9.2
NGC 7009	800	2.1
NGC 6210	700	1.6

Besides the graphical method, the emission measure EM can be determined even more precisely using the observed fluxes $F(\nu_1)$ and $F(\nu_2)$ at long, $\tau(\nu_1) > 1$, and short wavelengths, $\tau(\nu_2) < 1$. From (8),

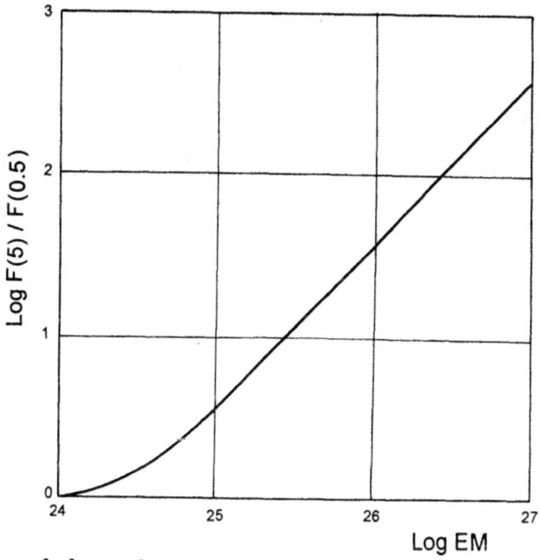

Fig. 14.4. Calculated dependence of the emission measure EM on the ratio of radio fluxes $F(5)/F(0.5)$ at frequencies of 5 GHz and 0.5 GHz

$$\frac{F(\nu_1)}{F(\nu_2)} = \left(\frac{\nu_1}{\nu_2}\right)^2 \frac{1 - e^{-\tau(\nu_1)}}{1 - e^{-\tau(\nu_2)}} . \tag{13}$$

For $\nu_1 = 5\,000$ MHz $= 5$ GHz and $\nu_2 = 0.5$ GHz, we obtain

$$\frac{F(\nu_1)}{F(\nu_2)} = 100 \frac{1 - \exp(-3.60 \times 10^{-27} EM)}{1 - \exp(-4.50 \times 10^{-27} EM)} , \tag{14}$$

where $\tau(\nu)$ is given by (3) and $T_e = 10\,000$ K. The left hand side of (14) is obtained from observations, and hence EM in units of cm^{-5}. The dependence of EM on the ratio $F(\nu_1)/F(\nu_2)$ is given in Fig. 14.4. Once we have the values of EM, it is not difficult to obtain $\tau(\nu_1)$ and $\tau(\nu_2)$ from the expressions

$$\tau(5 \text{ GHz}) = 3.60 \times 10^{-27}\ EM ,$$

$$\tau(0.5 \text{ GHz}) = 4.50 \times 10^{-25}\ EM ,$$

where EM is in units cm^{-5}.

No reliable observational facts exist that support any non-thermal, e.g. synchrotron, mechanism of radio emission in PNs.

14.5 Radio Recombination Lines

Free electron recombination with protons gives an atom of some discrete level n. The subsequent spontaneous transitions from that level cause the emission of *recombination lines*. For the hydrogen atom the frequency of such a line $n_{nn'}$ arising at the transition from level n to lower level n' is given by the relationship

$$\nu_{nn'} = R_{\rm H}\left(\frac{1}{n'^2} - \frac{1}{n^2}\right), \tag{15}$$

where $R_{\rm H} = 3.290 \times 10^{15}\ {\rm s}^{-1}$ is the Rydberg constant. Obviously, the discrete levels become closer with the increase of quantum number n. Low frequency transitions between close levels corresponding to the radio region cause *radio recombination lines.*

The line H 109 α is an example of a radio recombination line of hydrogen. This line arises at the transition from the level $n = 110$ to $n' = 109$; its frequency is $\nu \approx 5\,009$ MHz or $\lambda \approx 6$ cm. The difference between levels $\Delta n = n - n' = 1$ is denoted by the subscript α, $\Delta n = 2$ by β, $\Delta n = 3$ by γ, etc. The lowest frequency is at $\Delta n = 1$; the frequency increases with the increase of Δn, i.e. in the sequence α, β, γ etc.

It follows that radio recombination lines of hydrogen alone will infinite in number. In practice, however, their number is essentially limited by observation conditions. The possibility of detecting hydrogen recombination emission lines excited in the HII regions has been predicted by Kardashev (1959). These were discovered in 1964 (Dravskikh, Dravskikh, 1964). A number of such lines from $n = 54$ up to $n = 274$, i.e. from a frequency of 36 500 MHz up to 318 MHz were discovered later, for example in Orion A and M 17. Radio recombination lines of helium, He 109α and He 137β, and carbon, C 109α, have also been detected. The hydrogen line H 109α has been detected most often, in over a hundred objects: HI regions, nebulae, etc.

The situation concerning the detection of radio recombination lines in PN was different. In PNs these lines, according to calculations, should be very weak. This is the reason they were not reliably detected in PNs for a long time. Prior to 1978, the lines H 76α, H 85α, H 94α, H 109α, were detected and measured in only four or five PNs, NGC 7 027, NGC 6 543, NGC 7 662, IC 418 and M 1-78,and the lines H 113β and HeII 121α are fixed only in NGC 7 027. In Fig. 14.5, the recordings of the profiles of four lines, H 76α, H 85α, H 90α and H 109α, are given for NGC 7 027 (Terzian, 1976, 1978).

What kind of information may be deduced from the observations in the recombination lines of PNs? Exactly the same, in no cases any more, than for ordinary emission lines. Indeed, the strength and shapes of recombination lines are indicators of the electron temperature and velocity field of particles in an ionized gas. For a transparent medium the ratio $T_l/T_{\rm c}$ can be represented in the approximate form

$$-\frac{T_{\rm e}}{T_{\rm c}} \approx \frac{\tau_l}{\tau_{\rm c}},$$

where τ_l and $\tau_{\rm c}$ are the optical depths of the medium in a given radio line and in the continuous spectrum, respectively. Here $\tau_{\rm c}$ is determined by free–free transitions of electrons, and τ_l by Doppler broadening due to thermal motion, turbulence, expansion, etc. Then, we have for He/H = 0.10

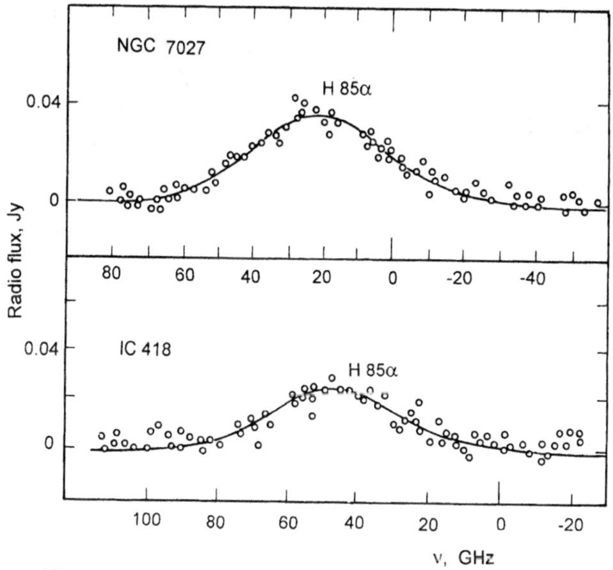

Fig. 14.5. Profiles of four radio recombination lines of hydrogen, H 76α (2.06 cm), H 85α (2.86 cm), H 90α (3.37 cm) and H 109α (6.00 cm), in the spectrum of NGC 7027

$$\frac{T_e}{T_c} = \frac{5.97 \times 10^{-11} \nu^2}{\Delta\nu_D \, T_e \ln(4.95 \times 10^7 \, T_e^{3/2}/\nu^2)}, \tag{16}$$

where $\Delta\nu_D$ is the half-width of the radio line in Hz. Here the values of T_l/T_c and $\Delta\nu_D$ are determined by direct measurements. We can then obtain the electron temperature and kinematic parameters of the emission medium.

Low velocity internal motions within the emission medium, usually unreachable by means of optical spectroscopy can, in principle, be studied simply with the help of radio recombination lines. However, even in view of this, one can hardly overestimate the importance of radio recombination lines in revealing the physics of cosmic emission sources in general, and of PNs, in particular.

14.6 The Planetary Nebula NGC 7027

In Fig. 14.6 the macrostructure of the continuous spectrum is shown for of a high excitation PN, NGC 7027. This spectra is typical for several reasons. It covers a broad range of wavelengths of visible, 3 000–10 000 Å, near and far infrared, 1–300 μm, and centimetre–decimetre, 1–200 cm, regions (Terzian, 1989). The flux interval, the ordinate, also covers an enormous range – four orders of magnitude. Only the near (2 000–3 000 Å) and far (1 000–2000 Å) ultraviolet bands are absent here. However, the structure of the continuous

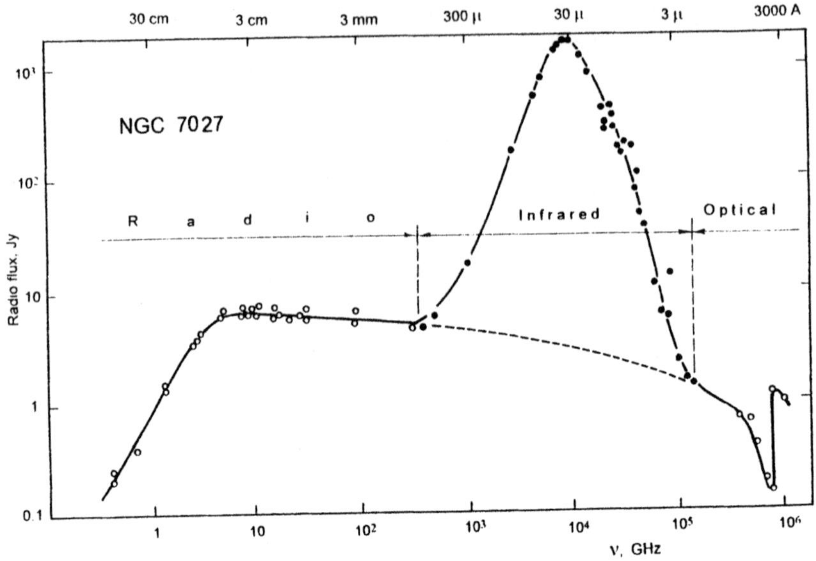

Fig. 14.6. The radio, infrared and optical spectra of NGC 7027. The continuous spectrum (*circles*) from the optical region (longer than 3 000 Å) to the radio through the broken line (~ 3 μm up to ~ 1 mm) is determined by thermal bremsstrahlung of the electrons of the gaseous component of the nebula with an electron temperature $T_e \sim 13\,000$ K. The infrared spectrum (*dots*) with a maximum at 30 μm and above the broken line is caused by black-body radiation of dust of effective temperature $T_d \sim 100$ K

spectrum in the ultraviolet is familiar to us (see Chap. 7), i.e. the Balmer continuum of hydrogen (shorter than 3 600 Å), the two-photon emission of hydrogen (longer than 1 216 Å) and the weak continuum of various ions. The nature of the latter is common, i.e. free–free and free–bound transitions of thermal electrons.

Figure 14.6 reveals the two-component structure of the continuous spectrum of this object. One has a purely atomic origin: it covers from the optical region (3 000–10 000 Å) radio (3 μm up to ~ 1 mm) (the circles). It is formed by a continuum of hydrogen, helium and other ions, as well as by the thermal bremsstrahlung of electrons of temperature $T_e \sim 130\,000$ K in the millimetre, submillimetre and radio regions. The second component, the infrared spectrum (the dots), with a maximum at ~ 30 μm, within the interval from ~ 1–2 μm to ~ 1 mm, is superposed over the first one. This belongs to the dust component of the PNs and is caused by black-body radiation of dust particles of effective temperature $T_d \sim 100$ K. In both cases the source of heating is the same – hard (ionizing) radiation of the central star (see Chap. 10).

A high resolution (~ 2″), radio image of this PN at the wavelength 3.71 cm (8 100 MHz) is shown in Fig. 14.7. *The bipolar* structure of this nebula is

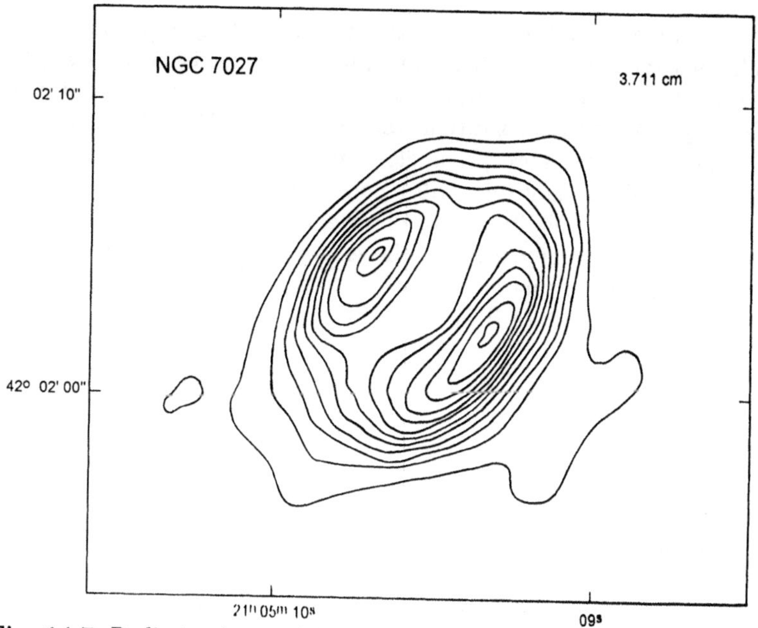

Fig. 14.7. Radio isophotes of the planetary nebula NGC 7027 at a frequency of 8 100 MHz (3.71 cm). Angular resolution ~ 2″. The bipolar structure is visible clearly seen here (Chap. 17), even more outlined than in optical images. This example demonstrates that in certain cases the radio images can be crucial for morphological studies of PNs.

14.7 Neutral Hydrogen in Planetary Nebulae

There is always a certain probability that some fraction of hydrogen in any PN can be in a neutral state, irrespective of the size or age. In the initial phase of the formation of the nebula when the hydrogen concentration is large, the ionizing radiation being absorbed in the inner regions of the nebula cannot reach its outer layers. In the late stages the ionization again cannot be complete because of the enormous size of the nebula and its strong rarefication. In both cases some part of the hydrogen remains neutral.

Indeed, there are indirect indications as well: in some nebulae we observe emission lines of neutral oxygen, most often of the doublet 6 300 + 6 363 [OI] or of the triplet 1 305 [OI]. So long as the ionization potential of both oxygen and hydrogen is the same, the neutral oxygen can exist only in the HI zone. Existing astrophysical methods enable one to obtain the mass of only the ionized part of the nebula, while to obtain a correct estimate of its total mass, it is necessary to evaluate also the mass of its neutral part. The problem of neutral hydrogen in PNs therefore arises.

The real sources of information on the neutral hydrogen are monochromatic radio observations in the 21 cm radio line. This line is caused by spontaneous transitions between sublevels of superfine structure of ground state of neutral hydrogen. In the PN conditions the 21 cm (1 410 MHz) line can originate in both emission and absorption. However, it is not difficult to convince oneself that the predominant ones will be absorption lines in the continuous radio emission of the nebula itself. The observations of PNs in the 21 cm line are rather difficult for many reasons: the extremely small expected fluxes, hundreds or thousands of times smaller than those of the Galactic HI regions and the small angular sizes of PNs.

These difficulties explain the appearance only in the 1980s of the first reliable observational results (Rodriguez, Moran, 1982; Schneider et al., 1987; Altschuler et al., 1986; Taylor et al., 1989), though attempts had been made since 1960s. The observations have been realized mainly by the 300 m Arecibo radio telescope and the VLA radio interferometer for PNs with large radial velocities, so that the 21 cm line was reliably distinguished at the background of powerful Galactic 21 cm emission. The results are as follows. No PN was detected with a 21 cm emission line. The absorption line was observed, for example, for NGC 6 302, NGC 6 970 and IC 4 997 as well as for IC 418; the radio isophote maps in this line had an angular resolution of 10″.

From the profiles of the 21 cm line, the value of $\int \tau \mathrm{d}V$ can be estimated. From here we obtain the mass of neutral hydrogen $\mathcal{M}_{\mathrm{HI}}$ for the given nebula in solar units

$$\mathcal{M}_{\mathrm{HI}} = 2.1 \times 10^{-6} (\varphi D)^2 T_{\mathrm{ex}} \int \tau \mathrm{d}V \,, \tag{17}$$

where D is the distance of the nebula in kpc, φ is its angular size in arcsec and T_{ex} is the effective temperature of the zone of neutral hydrogen, i.e. of the order of 100 K. The magnitudes of $\mathcal{M}_{\mathrm{HI}}$ obtained for the three observed PNs are presented in the sixth column of Table 14.1 along with the initial data. In the seventh column their ionized masses, $\mathcal{M}_{\mathrm{HI}}$, are given, and in the last column the ratio of both masses, $\mathcal{M}_{\mathrm{HI}}/\mathcal{M}_{\mathrm{HII}}$. In all cases this ratio is of the same order of magnitude, i.e. ~ 0.5; hence the masses of ionized and neutral hydrogen are about the same.

Table 14.1. Mass of neutral hydrogen, $\mathcal{M}_{\mathrm{HI}}$, in three planetary nebulae derived from data from the 21 cm radio line

Planetary nebula	φ''	D [kpc]	T_{ex} [K]	$\int \tau \mathrm{d}V$ [km s^{-1}]	$\frac{\mathcal{M}_{\mathrm{HI}}}{\mathcal{M}_{\odot}}$	$\frac{\mathcal{M}_{\mathrm{HII}}}{\mathcal{M}_{\odot}}$	$\frac{\mathcal{M}_{\mathrm{HI}}}{\mathcal{M}_{\mathrm{HII}}}$
NGC 6 302	4″.5	0.5	100	3.5	0.0037	0.0053	0.70
NGC 6 970	1.0	2.0	100	2.4	0.0015	0.0037	0.50
IC 4 997	0.8	2.6	185	2.1	0.0035	0.0049	0.71

The reliability of this conclusion is supported in the case of IC 4 997 by its youth and high density, in the case of NGC 6 302 by the existence in its spectrum of strong emission lines of neutral oxygen, both in the visible region, 6 300+6 363 [OI], and in the ultraviolet, 1305 [OI]. In the latter object there appears to be a torus-like zone enveloping the central ionizing zone HII.

The case of IC 418 is more peculiar. The 21 line emission is detected from a ring-like zone with dimension $> 180''$, i.e. far exceeding the optical sizes of the PN. Do we have an indication here of the phenomenon of *circumnebular clouds* analogous to circumstellar clouds? Judging by the negative results for numerous searches of the 21 cm absorption line, the fraction of neutral hydrogen in the majority of PNs must be smaller than in those discussed above. Can the lower limit of the fraction of neutral hydrogen be close to zero? Presumably other attractive methods, e.g. spectral observations in the ultraviolet are required to answer this question (see next chapter, Sect. 15.8).

The problem of neutral hydrogen in PNs remains open.

14.8 Interstellar Extinction from Radio Data

Radio observations of PNs are a powerful tool for determining of the magnitude of interstellar extinction.

Both the continuous radio emission and the emission in any hydrogen line are induced by the same thermal electrons. Also common is the effective volume of generation of both emissions in the radio continuum and in any hydrogen line, such as H_β. At last, as we have seen above, there is a definite region of radio frequencies for which all PNs, without exception, are transparent: higher than 3 000 MHz. The nebulae are completely transparent also in the H_β hydrogen line. The ratio $R(\nu)/F(H_\beta)$ can be easily obtained, where $R(\nu)$ and $F(H_\beta)$ are the fluxes of the continuous radio spectrum at a frequency ν and of the H_β line, respectively. This ratio depends neither on the nebula's size nor its electron concentration; it only weakly depends on the electron temperature T_e.

The interstellar medium is transparent for radio frequencies, but it is not transparent for the H_β line. As a result, the observed ratio $R(\nu)/F(H_\beta)$ must differ from its theoretical magnitude. Then, by a simple comparison of both observed and theoretical ratios, the magnitude of the interstellar extinction in the direction of a given nebula can be estimated. This is the main idea of the method.

The volume coefficient of radio emission $4\pi R(\nu)$ by free–free transitions of electrons in an optically thin medium of ionized hydrogen (HII zone) of electron temperature T_e, electron and ion concentrations n_e and n_i per unit interval of radio frequencies (Hz) is given by the following expression (Oster, 1961):

$$4\pi R(\nu) = n_\mathrm{e} n_i \frac{32\sqrt{\pi}\, e^6}{3m^2c^3}\left(\frac{m}{kT_\mathrm{e}}\right)^{1/2} \ln\left[\left(\frac{2kT_\mathrm{e}}{m}\right)^{3/2} \frac{m}{Ze^2}\right], \tag{18}$$

where Z is the charge of the ion, k is the Boltzmann constant, and m and e are the mass and charge of electron, respectively.

Taking into account only the contributions of H^+, He^+ and He^{++}, and assuming that the hydrogen is completely ionized, we will have from (18):

$$4\pi R(\nu) = 3.75 \times 10^{-40} t^{-1/2} n_\mathrm{e} \left\{ [n_i(\mathrm{H}^+) + n_i(\mathrm{He}^+)] \ln(4.95 \times 10^{13} t^{3/2}/\nu) + 4n_i(\mathrm{He}^{++}) \ln(2.47 \times 10^{13} t^{3/2}/\nu) \right\} \text{ ergs cm}^{-3}\text{ s}^{-1}\text{ Hz}^{-1}, \tag{19}$$

where $t = T_\mathrm{e}/10^4$ and $n_i(A)$ is the concentration of the ion A. On the other hand, we have for the volume emission coefficient in the H_β line:

$$F(\mathrm{H}_\beta) = 1.24 \times 10^{-25} n_\mathrm{e} n_i(\mathrm{H}^+)\, t^{-0.9} \text{ ergs cm}^{-3}\text{ s}^{-1}. \tag{20}$$

Assuming that the nebula is transparent at the radio frequency ν = 5 000 MHz = 5 GHz, we have for the ratio of both radio and H_β line emissions

$$\frac{R(5\,\mathrm{GHz})}{F(\mathrm{H}_\beta)} = 3.05 \times 10^{-18} t^{0.4} \ln(9\,900\, t^{3/2})[1 + (1-x)y + 3.7xy], \tag{21}$$

where R(5 GHz) and $F(\mathrm{H}_\beta)$ are in units of W m^{-2} Hz^{-1} and ergs cm^{-3} s^{-1}, respectively, y =He/H and x = He^{++}/He. According to the definition, we have for the interstellar extinction c

$$c = \log \frac{F(\mathrm{H}_\beta)_\mathrm{cal}}{F(\mathrm{H}_\beta)_\mathrm{obs}} = \log \frac{F(\mathrm{H}_\beta)_\mathrm{cal}}{R(5\,\mathrm{GHz})_\mathrm{cal}} \frac{R(5\,\mathrm{GHz})_\mathrm{cal}}{F(\mathrm{H}_\beta)_\mathrm{obs}}. \tag{22}$$

Substituting here the theoretical value of the ratio $R(5\ \mathrm{GHz})/F(\mathrm{H}_\beta)$ from (21), and bearing in mind that $R(5\ \mathrm{GHz})_\mathrm{cal} = R(5\ \mathrm{GHz})_\mathrm{obs}$ and $F(\mathrm{H}\beta)_\mathrm{obs} = F(\mathrm{H}_\beta)$, we obtain

$$c = \log \frac{3.28 \times 10^{-9} t^{-0.4} R(5\,\mathrm{GHz})}{[1 + (1-x)y + 3.7xy] \ln(9\,900\, t^{3/2})} - \log F(\mathrm{H}_\beta), \tag{23}$$

where R(5 GHz) is now the observed radio flux at the frequency 5 GHz in units of mJy, and $F(\mathrm{H}_\beta)$ is the observed flux in the H_β line in units of ergs cm^{-2} s^{-1}, as before. As we see, the dependence of c on T_e is weak. Therefore, we can adopt $t = 1$, i.e. $T_\mathrm{e} = 10\,000$ K, for all PNs. As for x and y, their numerical values are not known for all PNs. Taking on average $x = 0.50$ and $y = 0.11$, we get from (23), for $c = c_\mathrm{R}$,

$$c_\mathrm{R} = -9.546 + \log R(5\,\mathrm{GHz}) - \log F(\mathrm{H}_\beta). \tag{24}$$

Using this expression we can obtain the magnitude of interstellar extinction c_R in the direction of any planetary nebula by known observational magnitudes of their radio flux $R(5\ \mathrm{GHz})$ at a frequency of 5 GHz and that of the H_β line in units of mJy and ergs $cm^{-2}\ s^{-1}$, respectively. The subscript R indicates that this coefficient is obtained by the radio frequency method.

For the above values of x and y and at $T_e = 10\,000$ K we have for the ratio of both radio and H_β line fluxes:

$$\frac{R(5\,\mathrm{GHz})}{F(\mathrm{H}_\beta)} = 3.53 \times 10^{-14}\,. \tag{25}$$

In view of the weak dependence of this ratio on the radio frequency ν, with this quantity (3.53×10^{-14}), one can relate the radio fluxes in the centimetre and decimetre ranges.

The method just described for the determination of interstellar extinction c_R has been applied for a large number of PNs, more than 100 in the northern sky (Terzian, 1968), as well as for 125 PNs in the southern sky (Milne, Aller, 1975; Thompson, 1968). In the majority of cases, the values of c_R obtained by the radio methods for individual nebulae are in good agreement with the values of extinction c_{opt} obtained earlier by *optical* methods. This constant is confirmed by data for the following sample of PN

	NGC 7027	NGC 7293	NGC 6720	NGC 7026	J 900	BD+30
c_R	1.44	0.01	0.13	0.78	0.80	1.60
c_{opt}	1.34	0.02	0.15	0.87	0.75	0.76

As we can see, the largest difference between c_R and c_{opt} occurs only for the PN BD+30°3639.

15. Anomalous Planetary Nebulae

15.1 Superdense Planetary Nebulae

For ordinary PNs the ratio $E(N_1{+}N_2)/E(4\,363)$, as we have seen in Chap. 6, is very large, of the order of 100–300. These are objects with electron temperatures T_e of the order of 10 000–15 000 K and electron concentration n_e of the order of 10^3–10^4 cm^{-3} or smaller. However, for a group of just a few PNs, of star-like in form, an anomalously low value for this ratio has been discovered – of the order of ten or even a few units, i.e. nearly 100 times smaller than its typical value.

Apply the usual formula (see Chap. 6)

$$\frac{E_{N_1+N_2}}{E_{4\,363}} = 8.74\, e^{33\,000/T_e} \tag{1}$$

to IC 4 997, for example, for which the ratio $\delta = E_{N_1+N_2}/E_{4\,363}$ corrected for interstellar extinction is equal to 17.8. We obtain $T_e = 46\,000$ K, while for NGC 6 833 with $\delta = 9.15$ we have $T_e = 71\,300$ K (!) – obviously absurd results in both cases. Observations often give $\delta < 10$, which means that (1) cannot be applicable in principle, since the minimum value $\delta_{min} = 8.74$ is reached in accordance with that formula at $T_e \to \infty$.

The small δ signify the suppressing role of the second kind of electron collision when the transitions from the metastable levels are realized without the escape of photons in the forbidden line. Since the electron temperatures in PNs with anomalously small δ are the same as in ordinary ones, the efficiency of the second kind of electron collision can be increased only by high electron concentrations.

We arrive at an important conclusion: small δ indicate dense or superdense nebulae.

The small δ also indicates that the rate of spontaneous (radiative) transitions is comparable with the rate of transitions induced by the second kind collision. Calculations show that this is possible at electron concentrations of 10^6–10^7 cm^{-3}. More exact values of the electron concentration can be derived from the expression

$$\frac{E_{N_1+N_2}}{E_{4\,363}} = F\,(n_e,\, T_e)\, e^{33\,000/T_e}\,, \tag{2}$$

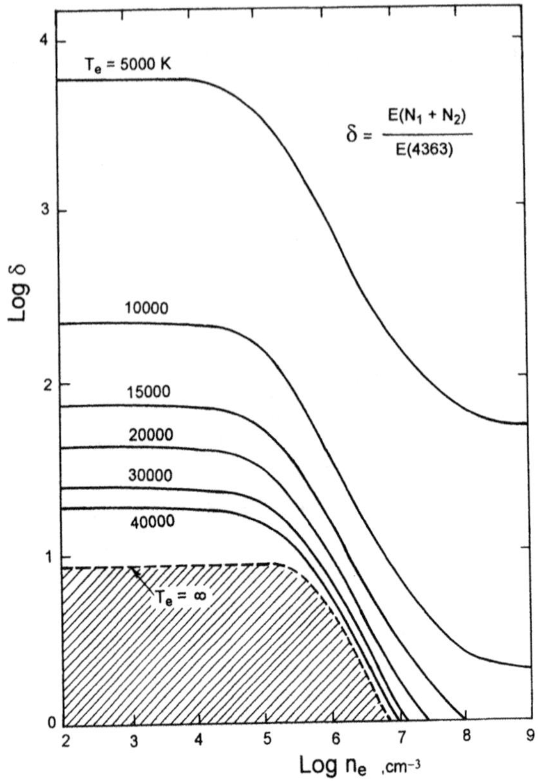

Fig. 15.1. Dependence of $\log n_e$ versus $\log \delta$ for the electron concentrations n_e from 10^2 up to $10^9\ \mathrm{cm}^{-3}$ and for various values of the electron temperature T_e

which is valid for all values of T_e and n_e (see Chap. 6).

Consider two limiting cases:

(a) As $n_e \to 0$ (actually already at $n_e < 10^3\,\mathrm{cm}^{-3}$) the function $F(n_e, T_e)$ given by (6.30), depends on n_e weakly and is equal to 8.76 at $T_e \sim 10^4$ K. Then, (2) coincides with (1).

(b) As $n_e \to \infty$, i.e. for dense or superdense nebulae (in fact already at $n_e > 10^4\ \mathrm{cm}^{-3}$), we have $F(\infty, T_e) = 0.0753$. Then instead of (2), we have

$$\delta = \frac{E_{\mathrm{N_1+N_2}}}{E_{4\,363}} = 0.0753\, e^{33\,000/T_e} . \qquad (3)$$

It follows, for a superdense nebula that δ can be equal, for example, to 2.04 and 0.68, at electron temperatures of 10 000 K and 15 000 K, respectively. The minimum possible value of δ for a superdense nebula can be reached as $T_e \to \infty$: $\delta_{\mathrm{min}} = 0.075$.

As we can see, the observed small values of δ reflect a real physical situation, possible, however, only in superdense, compact nebulae.

Equations (6.2) and (6.30) are plotted in Fig. 15.1 in the form of dependencies of $\log \delta$ on $\log n_e$ for T_e from 5 000 K up to 40 000 K. These curves are rather informative. They indicate, first, that at a given δ the electron concentration n_e decreases inversely with the electron temperature T_e. Second, at relatively small n_e ($< 10^5$ cm^{-3}) the minimum value $\delta_{min} = 9.17$ ($\log \delta_{min} = 0.96$) is reached as $T_e \to \infty$. In other words, no combination of n_e and δ corresponds to a real value of T_e (shaded region in Fig. 15.1). Third, δ_{min} corresponds to a higher electron concentration compared with the usual densities of nebulae. For example, at $\delta \approx 5$ we have $n_e \approx 10^6$ cm^{-3}, if $T_e \to \infty$, while since the real $T_e \ll \infty$, then it must always be the case that $n_e > 10^6$ cm^{-3} for $\delta \approx 5$.

The last conclusion is of special interest. It states that in all cases when δ is of the order of few a units, the concentration must be of the order of 10^6 cm^{-3} and higher *independent of the electron temperature.* Since in real cases $T_e \sim$ 10 000–15 000 K, one should expect that $n_e \sim 10^6$–10^7 cm^{-3} for $\delta \sim$ 5–10.

A sample of PNs and compact nebular objects is given in Table 15.1, in order of decreasing values of $\delta - I(N_1+N_2)/I(4363)$, corrected for interstellar extinction. The smallest $\delta = 2.0$ is recorded for M 1-2, apparently the densest PN.

Table 15.1. Electron concentrations n_e in dense ($\sim 10^5$ cm^{-3}) and superdense ($\sim 10^6 - 10^7$ cm^{-3}) planetary nebulae and two possible planetary nebulae V 1 329 Cyg and V 1 016 Cyg

Planetary nebula	$\delta = \frac{I(N_1+N_2)}{I(4363)}$	T_e [K]	n_e [10^6cm^{-3}]	$\frac{H_\alpha}{H_\beta}$	$\frac{H_\gamma}{H_\beta}$	Ref*
NGC 6 970	95	11 600	0.20	3.04	0.51	1
NGC 7 027	70	13 000	0.16	7.37	0.31	2
Vy 2-2	62	10 000	0.50	9.09	0.326	2
IC 5 117	28	12 000	1.2	2.73	0.51	1
Hu 1-2	23.8	11 900	1.2	7.47	0.345	2
Hb 12	23.6	11 500	1.3	7.52	0.339	3
IC 4 997	17.8	15 700	1.0	4.41	0.426	2
NGC 6 833	9.15	12 200	4.0	3.29	0.420	3
V 1 329 Cyg	7.5	15 000	2.8	5.25	0.390	2
V 1 329 Cyg	–	–	–	6.92	0.309	4
M 3-27	5.70	16 400	3.0	6.33	0.481	3
M 3-27	5.57	16 400	3.1	5.50	0.461	2
V 1 016 Cyg	4.37	15 100	5.6	4.04	0.441	2
V 1 016 Cyg	6.34	15 100	3.2	7.20	0.414	4
M 1-2	2.0	13 000	50	5.57	0.443	2

* 1. Kaler, 1976a, 2. Ahern, 1978, 3. Barker, 1978, 4. Blair et al., 1983.

The values of T_e estimated by various methods are given in the third column of Table 15.1. From known values of δ and T_e, the electron concentrations n_e were obtained using the graphs in Fig. 15.1; the results are presented in the fourth column. The smallest $n_e \sim 10^5$ cm^{-3} is obtained for NGC 7027 and NGC 6790, while in the remaining cases n_e is of the order of 10^6 cm^{-3}, and only in one case, for M 1-2, is $n_e \sim 10^7$ cm^{-3}. Note the stability of the obtained value $n_e \sim 10^6$ cm^{-3} with the variation of δ from 5 up to 25.

In fact one more superdense PN could be included in the Table 15.1; it is located in the Small Magellanic Cloud and denoted P23 (Boroson, Liebert, 1988). Its superdense nature is confirmed by the following additional facts:

(a) in the spectrum of this nebula the emission lines 3727 [OII] and 6731 [SII] are absent, for which the critical electron concentration is of the order of 3×10^3 cm^{-3};
(b) the ratio $N_1 + N_2/H_\beta = 0.27$, i.e. an order of magnitude less than we have for PN;
(c) the parameter $\delta = 1.07$, i.e. is extremely small;
(d) the line 5007 [OIII] is weaker than 6548 [NII] for which the critical concentration is 10^5 cm^{-3};
(e) the detected line 3869 [NeIII] has a higher critical concentration, $\sim 10^7$ cm^{-3}. Assuming $T_e \sim 10000$ K for SMC 23, we will $n_e = 8 \times 10^5 \sim 10^6$ cm^{-3}.

The radii of nebulae for which $n_e \sim 10^6$–10^7 cm^{-3} should be of the order of a few hundred a.u. and in all cases less than a thousand a.u., i.e. 10^{16} cm.

Note that T_e has a minimum value for a given δ when $n_e \to \infty$. The asymptotic expression for the determination of T_e(min) has the form:

$$T_e(\text{min}) = 14\,300\,(1.123 + \log\delta)^{-1}\,.$$

For NGC 7027 (δ = 70), M 3-27 (δ = 5.7) and M 1–2 (δ =2.0) the calculated values of T_e(min) are, 4800, 7600 and 10000 K, respectively.

15.2 M 1-2: The Densest Planetary Nebula

M 1-2 was discovered by Minkowski (1946) as a star-like ($< 0''.5$) emission object of $\sim 14^m$ visual magnitude. Soon after this the structure of its spectra, typical for a PN, was revealed: the forbidden lines N_1, N_2 [OIII], [OII], [SII], the lines of neutral and ionized helium, the strong lines of Balmer series of hydrogen etc. (O'Dell, 1963). However, the relative intensities of the lines were not ordinary: the extremely weak doublet 3727 [OII], the doublet 6548 + 6584 [NII] located on both sides of the line H_α was at the limit of discovery, etc.

This nebula differs from other objects with respect to at least three peculiarities of its spectrum. First, the parameter δ is small, of the order of 2–3, indicating quite a high electron concentration. Second, the unusual behaviour of the Balmer decrement is the indicator of an opaque medium in these lines. Third, the highest numbers of the Balmer lines, up to H 24, were detected.

These peculiarities have even cast some doubt on whether this object can belong to the category of PNs. The first reaction was to attribute M 1-2 to the symbiotic systems (Feibelman, 1983). Hence, could M 1-2 be a double system, although it reveals constant brightness levels and no signs of variations of radial velocities?

Two circumstances may throw light on its nature. First, its *IUE* ultraviolet spectrum includes an event of exceptionally rarity – the inclusion of all three categories of emission lines according to their degree of excitation: lower – 2 800 MgII, 1 335 CII, 1 260 SiII, 2 852 MgI, medium – 1 550 CIV, 1 909 CIII], 1 400 SiIV, 1 663 OIII], 1 640 HeII etc., and high – 1 240 NV, 1 371 OV, 1 575 [NeV] etc.

It is clear that all four low excitation lines of SiII, CII, MgII and MgI should be connected with the star G2 I, more precisely, with its chromosphere. The emission lines of the medium and high excitation groups, both in terms of their structure and relative intensities correspond to those of ordinary PNs of medium and high excitation classes. Incidentally, the excitation class of M 1–2 is $p = 10$.

The parameter δ is, as was mentioned above, of the order 2–2.5 (Table 15.2) or $\log \delta = 0.3$–0.4, which corresponds to an electron concentration of 3–5×10^7 cm^{-3} at $T_e = 13\,000$ K (Fig. 15.1). This is the largest electron concentration ever registered for PN. At angular sizes $\sim 0.''5$ the question arises whether this object belongs to a group of compact superdense PNs. This conclusion is supported also by the striking similarity of the *IUE* spectrum of M 1-2 in the region 1 200–1 900 Å (Fig. 15.2), and that of IC 4997, a well-known high density PN with $n_e = 1.0 \times 10^6$ cm^{-3} at $T_e = 15\,700$ K.

The second fact concerns the Balmer decrement of M 1-2, or more precisely, its location on the Balmer decrement diagram. Three values of the decrement of M 1-2 are collected in Table 15.2 (columns 5 and 6), and are plotted in Fig. 15.3. The changes during a ten year period are evident. Over this period a drift has occurred in the direction of decrease of the optical

Table 15.2. Electron concentration n_e and Balmer decrement H_α/H_β and H_γ/H_β at various periods of observations of the compact planetary nebula M 1-2

Observ. year	$\delta = \frac{I(N_1+N_2)}{I(4363)}$	T_e [K]	n_e [10^7cm^{-3}]	$\frac{H_\alpha}{H_\beta}$	$\frac{H_\gamma}{H_\beta}$	Ref.
1970	2.42	–	3	6.59	0.482	Barker, 1978
1974	2.00	13 000	5	5.57	0.443	Ahern, 1978
1981	2.53	16 000	1.2	8.10	0.388	Blair et al.,1983

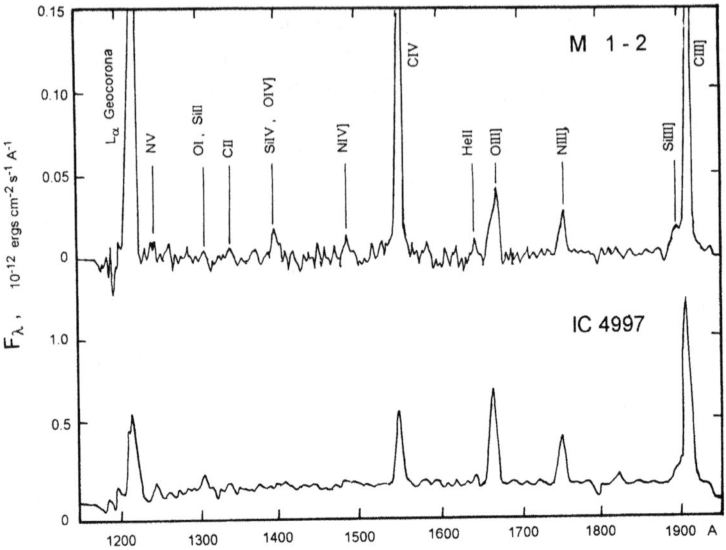

Fig. 15.2. *IUE* recordings of the ultraviolet spectra (1 200–1 900 Å) of two superdense planetary nebulae, IC 4 997 ($n_e = 10^6$ cm^{-3}) and M 1-2 ($n_e = 3 \times 10^7$ cm^{-3}). The similarity of both spectra is obvious. The flux scale in the case of IC 4 997 (V = 12) is 10 times larger than for M 1-2 (V = 14)

depth from $\tau_\alpha \sim 17$ up to $\tau_\alpha \sim 10$, which is possible especially with unchanged extinction c = const.

However, the drift of 1974-81 is hard to explain. Formally, this may be divided into two components: Δc and $\Delta\tau_\alpha$. The second one, $\Delta\tau_\alpha$, is related to the nebula itself and can be understood as a real but not large rise in optical depth, from $\tau_\alpha \sim 10$ up to $\tau_\alpha \sim 12$, which is quite possible in principle, say, as a result of the ejection of gaseous matter from the nebula's nucleus.

As for the first component, Δc, this change is related to the interstellar extinction, i.e. during this period (1974-81) a change in interstellar extinction from the value of $c \sim 0.5$ up to $c \sim 0.9$ has took place. However, physically this is absolutely impossible! Physically, the drift of any emission object on this diagram is possible only in the direction parallel to the line SA, "self-absorption", i.e. a change in the values of H_α/H_β and H_γ/H_β only; however, at a constant magnitude for this nebula of c – the parameter of interstellar extinction. Therefore the drift part in 1974-81 in the case of M 1–2 should be estimated as an annoying misunderstanding, arising, perhaps, through observational errors.

The spectrum of M 1-2 also shows the lines of ionized helium 1 640 HeII and 4 686 HeII (Barker, 1978) and of high potential ions NV, OV and NeV as well. To excite these lines, the temperature of the central star must be not less than 65 000 K. Can a binary system of a cool star and an O type subdwarf, or even an accreting dwarf, i.e. an *active binary*, exist at the centre of M 1-2?

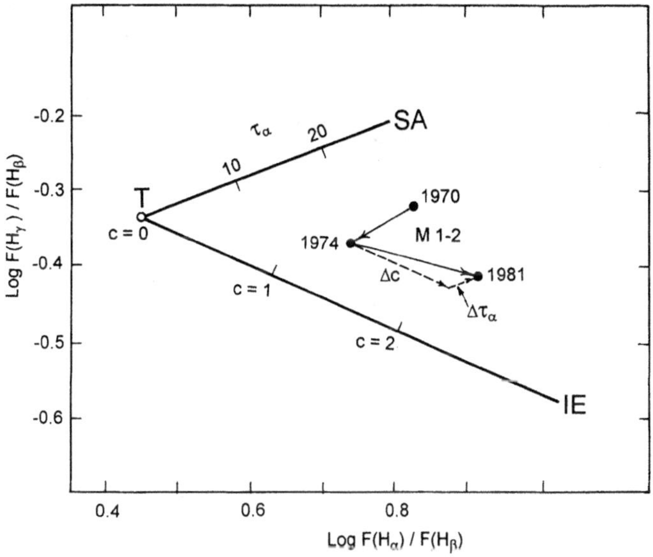

Fig. 15.3. Three positions (1970, 1974, 1981) of the superdense nebula M 1-2 on the decrement diagram. The directions of drift are shown (*arrows*) as well as the components of drift: Δc (interstellar extinction) and $\Delta \tau_\alpha$ (self-absorption)

At least some data, e.g. the high ratio $F(2\,800\ \mathrm{MgII})/F_{\mathrm{bol}} = 3 \times 10^{-3}$, etc., can support the association with, say, RS CVn type systems (Gurzadyan, Cholakyan, 1995).

15.3 Superdense Nebulae on the Decrement Diagram

In the fifth and sixth columns of Table 15.1 the *observed* values of the Balmer decrement, i.e. the relative intensities of H_α and H_γ lines in units of H_β are given for our sample of dense and superdense nebulae. These results are plotted in Fig. 15.4, on the decrement diagram. As we can see, the majority of the nebulae are located near the line of interstellar extinction IR; these objects are NGC 7 026, IC 4 997, NGC 6 833, Hb 12 and V 1 329 Cyg. Obviously, for them the discrepancy between the observed decrements and the theoretical one, i.e. from the point T, where $c = 0$ and $\tau_\alpha = 0$, is almost completely due to interstellar extinction. Here the self-absorption is practically absent.

Three objects, M 3-27, V 1 016 Cyg and M 1-2, are clearly far from the line IE. Note that all three objects are the densest among all others. Here we have the simultaneous action of both interstellar extinction and self-absorption; the latter corresponds to $\tau_\alpha \sim 10$–20 for all 3 objects.

The overview given by the decrement diagram in Fig. 15.4 enables us to draw remarkable conclusions. Indeed, the presence of objects on the line IE indicates that the dense nebulae can still be transparent in the lines of the

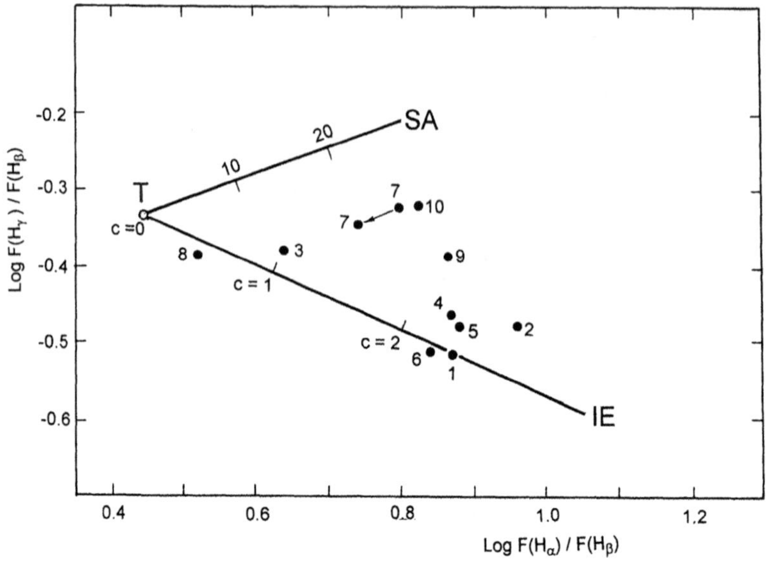

Fig. 15.4. Dense and superdense planetary nebulae on the decrement diagram. Numbers denote the following: 1 – NGC 7027, 2 – Vy 2-2, 3 – IC 4997, 4 – Hu 1-2, 6 – V 1329 Cyg, 7 – M 3-27, 8 – NGC 6823, 9 – V 1016 Cyg, 10 – M 1-2

Balmer series of hydrogen owing to a convenient combination of the electron concentration and temperature. In other words, the high density of a nebula still does not mean that has to be opaque in the mentioned lines.

This diagram also defines the direction of the evolution of a superdense nebula, i.e. its drift in the direction of the *IE* line with the final aim of moving onto it, along with the preservation of the parallelism with the self-absorption line *SA*, i.e. at constant *c*.

What about observations? For three objects, V 1329 Cyg, M 3-27 and V 1016 Cyg, we were able make to two observations for each. It can be seen at once that only in one case, that of the nebula M 3-27, does the drift take place strongly according to the above scenario (number 7 in Fig. 15.4, for the data of 1970 and 1974). In the other two cases, V 1329 Cyg and V 1016 Cyg (numbers 6 and 9), the directions of drift do not correspond to the expected picture; in both cases a component of drift appears parallel to the line *IE*, which means a rise in absolute magnitude of *c* over time, which is absolutely impossible. The problem remains open.

15.4 Nebulae Without Hydrogen: A New Class of Nebulae?

The nebula A 30 is extraordinary in many respects. The first is its unusual morphological structure: it is a two-envelope nebula with extended and faint

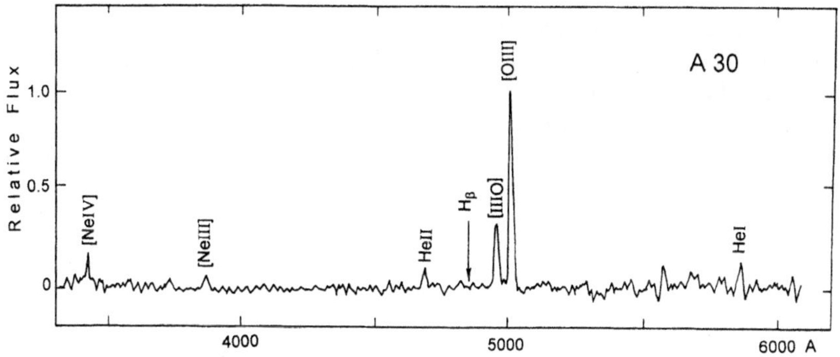

Fig. 15.5. The spectrum of the planetary nebula A 30 (condensation no. 3). The strongest lines are 4 959 + 5 007 [OIII]

outer envelope in the form of an exact circle of radius 64″. The central star, $V = 14^m.3$, is unusually bright for a nebula of this size. The main mass of the PN, however, is in the form of four condensations symmetrically situated with respect to the central star within a circle of diameter ~ 10″.

However, the main peculiarities are associated with the spectra of these condensations (Hazard et al., 1980): strong traditional lines N_1, N_2 [OIII], 4 686 HeII, 3 868 [NeIII], 3 727 [OII] etc., and without, or almost without signs of H_β in emission (see Fig. 15.5) (Jacoby, Ford, 1983).

The long wavelength parts of the spectra are represented by the emission lines 5 411 HeII, 5 876 HeII as well by the doublet 6 548 + 6 584 [NII]. H_α is absent in the spectrum of condensation no. 4 and is hardly noticeable in no. 3.

The condensations are seen to expand with a velocity of ±25 km s^{-1}, while the outer spherical envelope expands at 40 km s^{-1} (Reayet et al., 1983). At the distance of this nebula, 1 300 pc, this gives ~ 1500 years for the age of the condensations and ~ 10 000 years for that of the outer envelope.

The strongest emission lines in the *IUE* spectrum of this nebula are as follows (in units of J(1 640 HeII) = 1.00) (Harrington, Feibelman, 1984):

	1 240 NV	1 402 OIV	1 550 CIV	1 909 CIII	2 423 [NeIV]
$J(\lambda)$	4.77	1.37	9.00	0.80	1.63

The line 1 550 CIV is an order of magnitude stronger than 1 640 HeII. The line 1 240 NV is also unusually strong – nearly five times stronger than 1 640 HeII, which is rather unusual even in high excitation nebulae.

The electron temperature can be obtained from the known intensities of HeII lines with the help of the following relationship (Seaton, 1978)

$$\frac{J(1\,640\ \mathrm{HeII})}{J(4\,686\ \mathrm{HeII})} = 6.25 \left(\frac{T_e}{10\,000}\right)^{0.19} . \tag{4}$$

We have for the observed ratio $J(1\,640)/J(4\,686) = 7.1$, which gives $T_e = 19\,000$ K. The $N_1 + N_2$ [OIII]/4 363 [OIII] method gives $T_e = 16\,400$ K and the [NeIV] line method gives $T_e = 18\,300$ K.

The *IUE* recordings of the central star of this nebula are also interesting. On a fragment, shown in Fig. 15.6, the three lines 1 241 NV, 1 371 OV and 1 549 CIV are seen with sharply expressed P Cyg type profiles, probably indicating the matter outflow with quite a high velocity, 4 000 km s^{-1}, i.e. obviously a stellar wind. At such velocities, a shock wave is inevitable.

However, the absence of hydrogen emission lines A 30 should be considered the most interesting fact of the spectrum of A 30.

Since the lines 4 686 HeII and 1 640 HeII have a recombination origin, the ratio of their intensities must be proportional to He^{++}/He^{+}. From observations we have $J(4\,686\ \mathrm{HeII})/J(H_\beta) = 27$, which gives $N(\mathrm{He})/N(\mathrm{H}) \approx 10$, i.e. a hundred (!) times larger than its universal abundance (0.1).

Assume a nebula consisting only of helium atoms, i.e. with an absence of hydrogen atoms and a low abundance of heavy elements. The heating of such a nebula is realized, of course, by photoionization of He^{+}, and the kinetic energy of photoelectrons will be used for the excitation of forbidden or permitted lines by inelastic electron collisions. We can find the mean energy ε of free electrons, all "helium-born", from the condition of equality of the total energy emitted by the nebula in the emission lines excited by electron collisions and total energy of the photoelectrons. In this case the total intensity of all such lines will be expressed in units of the intensity of the line 1 640 HeII.

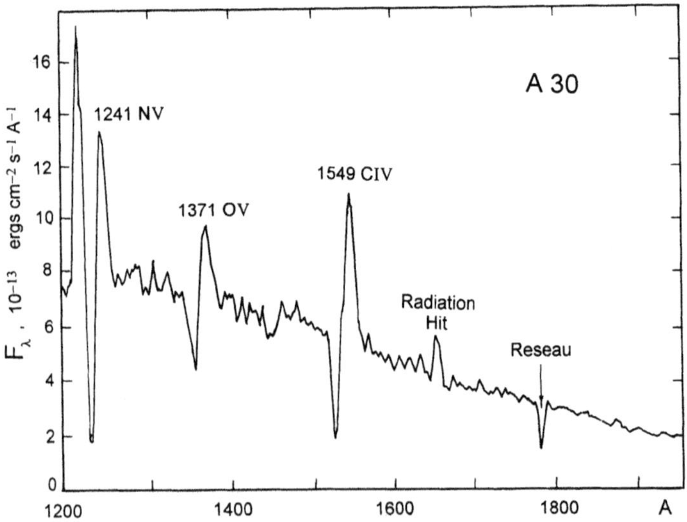

Fig. 15.6. A fragment from the *IUE* spectrum of the central star of the planetary nebula A 30. The profiles of all three lines, 1 241 NV, 1 371 OV and 1 549 CIV, are of P Cyg type

The number of recombination acts of free electrons with ions He^{++} per unit volume and per second will be equal to the total number of photons emitted by the nebula in the line 1 640 HeII

$$\frac{J(1\,640)}{h\nu_1} = n_e n\,(\mathrm{He}^{++})\,\alpha_{\mathrm{eff}}\,(1\,640\,\mathrm{HeII})\,, \tag{5}$$

where $h\nu_1$ is the energy of a photon at 1 640 Åand α_{eff} is the recombination coefficient depending on the optical depth of the nebula $\tau_c\,(\mathrm{He}^+)$ at the ionization frequency of He^+. In the case of optically thin nebulae we have $\alpha_{\mathrm{eff}} = \alpha_A\,(T_e)$, and, for optically thick nebulae, $\alpha_{\mathrm{eff}} = \alpha_B\,(T_e)$; these coefficients are as follows (in units of $10^{-12}\,\mathrm{cm}^3\,\mathrm{s}^{-1}$):

T_e, K	5 000	10 000	20 000	40 000
$\alpha_A(T_e)$	3.48	2.20	1.136	0.836
$\alpha_B(T_e)$	2.56	1.544	0.908	0.520

We have from the condition of ionization equilibrium

$$Z\,(\mathrm{He}^+) = n_e n\,(\mathrm{He}^{++})\,\alpha\,(\mathrm{He}^{++})\,, \tag{6}$$

where $Z(\mathrm{He}^+)$ is the rate of photoionization of He^+ and α (He^{++}) is the complete recombination coefficient for He^{++}. Setting here $n_e n(\mathrm{He}^{++})$ from (5) we get

$$Z(\mathrm{He}^+) = \frac{\alpha(\mathrm{He}^{++})}{\alpha_{\mathrm{eff}}(\mathrm{He}^{++})}\,\frac{J(1\,640\,\mathrm{HeII})}{h\nu_1}\,. \tag{7}$$

We can write for the mean energy $\bar{\varepsilon}$ of a photoelectron appearing as a result of the photoionization of helium

$$\overline{\varepsilon_{\mathrm{eff}}} = \frac{\sum F_i}{Z(\mathrm{He}^+)}\,, \tag{8}$$

where $\sum F_i$ is the summed energy of emission lines excited by electron collisions. From (7) and (8) we obtain, substituting also the value of recombination coefficients for the mean energy of free electrons $\overline{\varepsilon_{\mathrm{eff}}}$ (in eV),

$$\overline{\varepsilon_{\mathrm{obs}}} = 4.0\left(\frac{T_e}{10\,000}\right)^{-0.11}\frac{\sum F_i}{J(1\,640)}\,, \tag{9}$$

for optically thick nebulae, $\tau_c(\mathrm{He}^+) > 1$, and

$$\overline{\varepsilon_{\mathrm{obs}}} = 2.8\left(\frac{T_e}{10\,000}\right)^{-0.19}\frac{\sum F_i}{J(1\,640)}\,, \tag{10}$$

for optically thin nebulae, $\tau_c(\mathrm{He}^+) < 1$.

Thus, the mean energy of free electrons $\overline{\varepsilon_{\mathrm{obs}}}$ can be obtained without any difficulty direct from observations, through $\sum F_i$.

However, the mean energy of free electrons also depends on the temperature T_* of the central star, and if we can obtain this dependence, i.e. the function $\overline{\varepsilon} = \overline{\varepsilon(T_*)}$, then we can obtain the desired T_*.

Assuming that the photoionization of helium ions, He^+, is realized under the action of black-body radiation of the central star of effective temperature T_*, we can write, in analogy with (9) and (10), for the mean energy of a photoelectron that has escaped from He^+

$$\overline{\varepsilon(T_*)} = kT_* \left[\frac{\int_{x_0}^{\infty} (e^x - 1)^{-1}\, dx}{\int_{x_0}^{\infty} x(e^x - 1)^{-1}\, dx} - x_0 \right] \tag{11}$$

for $\tau_c(He^+) < 1$, and

$$\overline{\varepsilon(T_*)} = kT_* \left[\frac{\int_{x_0}^{\infty} x^3(e^x - 1)^{-1}\, dx}{\int_{x_0}^{\infty} x^2(e^x - 1)^{-1}\, dx} - x_0 \right] \tag{12}$$

for $\tau_c(He^+) > 1$.

In these expressions $h\nu_0 = x_0 kT_*$, and $x_0 = 631\,000/T_*$ corresponds to the ionization frequency of HeII. The dependence of $\overline{\varepsilon(T_*)}$ on T_* calculated via (11) and (12) is represented in Fig. 15.7 for both cases, $\tau_c(He^+) < 1$ and $\tau_c(He^+) > 1$.

Note that the $\overline{\varepsilon}$ of photoelectrons appearing as a result of the photoionization of hydrogen (Chap. 4), and of those appearing as a result of the photoionization of He^+ (Fig. 15.7), differ almost two-fold.

Now apply the expressions derived above to the nebula A 30. We have from observations

$$\frac{\sum F_i}{J(1\,640)} \approx 16\,,$$

which gives, at $T_e = 20\,000$ K, $\overline{\varepsilon} \approx 40$ eV or $\log \overline{\varepsilon} = 1.60$, for an optically thin nebula, and $\overline{\varepsilon} \approx 60$ eV or $\log \overline{\varepsilon} = 1.78$ for an optically thick one. This corresponds, as follows from Fig. 15.7, to $T_* = 380\,000$ K, in the first case, and $T_* = 880\,000$ K, in the second. At such high temperatures, one can hardly have a nebula that is optically thick at the frequencies of ionization of He^+. Therefore, the estimation $T_* = 880\,000$ K seems to be closer to reality.

Such exceptionally high temperatures were not discovered earlier. Presumably, this also characterizes the nebula A 30 as an extraordinary event.

Obviously, the unusual chemical composition of A 30 will be of particular interest from the point of view of stellar evolution and stellar cosmogony. Do we see here consequences of almost all of the hydrogen being transformed into helium?

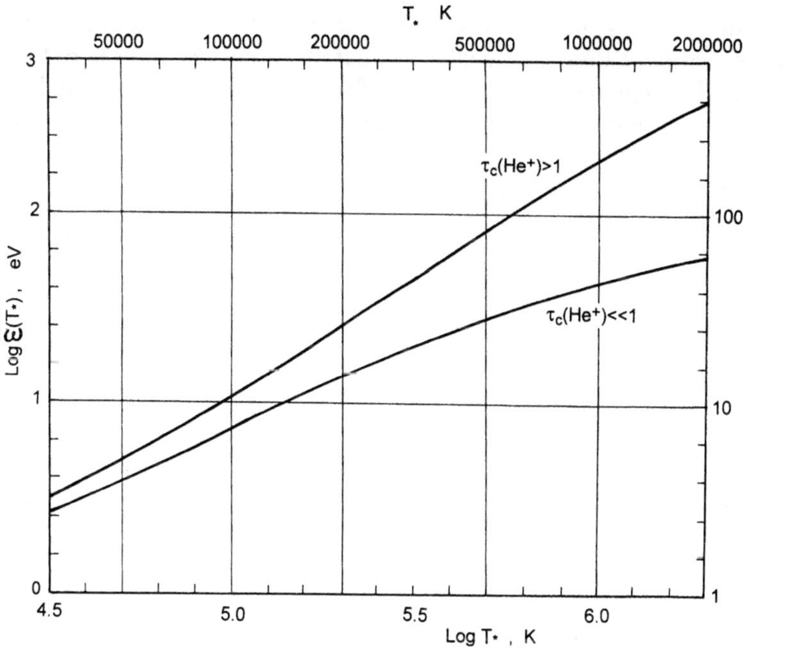

Fig. 15.7. Theoretical dependence of the mean energy of a photoelectron $\varepsilon(T_*)$ in the ionized helium medium on the central star's temperature T_* for the optically thin, $t_c(He^+) < 1$, and optically thick, $\tau_c(He^+) > 1$, nebula

The PN A 30 is not the only exception. With respect to its main properties 78 is rather similar to A 30. Thus, A 78 is also a two-envelope nebula with a diameter of outer weaker envelope of $\sim 130''$, the spectrum of its inner envelope repeats the spectrum of A 30 in detail, and again no trace of the hydrogen emission lines (Jacoby, Ford, 1983) are seen. The *IUE* spectrum of the central star of A 78 is also identical, i.e. the same lines 1 240 NV, 1 550 CIV and 1 360 OV with P Cyg type profiles indicating the mass outflow with a velocity of 3 400 km s^{-1} (Heap, 1979).

Both hydrogen-poor nebulae A 30 and A 78 have been observed with the help of the *the Hubble Space Telescope* (Borkowski et al., 1993b). On the high resolution images of these nebulae wind-blown cometary structures have been detected.

Another PN without hydrogen emission lines has been discovered in the Large Magellanic Cloud – LMC N26 (Boroson, Liebert, 1988). In the spectrum of this nebula all forbidden lines, as well as the line 4 686 HeII, are present, with a complete absence of hydrogen emission lines.

To this group of PNs should be added the highest excitation class, $p = 12^+$, for example, PN K 1-27 with a quite high intensity emission line 4 686 HeII, equal to 1.38 ($H_\beta = 1.00$), and with a hot hydrogen-deficient central star (Köppen, Werner, 1994).

Thus, there is sufficient reason to talk about the existence of a new independent class of PNs, i.e. *nebulae without hydrogen*. The objects of this class presumably represent the final phase of the evolution of nebulae and of their central star. Since the life-time of a helium burning star-nucleus cannot be large – less than 0.1 of the life of the nebula, i.e. 2 000–3 000 years – the total number of this class of objects cannot be large.

The main problems associated with nebulae without hydrogen are as follows:

- the combination of exceptionally high velocities of matter outflow with the low velocities of expansion of the nebula;
- the nature of formation of local condensations of gaseous matter;
- the origin of the second (outer) envelope;
- the role of shock waves in the dynamics of the nebula.

Infrared observations of this class of PNs (e.g. A 30 and A 78 are infrared sources) seem of special interest in revealing their morphological properties (Dinerstein, 1984).

15.5 Nebulae with Contrasting Nuclei

Apparently, one cannot attribute all nebulae that are large in absolute size to a single group to an identical evolutionary stage. For example, the PNs NGC 246 and NGC 3 587 are both the same size; the same holds for the temperatures of their nuclei which are higher than 100 000 K. However, the first one, NGC 246, is much more strongly excited, as follows already from the strength of the emission line 4 686 HeII: it is almost ten times stronger in NGC 246 compared with NGC 3 587. Moreover there are reasons to assume that NGC 246 is optically thick at the frequencies of the Lyman continuum of hydrogen, while NGC 3 587 is optically thin (Pottasch, 1980).

Meanwhile, the available data seem to indicate the existence of two types of objects assumed to be at the same stage of evolution because of their similarity in absolute size. Representatives of these types may be the PNs A 43 and A 50 (see Table 15.3) (Kaler, Hartkopf, 1981). A 43 is a high excitation nebula ($p = 12$), optically thin in the Lyman continuum, from the flat subsystem of the Galaxy. A 50 is a medium excitation optically thick nebula ($p = 9$), and belongs to the spherical subsystem – its distance from the Galactic plane is 1 200 pc. At the same time, both nebulae are equally large in absolute size – 0.41 pc and 0.27 pc – and located at distances from the Earth of 2 100 pc and 4 000 pc, respectively.

The attribution of the nebula to optically thin or thick types is established by the comparison of two temperatures of their nuclei obtained by two methods, those of hydrogen lines, $T_*(\mathrm{HI})$, and lines of ionized helium, $T_*(\mathrm{HeII})$. When

$$T_*(\mathrm{HI}) = T_*(\mathrm{HeII}), \tag{13}$$

Table 15.3. Basic parameters of the planetary nebulae A 43 and A 50

Parameter	A 43	A 50	Parameter	A 43	A 50
Excit. Class, p	12	9	T_*(HI), K	24 000	104 000
Distance, pc	2 100	4 000	T_*(HeII),K	68 000	125 000
Diameter, pc	0.41	0.27	$L_*(HI)/L_\odot$	73	55
z, pc	630	1 200	$L_*(HeII)/L_\odot$	1 020	91
n_e, cm^{-3}	30	90	$R_*/R_\odot$	0.23	0.020
T_e, K	14 000	10 000	τ_c	0.07	2.50
m_B	14.5	19.4			

then the nebula is optically thick at the frequencies of the Lyman continuum of hydrogen ($\tau_c \gg 1$). In the case where

$$T_*(\mathrm{HI}) < T_*(\mathrm{HeII}), \tag{14}$$

we have an optically thin nebula ($\tau_c \ll 1$).

Judging by the data in Table 15.3, condition (13) is fulfilled only for A 50, where both temperatures are of the same order, while for A 43 condition (14) is fulfilled, since T_*(HI) is almost three times smaller than T_*(HeII). The value of τ_c can be determined from known T_*(HI) and T_*(HeII) from the following expression:

$$\tau_c = 1 - \frac{J\left[T_*(\mathrm{HI})\right]}{J\left[T_*(\mathrm{HeII})\right]}, \tag{15}$$

where

$$J(T_*) = \int_{x_0}^{\infty} \frac{x^2\,dx}{e^x - 1}, \tag{16}$$

$$x_0 = 157\,000/T_* \,. \tag{17}$$

The obtained value of τ_c is given in the last line of Table 15.3.

Observations give for the intensities of the main emission lines of A 43 and A 50 (in units 4 861 H_β = 1.00),

	3 727 [OII]	4 363 [OIII]	4 686 HeII	4 959 [OIII]	6 563 H_α	4 340 H_δ	4 101 H_γ
A 43	–	0.13:	0.94	2.77	3.82	0.417	0.138
A 50	1.05	0.11	0.36	4.20	2.66	0.451	0.272

We see that the Balmer decrement H_α/H_β for A 50 is almost the same as the theoretical one (2.85); hence this nebula is not influenced by interstellar extinction, $c = 0$, while for A 43 $H_\alpha/H_\beta = 3.82$, i.e. $c = 0.38$. This discrepancy in c obviously indicates that these nebulae, though being at the same distances from us, belong to different Galactic populations.

The magnitudes of L_* of this nebulae differ 10 times, and this fact appears to be crucial for A 50. Indeed, the radius of the nucleus of A 50 is close to

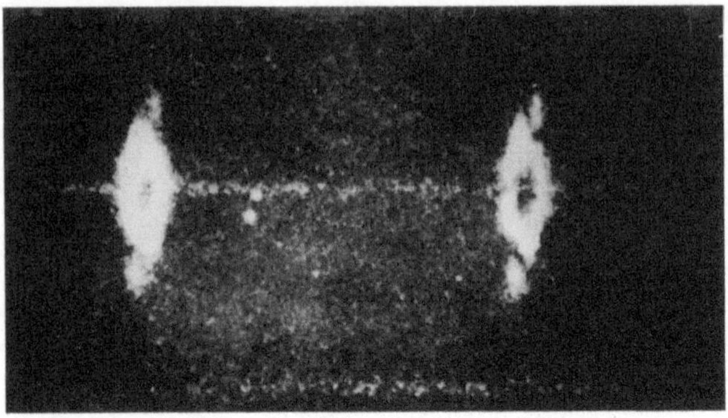

Fig. 15.8. Images of the emission lines H_α and 6 583 [NII] in the spectrum of the ring-like planetary nebula NGC 6 751. The main ellipsoidal form of the line corresponds to the space expansion of the nebula with a velocity of 40 km s^{-1}. Two condensations are situated symmetrically with respect to the nebula's axis and indicate bipolar outflow jets of velocity $\pm$ 30 km s^{-1}

that of a white dwarf ($R_*/R_\odot = 0.020$), almost an order smaller than the radius of the nucleus of A 43 ($R_*/R_\odot = 0.23$). Obviously, these nuclei are in different evolutionary states.

Thus, the examples of A 43 and A 50 demonstrate the principal differences existing between PNs with large absolute sizes.

15.6 Bipolar Outflow

Figure 15.8 shows a remarkable image of the emission lines H_α and 6 584 [NII] of the well known PN NGC 6 751 (Gieseking, Solf, 1986).

Note the two elongated condensations which are exposed only at certain positions along spectrograph's slit, at $\varphi = 100°$ in projection on the nebula's image; at six other positions these condensations are absent. Hence, these condensations should imply gaseous matter jets or streams.

Both condensations located symmetrically to the large axis of the ellipse are equally shifted in opposite directions; i.e. the two streams are directed opposite to each other.

The isolation of both condensations streams far outside the visible nebula of radius 10″is remarkable; the outer boundaries of the streams are located up to 16″from the nebula's centre. One could suppose that a bipolar outflow started at some period t_s, continued during Δt, and suddenly stopped at t_e. From the known linear size of the nebula, $R = 0.130$ pc (Daub, 1982), one can obtain t_s, t_e and $\Delta t = t_s - t_e$, at a constant velocity of outflow 30 km s^{-1}. The results are: $t_s \sim 7\,000$ years, $t_e \sim 3\,400$ years and $\Delta t \sim 3\,600$ years.

The phenomenon just described was first observed for the "Eskimo" PN, NGC 2 392 (Gieseking et al. 1985). For this object the velocity of symmetrical outflow is 185 km s^{-1}, i.e. three times higher than the expansion velocity of the nebula itself ($\sim$ 60 km s^{-1}). "Eskimo" ($p = 10$) is also a two-envelope PN, but both streams are located inside the outer envelope. The parameters of outflow are: $t_s = 1\,500$ years, $t_e = 400$ years and $\Delta t = 1\,100$ years. From [SII] observations, the electron concentration for these condensations was obtained: $\sim$ 200 cm^{-3}, an order smaller than in the main envelope ($\sim$ 2 000 cm^{-3}).

Among objects with bipolar outflow the "Saturn" PN, NGC 7 009 (Plate VIII), should be included as well. The difference from the two objects described above is in the spatial orientation of the outflow; in the case of NGC 2 392 the outflow takes place almost along the line of sight; for NGC 7 009, on the other hand, the outflow lies in the plane perpendicular to the line of sight. The direction of the outflow in NGC 6 751 has an intermediate orientation.

The phenomenon of bipolar outflow from the nuclei of PNs, obviously poses a number of unsolved problems. These concern the mechanism of outflow with velocities much less than those, for example, of stellar winds from the nuclei with velocities of 1 000–2 000 km s^{-1}, the stability of the condensations, etc. Indeed, a condensation, say, with $n_e \sim 1$ cm^{-3} should be dissolved in next 2 500 years; in fact, it will become unobservable much earlier. Hence, either we observe configurations that have just been born or there exist mechanisms preserving their properties.

15.7 Gaseous Envelopes with a Mixture of Dust

Planetary nebulae do contain dust particles (Chap. 10). Dust particles should be present also in the neutral outer envelopes of low excitation nebulae. In this connection, the radiative transfer problem through such a gas–dust medium arises in certain emission line. Below we shall confine ourselves to an examination of this problem with respect to the k and h MgII emission lines.

The behaviour of k and h Mg II photons in a pure gaseous medium without dust particles is analogous to that of L_α photons of hydrogen, i.e. multiple scatterings and subsequent escape from the nebula. Moreover, unlike L_α photons which can die after a certain number of scatterings via transformation to $2q$ photons, k and h MgII photons are forbidden such a fate theoretically.

On pure scattering in a medium free of dust particles, the depth r_0 in the centre of an absorption line is given by the expression (in the Schwarzschild–Schuster approximation)

$$r_0 = \frac{J(t_0)}{J(0)} = \frac{1}{1+t_0}\,, \qquad (18)$$

where t_0 is the effective optical depth in the centre of a given line.

The situation is different in a gas–dust medium. In this case the photons k and h MgII can die in collisions with the dust particles.

Thus, in a gas–dust medium MgII photons may simultaneously undergo two processes: pure scattering at Mg^+ ions and true absorption on dust particles. In this case the depth in the centre of the line is not described by (18). Now another relationship for the profile of the line is valid (Gurzadyan et al., 1991a):

$$\frac{J(t_\lambda)}{J(0)} = \frac{2\gamma\left[(1+\gamma)+(1-\gamma)\,e^{-2k(t_0-t_\lambda)}\right]\,e^{-kt_0}}{(1+\gamma)^2-(1-\gamma)^2\,e^{-2kt_0}}\,, \tag{19}$$

where

$$k = 2\sqrt{\eta\,(1-\eta)}\,, \tag{20}$$

$$\gamma = \sqrt{(\eta-1)/\eta}\,. \tag{21}$$

Here η is the "dustiness" factor,

$$\eta = 1 + \alpha/\sigma\,, \tag{22}$$

where α is the volume absorption coefficient on dust particles, σ is the volume scattering coefficient on Mg^+ ions, and t_0 is the full optical depth of the medium for the processes of pure scattering on Mg^+ ions. In deriving (19), $t=0$ in the inner boundary of the gas–dust envelope, and $t=t_0$ in the outer, is assumed.

Setting $t_\lambda = t_0$ in (19), we obtain for the relative flux of magnesium emission escaping from the nebula at the wavelength λ, or what is the same, for the spectral absorption line profile $r_{0\lambda}$ in the presence of both absorbed dust particles and scattering Mg^+ ions

$$r_{0\lambda} = \frac{J(t_0)}{J(0)} = \frac{4\gamma\,e^{-kt_{0\lambda}}}{(1+\gamma)^2-(1-\gamma)\,e^{-2kt_{0\lambda}}}\,, \tag{23}$$

where $t_{0\lambda}$ is now the full optical depth of the medium at the wavelength λ.

The value $\eta = 1$ corresponds to pure scattering ($\alpha = 0$). However, the transformation from (23) to (18) cannot be realized by merely setting $\eta = 0$, in view of the approximate character of (23).

The absorption of the line itself is determined by the damping process. Then, in view of (20), we have

$$kt_\lambda = \frac{2a}{\sqrt{\pi}}\,\frac{t_0}{x^2}\,\sqrt{\eta(\eta-1)}\,, \tag{24}$$

where t_λ is the total optical depth of the medium (envelope) in the centre of the line, $\lambda = \lambda_0$, and

$$x = (\lambda-\lambda_0)/\Delta\lambda_{\rm D}\,, \qquad \alpha = \Delta\lambda_{\rm E}/\Delta\lambda_{\rm D}\,,$$

where $\Delta\lambda_{\rm D} = \lambda(v/c)$ is the Doppler half-width of the line (~ 0.1 Å) caused by inner (thermal) motions of atoms, v is the thermal velocity of the gas and $\Delta\lambda_{\rm E}$ is the "natural" half-width of line, (~ 0.001 Å).

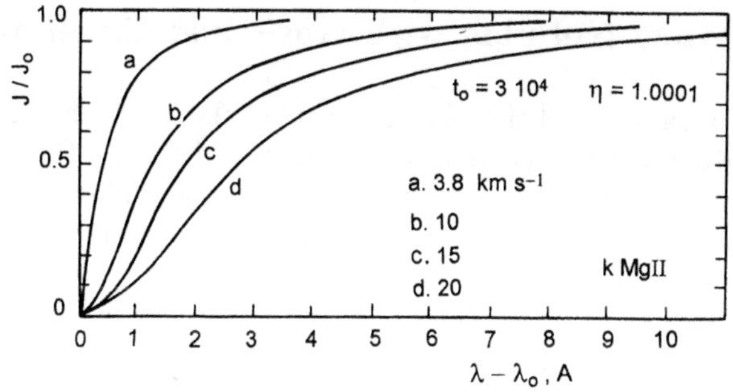

Fig. 15.9. Calculated profiles for the absorption line 2 795 *k* MgII for four values of the velocity of inner motions $v_0 = 3.8$, 10, 15 and 20 km s^{-1}

As an example, in Fig. 15.9 the profiles of *k* (*h*) MgII lines are shown, calculated using (23) and (24) with $a = 2.5 \times 10^{-2}$, $\eta = 1.0001$, $t_0 = 10^4$ and for four values of the velocity v: 3.8 km s^{-1} (corresponding to the thermal velocity of atoms at $T_e = 10\,000$ K), and 10, 15 and 20 km s^{-1}. By means of these and similar profiles, we calculate the equivalent widths $W(k$ MgII) for the line *k* (*h*) as a function of t_0,

$$W\,(k\,\mathrm{MgII}) = \frac{1}{J_0} \int_0^{\infty} J(t_\lambda)\,\mathrm{d}\lambda \tag{25}$$

The results are presented in Table 15.4 for three values of t_0: 10^4, 3×10^4 and 10×10^4, and for two values of the "dustiness" factor η: 1.0001 and 1.01. With the help of the curves $W(k$ MgII) $\sim t_0$ based on the data of Table 15.4, one can obtain t_0 numerically from the observed magnitude of $W(k$ MgII).

The results obtained in the present section, have been used to study the structure of the resonance doublet 2 800 MgII and of the resonance line 2 852 MgI in the spectrum of PNs possessing neutral gas-dust outer envelopes.

Table 15.4. Calculated equivalent widths $W(k$ MgII) (in Å) of the line *k* MgII for three values of the optical depth t_0: 10^4, 3×10^4 and 10×10^4, and four values of the inner velocity v_0: 3.8, 10, 15 and 20 km s^{-1}, and two values of the "dustiness" factor η: 1.0001 and 1.01

t_0, 10^4	v_0, km s^{-1}				t_0, 10^4	v_0, km s^{-1}			
	3.8	10	15	20		3.8	10	15	20
	$\eta = 1.0001$					$\eta = 1.01$			
1	0.80	2.42	3.52	4.80	1	0.37	0.95	1.32	1.85
3	1.48	3.84	5.22	7.92	3	1.16	3.10	5.58	6.16
10	2.90	7.15	9.64	16.1	10	3.88	10.2	12.9	18.3

15.8 Planetary Nebulae with Gas–Dust Envelopes

In Fig. 15.10 a fragment of the *IUE* spectrum of the low excitation class ($p = 3$) PN NGC 6 369 is given (Gurzadyan et al. 1991 b). Note the simultaneous presence of two groups of spectral lines – emission and absorption.

The strongest line in the first group is the doublet k and h MgII blended with a powerful 2 800 MgII line in emission. Its relative intensity $F^*(\mathrm{MgII})/F(\mathrm{H}_\beta) = 0.33$, corrected for the interstellar extinction, is rather high; it is equal to that of NGC 7 027, an object, however, of the high excitation class ($p = 10$).

In the second group, the resonance *absorption* line of neutral magnesium 2 852 MgI with an equivalent width of 6.6 Åis the strongest. The line 2 852 MgI has been detected in the spectrum of Cn 3-1 (see the next section) and with much lower intensity in IC 418 (Harrington et al., 1980), BD+30°3639 (Pwa et al., 1986) and LoTr 5 (Feibelman et al., 1988).

It is not difficult to ensure that such a powerful absorption line as 2 852 MgI cannot be caused within the nebula owing to self-absorption processes in the zone of the ionized magnesium Mg*. The only possibility is the weak emission in 2 852 MgI. Such powerful absorption may occur only in a medium where magnesium is almost completely in a neutral state. It cannot occur in the interstellar medium; its role in the absorption line 2 852 MgI might be less than 0.05 Å at the distance of this nebula of 0.42 kpc and at the specific equivalent width of the interstellar component of 0.1 Å per 1 kpc (Gurzadyan 1984). In our case this absorption line can originate only in the zone MgI immediately surrounding the nebula.

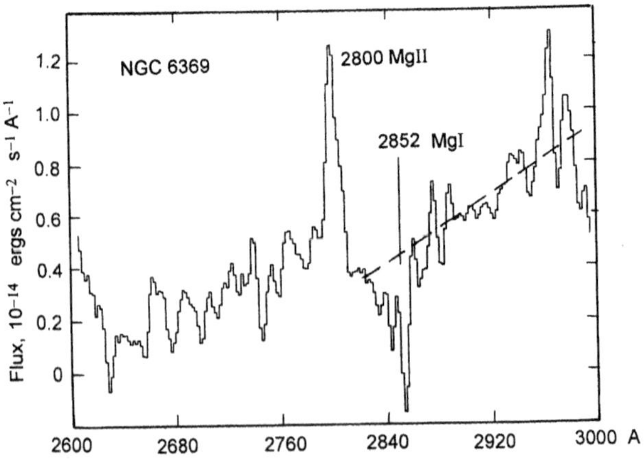

Fig. 15.10. A fragment from *IUE* ultraviolet spectra of the planetary nebula NGC 6 369 in the interval 2 600–3 000 Å. The powerful doublet 2 800 MgII in emission and 2 852 MgI in absorption is seen. The broken line indicates the adopted level of the continuous spectra

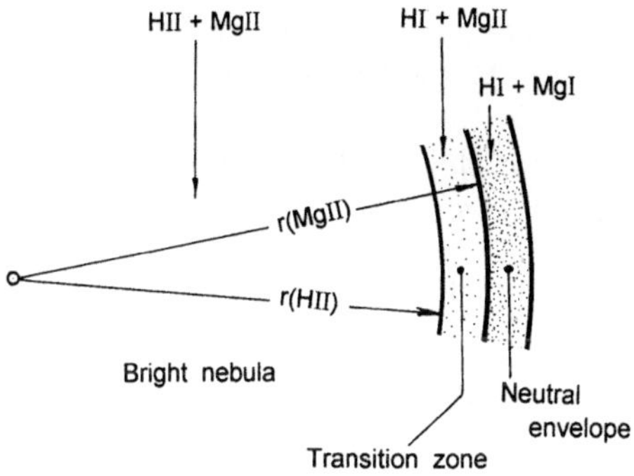

Fig. 15.11. The "three-envelope" structure of a planetary nebula

Thus, we arrive at the conclusion of the possible existence of a zone of neutral magnesium MgI around the luminous part of NGC 6 369.

On the other hand, it can be shown that the linear sizes of the ionization zone for Mg^+, r(MgII), are always larger than those of the ionized hydrogen, r(HII) (Gurzadyan, 1977). Therefore, we have the following three zone structure in the nebula NGC 6 369:

Visible part:	zone	HII + MgII
Transition part:	zone	HI + MgII
Neutral envelope:	zone	HI + MgI

In the first, central zone, the "visible nebula", both the hydrogen and the magnesium are ionized – HII + MgII; in the second, the "transition part", the hydrogen is neutral but the magnesium is ionized – HI + MgII, in the third, the "neutral envelope", both the hydrogen and the magnesium are neutral – HI + MgI. This stratification is shown schematically in Fig. 15.11.

The radius of the first zone is known, being the visible radius of the nebula, $r(\mathrm{HI}) = R_n$. The dimensions of the second and third zones have to be determined.

We start from the first zone. The energy E_{cal} (MgII) of emission in the doublet $k + h$ MgII, generated in the first zone, and falling onto the inner boundary of the transition zone, will be

$$E_{\mathrm{cal}}\,(\mathrm{MgII}) = \mathcal{E}\,(\mathrm{MgII})\,\mathrm{V\,ergs\,s^{-1}}\,, \tag{26}$$

where $\mathcal{E}$(MgII) is the emission per unit volume generated by electron collisions in the inner zone and V is the volume of the visible (ionized) part of the nebula.

For the magnesium emission coefficient we have

$$\begin{aligned}\mathcal{E}(\mathrm{MgII}) &= n_{\mathrm{e}} n(\mathrm{Mg}^+)\, b_{12}(T_{\mathrm{e}}) h\nu_{12} \\ &= 1.57 \times 10^{-20} n_{\mathrm{e}}^2 T_{\mathrm{e}}^{-1/2} \exp\left(-\frac{51\,400}{T_{\mathrm{e}}}\right) \mathrm{ergs\,cm^{-3}s^{-1}}\,.\end{aligned} \tag{27}$$

For NGC 6369 we have: $n_e = 7 \times 10^3\ \mathrm{cm}^{-3}$ and $T_e = 12\,500$ K. Using also $V \approx 4\,R_n^3$, at $R_n = 0.95 \times 10^{17}$ cm, we obtain: $E_{\mathrm{cal}}(\mathrm{MgII}) = 3.87 \times 10^{35}\ \mathrm{ergs\ s^{-1}}$.

On the other hand, for the observed magnesium emission corrected for interstellar extinction we have

$$E_{\mathrm{obs}}(\mathrm{MgII}) = 4\pi r^2 \frac{F^*(\mathrm{MgII})}{F(\mathrm{H}_\beta)} F(\mathrm{H}_\beta) \tag{28}$$

and for $r = 422$ pc, $F^*(\mathrm{MgII})/F(\mathrm{H}_\beta) = 0.33$ and $F(\mathrm{H}_\beta) = 5.0 \times 10^{-12}$ ergs $\mathrm{cm^{-2}\ s^{-1}}$, we obtain: $E_{\mathrm{obs}}(\mathrm{MgII}) = 3.50 \times 10^{31}\ \mathrm{ergs\ s^{-1}}$. Hence

$$\frac{E_{\mathrm{obs}}(\mathrm{MgII})}{E_{\mathrm{cal}}(\mathrm{MgII})} = 9 \times 10^{-5} \tag{29}$$

Thus, the magnesium emission reaching the observer is four orders of magnitude smaller than is actually generated in the central zone of the nebula. Such strong weakening can occur only in the transition zone.

The planetary nebula NGC 6369 is included in the *IRAS* catalogue, thus indicating the existence of *dust* particles which can absorb MgII photons.

The relative flux of the magnesium doublet emission from the "dusty" transition zone can be determined from (23), if we take into account that $2kt_0 \gg 1$ and η only slightly differs from unity. Then we have

$$\frac{E_{\mathrm{obs}}(\mathrm{MgII})}{E_{\mathrm{cal}}(\mathrm{MgII})} = \frac{J(t_0)}{J(0)} = 4\sqrt{\frac{\eta-1}{\eta}}\, e^{-2\sqrt{\eta(\eta-1)}t_0} . \tag{30}$$

The left hand side of this relationship is defined by (29). Then from (30) we determine, first, t_0 for several values of the "dustiness" factor η, after which we get for the column density of hydrogen atoms $N(\mathrm{H})$

$$N(\mathrm{H}) = 4 \times 10^4 \frac{t_0}{s_0} , \tag{31}$$

where we have taken, $\mathrm{Mg/H} = 2.5 \times 10^{-5}$. In (31) $s_0 = 0.95 \times 10^{-13}\ \mathrm{cm}^2$ is the coefficient of selective absorption of photons of k or h MgII per Mg^+ ion.

Once $N(\mathrm{H})$ is known, it is not difficult to obtain the relative linear thickness of transition zone, $\Delta\mathrm{H}$, from the relationship

$$\frac{\Delta \mathrm{H}}{\mathrm{H}} = \frac{N(\mathrm{H})}{n_{\mathrm{H}} R_n} . \tag{32}$$

Finally, for the mass of transition zone, $\mathcal{M}(\mathrm{HI})$, we have

$$\mathcal{M}(\mathrm{HI})/\mathcal{M}_\odot = 4\pi R_n^2 N(\mathrm{H}) m_{\mathrm{H}} = 3.6 \times 10^{-5} t_0 , \tag{33}$$

where m_{H} is the mass of the hydrogen atom. The numerical values of t_0, $N(\mathrm{H})$, $\Delta\mathrm{H}/R_n$ and $\mathcal{M}(\mathrm{HI})/\mathcal{M}_\odot$, obtained in the way described are given in Table 15.5 for a number of values of η and at $n_{\mathrm{H}} = 10^4\ \mathrm{cm}^{-3}$. In the last

Table 15.5. Calculated parameters of the transition zone around the visible part as a function of the NGC 6 369 in dependence of "dustiness" factor η ($n_e = 10^4$ cm^{-3})

η	t_0	$N(\mathrm{H})$ [cm^{-3}]	$\frac{\Delta \mathrm{H}_n}{R_n}$	$\frac{\mathcal{M}(\mathrm{HI})}{\mathcal{M}_\odot}$	t_d
1	10 000	4.2×10^{21}	4.4	2.35	0
1.0001	304	1.3×10^{20}	0.14	0.0012	0.030
1.001	114	4.8×10^{19}	0.050	0.0042	0.114
1.01	42	1.8×10^{19}	0.019	0.0016	0.42
1.10	15	6.3×10^{18}	0.0066	0.00056	1.50

column is given the optical depth t_d for the dust component of this zone, calculated according to the expression $t_d = (\eta - 1)t_0$.

Obviously, a transition zone with $\eta = 1$, i.e. without dust particles and pure scattering on Mg^+ ions, is also excluded because of the unrealistic mass of the transition zone – 10–100 times greater than the mass of normal PNs. At the same time, the existence even of a negligible fraction of dust particles, $\eta \sim 1.0001$, leads to quite acceptable parameters for the transition zone.

Note the relatively small linear depth of the transition zone – of the order of 0.01–0.001 of the nebula's radius (at $\eta < 1.001$). As a matter of fact, the transition zone in the nebula NGC 6 369 appears as a very thin formation, something like an egg-shell, enveloping the visible nebula.

Consider now the neutral zone, HI + MgI. Here both magnesium and hydrogen are neutral (Fig. 15.11). The neutral zone is located immediately after the transition zone and is responsible for the formation of *the absorption line 2 852 MgI.* The elementary process here is the selective scattering of photons of the continuous spectra of both the nebula and the central star, on the neutral magnesium atoms. No sources of emission exist in this neutral envelope.

One can determinate the optical depth $t(\mathrm{MgI})$ in the line 2 852 MgI using the following approximate expression

$$\frac{J}{J_0} = e^{-t(\mathrm{MgI})} , \tag{34}$$

where J_0 is the level of continuous spectra near 2 852 Åand J is the residual intensity in the centre of the absorption line 2 852 MgI. At $J/J_0 \sim 0.01$ (Fig. 15.11) (34) yields $t(\mathrm{MgI}) = 4.6$.

We also have $t(\mathrm{MgI}) = s_0 N(\mathrm{MgI})$, where $N(\mathrm{MgI})$ is the column concentration of neutral magnesium atoms in the neutral envelope and s_0 is the absorption coefficient in the centre of the line 2 852 MgI. For an oscillator strength $f = 1.902$ and the velocity of motions within the neutral envelope of 40 km s^{-1}, we have $s_0 = 2 \times 10^{-13}$ cm^2. Then we obtain $N(\text{MgI}) = 1.5 \times 10^{13}\,\text{cm}^{-2}$ or $N(\text{HI}) = 4 \times 10^4 N(\text{MgI}) = 6 \times 10^{17}\,\text{cm}^{-2}$, for the column concentration of hydrogen atoms in the neutral envelope.

Finally, we have for the mass of the neutral envelope: $\mathcal{M}(\text{HI}) = 4\pi R_0^2$ $N(\text{HI})\, m_\text{H}$, where R_0 is the radius of the neutral envelope Adopting $R_0 = R_n$, we have $\mathcal{M}(\text{HI}) = 5 \times 10^{-5} \mathcal{M}_\odot$. The mass of the ionized (luminous) part of this nebula is known more reliably: $\mathcal{M}(\text{HII})$–0.028 $\mathcal{M}_\odot$. Hence, the ratio of the two masses will be: $\mathcal{M}(\text{HI})/\mathcal{M}(\text{HII}) \approx 2 \times 10^{-3}$.

For a nebula like NGC 6 369, i.e. enveloped by a neutral zone, we can write

$$\tau_\text{c} = k_0\, n_1\, (\text{HI})\, R_n = 1\,, \tag{35}$$

the outer boundary of the visible nebula, where τ_c is the optical depth of the luminous part in the Lyman continuum of hydrogen, k_0 is the continuous absorption coefficient for one hydrogen atom, and $n_1(\text{H})$ is the concentration of neutral hydrogen atoms.

The next condition is the identification of the visible size of the nebula R_n with the size of the ionization zone r_0, i.e.

$$r_0 = R_n = 9.6 \times 10^{16}\ \text{cm}\,. \tag{36}$$

The radius of the hydrogen ionization zone r_0 can also be determined from the condition of equality between the number of ionization and recombinations

$$\int_{x_0}^{\infty} \frac{L_\nu}{h\nu}\, \mathrm{d}\nu = \frac{4\pi}{3}\, r_0^3 n^+ n_\text{e}\, C(T_\text{e})\,, \tag{37}$$

where $C(T_\text{e})$ is the recombination coefficient, and L_ν is the luminosity of the central star of radius R_* and effective temperature T_* at the frequency ν:

$$L_\nu = 4\pi\, R_*^2 B_\nu(T_*)\,. \tag{38}$$

From these relationships we obtain

$$r_0 = W\, \frac{24}{c^2} \left(\frac{kT_*}{h}\right)^3 \frac{1}{n_\text{e}^2\, C(T_\text{e})} \int_{x_0}^{\infty} \frac{x^2\, \mathrm{d}x}{e^x - 1}\,, \tag{39}$$

where W is the dilution coefficient, and $x_0 = h\nu_0/kT_*$.

Combining (39) and (36), we obtain $W = 5.4 \times 10^{-14}$ which gives for $n_\text{e} \simeq 10^4$ cm^{-3} for the star's radius $R_* = 2R_n\, W^{1/2} = 0.63\, R_\odot$, in agreement with the model computations for the PN nuclei.

Combination the first condition in (35) with the ionization formula (4.7) for hydrogen leads to the following expression:

$$T_*\, \exp\left(-\frac{158\,000}{T_*}\right) = 1.18 \times 10^{-35}\, \frac{R_n n_\text{e}^2}{W}\,. \tag{40}$$

Setting the values of R, n and W, we obtain for the temperature of the central star of the nebula NGC 6 369 $T = 50\,000$ K, a reasonable quantity for a nebula of excitation class $p = 3$. This result is remarkable because the temperature of the nucleus of this nebula, unknown before, is also determined.

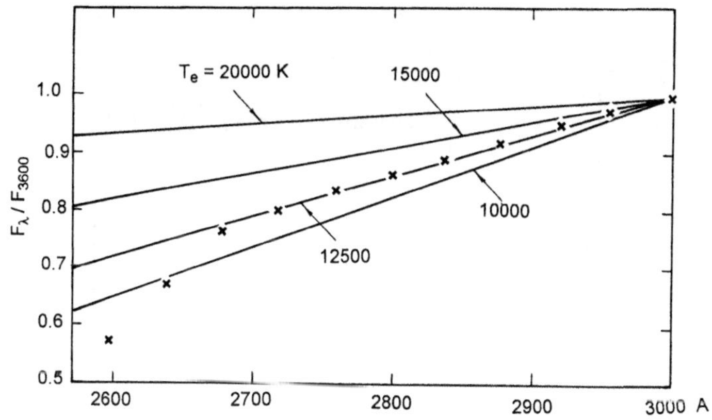

Fig. 15.12. The interpretation of the continuous spectra of NGC 6 369 in the region 2 600–3 000 Å. Solid lines are theoretical models for a line of electron temperature T_e; crosses indicate observations corrected for interstellar extinction

15.9 A Method for Determining the Electron Temperature

In Fig. 15.10, the continuous spectrum of the nebula NGC 6 369 is given with a visible excess in intensity at 3 000 Å and monotone fall in the direction of short wavelengths up to 2 000 Å. The following processes might contribute to this continuous spectrum: (a) the Balmer continuum of hydrogen; (b) the continuum of neutral helium; (c) the continuum of ionized helium; (d) the two-photon emission of hydrogen.

Processes (b) and (c) are excluded because of the absence of helium emission lines. Two-photon emission, though possible in principle, is hardly the case owing to the absence of signs of the continuous background near 2 000 Å. Thus, the continuous spectrum in Fig. 15.10 is almost represents the Balmer continuum. Figure 15.12 shows the calculated curves are for four values of the electron temperature T_e – 10 000, 12 500, 15 000 and 20 000 K (Chap. 4), along with observations (crosses) at 2 600–3 000 Å. We see that the observations favour the curve for T_e = 12 500 K.

As a matter of fact, this is the first attempt to determine the electron temperature of a nebula from the Balmer continuum in the ultraviolet, shorter than 3 000 Å.

15.10 Planetary Nebulae with Absorption Lines of MgII and MgI

Low excitation PNs have not been studied frequently, and can still lead to a number of surprises. This concerns, in particular, the star-like PN Cn 3-1

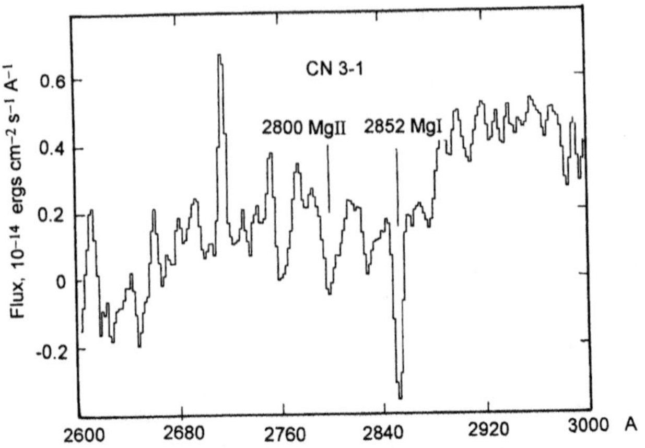

Fig. 15.13. A fragment of an *IUE* recording of the ultraviolet spectrum for Cn 3-1 for the region 2600–3000 Å. Both the resonance doublet 2800 MgII and the resonance line 2852 MgI are in absorption

of the lowest excitation class $p = 1$. Its *IUE* spectrum, given in Fig. 15.13, shows that both, the resonance doublet 2800 MgII and the resonance line 2852 MgI are strong in absorption (Gurzadyan et al., 1991).

Cn 3-1 reveals at least three anomalies.

The first follows from its visible spectrum. This contains the emission line 4686 HeII, though this is of small relative intensity 4686 HeII/$H_\beta = 0.0775$, but is nevertheless inconsistent with the excitation class of the PN. Indeed, recall Fig. 5.3 representing the empirical relationship established for typical PNs between the ratios $F(4686\ \mathrm{HeII})/F(H_\beta)$ and $F(N_1 + N_2)/F(4686\ \mathrm{HeII})$, while the position of Cn 3-1 in this diagram with log 4686 HeII/$H_\beta = -1.10$ and $\log(N_1 + N_2)/4686\ \mathrm{HeII} = 0.47$ lies far outside the zone of typical PNs.

This crucial fact indicates that the emission line 4686 HeII has no relation to the nebula and, therefore, has to be associated with its nucleus, according to some data, which is a Wolf–Rayet or Of type star. As indirect evidence, we can quote the absence of forbidden lines 3868 [NeIII] and 3426 [NeV] in the spectra of Cn 3-1 which usually accompany the line 4686 HeII.

The second anomaly concerns the temperature of the nucleus of this nebula. The temperature obtained via the ratio of 4686 HeII/H_β according to (5.18), yields 60000 K. However, the application of the [OIII]/[OII] method, leads to quite different result: $T_* = 26000$ K, in accordance with alternative estimates of 25100 K (Kaler, 1986) and 31600 K (Pottasch, 1983). The latter estimates $T_* \sim 30000$ K seem convincing, given the extremely faint lines $N_1 + N_2$ compared with 3727 [OII]. At the same time, note that helium lines cannot be excited at temperatures $T_* \sim 30000$ K.

Thus, we have two temperatures for the nucleus of this nebula – ~ 60000 K and ~ 30000 K. Which one is the true one? The peculiarity of the situation is that both determinations can be true.

We have two possibilities:

(a) The nucleus of Cn 3-1 is a binary system, one of the components of which is a WR star of effective temperature 60 000 K, and the second a star of $\sim$ B0 type of effective temperature $\sim$ 300 000 K. Presumably the gaseous envelope of the WR star is completely opaque for the continuum of He^{++}, O^{++}, etc. As a result we observe the WR star with its usual set of emission lines. The nebular emission is completely due to the ionizing radiation of the second component of the system, the B0 star.

(b) The nucleus of Cn 3-1 is a single star, the same WR star as above; however, its gaseous envelope is completely opaque for He^{++} the continuum and partly opaque in continuum of H^+ and O^+. Then He^{++} might be formed in the envelope of the central WR star, while the lines $N_1 + N_2$ [OIII], 3 727 [OII], hydrogen emission lines, etc., might be formed in the nebula. In this case, the amount of WR star radiation in the O^{++} continuum reaching the nebula, should be equivalent to radiation from an imaginary star of effective temperature $\sim$ 30 000 K.

It can be shown by direct calculations that in the latter case the contribution of both sources in H_β emission, of the WR star and the nebula, should be equivalent.

Then, (i) the emission lines of helium are due to the WR central star and are formed in its envelope, (ii) the forbidden lines N_1, N_2 [OIII] and 3 727 [OII] belong to the nebula, and (iii) the hydrogen lines are excited both in the envelope of the WR star and in the nebula.

The third anomaly is related to the absorption lines 2 800 MgII and 2 852 MgI. Both lines are visible on the background of some continuous radiation which obviously does not belong to the central star and most likely has to be identified with the Balmer continuum of the nebula; without such a background no absorption lines can be observed.

The problem is to understand where and how these absorption lines can be formed. Clearly, to create such lines, it is necessary to have a rich medium of Mg^+ ions in the first case, and of neutral atoms of magnesium in the second. One can imagine three possibilities : (a) the interstellar medium; (b) interstellar clouds; (c) envelopes of Mg^+ and Mg^o around the visible part of the nebula.

The first two possibilities, (a) and (b), should be excluded owing to unreasonably high densities of neutral hydrogen atoms – tens and hundreds of atoms per cm^3 required for the explanation of observations.

The only possibility seems to be to suppose the existence of an envelope, i.e. of a transition zone, similar to NGC 6 369, consisting of Mg^+ ions and neutral hydrogen atoms surrounding the luminous part of the nebula. Then the visible part of the nebula should coincide with the hydrogen ionizing zone; on its boundary the condition $\tau_c = 1$ should be fulfilled, while a B0 star of $T_* = 30\,000$ K should act as the source of ionization. Then identifying the left hand side of (39) with the radius of the visible nebula, $r(\text{HII}) = R_n =$

1.17×10^{17} cm and $n_e = 7.2 \times 10^3$ cm^{-3}, $C(T_e) = 5.4 \times 10^{-13}$ cm^3 s^{-1} at $T_e = 7\,500$ K, we obtain $W = 5.2 \times 10^{-12}$, and have for the star's radius $R_* = 5.8\,R_\odot$, which is in good agreement with the radii of B0 class stars.

Above we have found that $W(k$ MgII$) = 5$ Å, which in our model corresponds to an internal velocity V greatly exceeding 5 km s^{-1}. However, we are looking for a model with the least mass of the envelope HI + MgII, which is possible only at large velocities. For example, for $V = 20$ km s^{-1}, we find from the last curve on the right in Fig. 15.9 that $t_0 \approx 10^4$. The remaining parameters will be $N(\mathrm{Mg}^+) = t_0/s_0 = 10^{17}$ cm^{-2} or $N(\mathrm{HI}) = 4 \times 10^{21}$ cm^{-2} at $s_0 = 1.0 \times 10^{-13}$ cm^2. For the mass of the transition zone, i.e. of the gas–dust envelope, HI + MgII, we obtain $\sim 0.4\mathcal{M}_\odot$. This result may seem too large. Note, however, that for another low excitation PN, IC 418 ($p = 1$), 21 cm radio observations yield for the mass of the circumstellar neutral hydrogen envelope 0.35 $\mathcal{M}_\odot$ (see Chap. 14).

The model of a third–neutral envelope, HI + MgI, is similar to that described above for NGC 6 369, when we had a completely neutral envelope of neutral hydrogen and neutral magnesium in the presence of dust particles.

Note that this model requires high enough velocities of internal motions – 20 km s^{-1} – as a necessary condition for the existence of both the transition zone and the neutral envelope. The problem of the nature and character of these motions and, especially, of the source of their energy, therefore arises.

PNs with neutral envelopes are commoner events than it may seem at first glance. Indeed, neutral envelopes were discovered: with neutral hydrogen, HI, in ten PNs (from 21 cm radio line observations), with H_2 molecules in over thirty PNs (2.1 μm infrared observations), with CO molecules – over forty PN (1.3 mm and 2.6 mm observations), etc. (Taylor et al., 1990; Gussie, 1992; Zuckerman et al., 1990; Huggins et al., 1992; Graham et al., 1993; Kastner et al., 1994; Schild, 1995). The observations reveal that many PNs are surrounded by envelopes of neutral gas with masses even exceeding those of the ionized nebula. They also provide valuable information on the physical properties of the envelopes, their structure, and kinematics. The neutral envelopes firmly link the formation of PNs with the mass loss by AGB stars (see, the Chap. 18), and can play an essential role in the subsequent evolution of PN (Huggins, 1993).

15.11 A "Cross-Like" Planetary Nebula

Plate XIV shows the cross-like shape of the PN Wray 17-1. Its spectrum reveals the emission line 4 686 HeII of too high intensity, 4 686 HeII/H_β = 0.76, along with $N_1/H_\beta = 12.75$, which determines that it belongs to a high excitation class, $p = 10$.

This peculiar object, obviously bipolar, has two bright condensations on a single axis as is typical for bipolar nebulae. Another fainter pair of condensations exists on an axis perpendicular to that of the first pair. Also, a narrow arc indicating a fragment of an outer, second envelope is visible.

The formation of a bipolar structure as well as of the outer envelope takes place, presumably, according to the scheme of origin of two-envelope and bipolar nebulae (Chaps. 13 and 17). As for the second, polar, pair of condensations, the situation here is different.

One thought is that the origin of the polar pair of condensations might be connected with the outflow processes occurring from the polar regions of the central star – of its nucleus. Usually the magnetic axis of a star forms an angle with the rotation axis. If this is so, then the gaseous stream should occupy a conical angle between both magnetic and rotational axes. As a result, one should observe two conical formations located symmetrically with respect the to star. Judging from the absence of the visible contact of the conical configurations with the central star, the outflow process should have ceased a few thousand years ago.

This object, of course, requires more detailed study, particularly via spectroscopic observations, to decipher its real nature.

16. Dynamics of Planetary Nebulae

16.1 Light Pressure in Nebulae

The high density of radiation in the L_α line of hydrogen should be evaluated as the most characteristic property of a PN. This is one of the basic conclusions brought out by the radiative transfer theory for PNs (Chap. 11). As for the flux of $L\alpha$ radiation, it is different in various layers of the nebula. In this connection, the problem of the role of light pressure from L_α radiation in the dynamics of PNs, arises.

The magnitude and character of the flux distribution within the nebula depend strongly on what kinematics of the nebula are under consideration. In a static nebula the flux of L_α radiation is directed outward and increases from the inner to the outer parts of the nebula. In this case the L_α flux at the inner boundary is zero and at the outer boundary equals $+\pi S$ (Fig. 11.1). If the nebula expands with a velocity gradient, the flux at the inner boundary is directed inward and is equal to $-\pi Sq/\beta^{1/2}$, while at the outer boundary it is directed outward, being approximately $\pi Sq/\beta^{1/2}$. Since the factor $q/\beta^{1/2}$ is of the order of 10^{-2}, the flux at the outer boundary of the nebula must be about one hundred times smaller than in the static case. Nevertheless, in both cases the L_α flux at the outer boundary is quite large and consequently one must expect that the radiation pressure caused by L_α radiation will also be large. The problem is, by how much will the radiation pressure exceed the gravitational attraction of the central star?

Consider the ratio of L_α radiation pressure to the force of gravity in the case of a static nebula, within a unit volume near its outer boundary. Obviously, only atoms absorbing L_α photons, i.e. only of the ground state of concentration n_1, will directly experience the force of the L_α radiation. Subsequently this force is distributed by collisions and absorption among all particles.

The radiation pressure R due to L_α radiation acting on unit volume will be equal to

$$R = \frac{n_1\, k_\alpha}{c}\, \pi S\, h\nu_c\,, \qquad (1)$$

where $\pi\, h\nu_c$ is the intensity of L_c radiation onto 1 cm^2 of the inner boundary of the nebula and k_α is the coefficient of selective absorption in the line L_α per neutral hydrogen atom.

For the total number of L_c photons, πS, at the inner boundary of the nebula we have

$$4\pi r_n^2\,\pi S = 4\pi\,r_*^2\,\frac{2\pi}{c^2}\int_{\nu_0}^{\infty}\frac{\nu^2\,d\nu}{e^{h\nu/kT_*}-1}\,. \tag{2}$$

Consequently we obtain for R

$$R = \left(\frac{r_*}{r_n}\right)^2 \frac{2\pi\,n_1 k_\alpha}{c^3}\,h\nu_c \int_{\nu_0}^{\infty}\frac{\nu^2\,d\nu}{e^{h\nu/kT_*}-1}\,, \tag{3}$$

where r_* and r_n are the radii of the central star and of the nebula, respectively.

As distinct from the radiation pressure, which acts only on neutral atoms n_1, the gravitational force G of the central star acts on both neutral and ionized atoms, the concentration of which is n^+. Consequently we have for the gravitational force

$$G = \left(\frac{r_*}{r_n}\right)^2 g_*(n_1 + n^+)\,m_{\rm H}\,, \tag{4}$$

where g_* is the acceleration of the gravitational force on the surface of the star and $m_{\rm H}$ is the mass of hydrogen atom.

From (3) and (4) we have

$$\frac{R}{G} = \frac{2\pi K_\alpha\,h\nu_c}{m_{\rm H}c^3(1+n^+/n_1)g_8}\int_{\nu_0}^{\infty}\frac{\nu^2\,d\nu}{e^{h\nu/kT_*}-1}\,. \tag{5}$$

In such a representation the ratio R/G is independent of both radii r_* and r_n.

The value of k_α is determined by the Doppler half-width of the L_α line for thermal velocities of the order of 10 km s^{-1}. For the majority of PNs we have $T_* > 50\,000$ K up to $100\,000$ K and higher (see Table 5.4). Adopting $T_* = 70\,000$ K and $n^+/n_1 = 5\,000$, we find from (5) that

$$\frac{R}{G} = \frac{10^{10}}{g_*}\,. \tag{6}$$

It is hard to believe that the gravity of the nuclei of PNs can reach 10^{10} cm s^{-2}; if it did not, their masses should exceed the solar mass by over thousand times. Hence, undoubtedly for the nuclei of PNs we have $g_* \ll 10^{10}$ cm s^{-2}, so that $g_* \gg 1$. Thus, we conclude that the outer layers of PNs cannot be in static equilibrium and have to expand under the action of radiation pressure.

This result, however, concerns a stationary, motionless nebula. For a real nebula, expanding with a velocity gradient, the L_α flux at its outer boundary is two orders of magnitude less than for a static nebula. Therefore, (6) in this case will take the form

$$\frac{R}{G} = \frac{10^8}{g_*} . \tag{7}$$

A surface gravity of 10^8 cm s^{-2} for the nucleus, though theoretically possible, is improbable. In any case, in a real nebula expanding with a velocity gradient, the L_α radiation pressure no longer has the major role that it does for a motionless nebula.

The undoubted observational fact that PNs do expand with velocities noticeably exceeding the thermal velocity of atoms does not contradict the idea that radiation pressure plays a vital role in them. However, the initial force, leading to the evolution, expansion and separation of the outer layers of the central star and to the formation of the early phase of the nebula, may well be of a different nature.

In the analysis of the L_α field within a PN (Chap. 11), it was assumed that the parameter β is constant throughout the nebula. The constancy of β means the constancy of the degree of ionization. However, the latter not only varies within a nebula and at its boundaries, but also changes as the nebula expands. As a result, we shall have different values for R/G for various periods of evolution of the nebula. If this is the case, then one can expect a definite period in the nebula's life when the role of light pressure is the greatest. This stage can be associated with the appearance of the second (outer) shell around the primary one (Chap. 13).

16.2 Expansion of Nebulae: Theory

In the preceding section we saw that PNs cannot remain in a state of mechanical equilibrium and have to expand under the influence of L_α radiation pressure. This conclusion, although still qualitative, is in agreement with observations: the measured expansion velocities of nebulae – of the order of a few tens of km s^{-1} – considerably exceed, by a few orders of magnitude, the escape (parabolic) velocity of the central star at the distances of the order of a nebula's size.

Obviously, to estimate the real effect a detailed quantitative analysis is required. Besides this, the nebula will also experience the pressure of direct L_c radiation from the central star. Therefore, one has to discover the separate contributions of both L_c and L_α radiation..

Expansion under the Action of L_c radiation. Recall that L_c radiation is that in the Lyman continuum of hydrogen, i.e. in the region shorter than 912 Å. Under the action of direct L_c radiation from the central star, the nebula gains some momentum directed outward and has to increase its expansion velocity. The permanently acting diffuse L_c radiation of the central star will lead an accelerated expansion of the nebula. Our problem is to analyse this process theoretically and to compare the results with observations.

If E_c is the luminosity of the nucleus in the Lyman continuum, then for black-body radiation of an effective temperature T_*, we will have

$$E_c = 8\pi^2 r_*^2 T_*^4 \frac{k^4}{c^2 h^3} \int_{x_0}^{\infty} \frac{x^3\, dx}{e^x - 1}, \tag{8}$$

where $x_0 = h\nu_0/kT_*$, and ν_0 is the frequency of the Lyman boundary.

From this amount of L_c radiation, a fraction $E_c(1 - e^{-\tau_c})$ should be absorbed, where τ_c is the optical depth of the nebula at frequencies of the L_c radiation. The momentum arising as a result of absorption of that fraction of L_c radiation, will pass to the nebula envelope of mass $\mathcal{M}$. Therefore we can write for the equation of motion of such a nebula envelope

$$\mathcal{M}\frac{d^2 r}{dt^2} = \mathcal{M}\frac{dV}{dt} = \frac{E_c}{c}\left(1 - e^{-\tau_c}\right), \tag{9}$$

where dV/dt is the acceleration of the nebula envelope. So long as $dV/dt = V dV/dr$, we will have instead of (9)

$$V\frac{dV}{dR} = \frac{E_c}{c\mathcal{M}}\left(1 - e^{-\tau_c}\right), \tag{10}$$

where R is the nebula's radius at a given moment of its life.

Assuming a constant degree of ionization n^+/n_1 of hydrogen within the nebula, we will have for the optical depth τ_c

$$\tau_c = \int \kappa n_1\, dR = \frac{1}{4\pi}\frac{\kappa\mathcal{M}}{R^2 m_H}\frac{n^+}{n_1}, \tag{11}$$

where κ is the coefficient of continuous absorption per neutral hydrogen atom at the boundary frequency ν_0 of the Lyman series and m_H is the mass of hydrogen atom.

For the overwhelming majority of PNs, $\tau_c < 1$. Then, solving (10), we find for the dependence of the expansion velocity $V_c(R)$ on the nebula's radius R (Vitiasev, 1969)

$$V_c(R) = \left[V_0^2 + f\sqrt{\tau_c}\left(1 - \frac{R_0}{R}\right)\right]^{1/2}, \tag{12}$$

where V_0 is the velocity of expansion at the nebula's radius $R = R_0$, and f denotes

$$f = \frac{E_c}{c}\left(\frac{\kappa}{\pi m_H}\frac{n_1}{n^+}\right)^{1/2}\mathcal{M}^{-1/2}. \tag{13}$$

For a nebula's mass $\mathcal{M} = 0.01\mathcal{M}_\odot$ and $n^+/n_1 = 10^4$, we have from (13)

$$f = 0.158 \times 10^{-15} E_c, \tag{14}$$

i.e. f directly depends on E_c – on the L_c luminosity of the nucleus.

Equation (12) gives the sought for dependence between the expansion velocity $V(R)$ of the nebula and its radius R. This velocity, quite quickly, even at a radius $R/R_0 \sim 10$, approaches its largest value $V(\infty)$, given by the expression

$$V_c(\infty) = \sqrt{V_0^2 + f\sqrt{\tau_c}}\ . \tag{15}$$

The appearance of a limiting velocity is obvious: with the expansion of a nebula its optical depth τ_c decreases, and hence, the number of absorbed L_c photons drops sharply, as does the total momentum gained by the nebula from L_c radiation.

It is worth paying attention to the structure of (12) and (15). The second term under the square root is the squared additional velocity ΔV_c gained by the nebula as a result of the action of L_c radiation

$$\Delta V_c^2(R) = f\sqrt{\tau_c}\left(1 - \frac{R_0}{R}\right), \tag{16}$$

or for the limiting additional velocity $(R/R_0 \to \infty)$,

$$\Delta V_c(\infty) = (f\sqrt{\tau_c})^{1/2}\ . \tag{17}$$

The magnitude of $\Delta V_c(\infty)$ strongly depends on the nucleus temperature T_*, as follows from the results given below (at $r_* = 3 \times 10^{10}$ cm, $\kappa = 6.3 \times 10^{-18}$ cm^2):

T_*, K	50 000	70 000	100 000	120 000
$\Delta V_c(\infty)$, km s^{-1}	3.0	7.0	15.3	26.4

These results, as we can see, are comparable with the observed expansion velocities of PN and, hence, there is sense in insisting on the role of L_c radiation in the accelerating expansion of nebulae.

As for the role of the nebula's mass M on the magnitude of $\Delta V_c(\infty)$, it is not strong as follows from (17) and (13): $V_c \sim f^{1/2} \sim \mathcal{M}^{-1/4}$ when $V_0 = 0$. In particular, with a ten-fold increase of the nebula's mass, i.e. when $\mathcal{M} = 0.1\mathcal{M}_\odot$, $V_c(\infty)$ decreases only 1.8 times compared with its magnitude at $\mathcal{M} = 0.1\mathcal{M}_\odot$.

However, the most important result seems to be the dependence of $V_c(R)$ on the initial velocity V_0. In particular, we will have a sufficiently high velocity of expansion $2V_c$, up to 40–60 km s^{-1}, even in the case when the initial velocity is zero. In this case we will have infinite number of curves $V_c(R)$ depending on the size (radius) of the nebula R and initial velocity $2V_0$. In Table 16.1 the magnitudes of $V_c(R)$ are presented as a function of R/R_0, computed by means of (12) at nucleus temperatures T_* of 50 000, 70 000, 100 000 and 130 000 K and three values of initial velocity V_0: 0, 10 and 20 km s^{-1}. These results, however, are at $\tau_c = 0.5$. Figure 16.1 also shows the macrostructure of the dependence of $V_0(R)$ on R.

Table 16.1. Computed magnitudes of the expansion velocity $V_c(R/R_*)$ under the action of the L_c radiation of the nucleus of the nebula, as a function of the nebula's radius R/R_* for three values of the nucleus temperature T_*, 50 000, 70 000 and 130 000 K, and three values of the initial velocity V_0, 0, 10 and 20 km s^{-1}

$\frac{R}{R_0}$	50 000 K			70 000 K			130 000 K		
	0	10	20	0	10	20	0	10	20
1	0	10	20	0	10	20	0	10	20
2	2.1	10.2	20.1	4.9	11.1	20.6	18.6	21.2	27.3
4	2.6	10.3	20.2	6.1	11.7	20.9	22.8	24.9	30.4
6	2.8	10.4	20.2	6.4	11.9	21.0	24.1	26.1	31.3
∞	3.0	10.4	20.3	7.0	12.2	21.2	26.4	28.2	33.1

Each curve in Fig. 16.1 corresponds to a definite temperature of the nucleus, assumed to be constant during the life of the nebula. Meanwhile, there are serious indications that the nucleus temperature increases with the evolution of the nebula (Chap. 18). If this fact is taken into account, i.e. when T_* can vary, say, from 50 000 K to 130 000 K, there will be an increase in the rate of expansion velocities, as follows from Fig. 16.1.

Thus we arrive at an unexpected conclusion: the velocity field $V(R)$, should be covered by a dense set of curves $V_c(R) \sim (R, V_0, \tau_c, \mathcal{M}, r_*, n^+/n_1)$, including all possible combinations of V_0 and T_*. In other words, though the curves $V_c(R) \sim R$ grow in the direction of increasing size (radii) for each nebula, one cannot expect to pick out a single curve in the relationship $V_c(R) \sim R$.

Up until now we were considering optically thin nebulae, i.e. when $\tau_c < 1$. In the case when $\tau_c > 1$, i.e. when the nebula absorbs all L_c radiation emitted by the central star, we will have for the determination of the expansion velocity V, instead of (10)

$$\frac{\mathrm{d}V^2}{\mathrm{d}R} = \frac{2\,E_c}{c\,M} \,. \tag{18}$$

The increase of the expansion velocity will be

$$V^2 - V_0^2 = R\frac{2\,E_c}{c\,M}\left(1 - \frac{R}{R_0}\right) . \tag{19}$$

Calculations carried out with the help of this relationship give the growth of the expansion velocity to be about 10 km s^{-1}, in addition to the initial velocity when the nebula's radius approaches 10 000 a.u.

Another treatment of the dynamical role of L_c radiation in the nebula's expansion is based on the equality of the number of acts of ionization of hydrogen with the number of acts of recombination of electrons with protons per unit volume and time, i.e.

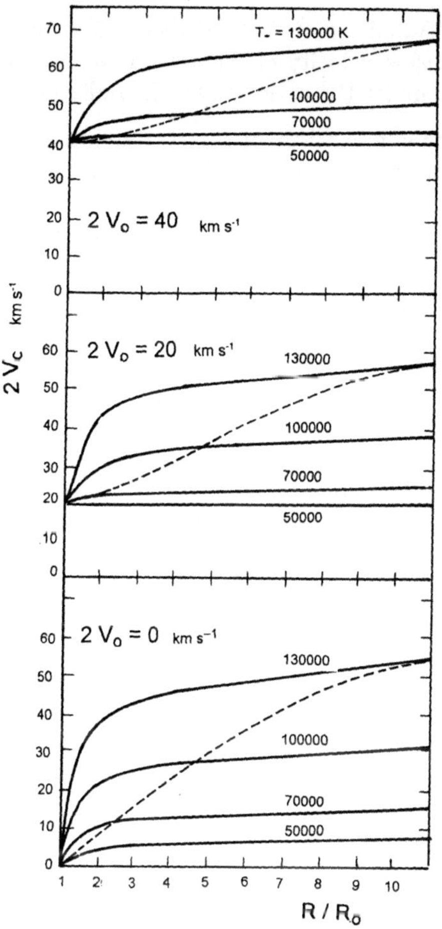

Fig. 16.1. Calculated curves of the dependence of the expansion velocity $2V_c$ for an optically thin nebula ($\tau_c = 0.5$) on its radius R/R_0 (*solid lines*) for three values of the initial velocity $2V_0$: 0, 10 and 20 km s^{-1}. The expansion is caused by the direct action of L_c radiation from the nucleus with temperatures T_* of 50 000, 70 000, 100 000 and 130 000 K

$$m_H n^+ \frac{dV}{dt} = n_e n^+ \frac{h\nu_0}{c} \sum C_n(T_e) \tag{20}$$

or

$$\frac{dV}{dt} = n_e \frac{h\nu_0}{c m_H} \sum C_n(T_e) . \tag{21}$$

We find that $\sum C_n(T_e) = 4.12 \times 10^{-13}$ cm^3 s^{-1}. Setting also the numerical values of other quantities, we get

$$\Delta V = 2.1 \times 10^{-10} n_e \, \Delta t \, \mathrm{cm\,s^{-1}} , \tag{22}$$

where $\Delta V = V - V_0$ and $\Delta t = t - t_0$. For $n_e = 5 \times 10^3$ cm^{-3}, we find from (22) that the expansion velocity of the nebula increases by 3.2 km s^{-1} over 10 000 years purely under the action of the L_c radiation of the nucleus.

Expansion Under the Action of L_α Radiation. A number of attempts were made to quantitatively estimate the role of L_α radiation in the expansion of PNs.

In one of these attempts, the initial point chosen was that the number of L_α photons born at the absorption of L_c photons at the outer boundary of the nebula is always smaller than the number of the same photons born at the inner boundary (Vitiasev, 1969). After multiple scatterings these L_α photons escape from either of the boundaries of the nebula. Clearly, the L_α flux escaping from the outer boundary will be smaller than from the inner boundary. The difference in the two fluxes directed inward, will result in the appearance of a flux force directed outward and acting to increase the acceleration of the expansion.

One can estimate the upper limit of the reactive force, neglecting the flux of L_α radiation directed outward:

$$F_{max} = \frac{H_\alpha}{2c}, \tag{23}$$

where H_α is the L_α flux directed inward. Since, in accordance with the theory, the number of L_α photons born is equal to that of absorbed L_c photons, we can write

$$H_\alpha = \frac{\nu_\alpha}{\nu_c} H_c = \frac{3}{4} H_c .$$

Therefore we have from (23)

$$F_{max} = \frac{3}{8}\frac{H_c}{c} = \frac{3}{8} P_c , \tag{24}$$

where P_c is the pressure of direct L_c radiation on the nebula. As follows from this expression, the pressure of the diffuse L_α radiation is even less than half (38%) of the pressure due to the direct L_c radiation of the nucleus in the Lyman continuum.

The largest additional velocity $\Delta V_\alpha(\infty)$ gained by the nebula under the action of diffuse L_α radiation can be represented, in analogy with (17), by the following approximate expression:

$$\Delta V_\alpha(\infty) = (0.4\, f \sqrt{\tau_c})^{1/2} . \tag{25}$$

For the resulting velocity of expansion under the simultaneous action of both L_c and L_α radiation, in analogy with (15) and (17), we get

$$V(\infty) = \sqrt{V_0^2 + \Delta V_c^2 + \Delta V_\alpha^2} . \tag{26}$$

The resulting maximum velocities of expansion will be, again for $V_0 = 0$,

T_*, K	50 000	70 000	100 000	130 000
$V(\infty)$, km s^{-1}	3.6	9.9	18.1	31.2

As we can see, the difference from that obtained above when considering the action of L_c radiation is not so large; nevertheless one should not completely neglect the contribution of the L_α field.

All these considerations, however, have an approximate character because of the number of factors still to be taken into account: e.g. the influence of the gradient of the expansion velocity, and the effect of the redistribution of L_α radiation at the frequency of multi-scattering processes, etc.

As calculations taking into account the complete redistribution of frequency indicate, the expansion velocity of the outer layers of the nebula will increase over 10 000 years by 18.7 km s^{-1} and 6.7 km s^{-1} at $t_0 = 10^4$ and $t_0 = 10^2$, where t_0 is the optical depth of the nebula at the centre of the L_α line (Zanstra, 1956). Since both effects, the redistribution in frequency and the velocity gradient, play almost equal roles in determining the L_α field, the results obtained above should be doubled.

16.3 Expansion of Nebulae: Observations

At the time the first theoretical results on the expansion of PNs appeared, the number of PNs with known expansion velocities $2V_0$ hardly reached two or three dozen. Obviously, at that time no unique conclusions could be derived, say, on the dependence of expansion velocities on the size of the nebulae.

By the late 1980s the number of PNs with more or less reliable measured values of expansion velocities $2V_0$ reached over 250 (Weinberger, 1989). Figure 16.2 shows a homogeneous results on the dependence of $2V_0$ on the radius of nebula R_n determined only with the help of lines of [OIII]; the values of $2V_0$ determined by HI or [NII] are systematically larger than those of [OIII].

Figure 16.2 is rather remarkable, since no definite correlation between the expansion velocity $2V_0$ and the nebula's radius R_n can be seen. The accumulation of the points at $R_n > 0.15$ pc is most likely to be explained by the expansion of the nebula, i.e. with the decrease of its surface brightness; some nebulae disappear earlier than others. The absence of points within the interval $2V_0 \sim 0$–10 km s^{-1} is due to the limiting accuracy of measurements of expansion velocities.

One can cover the points in Fig. 16.2 with the families of curves shown in Fig. 16.1 and corresponding infinite possible combinations between the initial parameters of the nebula and the nucleus. Then, obviously, without exception, all the points in Fig. 16.2 can be fitted by some curves. In other words, there exists a real path starting from a definite point on the axis $2V_0$, passing through the given nebula and predicting its further evolution.

As a result of the above analysis and the comparison carried out, the following conclusions may be drawn.

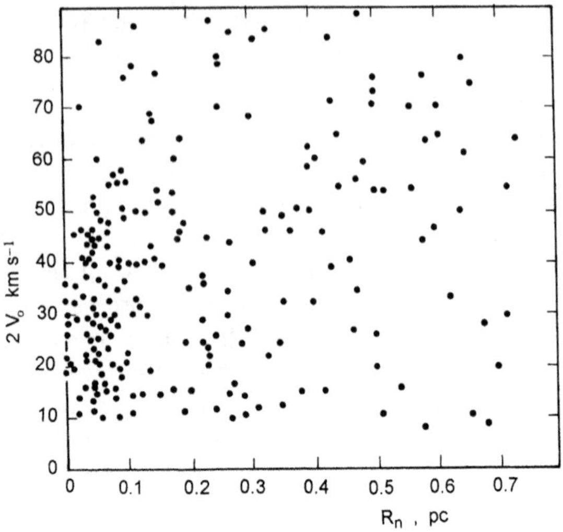

Fig. 16.2. Dependence of the observed expansion velocities $2V_0$ obtained by the lines of [OIII] on the linear radii R_n according to data for 250 planetary nebulae

(a) The initial expansion velocity, or more precisely, the velocity of ejection of the future envelope nebula from the star progenitor is not the same for all PNs.
(b) The ejection of the future nebula from the progenitor is also possible with zero initial velocity.
(c) The maximum initial velocity of ejection $2V_0$ does not exceed 40–45 km s^{-1}.
(d) No preferred velocity of ejection exists within the interval from zero to 40 km s^{-1}: for the considered sample of PNs the initial velocities $2V_0$, i.e. when $R/R_0 \sim 1$ (Fig. 16.1), are distributed uniformly within this interval of velocities.
(e) The radiation pressure plays an essential role in the acceleration of the nebula's expansion only in the case of high-temperature nuclei – higher than 70 000 K. In nebulae with relatively low temperatures, $\sim$ 50 000 K, the observed expansion velocities $2V_0$ should be identified with their initial ejection velocities.
(f) At various periods of the nebula's evolution different acceleration mechanisms are the dominant ones: in the early period, the pressure is due to L_c radiation (when the nebula is optically thick); in the late period, the pressure is due to L_α radiation.
(g) The initial radii R of ejection also are probably not the same, and large dispersion exists in the magnitudes of R_0.

Thus, at first glance the amorphous structure in Fig. 16.2 hides definite evolutionary regularities on the epoch of the birth and formation of PNs. The figure also reflects the essential role of both types of light pressure, L_c and L_α radiation, in the kinematic properties of nebulae.

It should be noted that for samples of PNs formed by definite criteria, e.g. according to the excitation class p, definite dependencies between $2V_0$, R_n and averaged values of $2V_0$ and R_n become evident (see Fig. 18.9).

16.4 Fast Winds from Central Stars

The safest way to ascertain if there is a stellar wind in a given PN nucleus is the registration of spectral lines of P Cyg-type profiles. In the optical region this method was applied, in particular, to the subordinate lines of hydrogen and helium H_α and 4 686 HeII.

The ultraviolet region is even richer. The *IUE* spectra of the nuclei of a number of PNs – NGC 40, NGC 2 371, NGC 6 210, NGC 6 543, NGC 7 009, IC 2 149, IC 3 568, A 30, A 78 etc. – all ultraviolet lines of high potential atoms, turned out to have P Cyg-type profiles (Heap, 1983). This is true first of all for the resonance lines 1 240 NV, 1 400 SiIV, 1 550 CIV, as well as for the subordinate lines 1 342 OIV, 1 371 OV, 1 579 NIV. The sign and magnitude of the shifts of these profiles correspond to the outflow of highly ionized matter with velocities of 1 000–3 000 km s^{-1}, i.e. to "fast stellar winds". The analysis carried out for a few objects reveals the outflow power: 4×10^{-9} up to $7 \times 10^{-7} \mathcal{M}_\odot/\mathrm{yr}$, i.e. a few orders smaller than for the usual WR type stars.

P Cyg type profiles have been registered by *IUE* in the nuclei of over 40 PNs (Perinotto, 1983) of all excitation classes:

Extremely high:	NGC 246 ($p = 12^+$), A 30 (12^+), NGC 1 360 (12);
High:	NGC 6 905 (11), NGC 2 392 (10), NGC 2 867(9);
Medium:	NGC 6 572 (7), NGC 7 293 (7), NGC 6 543 (5);
Low:	IC 3 568 (3), IC 2 149 (2), NGC 40 (1), IC 418 (1).

The fact that the nuclei of PNs are sources of stellar wind is rather remarkable. The data indicate that this fact may be one of the universal properties of all nuclei, while among these nuclei there are stars of various types: WR, Of as well as of Cont., and even O with no peculiarities.

The influence of the corpuscular stream on the nebula is two-fold: physical and dynamical. In the first case the fast wind may become a source of excitation and ionization of atoms by shock mechanism; protons rushing with a speed of the order of 1 000 km s^{-1} shall have an energy equivalent to the electron temperature of the order of 10^8 K (!). In the second case the corpuscular stream will affect the kinematics of the nebula; the nebula can gain additional momentum, with subsequent acceleration of the expansion. Below we will briefly discuss some of the possible dynamical consequences of the interaction of the corpuscular stream with the nebula.

Recall that no PN is known with an expansion velocity of 1 000 km s^{-1}. Already this sets a strong hint on the power of the corpuscular stream.

Let $\dot{\mathcal{M}}$ be the mass outflow rate and u is its velocity, and let both these parameters be preserved for some time Δt. The conservation of momentum will yield

$$\dot{\mathcal{M}}\, u\, \Delta t = \mathcal{M}_0 V_0 \,,$$

where V_0 is the expansion velocity of the nebula. From here we have for the acceptable limit of the outflow rate

$$\dot{\mathcal{M}} = \frac{\mathcal{M}_0 V_0}{u\, dt} \,.$$

For the nebula's mass $\mathcal{M}_0 = 0.01\mathcal{M}_\odot$, expansion velocity $V_0 = 10$ km s^{-1}, the velocity of the corpuscular stream $u = 1\,000$ km s^{-1} and duration of outflow commensurable with the age of nebula $\Delta t = 30\,000$ years, we have $\mathcal{M} = 3 \times 10^{-9}\mathcal{M}_\odot/yr$.

Note that this estimate requires the power of corpuscular stream to be preserved at a constant level – $10^{-9}\mathcal{M}_\odot/yr$ for all PN nuclei over their entire lifetime, i.e. for a few tens of thousands of years.

The interaction of the stellar wind with the gaseous envelope nebula with a velocity of the order of 1 000 km s^{-1} represents a complicated gas dynamic problem (see e.g. Kahn, 1983). However, there can be little doubt of the influence of stellar winds on the dynamics of the expansion of PNs. Recall the analysis of 834 *IUE* recordings of the spectra of 159 PN nuclei (Patriarchi, Perinotto, 1991). The existence of a stellar wind was confirmed in 60% of nuclei; for the remaining 40% no clear signs of stellar wind were observed. The latter are mainly nucleus stars possessing "absorption" spectra, among them many subdwarfs. The estimated parameters of the fast wind yield values for V_{ed} at "infinity" of 600 km s^{-1} to 3 500 km s^{-1}.

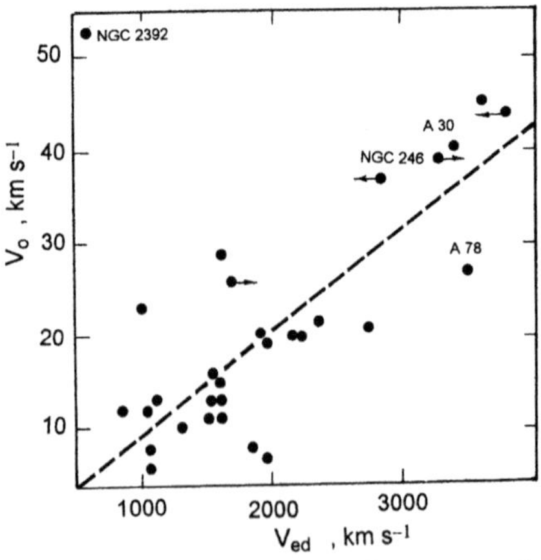

Fig. 16.3. Empirical dependence between the planetary nebula expansion velocity V_0 and the speed of the stellar wind V_{ed}

The comparison of $V_{\rm ed}$ with the expansion velocity V_0 of optical nebulae reveals the existence of a positive correlation between them; see Fig. 16.3, including the data for 30 PNs and their nuclei. Despite the significant scatter of points, an empirical law $V_0 = 0.01V_{\rm ed}$ can be followed for the power of outflow $10^{-7}\,\mathcal{M}_\odot/{\rm yr}$ during $t = 10\,000$ years for an initial mass of the nebula of $0.1\mathcal{M}_\odot$.

16.5 Shock Waves in Nebulae

The rapid and strong increase in pressure in a gaseous medium leads to so-called shock waves. The propagation of shock waves is accompanied by the heating of the medium.

As was mentioned above, *IUE* observations have shown that the central stars of PNs produce fast winds with speeds of 1 000 km $\rm s^{-1}$ and higher, probably common for high temperature nuclei of PNs. These winds carry a significant amount of momentum and energy, believed to play an essential role in the dynamical evolution of PNs, in particular, in the context of the multiple wind scenario mentioned above (Kwok, 1983; Kahn, Breitschwerdt, 1990; Marten, Schönberner, 1990)

Below, we will discuss in the simplest form the gas dynamic problem of the interaction of the fast wind and the slow envelope (for details see Kahn, 1989).

Assume a superwind with a thermal velocity u of a mass outflow rate $\dot{\mathcal{M}}$. A fast wind of velocity V, much exceeding u, is assumed from the nucleus of luminosity L. The latter wind carries energy at a rate $L_{\rm W} = \eta L$, and the star emits Lyman continuum photons at a rate $S_* = jL$ so that two shocks will develop as the result of the interaction between the winds.

The equations describing the motion are

$$\begin{aligned} &\frac{{\rm d}}{{\rm d}t}\left(2\pi r^3 P\right) + 4\pi r^2 P \frac{{\rm d}r}{{\rm d}t} = \eta L\,, \\ &P = \rho_0 \left(\frac{{\rm d}r}{{\rm d}t} - u\right)^2 + \frac{M_s}{4\pi r^2}\frac{{\rm d}^2 r}{{\rm d}t^2}\,, \end{aligned} \tag{25}$$

where

$$\begin{aligned} M_s &= \frac{\dot{M}}{u_0}(r - ut)\,, \\ \rho_0 &= \frac{\dot{\mathcal{M}}}{4\pi r^2 u}\,, \end{aligned} \tag{26}$$

where P is the pressure in the hot shocked stellar wind, M_s is the mass of the shell at a radius r, and ρ_0 is the density of superwind just after the outer shock.

The solution of (25) leads to the results

$$r = \lambda ut\,, \qquad \frac{dr}{dt} = V_0 = \lambda u\,, \qquad P = \frac{\dot{\mathcal{M}}(\lambda - 1)^2}{4\pi\lambda^2 ut^2}\,, \tag{27}$$

where λ satisfies

$$\lambda(\lambda - 1)^2 = \frac{2}{3}\frac{\eta L}{\dot{\mathcal{M}} u^2}\,. \tag{28}$$

The balance between the pressures of the fast wind and of the hot stellar wind shows that the inner shock should be at a radial distance

$$r_{\text{in}} = \frac{\lambda t}{\lambda - 1}\left(\frac{2uL_w}{\dot{\mathcal{M}}V}\right)^{1/2} \tag{29}$$

and therefore

$$\frac{r_{\text{in}}}{r} = \left(\frac{3V_0}{V}\right)^{1/2}\,, \tag{30}$$

where $V_0 = \lambda u$ is the expansion velocity of the nebula, typically 30 km s^{-1}, and the fast wind velocity V is 3 000 km s^{-1}, so that the ratio of the two radii, r_{in}/r is about 6, or the ratio of volumes is about 200.

As an illustration, we take the values of the main parameters $u = 10$ km s^{-1}, $\dot{\mathcal{M}} = 9 \times 10^{-5} \mathcal{M}_\odot$ yr^{-1}, $L = 6\,300 L_\odot = 2.5 \times 10^{37}$ ergs s^{-1}, $L_w = \eta L = 10^{35}$ ergs s^{-1}, $S_* = jL = 2.5 \times 10^{47}$ photons s^{-1}, consistent with a value $\lambda = 3$, and an expansion velocity $V_0 = dr/dt = 30$ km s^{-1} for the shocked shell, i.e. of the nebula itself.

The temperature T_s behind the shock front for a velocity V of the stellar wind is given by the relationship

$$T_s = 2\frac{\gamma - 1}{(\gamma + 1)^2}\frac{\mu m_p}{k}V^2\,, \tag{31}$$

where μ is the mean molecular weight of the gas, γ is the adiabatic index, and m_p is the proton mass.

For a hydrogen stellar wind, when $\mu = 1$ and $\gamma = 5/3$, we have from (31)

$$T_s = \frac{3}{16}\frac{m_p}{k}V^2$$

or

$$T_s = 0.22 \times 10^{-8}\,V^2,\ \text{K}\,, \tag{32}$$

when V is measured in cm s^{-1}, or

$$T_s = 22\,V^2,\ \text{K} \tag{33}$$

when V is in km s^{-1}.

For typical stellar wind velocities $V = 1\,000$ km s^{-1}, we have from (33) for the temperature behind the shock front: $T_s = 2 \times 10^7$ K, and $T_s = 2 \times 10^8$ K at $V = 1\,000$ km s^{-1} and $V = 3\,000$ km s^{-1}, respectively.

In relation to such high temperatures, the problem of the origin of strongly ionized atoms and, hence, of the excitation of high potential emission lines arises.

16.6 Thermal Expansion: Voids in Nebulae

There is still one more effect influencing the expansion of nebulae, i.e. the expansion of heated gas. Even a static, non-expanding nebula in time will increase its dimensions because of the expansion and diffusion of gaseous matter with the thermal velocity of hydrogen atoms: 10 km s^{-1}. Therefore, $V_0 = 10$ km s^{-1} will be the least value of expansion velocity. Though the real expansion velocity of nebulae is about 20–30 km s^{-1}, the existence of ring-like PNs, seems to be incompatible with relatively slow expansion.

The point is that, if the nebula is formed by a discrete outburst or the emission of gas from the nucleus on the form of a spherical shell, then there is no reason why thermal expansion will not also occur at the inner boundary of the nebula. As a result, the thickness of the shell must increase both from the inside and the outside. For an initial expansion velocity of a shell, say, of 10 km s^{-1}, the actual velocity at its outer layers would be 20 km s^{-1}, and zero at the inner layer. In other words, as the outer edge moves farther from the centre, the inner edge remains motionless. As a result one has not a ring nebula, but a nebula filling the whole volume inside its outer boundary.

In reality the expansion velocity of PN is larger than 10 km s^{-1}, and therefore a "void" should form in the centre of PNs. The point is that observations give a significantly higher ratio r_1/r_2 of the inner (r_2) and outer (r_1) boundaries of the ring than we might expect for "free" expansion.

Let V_0 be the observed velocity of the "void" nebula. Obviously, V_0 is the sum of the actual expansion velocity V and the thermal velocity u. Assume that V and u are constant within some time interval. Then the equations of motion of the outer and inner boundaries of the nebula relative to the centre, are

$$r_2 = (V + u)t + \Delta = V_0 t + \Delta ,$$

$$r_1 = (V - u)t = (V_0 - 2u)t ,$$

where Δ is the initial thickness of the shell. For large enough values of t, we obtain approximately

$$\frac{r_1}{r_2} = \frac{V_0 - 2u}{V_0} . \tag{34}$$

The electron temperature is practically the same for all PNs, while the expansion velocity V_0 can be different for various nebulae. Then, the theoretical relation between r_1/r_2 and V_0 will be given by (34). Figure 16.4 shows this relation for $u = 10$ km s^{-1} ($T_e \approx 10\,000$ K). In this figure the dots represent the data for 13 PNs with known expansion velocity V_0 and ratio r_1/r_2; the former data are from Table 1.3 and the catalogue of expansion velocities of PNs (Sabbadin, 1984), the latter are mainly from our measurements of images of ring nebulae (Kaler, 1974; Jewitt et al., 1986; Chu et al., 1987; Acker et al., 1992).

From Fig. 16.4 we see that the observations give significantly higher values for r_1/r_2, and therefore much smaller values for the relative thickness of the

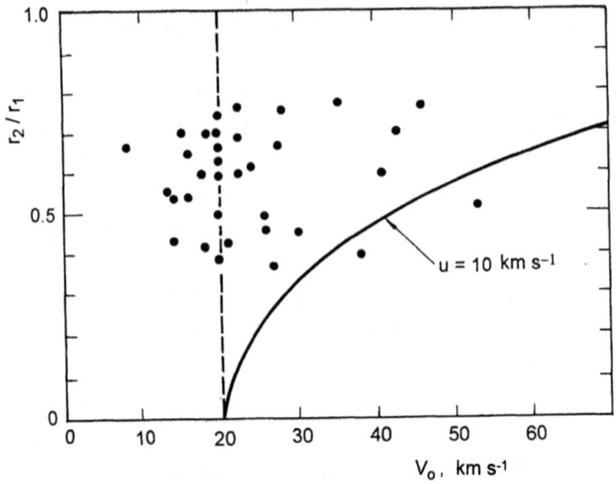

Fig. 16.4. Theoretical dependence (*solid line*) of the ratio r_2/r_1 for ring-like planetary nebulae on the expansion velocity V_0 for the "free" expansion of nebula. Dots are observations for particular nebulae

nebula than follows from the "free" expanding nebula: almost all dots lie far from the theoretical curve (solid line). Moreover, none of the nebula with an expansion velocity less than 20 km s^{-1} (almost 50%) possess a void in the centre.

One can attempt to explain the formation of a "void" by high electron temperatures in the central regions of the PN (see Sects. 6.4 and 6.5).

Assume that the boundary separating the void from the shell is in a state of equilibrium. This is possible only when the gas pressures on the boundary from both the shell and the void sides are equal. Neglecting the radiation pressure, we have at the boundary

$$n_1 T_{e1} = n_2 T_{e2} , \tag{35}$$

where the subscript 1 stands for the "void" and 2 for the shell. For the shell, T_{e2} is of the order of 10 000 K.

However, judging from the data, say, on the increase of brightness and therefore of the matter density inside the inner boundary in the majority of ring nebulae, the inner boundary is not in equilibrium, but moves outward. In other words, at this boundary we should have the condition

$$n_1 T_{e1} > n_2 T_{e2} . \tag{36}$$

Hence, we can write for the temperature of the void region

$$T_{e1} > \frac{n_2}{n_1} T_{e2} . \tag{37}$$

It can be easily shown based on the known isophotes for ring-shaped nebula that $n_2/n_1 \sim 3$. Then for the electron temperature in the central parts of the nebula we shall have

$$T_{e1} > 30\,000\ \mathrm{K}\,.$$

Thus, we have yet another indication of the high electron temperatures in the central parts of PN.

How can such high electron temperatures appear in the central parts of PN?

The optical depth of the void beyond the Lyman limit of hydrogen τ_c is very small – of the order of 0.001 – and, therefore, here the hydrogen will be ionized almost completely. The kinetic energy of the electrons born for photoionization cannot be used for the excitation of the usual forbidden lines of single and twice ionized oxygen, nitrogen, sulphur etc. In the void these elements will be at third or higher degrees of ionization. The excitation potentials of metastable levels of such highly ionized elements are much higher, and therefore the energetic losses of free electrons now will not be as important as earlier. As a result, the residual, "not used", kinetic energy of the electron gas will keep the electron temperature significantly higher than 10 000 K. In other words, the stronger the degree of ionization of the nebula, the less will be its cooling rate.

16.7 Deceleration of Nebulae

Planetary nebulae must also be influenced by decelerating forces during their expansion. Among such forces is that, for example, caused by the inward-directed L_α flux within the inner and central parts of the expanding nebula.

The problem of deceleration is of particular importance for the dynamics of PNs because of its direct relation to the problem of their origin. The possible alternatives are as follows:

(a) The initial velocity of ejection and the velocity of expansion of PNs are quantities of the same order of magnitude.
(b) The initial velocity of ejection is high, of the order of 1 000 km s^{-1}; however, subsequently it decreases owing to the deceleration forces acting on the nebula during its expansion.

Since the second alternative has attracted the most attention, we shall consider it first. Usually three possible deceleration mechanisms are considered: gravitational attraction, the drag of the interstellar medium and radiation pressure. We shall dwell on these mechanisms in some detail.

(1) Gravitational Deceleration. Obviously, the initial velocity of the ejected shell must exceed the escape velocity, i.e. the parabolic velocity at the surface of the star. Although the physical parameters of the nucleus of PNs are known with insufficient accuracy, there are reasons to assume its parabolic velocity to be high, of the order of 1 000 km s^{-1}. Thus, the velocity of the escaped shell has to be even higher.

At smaller distances from the star the velocity of the shell should decrease according to the law

$$V(r) = \left[V_0^2 - \left(1 - \frac{R_*}{r}\right) V_p^2\right]^{1/2}, \tag{38}$$

where $V(r)$ is the velocity of the shell at a distance r from the star, V_0 is the initial velocity, V_p is the parabolic velocity at the surface of the star and R_* is the radius of the star.

At larger distances, i.e. when $R/r \ll 1$, the velocity of the shell approaches a constant value $V(\infty)$:

$$V(\infty) = \sqrt{V_0^2 - V_p^2}\,. \tag{39}$$

Obviously, the difference $V_0 - V(\infty)$ is the maximum change in the velocity owing to the gravitational deceleration. If, for example, $V_0 = 1\,500$ km s^{-1} and $V_p = 1\,000$ km s^{-1}, then the least velocity of the shell will be equal to $1\,120$ km s^{-1}. Hence, although the initial high velocity is somewhat reduced, nevertheless it far exceeds the typical expansion velocities of the nebula. To have $V(\infty)$ of the order of 20 km s^{-1}, V_0 must be very close to V_p, which, as has been mentioned earlier, is quite unnatural. Consequently gravitational deceleration must be considered insufficient to explain the small expansion velocities observed in nebulae.

(2) Deceleration by the Interstellar Medium. The expansion of a gaseous shell evidently occurs in the interstellar medium which has a definite and far from negligible density. This medium will resist the moving shell and thereby decrease its velocity.

The mechanism of deceleration, first considered by Oort (1946), involves the following idea. As the shell expands, the particles of the interstellar medium fall onto it from all directions. These particles must be captured by the shell itself, since the linear thickness of the shell significantly exceeds the mean free path of the particles within it. At the same time, a part of the momentum of the shell is transformed to these new particles. This leads to a decrease of the expansion velocity of the shell and to the increase of its mass.

The condition of conservation of momentum for the shell can be written in the following form:

$$V_0 M_0 = V(r)\, M_0 + V(r)\, \frac{4\pi}{3}\, r^3 \rho_c\,, \tag{40}$$

where M_0 and V_0 are the original mass and expansion velocity of the shell, r_c is the density of the interstellar medium and $V(r)$ is the expansion velocity of the shell at a radius r. The first term on the right hand side of this equation refers to the shell, the second, to the additional mass received from the interstellar medium.

From (40) we have for $V(r)$

$$V(r) = \frac{V_0}{1 + \delta\, r^3}\,, \tag{41}$$

where

$$\delta = \frac{4\pi}{3}\,\frac{\rho_c}{M_0}\,. \tag{42}$$

Equation (41) defines the law of change (decrease) of the expansion velocity of the shell owing to the resistance of the interstellar medium. We see that a significant change in velocity can occur when $\delta\,\rho^3$ is comparable with unity. At what radius r_0 of the shell is this condition fulfilled? From $\delta\, r_0^3 = 1$, we obtain

$$r_0 = \left(\frac{4\pi}{3}\,\frac{M_0}{\rho_c}\right)^{-1/3}\,. \tag{43}$$

Taking $\rho = 5 \times 10^{-24}$ gr cm^{-3}, which corresponds to an upper limit for the density of the interstellar medium, i.e. $n(\mathrm{H}) = 3$ cm^{-3}, we obtain

$$\begin{aligned} r_0 &= 70\,000\ \text{a.u.} \qquad \text{for } M_0 = 0.01 M_\odot\,, \\ r_0 &- 140\,000\ \text{a.u.} \qquad \text{for } M_0 = 0.1 M_\odot\,. \end{aligned} \tag{44}$$

From here we are able to conclude that the expansion of gaseous shells (nebulae) is *really* decelerated by the interstellar medium, but it can be efficient only for nebulae of rather large dimensions – about 100 000 a.u. This mechanism has no role in the initial stage ($r \ll r)_0$) of ejection of gaseous shells.

(3) Deceleration by Radiation Pressure. For a long time L$_\alpha$ radiation pressure was considered the only reason for the deceleration of gaseous shells. In the central and inner parts of the expanding shell the pressure is directed inward, which can cause this deceleration.

However, as we have seen above, the acceleration due to L$_\alpha$ radiation pressure is comparable to the force of gravity. Consequently, its role will be of the same order as of the gravitational deceleration, i.e. again insignificant. This is true both for PNs and for shells of small dimensions. Therefore, we can conclude that the radiation pressure is not essential even in the initial stages of expansion of the nebula, so that velocities of the order of 1 000 km s^{-1} cannot be reduced by two orders of magnitude in this way.

One can show, meanwhile, that the relative role of radiation pressure in shells of small diameters is larger than for large sizes. This is seen from the following argument. For the flux of L$_\alpha$ radiation near the inner boundary of the nebula we have (11.47)

$$F_\alpha(0) = -\,2\pi\,\psi(0)\,\frac{q}{\sqrt{\beta}}\,, \tag{45}$$

where β as a function of r is expressed in the form

$$\beta = \frac{q}{2u}\,\frac{\mathrm{d}V}{k\,\mathrm{d}s}\,. \tag{46}$$

Neglecting the change of velocity along the radius of the nebula and setting $dV/ds = (2/3)V/r$ in (46), and in view of $\tau_c \sim r^{-3}$, we have

$$\beta = \frac{q}{2u}\frac{V}{\tau_0}\left(\frac{r}{r_0}\right)^3 , \tag{47}$$

where r_0 is the radius of the nebula when $\tau = \tau_0$.

As follows from (47), the effect of deceleration due to radiation pressure will be larger when the dimensions of the nebula are small.

Thus, although radiation pressure cannot decelerate a shell ejected with enormous velocities, it can, however, lead to the redistribution of the expansion velocities within the nebula along its diameters.

We conclude, that *the original velocity of ejection of a PN can in no case be of the order of 1 000 km* s^{-1}. Since the deceleration processes can lead only to small changes of expansion velocity, *the initial velocity of ejection of a nebula cannot be larger than 50–100 km* s^{-1}.

16.8 The Hydrodynamics of Planetary Nebulae

Up to now, while discussing the forces acting in PNs, we have mentioned the gravitational force of the central star and the radiation pressure of the nebula. Since we are dealing with a high temperature gas, of the order of 10 000 K, expanding with velocity of 10 km s^{-1}, it is clear that we should also consider the influence of the gas pressure on the nebula, along with all of its hydrodynamical consequences. Finally, since PNs, we believe, also possess magnetic fields (see the next chapter), the role of the magnetic force has to be considered as well.

In any nebula although these four types of force – gravity, radiation pressure, hydrodynamic and magnetic – can act simultaneously, one of them can become dominant. We cannot say *a priori* which of these forces will be dominant at any given stage of the development of the PN. However, the problem can be solved in a comparatively simple way. Since by their nature the four forces are intrinsically different, the form and structure of the nebula will change depending on which of the forces has the dominant role. Therefore, we can sometimes guess the nature of the primary force responsible for the shape and structure of the nebula simply by proceeding from the PN image. For example, such an empirical approach will lead us below to the conclusion that the dominant force responsible for the visible structure of bipolar nebulae is the magnetic force.

In certain cases it is possible to describe the dynamics of the nebula on the basis of *one* type of force; this greatly simplifies the treatment of the problem. Whether the hydrodynamic forces have to be taken into account in the presence of a magnetic field is determined from the inequality

$$\frac{H^2}{8\pi} < \frac{1}{2}\rho V^2, \tag{48}$$

where H is the strength of the magnetic field, V is the velocity of gaseous matter and ρ is its density. For PNs with $V_0 = 20$ km s^{-1}, $\rho = 1.6 \times 10^{20}$ gr cm^{-3} ($n_{\mathrm{H}} = 10^4$ cm^{-3}), we find that even at $H > 10^{-3}$ gauss the hydrodynamical forces can be neglected. The condition $H > 10^{-3}$ gauss cannot always be fulfilled in PNs. In nebulae with large linear dimensions, and also in extended regions of normal nebulae, we can expect $H < 10^{-3}$ gauss. In this case the hydrodynamical forces can play a certain role in the development of the structure of the nebula.

Finally, the inevitable interaction between the PN and the interstellar medium, where the expansion and motion of the nebula occurs, also requires the consideration of hydrodynamical effects on PN properties.

The strict quantitative elaboration of the problems of the hydrodynamics of PNs has serious difficulties associated with it.

The main point concerns the validity of the application of classical principles of hydrodynamics in the conditions of these cosmic objects. Even in the densest gaseous nebulae and Galactic clouds the density of the medium is negligible compared with the density of the medium for which the classical equations of hydrodynamics were deduced. Nevertheless, the usual laws of hydrodynamics can, under certain conditions, also be applied to cosmic objects – planetary and diffuse nebulae, gaseous envelopes ejected during the explosions of novae and supernovae, extended atmospheres of Wolf-Rayet stars, turbulent atmospheres of close binary systems, interstellar clouds, both cool and heated, etc. The determining factor is the ratio of the scale of the phenomenon or the size of the object – the "characteristic size" – to the mean free path of the particles. At least we can assert that as long as the distance d between two points with essentially different parameters of motion (e.g. the velocity) or of the medium (e.g. the density) remains larger than the mean free path l of the particles, i.e.

$$\frac{d}{l} \gg 1 \tag{49}$$

is satisfied, the usual laws and equations of hydrodynamics are valid.

In interstellar hydrogen clouds of much smaller density than PNs, the mean free path of particles is a few hundred a.u., while for PNs l is less than 1 a.u., i.e. much less than the dimensions of the nebulae.

No significant investigations into the various aspects of the hydrodynamics of PN have been performed. Of some interest may be some early results on the stability of the form of gaseous envelopes ejected by stars, among them PNs (Gurzadyan, 1953ab). We shall summarize these results briefly in the following section.

16.9 Stability of the Forms of Gaseous Envelopes. Statement of the Problem

It has been noticed that PNs preserve their shapes for a long time, while, say, the envelopes ejected by novae, in the majority of cases, during expansion lose their shape much more quickly. In other words, in one case the gaseous envelope keeps its shape, while in the other it is distorted and fragmented.

We will call the first type of envelopes "stable", and those distorted quickly "unstable". One can attempt to formulate the corresponding conditions for the kinematic and physical parameters of the envelopes, even to find out a kind "instability coefficient determining the fate of the ejected nebula.

Considering the reasons for the loss of stability, i.e. of the regular shape of gaseous envelopes, one obviously has to take into account their interaction with the interstellar medium. During the expansion the outer layers of the gaseous envelope experience the resistance action of the interstellar medium. This will lead to the redistribution of the initial density in the envelope, i.e. to its increase at the outer and decrease at the inner boundary of the envelope. The departure of the density distribution from the homogeneous, isotropic case owing to some force produces an acceleration within the nebula, i.e. related to the coordinate system of the moving envelope. We call it "internal acceleration" denoted by $g(t)$.

Thus, because of the resistance of the medium and the resulting compression of the envelope, it acquires the properties of an incompressible fluid with internal acceleration $g(t)$, varying over time t. It can be shown, however, that the internal acceleration is approximately equal to the deceleration of the envelope in the interstellar medium, i.e.

$$g(t) = \frac{\mathrm{d}V(t)}{\mathrm{d}t}, \tag{50}$$

where $V(t)$ is the expansion velocity of the envelope and is given by (41).

The $g(t)$ is directed opposite to the deceleration of the envelope, i.e. $g(t)$ acts in direction of expansion. Under these conditions, i.e. in the presence of an internal acceleration, the contacting surface between the nebula and the interstellar medium cannot remain stable for a long time. The smallest irregularity, caused for whatever reasons, can grow during the expansion of the nebula and lead to its total destruction.

The problem, therefore, is to establish under what conditions and at what stage of the expansion will the envelope become unstable with respect to original regular shape.

To simplify the problem, we confine ourselves to an examination of an isothermal medium; we neglect the effect of the compressibility of the gas, and also the fact that the expansion velocity of the envelope is supersonic. Then, using the well known methods of the study of gravitational waves on the surface of an incompressible fluid, we can write down an equation for the outer surface of the envelope as a function of time or of its radius. We can then determine the changes in the shape of the outer surface with time and the degree of stability of the envelope. In particular, we wish to know when the relatively quick increase of initial perturbations will occur and when the envelope will lose its equilibrium shape and, also, when this growth will be slow enough to keep the envelope stable.

The problem of the instability of the surface separating two liquids has been the subject of studies since the work of Rayleigh (1899), and later of Taylor (1950), who formulated what is now well known as the "Rayleigh–Taylor Instability".

The Rayleigh–Taylor instability has also been much studied experimentally. It is known that when two liquids of different densities are placed next to each other, an acceleration appears perpendicular to the surface of separation. This surface will be stable if the acceleration vector is directed from the lower density to the higher density fluid, and unstable when the vector is directed in the opposite way.

In our case the role of the denser fluid is taken by the envelope, so that the "internal acceleration" vector is directed from the envelope to the medium. Therefore, in general, we should expect the envelopes to be unstable. As we shall see later, the instability arising is really of Rayleigh–Taylor type; however, for envelopes with different kinematic and physical parameters, its consequences appear in different ways. In other words, the fate of various envelopes under equilibrium conditions is different.

To summarize, the question is, first, to determine the moment when the envelope becomes unstable and, second, to clarify the character of the instability.

16.10 Stability of the Forms of Gaseous Envelopes

Consider a coordinate system $x\,y$ comoving with the nebula so that the x axis is parallel to the surface of separation (the outer boundary of envelope) and the y axis is the normal vector to the surface, pointing outwards from the envelope, as is shown in Fig. 16.5. We assume that no motion exists along the z axis. The depth of the envelope will be considered large compared with small perturbations $\eta(x,\, y)$ of its outer boundary.

Our problem consists in establishing the form of the function $\eta(x,\, t)$ or $\eta(x,\, y)$, where r is the radius of the envelope (nebula), i.e. the form of the outer edge of the nebula at various times t or various distances of the edge r from the centre of the nebula.

It can be shown that in our case we are dealing with a potential flow, and therefore, we can use Eulerian coordinates V_x and V_y to describe the motion of disturbed particles relative to the coordinate frame $x\,y$. We have

$$V_x = -\frac{\partial \varphi}{\partial x}\,, \qquad V_y = -\frac{\partial \varphi}{\partial y}\,. \tag{51}$$

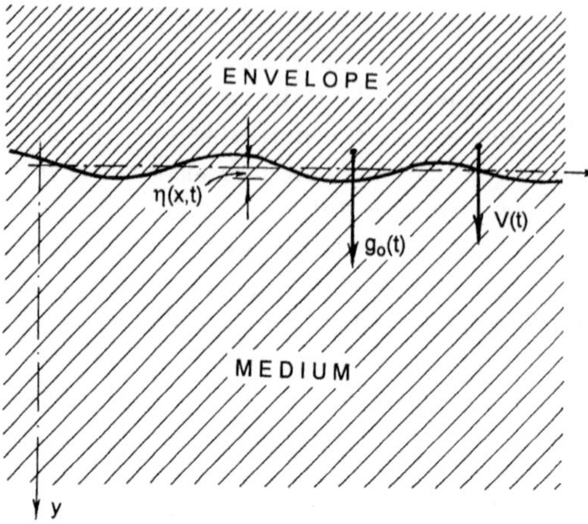

Fig. 16.5. The stability problem of the forms of gaseous envelope. $\eta(x,\, t)$ is a small perturbation, $V(t)$ the expansion velocity of the envelope, and $g_0(t)$ the inward acceleration of the envelope

The velocity field satisfies the equation of continuity, which for the potential flow of incompressible fluid can be represented by the Laplace equation

$$\frac{\partial^2 \varphi}{\partial x^2} + \frac{\partial^2 \varphi}{\partial Y^2} = 0\,. \tag{52}$$

The equation of motion is given by the Cauchy–Lagrange integral

$$V + P + \frac{1}{2}\left[\left(\frac{\partial \varphi}{\partial x}\right)^2 + \left(\frac{\partial \varphi}{\partial y}\right)^2\right] - \frac{\partial \varphi}{\partial t} = F(t)\,, \tag{53}$$

where V is the potential of external forces and P is the dynamical pressure. For V we have

$$V = g(t)\, y\,, \tag{54}$$

where $g(t)$, "inward acceleration", is determined by (50) and (41).

After some manipulation we find from (53) the following expression for the equation of the outer boundary of the envelope:

$$\eta = \frac{1}{g(t)} \left|\frac{\partial \Phi}{\partial y}\right|_{y=0}, \tag{55}$$

where $\Phi = \varphi - \int P(t)\, dt$. To find the function, we proceed from the condition that it satisfies Laplace's equation with the boundary conditions

$$\frac{1}{g(t)} \left|\frac{\partial^2 \Phi}{\partial t^2}\right|_{y=0} + \left|\frac{\partial^2 \Phi}{\partial t}\right|_{y=0} \frac{\partial}{\partial t}\frac{1}{g(t)} = -\left|\frac{\partial \Phi}{\partial y}\right|_{y=0}, \tag{56}$$

$$\left.\frac{\partial \Phi}{\partial y}\right|_{y=0} = 0\,, \tag{57}$$

$$\left.\frac{\partial \Phi}{\partial x}\right|_{y=0} = 0\,. \tag{58}$$

From (41) it follows that as long as $\delta r^3 < 1$, the motion of the envelope will be uniform, without acceleration, i.e. $g(t) = 0$. In this case, obviously, we will have

$$\eta(x,\, t) = \text{const}\,, \tag{59}$$

i.e. the surface of the envelope preserves its shape as long as it remains inside a sphere of radius r_0, determined by the condition $\delta r_0^3 = 1$:

$$r_0 = \left(\frac{4\pi}{3}\frac{\rho_c}{\mathcal{M}_0}\right)^{-1/3}\,, \tag{60}$$

where $\mathcal{M}_0$ is the mass of the envelope and ρ_c the density of the interstellar medium.

Outside the sphere of radius r_0 the condition $\delta r^3 > 1$ is satisfied. Then, from (41), we can write

$$V(r) = \frac{V_0}{\delta r^3} \tag{61}$$

for the expansion velocity of the envelope, and

$$r = \left(\frac{4V_0}{\delta}\right)^{1/4} t^{1/4} \tag{62}$$

for the equation of motion of the envelope. In this case $g(t)$ is different from zero, and is equal to

$$g(t) = -\frac{3}{16}\left(\frac{4V_0}{\delta}\right)^{1/4} t^{-7/4}\,. \tag{63}$$

It follows that $\eta(x,\, t)$ will no longer be constant outside the zone of radius r_0; the surface of the envelope will change its shape with time during the expansion of the envelope according to (62).

Substituting $g(t)$ from (63) into (56) and using (52), (57) and (58), we find the function $\Phi\,(x,\, y,\, r)$, and then the equation for the outer edge of the envelope $\eta\,(x,\, r)$ by means of (55). The final expression for $\eta\,(x,\, r)$ – for the small perturbations of the surface separating the two media – in linear approximation is

$$\eta(x,\, r) = \alpha_0 \left(\frac{r}{r_0}\right)^2 q(r) \cos kx\,, \tag{64}$$

where it has been assumed that the initial shape of the outer envelope when $r = r_0$ is given by the approximate equation

$$\eta(x,\, r_0) = \alpha_0 \cos kx\,, \tag{65}$$

and α_0 is the initial perturbation at $t = t_0$ or $r = r_0$. The latter is assumed to be small compared with the wavelength of perturbation λ, and $k = 2\pi/\lambda$.

The function $q(r)$ was already derived in 1953 (Gurzadyan, 1953, 1954). In a first approximation it can be presented in the form

$$q(r) \sim e^{\nu}\,, \tag{66}$$

where

$$\nu = 2\sqrt{6\pi}\left(\frac{r}{\lambda}\right)^{1/2}. \tag{67}$$

Equation (64) is the equation of "a tongue" of height given by the coefficient of cos kx, i.e. the magnitude of the perturbation at a given size of the envelope. We are interested, of course, only in the relative growth of the initial perturbations with time, $\delta(r)$, which we call the "instability coefficient". From (64) and (65) we have

$$\delta(r) = \frac{\eta(x,\, r)}{\eta(x,\, r_0)} = q(r)\left(\frac{r}{r_0}\right)^2, \tag{68}$$

where r_0 is determined by (60). For $r = r_0$ we have $\delta(r_0) = 1$, i.e. the envelope is stable. For $r > r_0$ the instability coefficient is greater than one, so that the envelope becomes unstable, more the inequality $\delta(r) > 1$ is fulfilled. This is Rayleigh—Taylor instability.

In Fig. 16.6 the development of the instability with increase of r is shown. The magnitude of the instability coefficient $\delta(r)$ is also indicated: envelope is stable of $\delta \sim 1$, unstable at $\delta > 1$ and strongly unstable at $\delta \gg 1$.

Thus, any envelope of a stable initial form will remain stable until the radius r_0 is reached (60) or up to the time t_0 approximately equal to r_0/V_0. After this the envelope undergoes Rayleigh–Taylor instability; the perturbed shape of the outer boundary will steepen exponentially as the radius r increases. The magnitude of r_0 is different for various envelopes and depends on their mass $\mathcal{M}_0$ and on the density of the interstellar medium ρ_0.

So, within this formulation of the problem of the stability of envelopes, we can conclude that for each envelope there exists a sphere of radius r_0, which we call the call it "sphere of stability", within which the envelope is stable; outside this sphere the envelope is unstable and after a certain time it becomes an amorphous nebula.

The Rayleigh–Taylor instability has been qualitatively studied by Spitzer (1954) in application to diffuse nebulae (see Sect. 16.15). Based on approximate qualitative considerations, an equation for $\eta(x,\, t)$ for medium size perturbations was derived for an incompressible fluid (Frieman, 1954):

$$\eta(0,\, t) \sim \exp(+\sqrt{gk}\, t)\,, \tag{69}$$

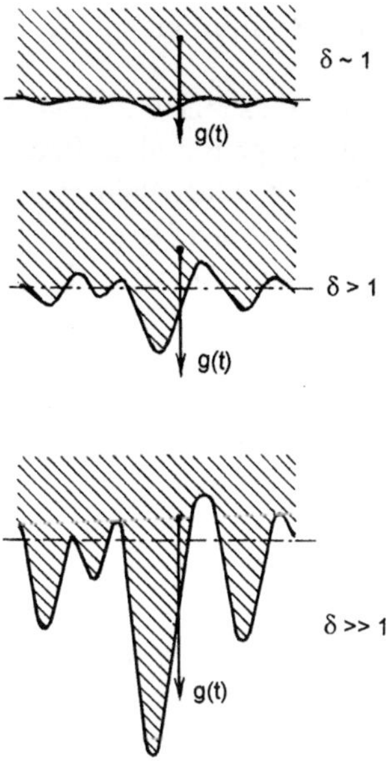

Fig. 16.6. The development of the Rayleigh–Taylor instability in a gaseous envelope. The magnitude of the instability coefficient δ is shown as well; $g(t)$ is the variable inward acceleration of the envelope

where g is the acceleration, and k is the wave number of the periodic perturbation. If we compare 69) with (64), we see that the development of instabilities for large η proceeds faster than for small values (i.e. for the initial phase of the loss of stability).

Further development of the "tongue" instability occurs with "free-fall" acceleration g:

$$V \sim \sqrt{g, \eta}\,. \tag{70}$$

This is the asymptotic behaviour of h for large values of perturbations.

A theoretical account of the Rayleigh–Taylor instability for incompressible liquids has been given by Chandrasekhar (1961). In a number of papers by Chevalier and Klein (1978) this type of instability is considered for the initial phases of formation of gaseous envelopes ejected in novae and supernovae explosions.

Below we give examples of gaseous envelopes for which the Rayleigh–Taylor instability criterion has been applied.

16.11 Stability of Planetary Nebulae

For a PN mass $\mathcal{M}_0 \sim 0.01$–$0.1\mathcal{M}_\odot$ and density of the interstellar matter $\rho_c \sim 5 \times 10^{-24}$ g cm^{-3} and 1.6×10^{-24} gr cm^{-3} ($n_H \sim 3$–1 cm^{-3}), we find from (60) for the radius of the equilibrium sphere r_0 100 000 a.u. and 150 000 a.u., respectively.

This result can be interpreted as follows: when the radius of a PN remains smaller than 1.5×10^5 a.u., its shape is stable. Planetary nebulae can develop a Rayleigh–Taylor instability only when their radii exceed the above value.

Since the radii of the majority of PNs are much smaller than 1.5×10^5 a.u. = 0.54 pc, they will be within the zone of stability and, therefore, will preserve their shapes. Observations, as we know, confirm this conclusion.

16.12 Stable Nova and Supernova Envelopes

The mass $\mathcal{M}_0$ of the envelope of a nova is of the order of $10^{-5}\mathcal{M}_\odot$. This gives for the radius of the sphere of stability for a density of the interstellar medium $n_H \sim 1\ \mathrm{cm}^{-3}$

$$r_0 = 10\,000\,\mathrm{a.u.} = 1.5 \times 10^{17}\,\mathrm{cm} = 0.05\,\mathrm{pc}\,.$$

As soon as the radii of nova envelopes exceed this value they undergo Rayleigh–Taylor instability. Envelopes reach this size very quickly – in a few tens of years at expansion velocities of the order of 1 000 km s^{-1}. This means that nova envelopes find themselves outside the zone of stability almost immediately and are quickly destroyed. The process of destruction is as follows: the initially small perturbation will grow, and as a result, the "tongues" appearing will disintegrate the envelope.

There are relatively few data on the detailed shapes and structure of gaseous envelopes ejected by fast novae. Even less is known about the way the shape changes during the expansion. Nevertheless, it can be considered established that in most cases the nova envelopes, starting a given stage of their expansion, enter into an unstable, i.e. amorphous, state.

As an example, consider the envelope produced by Nova Persei (1901). According to photographs published at various epochs, the envelope of this nova still had a fairly stable form of 10″in 1917. However, in later photographs it shows a partial loss of stability. Later the envelope of this nova completely lost its stability and was transformed into different fragments. This may have occurred because of the dynamical interaction between the envelope and the surrounding interstellar medium. We can even estimate an upper limit for the density of interstellar matter in the vicinity of Nova Persei from the available data on the mass of the envelope, $2 \times 10^{-5}\,\mathcal{M}_\odot$, its distance, approximately 600 pc, and the time when the envelope became unstable, 1917. This upper limit turns out to be 1.5×10^{-22} g cm^{-3}, i.e. fifteen times higher than the mean density of interstellar matter. This density equals that of an average diffuse nebula. Observations confirm this conclusion, that the ejection of this nova occurred in a dust cloud.

The Rayleigh–Taylor instability also seems to have occurred in the case of the Crab nebula. In numerous pictures it appears to consist of many filaments and elongated blobs, distributed radially relative to the centre of the nebula. Assuming that after the explosion of the the supernova (1 054) the ejected envelope had an initially more or less regular shape (cf. Supernova 1987A); we can attribute the later loss of stability to the interaction with the interstellar medium. From the fact that the envelope is highly disrupted we conclude that at present it is outside the sphere of stability. We can even make an attempt to determine the mass of the gaseous envelope of the Crab nebula from the known density of interstellar matter in the vicinity of this nebula. Unfortunately, data on this density are very poor. Nevertheless, from the typical value of the density of interstellar matter and more or less reasonable estimation of the radius of the sphere of stability, we conclude that if the disruption of the gaseous envelope of the Crab nebula is really due to its interaction with the interstellar matter, then the mass of its envelope should be less than $0.1\mathcal{M}_\odot$.

We also have cases of nova envelopes with regular spherical shapes, for example, the envelope of Nova Aquila discovered almost immediately after the explosion in October of 1918. Its diameter was then $0''.65$, and the expansion velocity 1 700 km s^{-1}. Within two months the diameter of the nebula had reached $1''.8$. In photographs of 1926 it represents a full circle of diameter $16''.4$. At Nova Aquila's distance of 360 pc, this corresponds to a linear diameter of 6 000 a.u. In 1930 its diameter was 8 000 a.u. and in 1940, 11 000 a.u. and the envelope was still stable.

The stability of the envelope of Nova Aquila can be explained by its location remaining within the sphere of stability. If we estimate this radius from the known mass of the envelope of Nova Aquila, which is $4.5\times10^{-4}\mathcal{M}_\odot$, we get from (60) $r_0 \approx 14000$ a.u. for $n_H = 3$ cm^{-3} and $r_0 \approx 20\,000$ a.u. for $n_H = 1$ cm^{-1}. The envelope will be able to retain its shape until its radius is 14 000 a.u. or 20 000 a.u. With the expansion velocity quoted above, this will occur some in 60–90 years after the ejection, i.e. between 1980 and 2010

As we can see, between these two categories of objects, PNs and nova and supernova explosions, there is almost nothing in common, not only from the cosmogonical point of view, but also with respect to the criterion of stability: gaseous envelopes of nova and supernovae lose their stability relatively quickly while PNs preserve them all their lives.

16.13 Stability of the Envelopes of Wolf–Rayet Stars

Sometimes it is assumed that PNs may be the result of a continuous flow of gaseous matter from non-stationary stars, e.g. Wolf–Rayet stars. If the star ejects matter at a rate of $10^{-5}\mathcal{M}_\odot$/yr, continuously for several thousand years, then at some distance from the star an accumulation of mass equal to the PN mass can occur. The ejected matter will be decelerated owing to the resistance of interstellar matter.

Aside from other possible shortcomings of this hypothesis, consider its validity from the point of view of the stability of the envelope: can the envelope produced by the continuous ejection of gaseous matter from a central star be stable and turn into a PN?

Let $\mathcal{M}_0$ be the initial mass of the ejected envelope. The mass will increase owing to the matter ejected by the star with a velocity V_0 and rate αV_0 plus the matter gained by the envelope from the interstellar medium, i.e.

$$\mathcal{M} = \mathcal{M}_0 + \frac{4\pi}{3}\rho_c r^3 + \alpha \int_0^t (V_0 - V)\,dt\,, \tag{71}$$

where ρ_c is the density of the interstellar medium, and r is the radius of the envelope at t.

By definition, the loss of stability of the envelope starts at $\delta r^3 = 1$, i.e. when the proper mass of the expanding envelope becomes comparable with the amount of interstellar matter inside a volume equal to that of the nebula. Therefore we can write for the critical mass and radius of a nebula that will still remain stable

$$\mathcal{M} = \frac{4\pi}{3} r^3 \rho_c\,. \tag{72}$$

To determine $\mathcal{M}$ or r from (71) and (72) we need to know the equation of motion of an envelope with variable mass and variable expansion velocity. The study of this problem led to the following expressions for r, $\mathcal{M}$ and V (Minin, 1955):

$$r = \frac{\mathcal{M}}{a} x\,, \tag{73}$$

$$\mathcal{M} = \mathcal{M}_0(1 + \alpha x^2)/y\,, \tag{74}$$

$$V = V_0(1 - y)\,, \tag{75}$$

where x and y are related through the equation

$$\frac{1-y}{y} = \frac{\left[2x + \alpha x^4/2\right]^{1/2}}{1 - \alpha x^3}\,, \tag{76}$$

and where

$$\alpha = \frac{4\pi}{3}\frac{\rho_c \mathcal{M}_0^2}{a^3}\,. \tag{77}$$

Using (73), (74) and (75), (72) can be written as

$$\frac{4\pi}{3}\frac{\rho_c \mathcal{M}_0^2}{a^3} = \frac{1}{x^3}\left[(1 + \alpha x^3) + \sqrt{2x + \frac{\alpha}{2}x^4}\,\right]. \tag{78}$$

From the known values of $\mathcal{M}_0$, ρ_c and a the parameter x, and then, $\mathcal{M}$ and r by means of (74), (73) and (75), can be obtained.

Adopting, for example, $\mathcal{M}_0 = 10^{-5}\,\mathcal{M}_\odot$/yr, $V_0 = 1\,000$ km s^{-1} and $R_c = 5 \times 10^{-5}$ g cm^{-3}, we find for the parameters of the stable, "non-destructed" nebula

$$\mathcal{M} = 0.0002\mathcal{M}_\odot, \qquad r = 18\,000 \text{ a.u.}, \qquad V = 870 \text{ km s}^{-1},$$

where r stands for the maximum size of the nebula. We see that a typical PN cannot be formed in this way, since its lifetime would be of the order of 100 years; after this the envelope has to disintegrate.

Thus, consideration of the dynamic stability of gaseous envelopes leads us to the conclusion that PNs cannot originate by continuous ejection of gaseous matter from non-stationary stars, from Wolf–Rayet stars, in particular.

An example of a gaseous envelope nebula formed probably as a result of the action of a Wolf–Rayet star may be the well known nebula (?) NGC 6 888 in Cygnus with a WR type star at its centre, which, however, has already lost its stability, transforming to a filamentary structure.

16.14 Application of the Theory to Diffuse Nebulae

The necessary condition for the Rayleigh–Taylor instability is the appearance of acceleration at the surface of separation of the two fluids. This acceleration can arise for various reasons in different situations. In the case of PNs and the gaseous envelopes ejected by novae and supernovae we have seen that it is due to a compression at the surface of separation between the envelope and the interstellar medium. An analogous situation can occur through the interaction of diffuse nebulae with the surrounding small gas dust clouds, with the only difference that in this case the acceleration originates in the thermal expansion of the cloud or nebula (Spitzer, 1954). During the expansion with a velocity of the order 10 or 20 km s^{-1}, the hot gas compresses the dense, cold gas (cloud). This pressure is equivalent to an acceleration at the surface of separation between the hot and cold gases, so that it can result in a Rayleigh–Taylor instability.

Spitzer discovered a region of Rayleigh–Taylor instability in the diffuse nebula M 16 (NGC 6 611). It had the appearance of a tongue or "elephant trunk", and could be easily distinguished against the bright background of the nebula. The "elephant trunks" are less luminous, and in some cases they are completely dark; they consist, obviously, of cold gas associated with a large amount of dust.

One should not forget the further possibility that perturbing acceleration, and even a Rayleigh–Taylor instability, can be caused by gravitation, when say, a relatively cold condensation of gas and dust can find itself within the nebula. However, for typical diffuse nebulae of dimensions of the order of a few parsecs and condensations of 50 atoms cm^{-3}, calculations show that this acceleration is negligible. Therefore, a Rayleigh–Taylor instability arising from gravitation will occur only in either a particularly dense or a small diffuse nebula.

Figure 16.7 schematically shows the development of a Rayleigh–Taylor instability due to any kind of acceleration. The heavy fluid (cold cloud) is

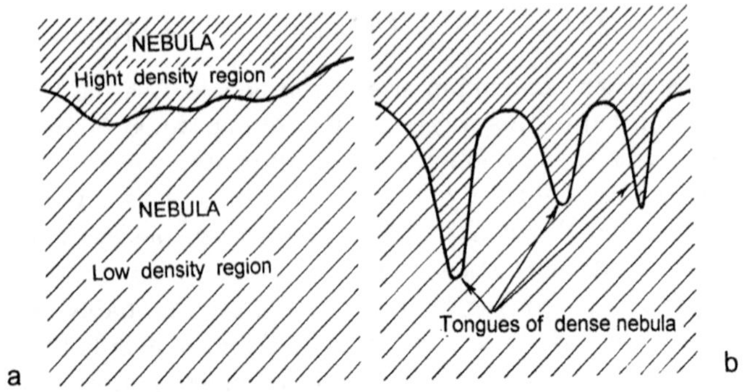

Fig. 16.7. Formation of Rayleigh–Taylor instabilities in diffuse nebulae. (**a**) Onset of the condition at the boundary between the low density cold nebula and the high density hot nebula. (**b**) Formation of tongues or "elephant trunks"

located above, and the light fluid (heated gas) is below. When the amplitude of the perturbation (irregularity) at the surface of separation of the two fluids is small compared with the wavelength of the perturbation, the amplitude grows according to the Rayleigh–Taylor theory, i.e. with an exponential law. For larger amplitudes, the theory no longer holds and the instability results in simple "free-fall" of filaments or "elephant trunks" of the heavy fluid into the light one.

An examination of the images of diffuse nebulae shows that Rayleigh–Taylor instabilities are quite common among them. Many dark tongues and "elephant trunks" can be seen in the diffuse nebulae NGC 2 264, NGC 2 237-9, NGC 7 000, IC 410, IC 1 848, S 193, etc. Especially remarkable in this respect are NGC 6 523 and IC 1 396. In the majority of cases the "elephant trunks" develop in the periphery of a nebular region, and, in particular, at its outer boundaries. Many clear "elephant-like" structures can be seen, for example, in large scale Hubble Space Telescope images of various regions of the diffuse nebula of Orion (NGC 1 976).

A number of clear manifestations of "elephant trunk" instabilities in PNs are shown in Fig. 16.7, in the case of NGC 6 543: note the boundary of its large halo with the surrounding interstellar medium.

16.15 Macroturbulence and the Stability of Envelopes

Up to now we have dealt only with the external factors which can cause the instability of a gaseous envelope, i.e. the resistance of the medium. But there is another, internal factor that can lead to the same consequences. We have in mind *macroturbulence* in the envelope, i.e. turbulence caused by the radial expansion of the nebula leading to velocity gradients between any two points within it. As far as microturbulence is concerned, as we have already mentioned in Chap. 1 (concerning Wilson's observations), it plays no significant role in PNs.

The regime of motion of a fluid is characterized by the dimensionless Reynolds number. For small values of this number the laminar flow is stable, while for large values the motion becomes turbulent. Between these values there is a transition region, where the motion changes from laminar to turbulent. The corresponding number is called *critical*. It can be determined experimentally and, for example, in conditions of terrestrial aerohydrodynamics, it is of the order of one thousand. However, its value in cosmic objects like PN, is unknown. Obviously, it is naive to simply assume that the critical Reynolds number for cosmic objects is the same as for the terrestrial conditions. Undoubtedly some role must be played, for example, by the magnetic interaction with the gas.

One can, however, attempt to estimate the order of magnitude of the critical Reynolds number from observational data. Indeed, among PNs there are those that show quite regular and stable shapes (IC 418, NGC 2 610 and others). This means that the Reynolds number in these nebulae is smaller or at least of the order of the critical number. It is not difficult to compute their Reynolds number, which turns out to be of the order of 10^5.

Thus, the Reynolds number in real PNs is one or two orders of magnitude larger than the critical Reynolds number of usual hydrodynamics. Nevertheless, turbulence is not present in nebulae. One must conclude that there exist forces or mechanisms which prevent the transition to the turbulent phase. Such a force or factor may be the magnetic field associated with the nebula (Chap. 17).

It is also of interest to determine how the Reynolds number varies as the nebula expands. First, we need to establish the dependence of the Reynolds number upon the parameters of the envelope.

The velocity gradient between two arbitrary points in an expanding nebula depends on the chosen points. Therefore, the Reynolds number will also change; its maximum value can be

$$R_e = \frac{2V_0\,D}{\nu}\,, \qquad (79)$$

where V_0 is the maximum velocity difference, D is the diameter of the envelope, and ν is the kinematic viscosity

$$\nu \approx ul\,, \qquad (80)$$

where u is the thermal velocity, and l the mean free path of the particles in the gas. We have $l \approx 1/\sigma n$, where n is the particle concentration and σ is the effective collisional cross-section. Finally, for the Reynolds number in the outer layers of an expanding envelope, we get

$$R_e = \frac{2V_0}{u}\,Dn\sigma\,. \qquad (81)$$

If we assume that the expansion velocity is constant and that the gas density decreases as the square of the radius, we have $n \sim D^{-2}$. Then, instead of (81), we obtain

$$R_e = \frac{2V_0}{u} \frac{n\sigma}{D}, \tag{82}$$

i.e. the Reynolds number decreases as the nebula increases in size. Thus, *if a gaseous envelope at some initial time is not in a turbulent state and is stable in shape, then later, with increasing size, it will certainly be stable.*

So, there are two hydrodynamic factors leading to the loss of stability of the envelope and eventual destruction that act in opposite directions. The first one, a resisting interstellar medium, plays a significant role only in envelopes of large dimensions, the second one, the inner turbulence, on the contrary, is significant only for small envelopes.

16.16 Motion of PNs in the Interstellar Medium

Now, consider briefly the possible influence of the interstellar medium on the geometry of PNs as they move through this medium.

The majority of PNs are distributed near the Galactic equator, where the mean density of the interstellar matter is high. The gaseous envelope of a nebula of giant dimensions and of relatively small mass must undergo drag opposing its motion through the medium. After a certain time, the central star should not appear at the centre of the nebula but will be somewhat shifted to the front edge. If such a phenomenon does indeed occur we should frequently observe PNs with shifted nuclei.

Actually, however, the nuclei of the majority of the nebulae are located quite symmetrically relative to the nebula. If any eccentricity is observed it never exceeds 0.1 or 0.2 of the radius of the nebula. Therefore we must conclude that for some reason the interstellar medium exerts no influence on the motion of the nebula.

In its motion through the interstellar medium only half of the nebula experiences drag. The envelope is decelerated in so far as part of its momentum is transferred to the interstellar matter gathered along the way and accreted by the envelope. It is quite probable that such a process does not occur because the linear thickness of the envelope of the nebula is many times larger than the mean free path of the gas particles. Therefore, neglecting variations in sizes of the nebulae, we can write

$$\mathcal{M}_0 \, w_0 = \left(\mathcal{M}_0 + \pi R^2 \rho_c s\right) \, w \,, \tag{83}$$

where w_0 is the initial relative velocity of the envelope and also of the nucleus, R is the radius of the nebula, and w is the velocity if the "frontier" of the nebula, i.e. the side facing the motion.

From (83) after the nebula has moved a distance s we have:

$$w = w_0 \left[1 + \frac{\pi \rho_c}{\mathcal{M}_0} R^2 s \right]^{-1} . \tag{84}$$

Substituting $w = ds/dt$ and integrating we find that

$$(s - R) + \frac{\pi \rho_c}{\mathcal{M}_0} R^2 \left(s^2 - R^2 \right) = w_0 t \, , \tag{85}$$

where the constant of integration has been evaluated from the condition that at $t = 0$ the nucleus is at the centre of our coordinate system and the front side of the nebula is at a distance R.

Equation (85) is actually an equation of motion for the front side of the envelope. Since the nucleus is not decelerated, its equation of motion is simply

$$s' = w_0 t \, . \tag{86}$$

As an example, let us calculate the time necessary to produce a "displacement" of the nucleus relative to the centre of the nebula (eccentricity) of one tenth of the radius of the nebula. We have

$$s = s_0 \, ,$$

$$s' = s_0 - 0.9R \, .$$

Comparing the above with (86) and also with (85), we find that the distance that the nebula must cover in order to reach the specified eccentricity is

$$s_0 = \left[R^2 + \frac{0.1 \mathcal{M}_0}{\pi \rho_c R} \right]^{1/2} , \tag{87}$$

and the required time is

$$t_0 = \frac{1}{w} \left[R^2 + \frac{0.1 \mathcal{M}_0}{\pi \rho_c R} \right]^{1/2} . \tag{88}$$

For $\mathcal{M}_0 = 0.01 \mathcal{M}_\odot$, $\rho_c = 5 \times 10^{-24}$ g cm^{-3} ($n_H \sim 3$ cm^{-3}), $w_0 =$ 30 km s^{-1}, and $R = 20\,000$ a.u., we get $t \approx 10\,000$ years.

Thus, the resistance of the interstellar medium can produce asymmetry in the position of the nucleus relative to the nebula of one tenth of its radius in a time of the order of 10 000 years; such an asymmetry in the position of central star is observed in the case of Lo Tr 4 by about 2″relative to the shell, which is nearly one-seventh of its radius (Rouch et al., 1996). Here we did not take into account the expansion of the nebula, which if included would lead to even longer time scales. Also, for calculations an obviously higher value (~ 3 cm^{-3}) for the density of the interstellar medium is assumed. Therefore, we conclude that in order to produce a significant displacement between the nucleus and the nebula under the influence of the drag of the interstellar medium, the required time scale is of the order of the lifetime of the nebula itself. The consequences of such an influence could only be observed in very large nebulae, and have indeed been observed (Zucker, Soker, 1993).

The fact that in few relatively large PNs can we observe an increase in brightness on one side shows that the drag caused by the interstellar medium does already occur. Examples of such nebulae are NGC 7139, A 16, as well as NGC 6888, which, however, is not included in catalogues of PNs. A special programme of deep narrow-band CCD imaging is performed through the interference filters, to investigate the mechanism responsible for the present shaping of a group of nine faint nearby (within 500 pc from the Sun) PNs of large angular extent (Xilouris et al., 1996). Most of these PNs were found to interact with interstellar medium. The morphologies which were originally produced by the initial ejection, have faded and become less detectable in most cases, allowing the "planetary nebula – interstellar medium" interaction to emerge as the dominant process. The interaction regions appear mainly as asymmetric brightness enhancements (A 62, A 7). In more advanced phases (A 74, HFG 1, S 188, IW 1, S 176) a strong ionization stratification is evident with the [OIII] emission, confirmed by the H_α emission. In late stages of expansion, the central star is substantially displaced from its original position (S 68, IW 1, IW 2, S 188). In one case, A 7, located far from the plane of the Galaxy, no significant evidence an on-going interaction has been found, as expected given the low density of the interstellar medium (Borkowski et al., 1992).

So, the dynamic interaction of nebular material with interstellar medium is a reality.

17. Magnetic Fields in Planetary Nebulae

17.1 Bipolarity of Nebulae

Bipolarity is one of the most common structural features of PNs. Objects of this type have two bright "beads", or regions of enhanced brightness, arranged symmetrically in relation to the nucleus. Examples of bipolar nebulae are given in Chap. 1 (Plate VII).

No PN with, say, one or three "beads" has been discovered; we always observe only two "beads", and their forms differ in different nebulae – from sickle or banana-like to segment-shaped. One nebula, as a rule, has two "beads", of similar types and brightness, and located strongly symmetrically to the nebula's centre. The highly pronounced bipolar structure of PNs can be considered evidence of their electromagnetic origin, including, in rare cases the possible contribution of the Galactic magnetic field.

17.2 The Role of the Galactic Magnetic Field

The regular magnetic field of the Galaxy may lead to the redistribution of almost completely ionized nebular matter, and as a result, it can become non-uniformly distributed in different sections within the nebula. If this hypothesis is true, a certain correlation will be expected between the direction of the Galactic magnetic lines and, say, of the direction of the major axis of a bipolar PN (the minor axis of a bipolar nebula passes through both "beads", and the major axis is perpendicular to the minor axis; see Chap. 1).

The directions of the Galactic magnetic force lines usually change from one region to another. However, in certain sufficiently large regions of the Galaxy the magnetic field is more or less homogeneous – the field lines are almost parallel to each other and are parallel to the Galactic equator. As for the strength of Galactic magnetic field, estimated by various means, it is of the order of 10^{-5}–10^{-6} gauss.

Is there any correlation between the orientation of the "beads" of bipolar nebulae and the direction of the Galactic equator? There is no definite opinion on this; everything seems to depend on the manner in which the analysis of observational material is carried out (Heiligman, 1980). Nevertheless, the answer to this question seems to be negative. The analysis of observational data does not reveal any correlation between the orientation of "beads" and the direction of the Galactic equator: PNs with their major axis parallel to the Galactic equator are as frequent as those with major axis perpendicular to the equator (Jewitt et al. 1986). For example, five PNs – NGC 6 153, Mz 1, Mz 2, Mz 3 and Sp 1 – four of which are bipolar, are clustered in a relatively small region of the southern sky, near the Galactic equator. One of them, Sp 1, is an ordinary fairly large ring nebula without any traces of bipolarity, while the major axes of the next four bipolar nebulae are neither parallel to each other nor correlated with the Galactic equator.

One more fact: bipolar PNs are sometimes located at rather high Galactic latitudes – a typical bipolar nebula NGC 3 587 has $b = +58°$ – and there the magnetic field is too weak to have a marked influence on internal motions of the nebula and on its structure.

So, at least the process of "bead" formation in PNs is independent of the Galactic magnetic field.

17.3 The Role of the Central Star

The magnetic field of the central star – the nebula's nucleus – cannot affect the ionized matter of the nebula either. As this is a dipole field, it decreases as the cube of the distance, so that the magnetic field strength in the region of the "beads" will be as low as 10^{-10} gauss, even if a field of 10^6 gauss at the surface of the central star is assumed.

17.4 The Intrinsic Magnetic Field of a Nebula

Thus, neither the regular magnetic field of the Galaxy nor the dipole magnetic field of the central star markedly affect the structure and the shape of a PN, but the idea of the electromagnetic origin of the bipolar structure of PNs cannot be easily rejected. If this is so, we have to conclude that the *nebulae possess an intrinsic magnetic field*, which is neither an extension of the field of the central star nor related to interstellar magnetic fields.

The magnetic field of a PN cannot be homogeneous, i.e. of constant magnitude and direction at all points of the nebula, otherwise, no redistribution of matter within the nebula might occur. For the same reason, this field cannot be random. The bipolar structure of PNs indicates, first, that the strength and the direction of the magnetic field should vary in a definite manner from one point to another, and second, that this field should be symmetrical with respect to the axis of the nebula.

A *dipole magnetic field* within a PN formally fulfils these requirements.

In the next sections of this chapter we will show that the hypothesis of a PN dipole magnetic field provides a satisfactory interpretation of the observed shapes and structures.

17.5 Equilibrium of a Planetary Nebula with a Magnetic Field

PNs are expanding gas spheres, much larger than stars, and can possess positive energy. How will a sphere of this kind behave in the presence of a magnetic field?

If the nebula has an intrinsic magnetic field, obviously, it also carries a certain amount of magnetic energy. Let $\bar{H}$ be the average magnetic field strength, averaged over the nebula. The specific magnetic energy, i.e. the energy per unit volume of the nebula, is then $\bar{H}^2/8\pi$. The total magnetic energy $\mathcal{E}$ of a homogeneous spherical nebula is thus given by

$$\mathcal{E} = \frac{1}{8\pi}\,\bar{H}^2 V = \frac{1}{6}\,\bar{H}^2\,R^3\,, \tag{1}$$

where V and R are the volume and the radius of the nebula. The magnetic energy, obviously, is a positive quantity, independent of the polarity of the field. Its contribution to the mechanical equilibrium of a gaseous sphere is therefore identical to that of kinetic energy: the magnetic energy opposes the negative potential energy and tends to destroy the equilibrium, i.e., to cause the sphere to expand. The problem is to estimate the field that can cause expansion in spheres of known mass and radius.

This problem can be solved by means of the virial theorem, derived in the case of a magnetic field by Chandrasekhar and Fermi (1953):

$$\frac{1}{2}\,\frac{\mathrm{d}^2 I}{\mathrm{d}t^2} = 2T + 3(\gamma - 1)\,U + \mathcal{E} + \Omega\,. \tag{2}$$

Here I is the moment of inertia of the gaseous sphere, T, U, $\mathcal{E}$ and Ω are, respectively, its kinetic, thermal, magnetic and potential energies, and γ is the adiabatic index.

The moment of inertia I varies as the nebula expands. Assume that the velocity of expansion V_0 is constant; then for a homogeneous nebula of mass $\mathcal{M}$ we have

$$I(t) = \int r^2\,\mathrm{d}m = \frac{3}{5}\,\mathcal{M}\,v_0^2 t^2\,, \tag{3}$$

where t is the time for the nebula to expand up to a radius $R\ (= v_0 t)$.

Inserting into (2) the expression for I from (3) and observing that $\mathcal{M}v_0^2/2 = T$, we find that

$$\frac{4}{5}T + 3(\gamma - 1)\,U + \mathcal{E} + \Omega\,. \tag{4}$$

On the other hand, the total energy of the nebula is, if we neglect the radiation energy,

$$E = T + U + \Omega . \tag{5}$$

Eliminating U from (4) and (5), we find that

$$E = -\frac{3\gamma - 4}{3(\gamma - 1)} \left(|\Omega| - \mathcal{E} \right) + \frac{15\gamma - 1}{15(\gamma - 1)} T . \tag{6}$$

The nebula will not expand, i.e. will remain in equilibrium, if its total energy is negative, $E < 0$. Thus, for $\gamma = 5/3$ (hydrogen), we have from (6)

$$\frac{6}{5} T + \mathcal{E} < |\Omega| . \tag{7}$$

If $T = 0$, the nebula is in equilibrium when

$$\mathcal{E} < |\Omega| . \tag{8}$$

PNs do expand, i.e. they are not in equilibrium. Assuming for the moment that the lack of equilibrium is due to the action of the magnetic field, we replace (8) with

$$\mathcal{E} > |\Omega| . \tag{9}$$

For constant spherical configurations we have

$$|\Omega| = \frac{3}{5} \frac{G\mathcal{M}^2}{R} . \tag{10}$$

Substituting (1) and (10) into (9), we find (Gurzadyan, 1957) that

$$\bar{H} > \left(\frac{18}{5} G \right)^{1/2} \frac{\mathcal{M}}{R} . \tag{11}$$

For a typical PN, we have $\mathcal{M} \approx 0.1 \mathcal{M}_\odot$ and $R \approx 10\,000$ a.u. Inserting these into (11), we find that

$$\bar{H} > 10^{-6} \text{ gauss} \tag{12}$$

This result is remarkable for outlining the possibly crucial role of the magnetic fields in the equilibrium of nebulae. Even with the absence of other forces responsible for the expansion – internal forces, gas and radiation pressures – *a magnetic field of* $\bar{H} > 10^{-6}$ *gauss will be enough to destroy the mechanical equilibrium of a nebula and to cause its expansion.*

A realistic field strength can be estimated, if one assumes that the motion of gaseous matter in certain parts of the nebula is controlled by magnetic fields. This implies that the magnetic energy in the relevant regions is comparable with the corresponding thermal energy of the gas. In other words, if the formation of "beads" in PNs is attributed to the internal magnetic field, we should have $U \sim \mathcal{E}$ or

$$\frac{1}{6}\bar{H}R^3 \sim \frac{\mathcal{M}u^2}{2}, \tag{13}$$

where u is the thermal velocity of the gas particles. At ~ 10 km s^{-1} we obtain for the average field strength (for the same mass and radius as above)

$$\bar{H} \sim 10^{-3} \text{ gauss}. \tag{14}$$

Given magnetic fields of this magnitude, PNs will no longer be in equilibrium and must therefore expand, even if no other forces are acting in this direction.

Here we have considered a method of determining the mean magnetic field in nebulae. Another, more precise method, which can be applied to particular PNs, is described below.

Table 17.1. Total energies of various kinds for a homogeneous PN with $\mathcal{M} = 0.1\mathcal{M}_\odot$ and $R_n = 1.5 \times 10^{17}$ cm

Form of energy	Energy in ergs
Kinetic T, $v_0 = 20$ km s^{-1}	4×10^{44}
Thermal U, $u_0 = 10$ km s^{-1}	10^{44}
Gravitational Ω	2×10^{40}
Magnetic $\mathcal{E}$ $H = 10^{-3}$ gauss	5×10^{44}
$H = 10^{4}$ gauss	5×10^{42}
$H = 10^{-5}$ gauss	5×10^{40}
Total energy of an average star (Sun)	5×10^{48}

Table 17.1 gives a summary of the contribution of various types of energy in PNs of average size and mass. The last line shows the total energy – gravitation, thermal, ionization etc. – for an average star (the Sun). Note that the gravitational energy of a nebula is three orders of magnitude less than the kinetic, the thermal and the magnetic energies, while the sum of the last three components is three orders of magnitude less than the total energy of the central star, if this is a solar type star. In other words, a PN contains only a negligible fraction of the total energy of the nucleus.

17.6 The Energy Balance in Planetary Nebulae

The picture is different if one considers the energy balance in PNs possessing a magnetic field. The possibility of radiative equilibrium was demonstrated in Chap. 11. Other types of energy are obviously also responsible for the equilibrium, otherwise the apparent pressure gradient will cause large scale motions. This is not supported by observations (see Chap. 1). We therefore conclude that energy balance is reached and that the relaxation time (time to reach that equilibrium) is significantly less than the lifetime of the PN.

These considerations, however, cannot be considered a rigorous proof. In particular, estimation of the relaxation time requires special consideration. So, we postulate that *equipartition does exist between all types of energies at all points within the nebula*, i.e.

$$\sum E_i = C\,, \tag{15}$$

where the sum is taken over all energy types. Below we shall show that this postulate, together with the hypothesis of symmetrical, dipole-like magnetic fields in PNs, provides a satisfactory interpretation of various features of their structure.

Both the kinetic energy of the thermal motion of gas particles and the magnetic energy are the basic components of total energy of the nebula: gravitational energy is negligibly small, and actually there is no turbulence. For simplicity we will consider a stationary, non-expanding nebula, although this is not a fundamental restriction; it only implies that the energy balance within the nebula is reached before the nebula can significantly change its size.

We can therefore *postulate* the condition of energy balance (15) in the form (Gurzadyan, 1958, 1969)

$$\frac{H^2}{8\pi} + 2nkT = \text{const}\,, \tag{16}$$

where H is the magnetic field at a point within the nebula, n is the concentration of protons (electrons) at that point, and T is the electron temperature. The first term on the left hand side of (16) is the density of magnetic energy, the second is the density of the thermal energy. Both are local quantities, varying with position. The factor 2 in the second term implies that the gas (hydrogen) in the nebula is almost completely ionized, the number of protons being equal to the number of electrons, so that the total number of free particles is twice the number of free electrons.

If the field strength varies along a line of force, the magnetic energy density will lead to a difference in magnetic pressure at various points. This will destroy any initial uniform gas distribution: the gas density will increase in the regions of low magnetic pressure, and vice versa. The drift of gas along the line of force will cease only when the magnetic pressure gradient $\nabla H^2/8\pi$ is balanced by the gas pressure gradient $2kT\nabla(n)$:

$$\nabla\left(\frac{H^2}{8\pi} + 2nkT\right) = 0\,. \tag{17}$$

Integrating this, we obtain (16).

In (16), H and n are the only position dependent quantities; the electron temperature t is generally determined by atomic absorption and emission processes and henceforth is assumed constant over the entire nebula. In functional notation (16), therefore, has the form

$$\frac{1}{8\pi} \mid H(x,\, y,\, z) \mid^2 + 2kTn(x,\, y,\, z) = C\,. \tag{18}$$

The constant C is determined from the conservation of mass along the line of force:

$$\oint n\,(x,\, y,\, z)\, \mathrm{d}l = n_0 \oint \mathrm{d}l = n_0 L\,, \tag{19}$$

where the integral is taken along the line of force; L is the length of the line, n_0 is the concentration of atoms of the "unperturbed" nebula, i.e. of the uniform density nebula without a magnetic field. Using (18), we find that

$$C = 2n_0 kT + \frac{1}{8\pi L} \oint \mid H\,(x,\, y,\, z) \mid^2 \mathrm{d}l\,. \tag{20}$$

Equation (18) is the basis of all our subsequent calculations. In each particular case the distribution of gas density, and therefore the distribution of brightness, over the nebula is determined by the character and the magnitude of the magnetic field $H\,(x,\, y,\, z)$. The apparent shape and structure of the nebula is governed by this factor.

Our problem is to find out the type of magnetic field that will fit the observed bipolar structure of the PN. The magnetic field of an extended dipole is found to fulfil these requirements fairly closely.

We start with a formal discussion of the character of the nebular structure in the case of magnetic fields of point and extended dipoles.

17.7 The Point Dipole

A magnetic dipole is a system consisting of two "charges" of opposite signs at a distance l from each other. Normally, in physics the term dipole is used only where l is much less than the distance r of the "charges" from the field point, i.e. when $r \gg l/2$. In PNs, however, we are dealing with dipoles of a size comparable with the radius of the nebula, i.e. $r \sim l/2$. We therefore speak of a "point" dipole (Fig. 17.1(a)), when $r \gg l/2$, i.e. when the dipole is in fact situated at the centre of the nebula, and of an "extended" dipole (Fig. 17.1(b)) when $r \sim l/2$, i.e. when the dipole "occupies" the entire nebula or, more precisely, when the "charge" or the poles of the dipole are at opposite ends of the nebula. The line through the "charges" (or the S and N poles) is the dipole axis or the magnetic axis of the nebula.

In this section, we shall consider the effect of a point dipole on a nebula and show that the dipole magnetic field does indeed produce bright "spots" situated symmetrically with respect to the nucleus.

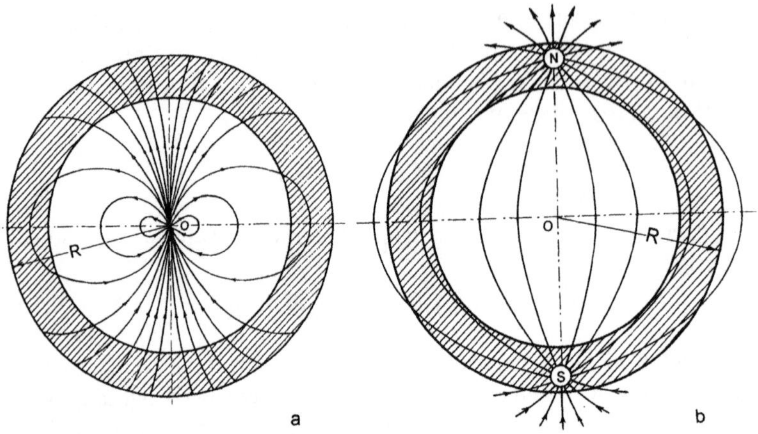

Fig. 17.1. (**a**) A point magnetic dipole in a ring-like nebula. (**b**) An extended magnetic dipole in a ring-like nebula

Consider a spherical nebula of radius R with a constant concentration of electrons and protons, n_0, at all points. Consider a section passing through the centre O of the nebula ("a central section") and replace there a point dipole with a moment a. The field strength at a given point (r, φ) owing to the point magnetic dipole in plane polar coordinates should be

$$H\ (r,\ \varphi) = a\eta(r,\ \varphi)\ , \tag{21}$$

where

$$\eta(r,\ \varphi) = \frac{1}{r^3}\sqrt{1 + 3\sin^2\varphi}\ . \tag{22}$$

The field vector is directed along the tangent to the line of force at an angle α with respect to the radius vector, defined as

$$\tan\alpha = \frac{1}{2}\cot\varphi\ . \tag{23}$$

Note that α is constant for all force lines, i.e. it is independent of r. This is a characteristic feature of the point dipole, violated, however, as we shall see below, in the case of an extended dipole.

Equation (21) shows that the magnetic field in the central section has gradients in both coordinates. The field gradient creates a magnetic pressure gradient which, as we have noted in the preceding, destroys the initial uniform gas density distribution in the nebula.

Inserting for $H(r, \varphi)$ in (18) its expression from (23), we obtain an equation for the relative concentration of gas at the point (r, φ):

$$\frac{n(r,\ \varphi)}{n_0} = 1 - \sigma\eta^2(r,\ \varphi)\,, \tag{24}$$

where

$$\sigma = \frac{a^2}{16\pi kTn_0} . \tag{25}$$

Since the density of kinetic energy n_0kT is conserved, we shall call σ an "energy parameter".

As follows from (24), first, the density distribution is symmetrical with respect to the magnetic axis of the nebula and, second, the concentration at a given distance from the centre has a minimum in the direction of the magnetic axis ($\varphi = 90°$) and a maximum in the magnetic equator ($\varphi = 0$). Moreover, the concentration of ions is relatively low near the centres, increasing outwards.

The volume emission of the nebula is proportional to the square of the electron or ion concentration. Therefore, even only a small difference between the concentrations at various points within the nebula may lead to a substantial difference in the surface brightness. The largest difference occurs between the brightness in the direction of the magnetic axis of the nebula (region of minimum brightness) and that on its equatorial axis (region of maximum brightness). The contrast is preserved when the actual space distribution obtained by rotating the central magnetic axis is projected onto the image plane perpendicular to the magnetic axis of the nebula. The resulting apparent image is therefore a nebula with two brightness maxima at points symmetrically located with respect to the centre, i.e. a *bipolar nebula.*

The problem is to define a kind of "avoidance" zone for which $\sigma\eta^2(r_0, \varphi_0) \approx 1$. Roughly this "avoidance" zone can be represented as a sphere with a radius

$$r_0 \approx \sigma^{1/6} = \left(\frac{a^2}{16\pi n_0 kT}\right)^{1/6} . \tag{26}$$

The hypothesis of a point dipole at the centre of the nebula thus qualitatively explains the formation of "beads", i.e., the formation of bipolar PNs. The results, however, are quantitatively inadmissible, if the actual size of the nebula is compared with the geometrical extent of a dipole, $l = (a/H)^{1/3}$, the magnetic field at the dipole should be of the order of 10^6 gauss, assuming a nebular field of 10^{-4}–10^{-5} gauss, and the dipole length of the order of a stellar radius.

This difficulty can be eliminated if the hypothesis of a point dipole is replaced by one of an extended dipole with a size comparable to that of the nebula.

17.8 The Extended Dipole

The magnetic field strength in a nebula at a point $(r,\ \varphi)$ caused by an extended dipole is

$$H(r,\ \varphi) = a\eta_1(r,\ \varphi) , \tag{27}$$

where $\eta_1(r,\ \varphi)$ has the form (Gurzadyan, 1957)

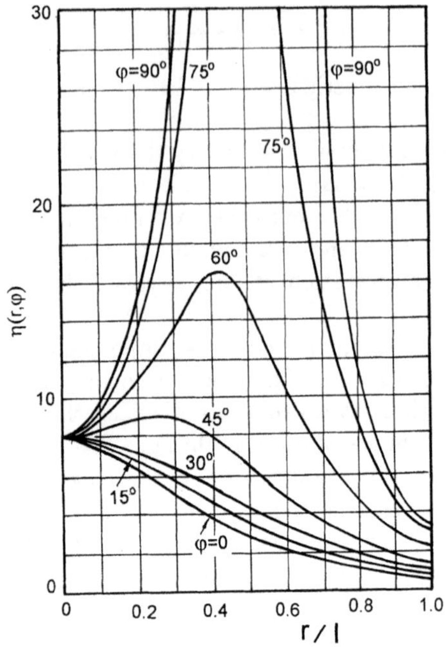

Fig. 17.2. Curves for the function $\eta\,(r,\,\varphi)$ as a function of r/l, for various values of φ

$$\eta\,(r,\,\varphi) =$$

$$2^{5/2}\kappa^2 \frac{\left\{(1+\kappa^2)^2 + 4\kappa^2\sin^2\varphi - (1-\kappa^2)\left[(1+\kappa^2)^2 - 4\kappa^2\sin^2\varphi\right]^{1/2}\right\}^{1/2}}{(1+\kappa^2)^2 - 4\kappa^2\sin^2\varphi}\,, \tag{28}$$

where $\kappa = l/2r$ is a measureless distance.

This is a general formula applicable to large and small distances. In particular, for $l/r \ll 1$, (28) reduces to (22). In practice, the dipole can be regarded as a point dipole for l/r of the order of 0.1. The function $\eta_1(r,\,\varphi)$ is plotted against r/l in Fig. 17.2 for various values of φ.

The distribution of the relative concentration of electrons in the central section of the nebula is given as in (24) by

$$\frac{n(r,\,\varphi)}{n_0} = 1 - \sigma\eta_1^2(r,\,\varphi)\,. \tag{29}$$

In the case of an extended dipole the geometry of the central section and of the shape of the nebula as projected onto the sky are markedly affected by the ratio l/R of the dipole length (l) to the outer radius of the nebula (R). This ratio also characterizes the relative "depth" of the magnetic poles in the nebula. On the other hand, the outer radius R of the nebula and the thickness h of the envelope enter the theory as independent quantities, unrelated to the magnetic field. We are therefore dealing with two parameters, l/r and l/h, which may take on various values in different nebulae.

Going back to (16), one should note that the application of this condition to the extended magnetic dipole and at all points of the nebula, strictly speaking, is wrong. First, the conjecture (16) is true only when the magnetic force lines within the nebula are parallel to each other, which is not true for an extended dipole and its macrostructure, and, second, in spite of the compensation of the strength of the magnetic field by the tension of the force lines, the dipole field cannot change the initial distribution of ionized gas in nebula. The condition on the constancy of the sum of the kinetic energy and the magnetic energy at all points of the nebula may be accepted therefore only as a working hypothesis, as a kind of postulate, to explain the phenomenon of PN bipolarity.

It is difficult to doubt the existence of a proper magnetic field connected with the PN itself and therefore the existence of a magnetic axis as well as a magnetic equator with an obvious gradient in the magnitude of the magnetic field while moving from one to another. As we will see below, by means of these two assumptions, namely (a) the existence of a dipole-like magnetic field and (b) the constancy of the sum of magnetic and kinetic energies at all points of the PN, one can explain the observed bipolarity and diversity of the PN forms. The magnetic fields within the PNs resemble a dipole field only with respect to the character of *the distribution* of the absolute magnitude of the field. In reality, however, these fields differ, perhaps essentially, from dipole fields in the *structure of the field lines.* In PNs one can imagine conditions that are favourable for the required behaviour of the lines, e.g. when the lines of the quasidipole field can be quasiparallel to each other, thus fulfilling condition (16).

The structure of the field in the regions of magnetic poles is also also far from clear. The field in the central regions of the nebula, at least, cannot be of dipole type as follows from the structure of monochromatic images of the central regions of PNs. For example, the main images of the PNs NGC 6 720, NGC 7 662 and IC 2 165 (Plates V, VIII) are bipolar in the lines N_1, N_2, H_β etc., while the central parts of the same PNs in the high excitation lines 4 686 HeII, 3 869 [NeIII], 3 426 [NeV] appear as disks or rings without noticeable signs of bipolarity (Wilson, 1990). One should look for an explanation in the character of the "avoidance" zones in the central parts of PNs, though the very problem of the equilibrium of ionized gas in the central parts of PNs under the presence of dipole field loses its sense.

In view of all these considerations, below, when we discuss the dipole or quasidipole field within PNs, we will have in mind simply the corresponding character of the magnetic field.

17.9 Brightness Distribution of a Nebula with a Magnetic Field

In this section the isophotes derived from the above theoretical computations are compared with the observed isophotes or shapes of individual PNs.

We will consider the particular case of a nebula with its magnetic axis orthogonal to the line of sight. The relative surface brightness $J(p,\ \psi)$ at the point $(p,\ \psi)$ of the PN is then given by

$$J(p,\ \psi) = C_1 \int_P^{P_1} \frac{n^2(r,\ \psi)}{\sqrt{r^2 - p^2}}\, r\, \mathrm{d}r\,, \tag{30}$$

where C_1 is a constant, $p_1 = (R^2 - p^2\ \cos\psi)^{1/2}$ and $n(r,\ \psi)$ is given by (29) with $\sin\varphi$ replaced by

$$\sin\varphi = \frac{p}{r}\sin\psi\,.$$

In (30) it is assumed that the nebula is transparent with respect to its own radiation. The integral in (30) is evaluated graphically for various values of the radius R of the PN, the thickness of the envelope h (both in units of l), and the parameter σ.

Figure 17.3(a) shows the plot of isophotes for a full nebula with $R = 2.25$ (i.e., $R/l = 2.25$), and $\sigma = 1$. The isophote numbers in this and the following figures denote the intensity in arbitrary units; the dashed line indicates the

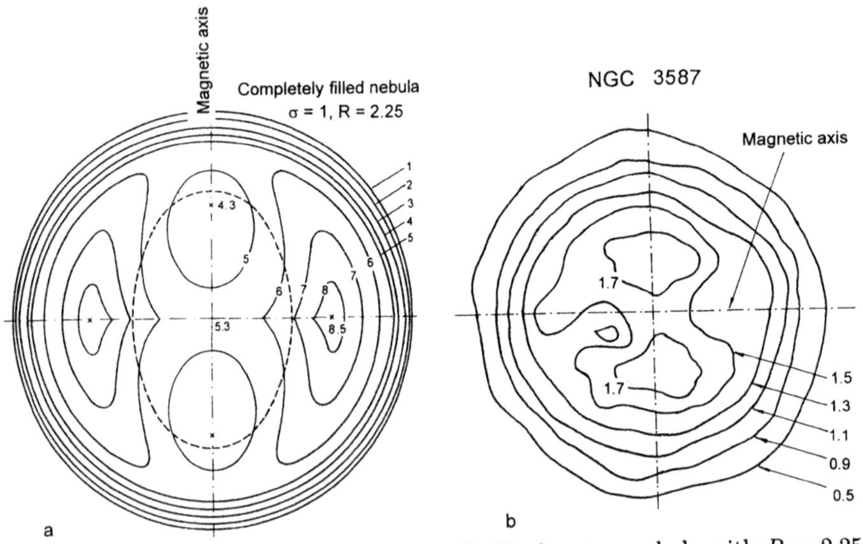

Fig. 17.3. (**a**) Theoretical isophotes of a "full" planetary nebula with $R = 2.25$ and $\sigma = 1$. (**b**) The observed isophotes of NGC 3587 (the "Owl" nebula) in red light

limits of the avoidance region. The two "beads" – regions of maximum brightness in the equatorial plane of the nebula located symmetrically with respect the centre – are rather prominent. The regions of minimum brightness lie on the magnetic axis.

The surface brightness distribution of NGC 3 587 (the "Owl" nebula, see Plate VI), is quite similar to the theoretical pattern of Fig. 17.3(a). The observed isophotes of this nebula are shown in Fig. 17.3(b); both qualitative and quantitative similarity is seen. For example, the ratio of the maximum – in the equatorial section of the nebula – to the minimum (on the magnetic axis) intensity for NGC 3 587 is approximately 1.9, while the theoretical ratio, according to Fig. 17.3(a), is 2.

Since the theoretical and the observed isophotes are identical, we can estimate the magnetic field strengths in various parts of NGC 3 587. In this case $\sigma = 1$, and we have from (25)

$$\sigma = \frac{a^2}{16\pi n_0 k T_e} = 1 . \tag{31}$$

We have for the electron concentration and electron temperature of NGC 3 587 $n_0 = 150\,\text{cm}^{-3}$ and $T_e = 100\,000$ K. Using these data, we find from (31) that $a \approx 10^{-4}$ gauss, and for the field at any point of the nebula we have from (29)

$$H(r,\ \varphi) = a\eta_1(r,\ \varphi) = 10^{-4}\eta_1(r,\ \varphi) \tag{32}$$

For $r = 1$, approximately half the PN's radius, we obtain from (28) or Fig. 17.2: $\eta_1(1,\ 0°) = 0.715$ and $\eta_1(1,\ 90°) = 3.56$. Hence, $H(1,\ 0°) = 0.7. \times 10^{-4}$ gauss and $H(1,\ 90°) = 4 \times 10^{-4}$ gauss. In the same manner, for $r = 2$, i.e. near the edge of the PN, we have $\eta_1(2,\ 0°) = 0.115$ and $\eta_1(2,\ 90°) = 0.286$, and we find, respectively, $H(2,\ 0°) = 3 \times 10^{-5}$ gauss and $\eta_1(2,\ 90°) = 3 \times 10^{-5}$ gauss. At the centre ($r = 0$) we have $\eta_1(0) = 8$ and, hence, $H = 0.8 \times 10^{-3}$ gauss. Note that in our computational scheme the magnetic field is not too sensitive to the electron concentration and electron temperature: $H \sim (n_0,\ T_e)^{1/2}$.

Figure 17.4(a) shows the theoretical isophotes for a full nebula with $\sigma = 1$, as before, but with $R = 1.67$. Here the "beads" are peaked inward, and the regions of maximum intensity are generally more rounded. The result is a "dumb-bell" nebula, similar to NGC 3 195, He 2–153. A good example of this kind is provided by A 70 (see Plate I).

Figure 17.4(b) is still another example of theoretical isophotes. It corresponds to $\sigma = 10$, $R = 2$ and $h = 0.1R$. Here the contrast between the "beads" is even greater. This brightness distribution is also observed, for example, for NGC 40, NGC 6 058, and others. Remarkable objects of this class are A 66 (Plate VIII) and A 19.

Figure 17.5(a) shows the theoretical isophotes for an envelope of thickness h (a hollow nebula), yielding 1/4 of the nebular radius and $\sigma = 1/12.7$. The external PN radii are $R = 1.34$.

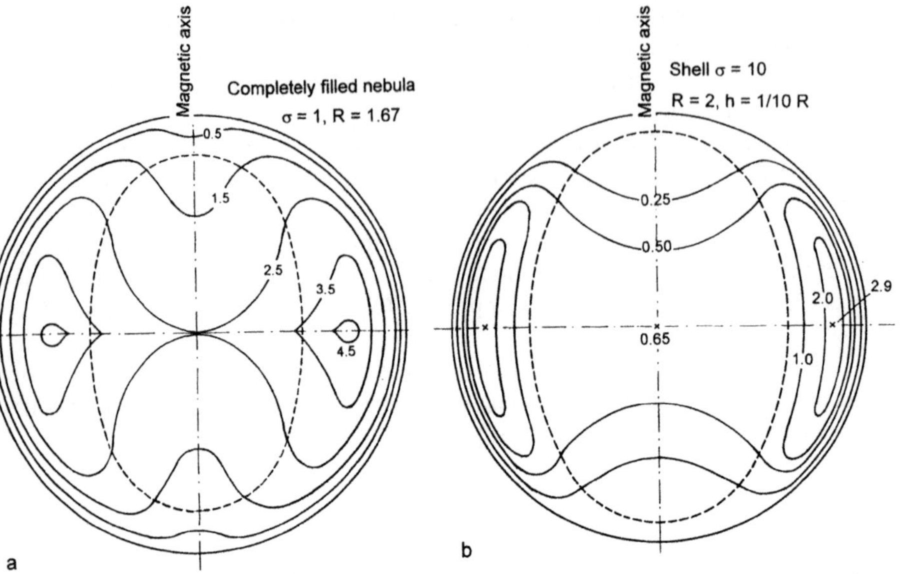

Fig. 17.4. (**a**) Theoretical system of isophotes for a "full" planetary nebula with $R = 1.67$ and $\sigma = 1$. (**b**) Theoretical system of isophotes for a hollow planetary nebula with $R = 2$, $h = 0.1R$ and $\sigma = 10$

The isophotes of Fig. 17.5(a) are structurally similar to those of a ring nebula, where, although no distinct "beads" are observed, the brightness increases towards the ends of the minor axis (NGC 6720, NGC 7293 etc., Plates V, I). From the observed ratio of maximum intensity to the central intensity for NGC 6720 is about 2.5–3. Almost the same ratio is obtained from the isophotes of Fig. 17.5(a). Thus we find that $H \approx 10^{-4}$ gauss for the magnetic field strength at a distance $r \sim 1$–1.3 in the region of enhanced brightness ($\varphi = O^0$), $H \approx 5 \times 10^{-4}$ gauss at the magnetic poles ($\varphi = 90°$), and $H \approx 10^{-3}$ gauss at the centre of the nebula.

Isophotes in Fig. 17.5(b) are in satisfactory qualitative agreement with the structure and the shape of the PN Mz 2 as well as of NGC 7662 (Plates VII, VIII). The "beads" of these PNs are "banana-shaped".

Thus, the conjecture on the existence of an extent dipole within the PN can explain the observed variety of forms and the main structural features of bipolar nebulae. Bipolar PNs may thus indicate whether the magnitude of the magnetic field within the PN follows the pattern of the field of an extended dipole.

According to an alternative point of view, the bipolarity of PNs can be the result of the processes occurring in the nebula's nucleus if the latter is a close binary system (Corradi, Schwarz, 1992; Han et al, 1995). In this case the bipolar structure of the nebula is treated as an event or process of purely gasdynamical origin, i.e. without the couplication of magnetic fields. In this

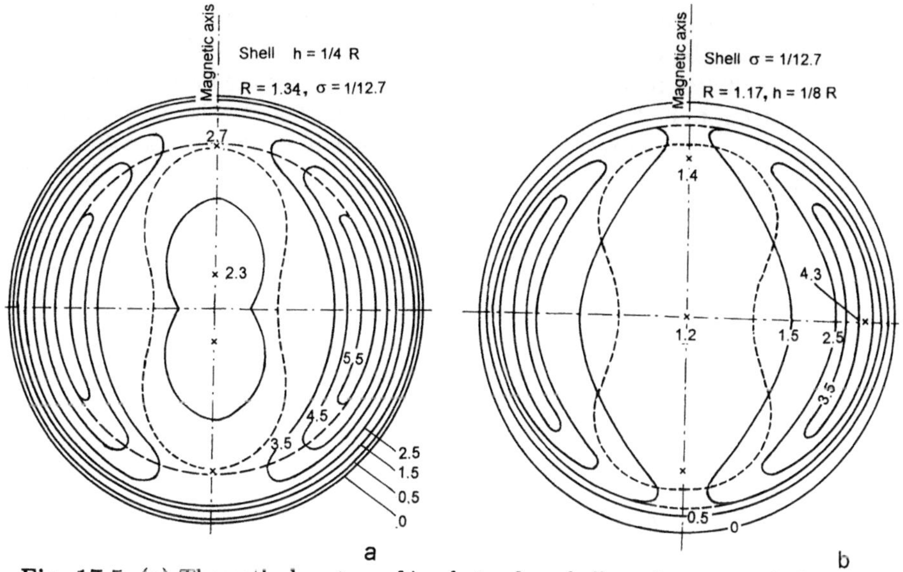

Fig. 17.5. (**a**) Theoretical system of isophotes for a hollow planetary nebula with $R = 1.34$, $h = 0.25R$ and $\sigma = 1/12.7$. (**b**) Theoretical system of isophotes for a hollow planetary nebula with $R = 1.17$, $h = 0.125R$ and $\sigma = 1/12.7$

case, however, one has to explain in a natural way the huge difference between the scales of the bipolarity phenomenon and the effects within the centre of the PN, while macromagnetic fields succeed in doing this quite easily.

17.10 Elongated Nebulae

The intrinsic magnetic fields within PNs affect their external structure, and in particular their apparent shapes. The question of the mechanical equilibrium of a nebula or a gaseous sphere is of crucial significance in this respect.

The role of the magnetic field with respect to the shape of a conducting fluid sphere in mechanical equilibrium was first considered by Chandrasekhar and Fermi (1953). They showed that a dipole magnetic field acting on the external parts of the sphere and a uniform magnetic field acting on the interior parts, would flatten the gaseous sphere in the direction of the magnetic axis. The relative flattening (oblateness) in this case is proportional to the square of the polar value of the magnetic field.

The situation is rather different in gaseous spheres that are not in a state of mechanical equilibrium. It has been shown (see Sect. 17.2) that a magnetic field leads to the expansion of the PN by passing additional kinetic energy to it. In other words in the presence of a magnetic field a nebula will expand even more rapidly than without it. If the magnetic field is homogeneous, its influence is isotropic and the external appearance of the nebula will remain unchanged; a nebula that is initially spherical will remain so while expanding.

The results of previous sections show, however, that the magnetic fields in PNs ought not to be homogeneous: the field strength along the magnetic axis is larger than in the direction of the magnetic equator. The nebula will therefore expand at different rates in various directions, and its initial spherical shape will be distorted.

Suppose that the unknown force causing the expansion acts uniformly in all directions, i.e., the gas maintains its spherical shape. The intrinsic kinetic energy T_0 density of the nebula will therefore remain constant in all directions during the expansion.

If a dipole magnetic field is now superimposed on the nebula, the density of total energy is modified by the magnetic energy:

$$T_0 + \mathcal{E}(r,\ \varphi)\ , \tag{33}$$

where $\mathcal{E}(r,\ \varphi)$ is the magnetic energy per unit volume; it is a function of direction, since the dipole field is not homogeneous. The additional energy will increase the rate of expansion $V(r,\ \varphi)$ of each volume element in a given direction. We may therefore write

$$\frac{\rho V^2(r,\ \varphi)}{2} = T_0 + E(r,\ \varphi)\ , \tag{34}$$

where ρ is the gas density.

The magnetic field is maximal at the poles and minimal on the equator. The rate of expansion of a nebula with a dipole magnetic field is therefore the largest along the polar axis and smaller towards the equator. An initially spherical nebula is thus gradually stretched in the direction of the magnetic axis. This prolate shape naturally increases with the ratio of the magnetic energy to the intrinsic kinetic energy.

In this way, the dipole magnetic field stretches the nebula along the magnetic axis distorting its spherical shape. This conclusion is consistent with observational data; the great majority of PNs are non-spherical and elongated to a certain degree.

We have discussed only the qualitative aspects of the phenomenon. A strict quantitative treatment requires consideration of the equations of magnetohydrodynamics for a moving medium for the given geometry of the magnetic field and the dipole size. In the conditions where there is no reliable information about the latter, we will restrict our analysis by an approximate solution of the problem, thus ignoring the higher order corrections.

The magnetic energy density at the outer boundary of a nebula is

$$\mathcal{E}(r,\ \varphi) = \frac{a^2}{8\pi}\,\eta_1^2(r,\ \varphi)\ . \tag{35}$$

Substituting this expression into (34), we find that

$$\left[\frac{V(r,\ \varphi)}{V_0}\right]^2 = 1 + \frac{a^2}{8\pi}\,\eta_1^2(r,\ \varphi)\ , \tag{36}$$

where V_0 is the rate of expansion without a magnetic field. Bearing in mind that at the centre of a nebula

$$\mathcal{E}_0 = \frac{a^2}{8\pi}\eta_1^2(0,\ 0)\,, \tag{37}$$

we obtain from (36)

$$\left[\frac{V(r,\ \varphi)}{V_0}\right]^2 = 1 + \frac{\mathcal{E}_0}{T_0}\left[\frac{\eta_1(r,\ \varphi)}{\eta_1(0,\ 0)}\right]^2 \tag{38}$$

or, writing

$$\frac{\mathcal{E}_0}{T_0}\,\frac{1}{\eta_1^2(0,\ 0)} = q\,, \tag{39}$$

we find for the relative rate of expansion of the nebula in the direction φ

$$\frac{V(r,\ \varphi)}{V_0} = \left[1 + q\eta_1^2(r,\ \varphi)\right]^{1/2}\ . \tag{40}$$

The extent of the nebula, or more precisely, of some thin layer a distance r from the centre, in the given direction, $R(\varphi)$, is clearly proportional to $V(\varphi,\ r)$. Therefore we may write for the outermost boundary of the nebula

$$R(\varphi) \sim \left[1 + q\eta_1^2(r,\ \varphi)\right]^{1/2}\ . \tag{41}$$

The shape of the nebula in the direction of a dipole magnetic field is thus a function of q, i.e., it depends on the ratio of the magnetic energy density at the centre to the intrinsic kinetic energy, and on the "depth" at which the magnetic poles are located, represented by the function $\eta_1(r,\ \varphi)$.

Figure 17.6 shows schematically the outermost lines of a PN according to (41) and (28) for various values of q, from $q = 0$ to $q = \infty$, and three values of the ratio r/l, 0.5, 1 and 2, corresponding to various "depths" of the magnetic poles. The case $q = \infty$ implies that the magnetic field is entirely responsible for the shape of the nebula, while for $q = 0$, the magnetic field is absent and the nebula is spherical. An increase in r/l signifies that the magnetic poles are situated deeper in the interior of the nebula (reckoning from the outer boundary, the radius of which is r), and the ratio of the two diameters is correspondingly smaller. The minimum diameter ratio for $q = \infty$ is 2; it is obtained in the limit $r/l \to \infty$ (point dipole).

Dipole magnetic fields are also responsible for the appearance of regions of enhanced brightness. These "beads" are situated on the equator symmetrically relative to the nucleus. In Fig. 17.6 the equator is represented by the horizontal, minor axis; the magnetic, major axis is along the vertical one. The brightness at the ends of the major axis falls off gradually, and no distinct boundary should be observed. At the ends of the major axis, on the other

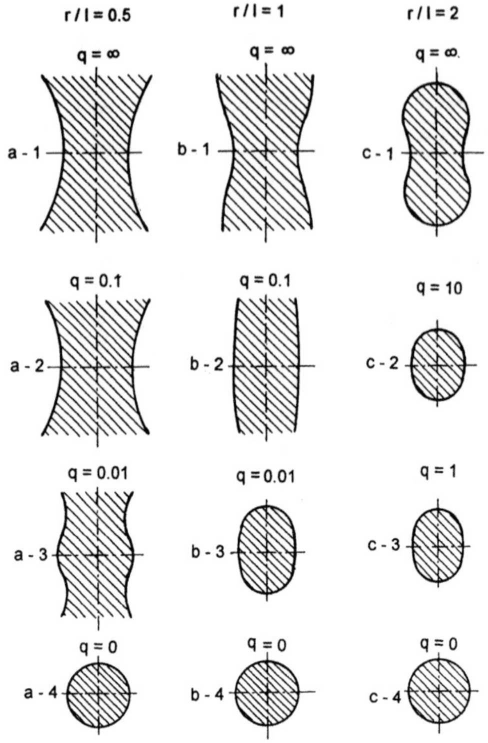

Fig. 17.6. Possible theoretical configurations of planetary nebulae with various parameters of the dipole magnetic field

hand, the brightness is maximal and the outer shape of the nebula (the edge of the "beads") is rather sharp. At first glance it would therefore appear as if some confining force operates in the direction of the minor axis, preventing the expansion of the PN. In fact, however, the nebula expands at a normal rate, as we have seen before, the rate of expansion increasing in the opposite direction from the equator.

Figure 17.6 clearly shows that the shapes of PNs with dipole magnetic fields may vary between wide limits depending on the field parameters. Though our calculations are rather approximate, it seems improbable that further improvement of the theory will modify this conclusion radically. PNs may possess a variety of shapes depending on the geometry and the relative strength of the field, including "hour-glass" (types $a-1$, $k-2$, $b-1$), "dumb-bell" (type $c-1$), "rectangular" ($b-2$), ellipsoidal ($b-3$, $c-2$) and, finally, almost spherical ($c-3$) nebulae.

PNs with these shapes are exclusively bipolar. We thus conclude that bipolar PNs may vary from almost spherical to almost rectangular. Examples of these extreme types are NGC 2474-5, a spherical nebula, and IC 4406 (Plate XII), a rectangular nebula (see below).

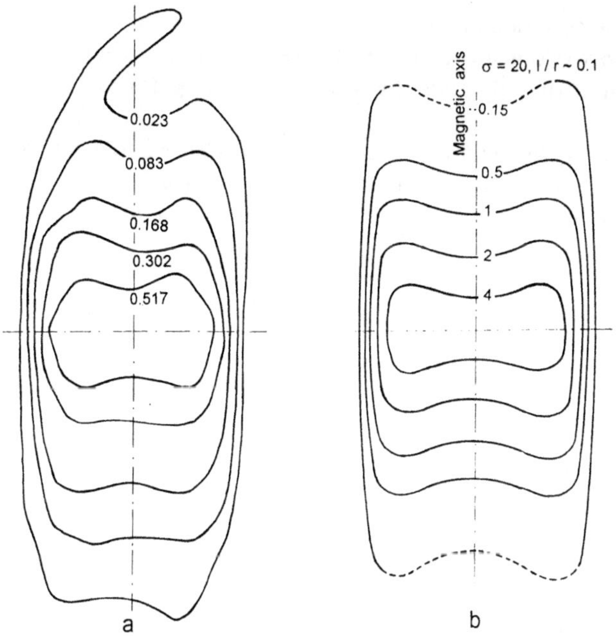

Fig. 17.7. (**a**) Observed isophotes of the planetary nebula IC 4406 in red light. (**b**) Theoretical isophotes of a "rectangular" nebula with parameters $l/r = 0.1$ and $\sigma = 20$

17.11 "Rectangular" Nebulae

A remarkable representative of this type is IC 4406 (Plate XII). Its shape is almost rectangular; more precisely, it has straight edges parallel to the major axis. Figure 17.7(a) shows its isophotes (in the H_α line of hydrogen); the numbers indicate intensities in relative units.

According to its shape, IC 4406 is classified as type $b-2$ of Fig. 17.6. This configuration is therefore of electromagnetic origin, arising from a dipole magnetic field of a certain type with the dipole axis pointing along the major axis of the nebula. We will try to plot the approximate theoretical isophotes for this "rectangular" nebula.

We again start with (18), introducing on the left hand side an additional term which allows the non-uniform expansion:

$$\frac{H^2(r,\,\varphi)}{8\pi} + 2nkT + \frac{\rho V_\varphi^2}{2} = C\ , \tag{42}$$

where V_φ is the translational velocity in the direction φ, and ρ is the mean gas density in the direction ($\rho = nm_{\rm H}$).

In principle, we should insert for V_φ in (42) its expression from (40). However, as we are dealing with a rectangular nebula, a simple function can be used. The equation of the straight edges parallel to one of the coordinate axes can be written in polar coordinates as $R(\varphi) \sim \sec\varphi$ instead of (41). Comparison with (40) gives for the outermost parts of the nebula $V_\varphi = V_0 \sec\varphi$.

Substituting this expression for $V\varphi$ into (42), we find, proceeding along the same lines as in (31), an expression for the relative gas concentration in a certain section of a non-uniformly expanding nebula:

$$\frac{n(r,\ \varphi)}{n_0} = \delta(\varphi)\left[1 + \left(\frac{V_0}{W}\right)^2 - \sigma\eta_1^2(r,\ \varphi)\right] , \tag{43}$$

where

$$\delta(\varphi) = \frac{1}{1 + (V_0/W)^2 \sec^2\varphi} , \tag{44}$$

and w is the thermal velocity of the gas particles, and V_0 the rate of expansion in the direction $\varphi = 0$.

Using (43) we find the gas density at any point of a central section of a rectangular nebula and hence, and from (32), we can compute its observed brightness distribution.

Figure 17.7(b) shows the isophote contours calculated by this method for a rectangular nebula with $W = 10$ km s^{-1}, and $\sigma = 20$. A point dipole was assumed ($l/r \sim 0.1$).

As we can see from the figures that despite the considerable simplification introduced in the theory, the two isophote plots, observed and computed, are similar. The similarity extends beyond the general shape of the isophotes to the almost identical brightness gradients, especially in the direction of the magnetic axis. Calculations show that the magnetic field strength near the centre of the PN should be of the order of 3×10^{-3} gauss (for $n_0 = 5 \times 10^3$ cm^{-3}, $T_e = 10^4 K$).

Despite the low probability of observing of such a PN, a number of objects are known to have a partially rectangular or close shape. The bipolar nebula

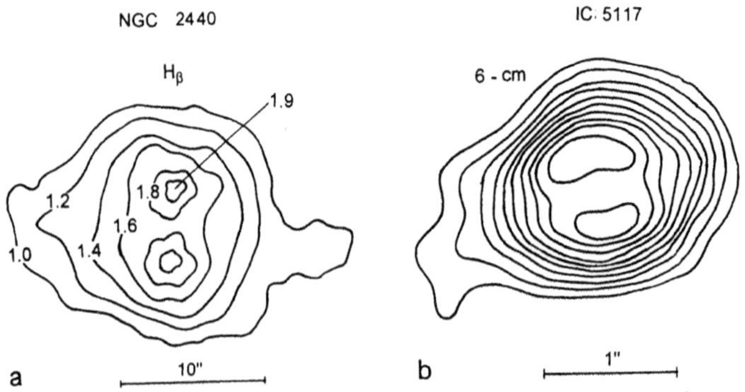

Fig. 17.8. (**a**) Monochromatic isophotes of NGC 2440 (H_β) a bipolar and almost rectangular PN. (**b**) Radio isophotes of PN IC 5117

NGC 2 440 is quite similar to IC 4 406; its isophotes are shown in Fig. 17.8. NGC 7 026 and Hb 5 have a noticeable rectangular shape (See Fig. 17.11, Plate XIII). Isophotes for NGC 40 may also be manifestations of the role of the magnetic field (Meaburn et al, 1996).

Among those PN which are rather rare in terms of their form and structure as well as their beauty are "hour-glass" PNs. In Plate XII three PNs of this type are shown: M 2-9, Hf 48 and NGC 6 302. All of then are bipolar. Simultaneously bipolar, rectangular and "hour-glass" type is NGC 650-1 (Bruce et al., 1996; Recillas–Cruz, Pişmiş, 1984; Pişmiş et al., 1993)

The "hour-glass" PNs seem to indicate exceptionally favourable and rare orientations of their major axes in relation to the observer, i.e. perpendicular or almost so to the line of sight.

The following PNs can be attributed to "rectangular" and "hour-glass" types:

"Rectangular"	"Hour-glass"
NGC 2 440	NGC 2 346
NGC 6 772	NGC 6 302
NGC 6 781	NGC 650-1
IC 4 406	He 2-119
A 1	M 2-9
Sh 2-71	Hf 48

Objects of "hour-glass" or "dumb-bell" type may well be cornerstones in our understanding of the processes of the origin and formation of PNs. On the other hand, one cannot even be sure whether these extraordinary objects are really classical PNs. In both cases their further study is of special interest.

17.12 The Magnetic Field of NGC 7 293

Can one obtain the actual character of the magnetic field within the given PN? It seems that if the real distribution of the electron concentration throughout different sections of the PN is known and if one assumes a balance of the magnetic and kinetic energies, i.e. the fulfilling of (16) at each point of the object. Consider one of the well known and the largest PNs, in terms of its visible size, in Aquarius, NGC 7 293, a typical bipolar object (Plate I). Its isophotes plotted from the plates in the H line obtained using the 40″ Schmidt telescope in Byurakan Observatory are shown in Fig. 17.9; the numbers indicate intensities in relative units.

Note the strongly expressed bipolarity of this object and the existence of two clearly visible "beads" symmetrically situated with respect to the central star. In accordance with our ideas about the role of the magnetic field one can conclude that the small axis I–I coincides with the magnetic equator while the magnetic axis II–II is perpendicular to I–I.

Our first problem is the determination of the distribution of the electron concentration along the different sections of the nebula using the observed

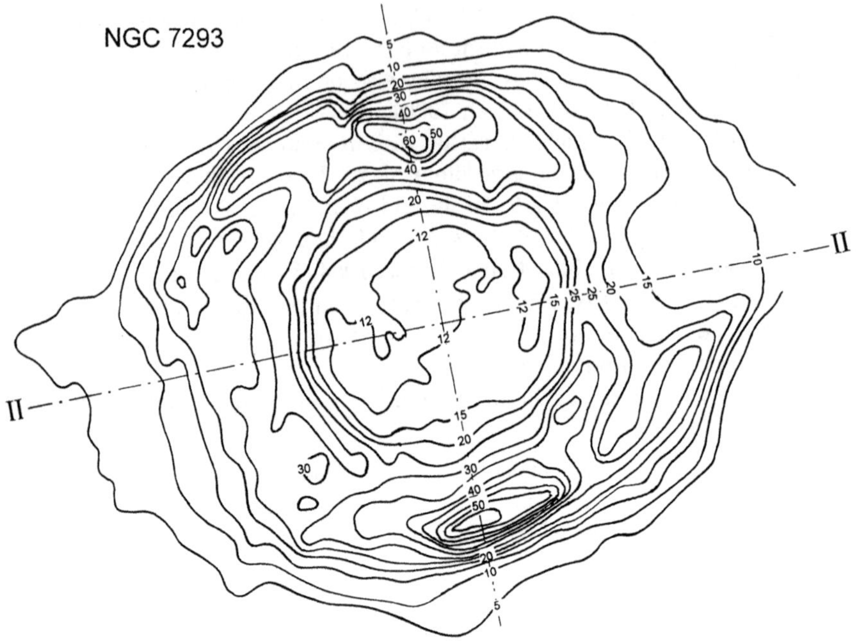

Fig. 17.9. Isophotes of NGC 7293 in H_α + [NII]. Figures indicate intensities in arbitrary units. I–I is the equatorial axis and II–II is the magnetic axis of the nebula

system of isophotes in Fig. 17.9. The solution of similar problems in the case of spherical or cylindrical configurations is usually obtained from the Abell equation. In our case we will use numerical methods, obtaining first the distribution of the volume emission coefficient $i(r,\ \varphi)$ along the PN radius and for various sections (different φ), and then the electron concentrations $n_e(r,\ \varphi)$, assuming $n_e \sim i^{1/2}$.

Applying the main conjecture concerning the energy balance and substituting for the constant in (16) the corresponding expression with $\varphi = 0$, we obtain

$$\frac{H_\varphi}{H_0} = \left[1 + \frac{16\pi n_e k T_e}{H_0^2}\left(1 - \frac{n_\varphi}{n_0}\right)\right]^{1/2}, \tag{45}$$

where H_0 and n_0 are the magnetic field and the electron concentration in the "beads" ($\varphi = 0$), respectively.

We proceed as follows. First, combining our system of isophotes for NGC 7 293 (Fig. 17.9) with the theoretical calculations (Figs. 17.3-5), we obtain the numerical value for σ that best matches the observed system of isophotes: $\sigma \approx 0.1$. Then, we obtain from (25), at $T_e = 10^4$ K, $a(\varphi) = 0.8 \times 10^{-5} n_e^{1/2}(\varphi)$. By mean of (30) or Fig. 17.2 we obtain, taking $l/r \approx 1$, for the strength of magnetic field $H(\varphi) = 0.8 \times 10^{-5} \eta(\varphi) n_e^{1/2}(\varpi)$. In the case of NGC 7 293 we have $n_e(90°) = 160\ \mathrm{cm}^{-3}$ and $n_e(0°) = 107\ \mathrm{cm}^{-3}$. We also have, from Fig. 17.2, $\eta(90°) \approx 4$ and $\eta(0) \approx 1$. Then we obtain for the strength of the magnetic field in the region of the "caps" of NGC 7293 $H(90°) = 4 \times 10^{-4}$ gauss and on the equator $H(0) = 0.9 \times 10^{-4}$ gauss, i.e. $H(90°)/H(0) \approx 4.5$.

Second, we find out the dependence of $H(\varphi)/H(0)$ as a function of φ, i.e. the observed variation of the magnetic strength from the polar regions to the equator. Omitting the details (Gurzadyan, 1963), we present the main conclusion: the observed decrease of the field is somewhat slower, at least for NGC 7293, than in the case of classical dipoles of either point-like or extended types. The reason is presumably to do with the various gradients of the field while moving from the magnetic equator to the poles.

17.13 Spiral Planetary Nebulae

By spiral or spiral-like PNs we mean those which have two and only two more or less developed spiral arms located symmetrically with respect to the core. Typical objects of this class are NGC 4361, NGC 6309, NGC 7026, A 79, Hb 5 some shown in Plate XIII.

Spiral structure has been established for at least the following PNs (Aaqust, 1993):

NGC 650-1	NGC 6853	He 2-186
NGC 2440	NGC 7009	K 3-35
NGC 4361	NGC 7026	PC 14
NGC 5882	NGC 7293	Hb 4
NGC 6210	J 320	Hb 5
NGC 6309	A 79	M 3-1

Some spiral PNs are also bipolar (NGC 650–1, NGC 7026, NGC 7293) or possess two envelopes (NGC 4361, NGC 7009, NGC 7293, Plates II, VIII, IX, XIV). The spiral arms vary both with respect to their forms and prominence. In some cases the spiral arms are clearly outlined and are fairly well developed (NGC 4361, NGC 6210) while in others they are thin and hardly visible (NGC 6309). Occasionally the spiral arms are reduced to mere projections (NGC 7026), or "pincers" (Hb 5, NGC 650-1, Fig. 17.11), but they are always arranged symmetrically with respect to the nebula. In one case (A 79, Plate XIII) the spiral arms are so outlined that the PN can easily be mistaken for a galaxy, while in another case, though no typical arms are apparent, the spatial orientation of the object with respect to the observer allows certain features to be interpreted as manifestation of spiral structure, e.g. the conical projections on both sides of NGC 7009 (Plates IX, XIV).

Thus, although the known spiral PNs are not so numerous, one can hardly doubt their reality.

The spiral arms of PNs can also have an *electromagnetic* origin, resulting from the *combined* action of the inhomogeneous intrinsic fields of the nebula and the regular Galactic magnetic field. Confining ourselves to an outline of this idea, let us again assume an extended dipole-like magnetic field in a PN.

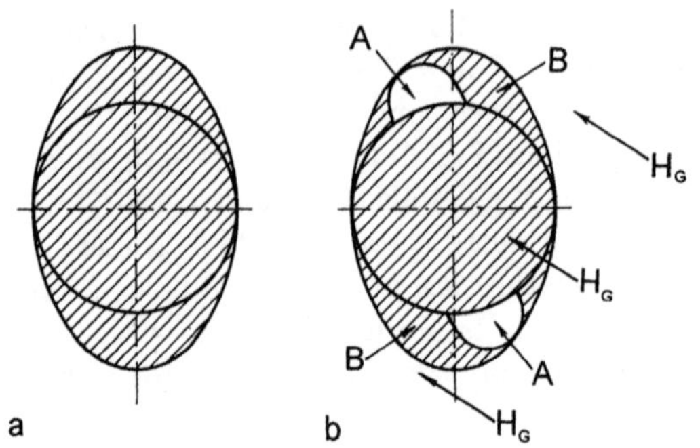

Fig. 17.10. Theoretical possible forms of a planetary nebula (**a**) without the regular Galactic magnetic field, and (**b**) with the Galactic magnetic field superimposed on the dipole field of the nebula

The expanding nebula with a dipole magnetic field is inevitably stretched in the direction of the magnetic axis. One might have the impression that the gaseous matter escapes from the poles of the PN in the form of a wide jet, projected on both sides. However, in reality, the PN spiral arms cannot have the same dynamical nature as in galaxies.

Indeed, consider the projection effects close to the magnetic poles where the dipole field is comparable with the magnetic field of the Galaxy, i.e., of the order of 10^{-5}–10^{-6} gauss, which can easily be assumed as homogeneous within the space occupied by the PN.

Consider two points A and B near the edge of a PN (Fig. 17.10), symmetrical with respect to the magnetic axes. The magnetic lines of an extended dipole passing through the point A should clearly mirror those passing through B. The dipole field $H(r, \varphi)$ at these points, A and B, is the same only in absolute magnitude, while the Galactic homogeneous field vector H_σ, will be the same in both magnitude and direction. As a result of the combined action of both magnetic fields, the field can be strengthened at one point, A, and weakened at the other, B. According to conjecture (16), on the energy balance, we can conclude that the matter density in the regions surrounding B must be greater than at A: $\rho_A > \rho_{PB}$.

Figure 17.10 shows a diagram of an elongated nebula when the Galactic magnetic field is neglected (a), and when it is taken into account (b).

Thus, the spiral arms in the PN are in fact geometrical areas, with extremes of the magnetic field, and in contrast to the galaxies, they are the merely static ornaments "adorning" the nebula.

NGC 7026 was listed above among the spiral PNs. Figure 17.11 shows the mean isophotes of this nebula, revealing its two or even three important properties: bipolarity (isophotes 1.1), and its spiral and rectangular form; the dashed line denotes the magnetic axis, the solid line the direction of the plane, parallel to the Galactic equator, the probable direction of the magnetic lines.

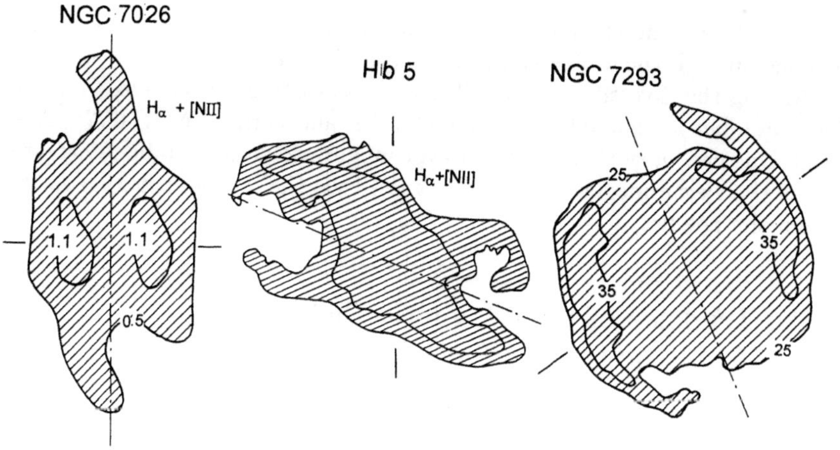

Fig. 17.11. The silhouette images of the planetary nebulae NGC 7 026, NGC 7 293 and Hb 5., all three of which are spiral-like. The dashed line is the magnetic axis; the solid line, the direction of the plane parallel to the equator of the Galaxy

In the same figure (17.11) the isophotes of two other PNs, Hb 5 and NGC 7 293, are shown. The most prominent feature in the first case, Hb 5 (Aller, 1956; Pişmiş et al., 1993), is the highly flattened, almost rectangular shape, evidence of the intrinsic magnetic field, as discussed above. The spiral arms at the ends of the magnetic axis are pincer shaped.

The presence of the signs of spiral arms in NGC 7 293 is interesting and unexpected. So, if we eliminate the isophotes marked No. 25 (see Fig. 17.9) we discover the presence of two small prominences located symmetrically with respect to the centre of the nebula. Within the general pattern of PNs these prominences look like spiral arms. It is interesting that in this case the Galactic equator lies almost perpendicular to the magnetic axis of the nebula – the most favourable orientation for the appearance of the spiral structure.

Note one more peculiarity of NGC 7 293: the apparent intersection of two envelopes (rings). In reality within the described picture only one envelope can exist at the ends of the major (magnetic) axis, influenced by both types of magnetic field – the dipole of the nebula and the homogeneous field of the Galaxy.

The fairly distinct spiral structure is also observed within the second (outer) envelope of Mz 2 (Acker et al., 1992), one of the most remarkable bipolar PNs (Plate VII).

In the double-envelope nebula NGC 7 009 ("Saturn", Plates IX, XIV) the spiral arms are oriented differently with respect to the observer: the arm plane lies nearly perpendicular to the image plane, as does the Galactic magnetic field. However, judging by the symmetrical bright spots in the second envelope, one can conclude that the latter field is slightly inclined ($\sim 30°$)

towards the magnetic axis of the nebula: the magnetic axis follows the line passing through the centre and the two prominences.

Among the diversity of spiral PNs one can distinguish *Z-shaped nebulae.* Their number is relatively small: NGC 2 452, and perhaps NGC 6 778, NGC 6 853. One can show that the Z-shaped configurations can also be interpreted within the electromagnetic scheme discussed via the magnetic field geometry within the nebula.

17.14 Transfer of the Magnetic Field from the Star to Nebula

In previous sections, the role of the static magnetic field within the PN has been examined. In spite of a number of simplifications, this concept alone enables one to easily explain the rather different but essential observational properties of PN forms and structure, namely:

(a) the absence of spherical-symmetry in the overwhelming majority of PN;
(b) the existence of a sharply expressed bipolar structure in many PNs;
(c) the existence of rare and exotic forms of PNs – spiral, Z-shaped, "rectangular", "hour-glass" etc.

Proposed originally by Gurzadyan (1962), the idea of the essential role of the magnetic field in PNs later became a matter for more detailed studies, in particular, by Pascoli (1985, 1987, 1990), involving the effects of gaseous matter outflow from the central star, thus transferring to the nebula the "frozen" magnetic field, or within the model of a toroidal magnetic field (Chevalier, Luo, 1994).

The main conclusion is as follows: the absence of spatial symmetry in the distribution of matter within the nebula is a direct consequence of the presence of an intrinsic magnetic field. The latter is particularly required to have the elongated shapes of PNs. Studies at least outline the importance of intrinsic magnetic fields for the formation of PNs with filamentary structure (see Sect. 17.17).

As for the origin of the magnetic field itself, the main process proposed is the transfer of magnetic energy from the central star via stellar wind or the stationary outflow of gaseous matter. Not everything seems clear here concerning the maintaining of the initial dipole and space orientation of the magnetic field of the central star, which is at the same time in a state of axial rotation during the mass transfer. Recall that the clearly expressed bipolar structure of PNs of rather different sizes shows that the invariance of the space orientation of bipolar structures and, hence, of the dipole nature of the field, is preserved presumably during the life of the nebula. In the framework of the hypothesis of the transfer of the magnetic field of the central star to the nebula, one has to assume, therefore, the strongly equatorial character of the mass outflow process.

17.15 Observational Evidence for the Presence of Magnetic Fields in PNs

Possible persuasive evidence for the presence of intrinsic magnetic fields within the PN can obviously be implied by the occasionally remarkable agreement between the theoretical predictions of the form and structure of nebulae and the observations, as discussed above.

However, other evidence may also can exist: (a) the synchrotron radiation, and (b) the polarization of the radiation.

In spite of the search for continuous emission of a non-thermal nature and even claims of finding it, no sufficiently reliable evidence can be considered to exist.

No less problematical are the results of polarimetric observations of PNs. Recall that the first attempt at such observations was carried out for NGC 7 026 by the author together with Razmadse in 1959, and evens then an unusually high – of the order of 5% degree – polarization had been observed in the continuous spectrum between 3 740 and 4 600 Å.

Latter on, polarimetric observations of a number of PNs were carried out, and polarization has been discovered in almost all cases in both emission lines and continuous spectra. The degree of polarization, for example, in the emission line 5 007 [OIII], lies from 0.1% (NGC 2 392, NGC 3 242, NGC 6 720) up to 1% (NGC 6 572, NGC 7 009). The degree of polarization in the continuous spectrum, in the interval 5 200–5 700 Å, is sufficiently high – of the order of 2–3% (NGC 6 572, NGC 7 009). Two remarkable facts are that the polarization is not constant within the area of the nebula, the minimum polarization being registered at the PN centre, and the polarization map in the majority of cases is symmetrical with respect to the nebula's centre.

The last two properties – the variation within the PN and the symmetrical character of the polarization – could be considered to support the presence of the intrinsic magnetic field. Note that the observed picture of polarization is a superposition of two effects – an intrinsic one within the nebula and polarization due to the interstellar medium or that the located in the immediate neighbourhood of the PN. After the reduction of the second component, both the above mentioned features become even more marked: the map of intrinsic polarization turns out to be more symmetrical, and with almost zero polarization within the centre of the nebula. The map in Fig 17.12 of the nebula NGC 3 242 is an interesting example.

Attempts to interpret these polarization data within the framework of electron scattering or scattering on the dust particles of the light have been made. Within the magnetic field paradigm all these observations can be naturally explained by synchrotron emission. First, the polarization maps, and therefore the magnetic field configurations, are consistent with the point-like or extended dipole conjectures.

Moreover, the theoretically predicted degree of polarization, in the case of synchrotron radiation, should be very high – almost 100%. Depending on the

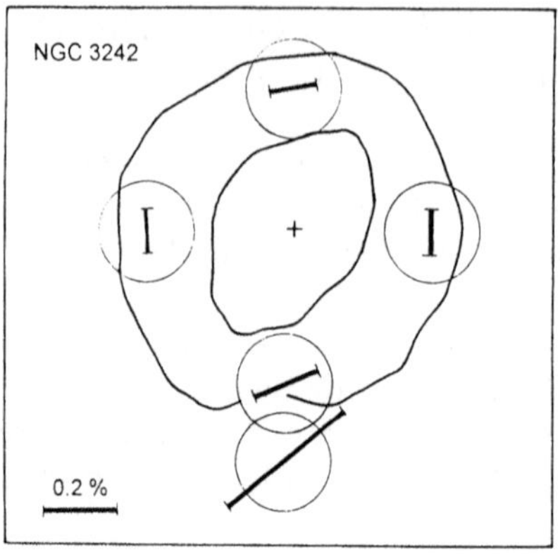

Fig. 17.12. The map of proper polarization of the planetary nebula NGC 3 242. Polarization in the centre of the nebula is zero. The symmetrical and tangential character of the polarization is apparent

magnetic field geometry, the resulting degree of polarization of non-thermal emission will be much less, say, of the order of 10–20 percent. Hence, one can readily estimate the fraction of synchrotron emission within the whole continuous radiation of recombination, i.e. of bremsstrahlung origin: less than of the 10%, which is too small to be discovered by usual means of observations.

17.16 Dissipation of Magnetic Energy

The dimensionless parameter γ introduced above, the ratio of the kinetic energy density $P_{\rm kin}$ that of the magnetic density $P_{\rm mag}$,

$$\gamma = \frac{P_{\rm kin}}{P_{\rm mag}} = \frac{4\pi\rho V^2}{H^2} \tag{46}$$

in the central regions and near the magnetic poles, γ, should be assumed to be smaller than unity. Near the "beads" γ must be of the order of unity. For $\gamma < 1$, the internal motions should be controlled predominantly by the magnetic field, while in the "beads" the thermal motion has a comparable contribution. The nebular structure is not influenced by the magnetic field only when $\bar{\gamma}$ is sufficiently larger than unity.

The average index γ is defined by

$$\bar{\gamma} = \frac{1}{V}\int \gamma \mathrm{d}V , \tag{47}$$

where V is the PN volume. Obviously the variation of γ during the PN expansion is indicative for the relative role of the magnetic field at various stages of the nebula's evolution.

Let $\mathcal{E}_{\text{mag}}$ be the total magnetic energy of the nebula at the time of its separation from the nucleus. As a starting point we assume that this energy remains constant as the nebula expands. The total kinetic energy $\mathcal{E}_{\text{kin}}$ of the nebula, however, behaves differently.

In general, $\mathcal{E}_{\text{kin}}$ is a sum of three components: the kinetic energy of expansion, the kinetic energy of the ionized component due to the thermal motions of gas particles at a temperature of 10^4 K, and the kinetic energy of the non-ionized component of the nebula, with the thermal motions of much colder gas, at about 100 K or 1 000 K. The rate of expansion is fairly constant regardless at the size of the PN; therefore the first component is virtually constant during the PN evolution. The third component is small and can be ignored. As for the second component, it is determined by the amount of ionized matter relative to the total mass of the PN. The total optical thickness τ_c at the frequencies of the hydrogen ionization of a small PN is markedly greater than unity, and therefore only a small fraction of the nebula is heated to 10^4 K. This fraction, however, increases as the nebula expands, and E_{kin} increases correspondingly. Therefore, in small nebulae, when $\tau_c \gg 1$, the magnitude of $\gamma = E_{\text{kin}}/E_{\text{mag}}$ will increase as the nebula expands until τ_c has reached unity.

For double-envelope PNs, τ_c is less than unity. Many of these PNs are bipolar as well, and their structure is therefore influenced by magnetic fields. If one assumes $\gamma \sim 1$ for double-envelope PNs, then γ should be much smaller than unity for small PNs with $\tau_c \gg 1$. In other words, the role of the magnetic field in PN dynamics is greater for small nebulae and decreases with expansion.

Consider how the situation develops when τ_c becomes less than unity. In accordance with the preceding arguments, γ remains constant, so that the relative contribution of the magnetic and internal forces is diminished. This conclusion, however, does not correspond to what really happens in PNs at this stage.

As the nebula expands, the magnetic energy is inevitably dissipated by at least two mechanisms: the first is connected with the expansion, and the second with the partial conversion of magnetic into thermal energy one, Joule heat (j^2/σ per unit volume).

There exists, however, a third source of dissipation, namely, turbulence. Although turbulence is able to suppress the field as well as the magnetic energy, the dissipative action of the turbulence hardly seems efficient in small PNs, becoming more important presumably in later stages of the life of the nebula, when it becomes relatively larger.

17.17 Filamentary Nebulae

Irrespective of the rate of dissipation of magnetic energy, Γ is clearly larger than unity at some stage of the PNs' life. At this moment the magnetic lines of force are stretched to the ultimate limit, so that the field dissolves into small fragments: the dipole field ceases and is replaced by local magnetic fields – magnetic energy fluctuations arise, and this in turn causes redistribution of the ionized nebular matter into smaller scales. In other words, fluctuations of surface brightness should be observed. Indeed, the outer envelopes of some double-envelope PNs (NGC 2392, NGC 6720, NGC 7662 etc.) are distinguished by considerable fluctuations in surface brightness, whereas the inner envelopes do not show any such effect. This clearly supports our hypothesis.

It should be mentioned, however, that there do exist certain bipolar and double-envelope PNs (NGC 7354, NGC 3587 etc.) in which neither of the envelopes show brightness fluctuations. This means that γ in both envelopes is smaller than unity and magnetic energy dissipation has not yet terminated.

Along with this, there exist PNs which are entirely covered by spots and filaments of non-uniform brightness: NGC 7094 (Plate IV), NGC 1501, A 43, A 72 (Plate XI), as well as NGC 40, NGC 2461, NGC 1514, A 56 etc. In these objects the dipole magnetic fields apparently disappeared a long time ago (if they ever existed) and only local fragmentary fields remain.

As early as Chapter 1 we pointed out one of the most remarkable facts, namely, that PNs with large brightness fluctuations do not reveal any signs of turbulence. We now see that this "anomaly" is readily removed if the brightness fluctuations, i.e. the non uniform gas density, are correlated with the magnetic field fluctuations.

Thus the brightness fluctuations of PNs may be caused not by the turbulence but by the fluctuations of the magnetic field.

From this point of view it is interesting to follow the story of the study of the diffuse Orion nebula begun many decades ago. Radial velocity measurements at 85 points of this nebula, carried out by Campbell and Moore in 1917, showed a considerable scatter which was naturally attributed to turbulence of the nebular gas. Collateral evidence in favour of this hypothesis was the clearly non uniform distribution of brightness over the nebula, with some large fluctuations. But if brightness fluctuations and radial velocity fluctuations are indeed caused by turbulence, they should obey the Kolmogorov law of locally isotropic turbulence for high Reynolds numbers. This law can be formulated as follows: the relative velocity V of two points at a distance l from each other is on average proportional to $l^{1/3}$, if l is large enough. Indeed, denoting by ρ the gas density, we have

$$V \sim (l\,/\,\rho)^{1/3}\,. \tag{48}$$

Van Hoerner in 1958 compared this law with Campbell and Moore's observations and concluded that the Kolmogorov relation was satisfied for the Orion nebula. Munch (1958) analyses far more complete data (50 000 radial velocity measurements, to mention one set of data alone!) and established that the Kolmogorov law is not fulfilled for the Orion nebula. The deviations from this law increased at smaller l and therefore the brightness and radial velocity fluctuations cannot be entirely dependent on the turbulence; some other factor may also have a role. Munch suggested explaining the velocity distribution deviations by the existence of velocity gaps in all directions throughout the nebula, caused by the interaction of hot and cold gases. This may be true, but the role of local magnetic fields in the nebula should not be disregarded.

Finally, let us mention that if the total magnetic energy of an average-sized PN is attributed to the gaseous mass detached from the nucleus, then the magnetic field strength of the nucleus at the time of the separation of the nebula's mass should be at least 10^5 gauss. Therefore either PNs are formed from objects with rather strong magnetic fields, or they possess an independent mechanism for generating or amplifying magnetic fields.

17.18 Interstellar Magnetic Field of the Intermediate System of the Galaxy

An expanding PN eventually dissolves within interstellar space. Its gaseous matter carries away the nebular magnetic field, which is dispersed within the interstellar medium, provided that the magnetic energy has not been fully dissipated by the time the nebula is finally dissolved. The magnetic fields of PNs in the final stages of their evolution should be highly irregular, or even *random*.

In view of the relatively high rate of PN production – about one per year – and their short lifetime – a few tens of millennia – the PNs can presumably make a certain contribution to the interstellar magnetic fields.

If, say, one PN is created in 10^9 years in a volume V pc^3 within the Galaxy, and assuming that no magnetic energy is dissipated in the interstellar medium, there is a resulting average magnetic field $\mathcal{H}$,

$$\mathcal{H} = \mathcal{H}_0 (r \,/\, r_0)^3 , \tag{49}$$

where $\mathcal{H}_0$ and r_0 are the mean field and the mean radius of the PN, $r \approx V^{1/3}$, respectively. Calculations yield $H \sim 10^{-6}$–10^{-7} gauss for $V = 125$ pc^3 and $H \sim 10^{-8}$–10^{-9} gauss for $V = 125\,000$ pc^3. In both cases, $r_0 = 10\,000$ a.u. and $H_0 \sim 10^{-3}$–10^{-4} gauss for a typical PN are assumed.

17.19 The Problem of Magnetic Fields in Planetary Nebulae

Thus, there is sufficient background to suppose that in PNs stationary magnetic fields of dipole or quasidipole type might exist. The mean magnetic field obtained in two independent ways – by the specific kinetic energy of thermal motions and by the comparison of theoretical isophotes with observed ones for a large number of objects – is of the same order, 10^{-3} gauss (see Sects. 5 and 7). By means of such a field of various configurations, all the observed diversity of the forms and structures of PNs turn out to be explained, including even rather exotic and anomalous ones – elongated, rectangular, "hour-glass", spiral-like nebulae etc.

All these fundamental conclusions claimed by us as long ago as the early 1960s, were confirmed soon after by a number of authors. For example, Menzel (1968), also considering the role of magnetic fields in PN dynamics, came to a conclusion confirming the above results, although leaving the problem of the origin of magnetic fields open. Pascoli (1985, 1987, 1990), as already mentioned, in his thorough studies confirming again the extreme importance of magnetic fields in the formation of the spatial structure and form of nebulae, outlines one of the consequences of this paradigm, namely, the nebular spiral structure as the result of the combined action of two fields – the dipole in the peripheries of nebulae and the Galactic regular field.

Aller and Liller (1968) also thoroughly analysed the arguments in favour of the existence of intrinsic magnetic fields in PNs and the origin of bipolar configurations as a result of the action of these fields. They especially emphasized the coincidence of the magnetic field (10^{-3}–10^{-4} gauss) obtained by the two methods mentioned above, namely, the energy balance and the comparison of theoretical and observed isophotes.

The reality of the presence of magnetic fields in PNs is also supported by the results of polarimetric observations (Sect. 17.13). The absence of synchrotron radiation at the radio frequencies should presumably imply the lack of high energy electrons.

The field of 10^{-3}–10^{-4} gauss is two to three orders larger than that of the Galactic interstellar medium, although it is much smaller than the magnetic fields on the surfaces of the central stars of PNs. However, the nebular magnetic field cannot be a simple continuation of the dipole magnetic field of the central star, otherwise the field at the star's surface would have to be of the order of 10^{13} gauss(!).

The magnetic field of a planetary nebula cannot also be primordial, remaining since the proto-PN embryo. The total magnetic energy for an ordinary star (the Sun) is hardly 10^{40} ergs, if we even assume that $H \sim 10^4$ gauss within the entire volume of the star. Meanwhile, the complete magnetic energy of a PN, as we have seen above (Table 17.1), is of the order of 10^{44} ergs.

It is clear, therefore, that the PN magnetic field itself should most probably form independently, during all of its evolution. The stellar wind may have an essential role as the carrier of the magnetic energy, passing it to the nebula. In any case the problem remains open, though there seems to be little doubt about the very existence of magnetic fields in PNs.

18. Origin and Evolution of Planetary Nebulae

18.1 The Problem of Progenitors

All known forms of ejection of gaseous matter by stars – either catastrophic explosions, resulting in speeds of the order of 1 000 km s^{-1} or continuous outflows of the same velocity – cannot produce PNs, first of all, because of the impossibility of reducing such high speeds by two orders of magnitude.

At the same time, the fact of the expansion of the nebulae and the permanent increase in their absolute sizes indicates that the gaseous envelopes separating from the central stars with small initial velocity – of the order of 20–30 km s^{-1} – are significantly smaller than the escape (parabolic) velocity of the stellar surface. This implies that the gaseous envelope could not have been ejected from the star's surface.

However, if one assumes that the forces acting in the interior of the central star are increased not rapidly but gradually over a long period of time, the outer layers of the star can inflate without separating from the star up to large distances. After some time, these layers, of course, can reach distances where the escape velocity at the given distance is much less. If the departure occurs at that radius, the outer layers of the star can leave the star by inertia.

As is well known, the parabolic velocity of escape $V_p(r)$ decreases slowly with distance, proportional to the inverse square root of the distance r:

$$V_p(r) = V_p(0)(R_*/r)^{1/2}, \qquad (1)$$

where R_* is the star's radius and $V_p(0)$ is the parabolic velocity on the star's surface, i.e. at $r = R_*$. We see, that in the case of the Sun, for which $V_p(0) =$ 618 km s^{-1} at $R_\odot = 7 \times 10^{10}$ cm, the parabolic velocity of the order of $V_p(r) = 20$ km s^{-1} one will have for the radius of the inflated star $r \approx 1\,000\,R_\odot = 7 \times 10^{13}$ cm, or nearly 5 a.u.

Can a star inflate in principle? It appears that it can. Inflation can occur at a certain stage of its evolution, i.e. after the complete burning of the hydrogen in the central regions of the star, when its nucleus has ceased to be the generator of nuclear energy. The only possibility then is the compression of the nucleus, the only source of the energy required being the potential energy. Compression of the nucleus will occur slowly at the beginning, and then this process will accelerate, up to the final collapse of the central regions of the star.

The condition of preservation of total (instantaneous) gravitational energy during the collapse without any loss of mass, and, hence, without loss of the kinetic energy of the star, will be fulfilled only via the departure of the outer layers or the inflation of the star.

So the compression of the star's nucleus has to be compensated by an increase in its size, i.e by its expansion. The last process implies that the star has quite soon to become a giant and then a supergiant. In this case the effective temperature of the star remains practically unchanged. This means that if the star initially has not too high a temperature, then most probably it will become a red giant.

Thus, we arrive at the conclusion that the progenitors or embryos of PNs should be searched for among the red giants.

18.2 Comparison with Observations

The process described above – the collapse of inner parts of the star and the expansion of its outer layers – allows a theoretical treatment. It appears that the rate of the process, and especially the basic parameters – the luminosity L and effective temperature T of the naked star, i.e. of the star with lost outer parts spent on the formation of the future nebula – depends strongly on the initial mass of the star $\mathcal{M}$.

The interrelation of the parameters L and T represents nothing other than the Hertzsprung–Russell diagram. The theory permits us to compute the tracks showing the dependence of L on T for various values of the initial parameter, the mass $\mathcal{M}$. The idea of checking the theory lies in the determination of observational values for the luminosity L and effective temperature T of the central star for each nebula separately, and in their comparison with the Hertzsprung–Russell diagram.

The results of one such analysis are presented in Fig. 18.1 (Pottasch, 1984). The solid lines indicate the theoretical tracks of dependence of L on T corresponding to the so-called helium flash model for two values of the mass of the central star, 0.8 $\mathcal{M}_\odot$ and 0.6 $\mathcal{M}_s$, derived by Paczynski (1971). The evolution of a star takes place from the right hand side and from top to bottom. The numbers on the curves denote the age of the central star (in 10^3 years), starting from the moment of loss of the outer part of the star's mass.

According to modern viewpoints, the early and late stages of the star's evolution are realized according to the following approximate scheme. For a star on the main sequence, the basic process is hydrogen burning with its transformation into helium. This starts at the earliest period of evolution, although the duration of the star's stay in this condition is quite long – of the order of 90% of the star's life. However, with the exhaustion of nuclear energy and the ceasing of all possible nuclear reactions, the process of compression of the central regions of the star – collapse – starts with the loss of gravitational energy. From this moment, the star leaves the main sequence and moves to

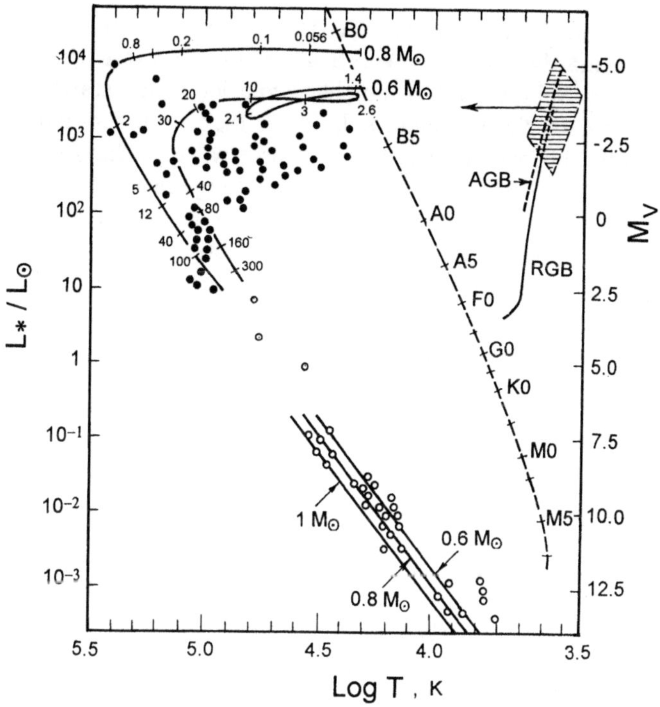

Fig. 18.1. Hertzsprung–Russell diagram, i.e. the luminosity (L) – effective temperature (T) relationship, for the central star nuclei of PNs. The theoretical evolutionary tracks are presented according to Paczynski for two values of the mass of the central star nucleus of the nebula, 0.8 $\mathcal{M}_\odot$ and 0.6 $\mathcal{M}_\odot$ (*solid lines*). Numbers on curves denote the age of the nebula (in 10^3 years), dots the nuclei of PNs with known values of L and T, circles white dwarfs, the broken inclined line from the right the main sequence (B0–M5), the RGB sequence the branch of red giants, the AGB sequence the asymptotic branch of giants, and the shaded area long period Mira Ceti type variables

the right and upward (Fig. 18.1). As a result, the first branch of the red giants – the branch of "RGB stars" arises (it is shown in Fig. 18.1 by nearly vertical solid line); this corresponds to the period of helium burning.

Quite quickly the helium in the central parts of the star is exhausted, and the compression of the star's nucleus, composed now mainly of carbon, oxygen and strongly degenerated electrons, starts. The compression is accompanied by strong heating, which in turn results in the expansion of an envelope. As a result, a new sequence of stars with an inert carbon–oxygen nucleus, the so-called "asymptotic branch of giants" – "AGB stars" – appears; it is shown in Fig. 18.1 by a broken line quite close to the RGB branch. After some time, these stars arrive at the upper right region, the shaded area corresponding to stellar luminosities of the order of $10^3 - -10^4\, L_\odot$. This region is also the region of Mira Ceti type long period variables.

It is necessary to mention that AGB stars are attracting more and more attention, primarily, from the theoretical point of view (see. Proc. IAU Symp. 155, Planetary Nebulae, 1993). Further evolution of the star now goes from the right hand to the left hand side, "starting" from the diamond-shaded area. Initially, this evolution proceeds almost without change of luminosity, though with a rapid increase of the effective temperature.

Thus, the determining factor of each of these sequences is the structure of the nucleus and the nature of the loss of interstellar energy throughout the main sequence – the hydrogen nucleus, burning of hydrogen and its transformation into helium: the RGB sequence – helium nucleus, without hydrogen, but with degenerate electrons and burning of helium and its transformation into carbon; and the AGB sequence – the carbon–oxygen nucleus without hydrogen and helium, and loss of gravitation energy.

18.3 Further Evolution

Consider now the further evolution of the star from the diamond-shaded area in Fig. 18.1. Here, some points are worth special attention.

First, the behaviour of these tracks is remarkable, i.e. there is a nearly tenfold rise in temperature at an almost constant level of luminosity during the initial period of the evolution, and then a too strong, thousand-fold falling of the luminosity with a relatively small decrease of the effective temperature.

Second, the existence of loops in these tracks, along with the points of reversal – the retrograde motion of the star from the left to the right – in the direction of low temperatures is no. In other words, the evolution is accompanied by periods of noticeable variation of the effective temperature of the star together with an almost unchanged level of its luminosity (for more details see Sect. 13.16).

The origin of the loops in the framework of the "helium flash" model of the star's nucleus is confirmed by results of model computations (Wood, Faulkner, 1986). It is important to mention that the loops and reversal points for the temperature on the evolutionary tracks exist only in the helium flash model: they are absent, for example, in the "hydrogen flash" model (Schönberner, 1981).

In fact, definite parts of each of the tracks in Fig. 18.1 correspond to the various phases of nucleus evolution: the loops to the "helium flash" phase in the layers, the horizontal part mainly to the "hydrogen flash" phase, again in the layers, and the inclined part to the phase of gravitation contraction.

In Fig. 18.1, the observational data for L and T are plotted for the central star nuclei for a sample of 70 well studied PNs with reliably determined distances and effective temperatures. As we can see, the observational points are scattered near the horizontal parts of theoretical tracks, which might indicate that these star nuclei are in the hydrogen flash phase. However, this is not so. To draw a final conclusion it is necessary to look for the observational confirmation of the presence or absence of loops in general. For example, if one could succeed in explaining the origin of three-envelope PNs, as described in Sect. 13.16, then this can be considered indirect support for the origin of the existence of loops.

Third, the variations in the brightness of the star nucleus during the life of the nebula are extremely small. Indeed, the star nucleus covers the horizontal part of the diagram for nearly 10 000 years and 30 000 years for masses of 0.8 $\mathcal{M}_\odot$ and 0.6 $\mathcal{M}_\odot$, respectively. The reason for such behaviour is clear: in this period of evolution, the rise in the temperature is compensated by the diminished size of the star, and as a result the star's brightness remains nearly constant.

18.4 The Final Phase of Evolution

The essential change – the fall in brightness of the star nucleus takes place on the inclined parts of the tracks in Fig. 18.1. However, large changes in brightness occur during rather long periods of time: from $10^4\,L_\odot$ up to $10\,L_\odot$ over 2×10^5 years in the case of $\mathcal{M} = 0.8\,\mathcal{M}_\odot$ and from $2 \times 10^3\,L_\odot$ to $10\,L_\odot$ over 2.7×10^5 years for $\mathcal{M} = 0.6\mathcal{M}_\odot$, which gives for the rate of brightness variations $0.00004\,L_\odot$ and $0.00002\,L_\odot$ per year, respectively. Such variations (weakening) cannot be discovered in practice, hence, during such relatively long periods, of the order of hundreds of thousands of years, the nucleus of the PN should exhibit a highly stable brightness.

However, the decline in the luminosity L occurs more rapidly than is expected by the $L \sim T^4$ law; hence, the decrease in both the luminosity and temperature is accompanied by a simultaneous diminishing of the star's radius.

In Fig. 18.1 nearly 30 known white dwarfs are plotted, as circles. They are all scattered in the lower-right part of the diagram and are in satisfactory agreement with theoretical tracks.

The conclusions from all these facts are as follows. First, the real nuclei of PNs form an isolated group located to the left of the main sequence (broken inclined line from M5 up to B0), i.e. in the region of high temperatures. Although in terms of brightness these nuclei are compatible with B5V–A5 stars of the main sequence, the central stars with collapsing nuclei have nothing in common with normal stars.

Second, the nuclei of real PNs are located (points) quite far from the zone of white dwarfs (circles) – the luminosities of the PN nuclei are two to seven (!) orders larger than the luminosities of classical white dwarfs. As for the effective temperatures, maximum temperatures of the order of 60 000 K have been registered in DA type white dwarfs (Fiege 24 and HZ 43) and at the same time, such temperatures are only a lower limit for PN nuclei.

The nuclei of PN are not typical white dwarfs, and, moreover, they even do not even form, during the period when they are enveloped by gaseous shells, a continuous sequence from classical PNs to classical white dwarfs. And although extremely faint formations have been discovered around a few hot H-rich DA white dwarfs (Tweedy, Kwitter, 1994), it is hard to accept these formations as being the remnants of former PNs.

Third, the former nuclei of PNs, starting from approximately, $L/L_\odot \sim 10$ and $T \sim 100\,000$ K, however, continue, their evolution to the right and downwards over a long period, of the order of a million years, before reaching the zone of classical white dwarfs, where $L/L_\odot < 0.1$ and $T < 30\,000$ K.

Fourth, although the majority of PN nuclei are scattered sufficiently below the theoretical track with $0.6\mathcal{M}_\odot$, from this it does not follow that the nuclei of these nebulae can be stars with masses less than $0.6\mathcal{M}_\odot$. This, most likely, should be considered the shortcoming of the theory, since the initial and final periods of evolution can differ owing to chemical composition.

18.5 Red Giants as Progenitors of Planetary Nebulae

The theoretical tracks in Fig. 18.1 show the star deviating from the main sequence, where it spent the first 90% of its lifetime, and for all that time the generation of energy was maintained by burning of hydrogen and its transformation into helium. The final 10% of the star's lifetime is spent leaving the main sequence with a simultaneous change of generation of energy from nuclear to gravitational by compression, first slowly and then rapidly with the collapse of its nucleus.

The main point, however, is that all these transformations are accompanied by the expansion of the outer layers of the star, with inflation up to enormous sizes. As a result, at some stage the dynamical separation of a part of the star's mass takes place; this mass then continues its independent evolution in the form of an envelope moving away – the future PN.

Hence, the important conclusion is that *PNs originate in the final phase of the star's life.* Thus, PNs are the products of the natural evolution of the star at the stage when the excess of the mass is lost and there is a switch from one type of energy generation – nuclear – to another – gravitation.

All these points concern stars whose masses and temperatures are not too great but which are fairly large as a result of the strong expansion of their outer layers. Red giants are just such objects. Thus, we arrive at the idea of red giants being the most probable progenitors of PNs (Shklovsky, 1956; Abell, Goldrich, 1966; Kwok, 1982; Blanko, McCarthy, 1981; Kahn, 1983; Renzini, 1983; Pottasch, 1984).

However, the red giants are not a homogeneous group of stars; they include cool giants, carbon stars, Mira Ceti type variables, OH masers, certain types of infrared object, etc. As for the protoplanetary objects and symbiotic stars, they are not direct progenitor embryos of PNs, but only a transition stage from the red giant phase to PNs.

In the forthcoming sections, additional arguments are presented supporting the view that PN progenitors do not form a homogeneous group of stars or star-like objects. The detailed analysis of this problem – the evolution from red giants to PNs, the evolution of the central stars themselves, the summary of theoretical models of stars and their drift along the "luminosity–temperature" diagram, the contribution of each type of progenitor at the formation of PNs are given in the monograph by Pottasch (1984).

18.6 Planetary Nebulae in the Galaxy

The search for possible progenitor ancestors of PNs among the non-ordinary objects of the Galaxy should be carried out, in particular, through a comparison of their statistical, morphological, kinematic, evolutionary and other parameters. To this end, we present here some of these parameters for the PN subsystem of the Galaxy.

Within 1 kpc around the Sun the number of PNs is about 50 (Maciel, 1984; Daub, 1982; Kagn et al., 1992). At the half-depth of the subsystem of PNs in Galaxy of $\beta = 230$ pc (Sect. 18.12) the spatial concentration of PN will be $n = 35$ kpc^{-3}. Assuming for the lifetime of the PNs stage $t = 20\,000$ yr, we have for the PN formation rate $\kappa = 1.8 \times 10^{-3}$ PNs per kpc^3 per year.

During the past 20 years over 15 attempts have been made to determine the main statistical parameters of PNs in the Galaxy, and Table 18.1 summarizes the main results. As we can see, at the half-depth of the subsystem of the order of $\beta = 200$ pc, the formation rate of PNs is of the order of 2×10^{-3} kpc^{-3} yr^{-1}. The formation rate of white dwarfs of 2×10^{-3} kpc^{-3} yr^{-1} (Wiedmann, 1977) is in surprising agreement with the above results for PNs.

Table 18.1. Planetary nebulae statistics in the Galaxy

	n [kpc^{-3}]	β [pc]	κ [10^{-3} kpc^{-3}yr^{-1}]	t [yr]	N
Cahn-Wyatt (1976)	80	115	5.0	16 000	38 000
Weidmann (1977)	36	150	1.8	20 000	20 000
Acker (1978)	48		3.0	16 000	25 000
Maciel (1981)	41	144	2.0	20 000	31 000
Daub (1982)	53	125	5.0		28 000
Pottasch (1984)	50	120	2.0	30 000	
Aller (1984)	50	125	2.0	30 000	15 000
Ishida-Weinberger (1987)	326	100	8.3		140 000
Gurzadyan (1996)*	35	230	1.8	20 000	16 000
Pottasch (1996)	50	160	3.0	25 000	24 000

* See Sect. 18.12.

Within the central 2° ($\sim$ 350 pc) of the Galaxy about 300 PN are known, which yields nearly 2 000 objects in 1 kpc^3, i.e. almost 40 times higher than in the neighbourhood of the Sun, and correspondingly the higher should be the formation rate: 8×10^{-2} kpc^{-3} yr^{-1}.

The total number of PNs in the Galaxy is expected to be not lower than $N = 2\beta \pi R^2 n = 16\,000$, for $R = 15$ kpc, $\beta = 230$ pc and $n = 50$ kpc^{-3}. However, as follows from Table 18.1 (last column), there are estimates of up to 40 000. The upper limit is more difficult to estimate, and may be over 10^5, depending on the adopted spatial density in the solar neighbourhood. It should be suggested that the total number of PNs may lie in the range from 15 000 to 100 000. Similarly their special density n and birth rate κ also have large uncertainties and fall in the range from 40 to 100 kpc^{-3}, and 1 per 3×10^{-3} PN kpc^{-3} yr^{-1}, though higher values have also been suggested as well (Maciel, 1989; Phillips, 1989).

The total number of stars in the Galaxy is estimated at 10^{11}. Hence, during 10^{10} years 2×10^{10} PNs should be formed. This means, that at least every fifth star of the Galaxy has been the ancestor of a PN, and perhaps every fifth dwarf star in our Galaxy is the former nucleus of a PN.

The mean dispersion of velocities has been obtained from the data of radial velocities for 300 PNs; it is equal to 140 ± 10 km s^{-1}. This dispersion reflects the differential rotation of the Galaxy and is a characteristic property of objects located far from the Sun.

18.7 Mirids and Maser OH/IR Sources as Ancestors of Planetary Nebulae?

It is hardly possible to insist that we have a clear notion about the transformation of red giants into PNs. Moreover, there is an impression that between red giants and PNs an intermediate stage should exist, i.e. long period variables of Mira Ceti type – Mirids – as well as cosmic OH/IR maser sources (Ortiz, Maciel, 1993). According to conventional viewpoint, the direct ancestors of PNs are just these objects, especially the Mirids at an early stage!

Mirids, i.e. Mira Ceti type variables, are long period variable giants of later types (M5e) with a period of brightness variation of 100–1 000 days, mean absolute brightness of $M_{\rm bol} = -3^{\rm m}.8$, high amplitude of brightness variations (up to $5^{\rm m}$ or more), extremely low effective temperature of 2 300–3 000 K and with characteristic emission spectra of the later classes.

One must bear in mind the following facts and considerations when relating Mirids and PNs.

(a) The atmospheres of these giants (radii $\sim 10^{16}$ cm) are unstable and are accompanied with intense mass loss.

(b) Mirids are located on the upper part of the Hertzsprung–Russell diagram, on its asymptotic giant branch (Fig. 18.1), i.e. in the area from which, presumably, the evolution of red giant starts in the direction of PNs.
(c) The dispersion of the velocity for Mirids is equal to 112 ± 17 km s^{-1}, i.e of the same order as in the case of PNs.
(d) The mean lifetime of the Mirids is estimated to be of the order of 7×10^5 years (Jewell et al.1980). In combination with their spatial distribution – nearly 250 per 1 kpc^3 in the solar neighbourhood (Barker, Tuchman, 1980) – this yields a formation rate of 4×10^{-4} Mirids in 1 kpc^3 per year. This rate, however, is 5 times smaller than the birth rate of PNs (2×10^{-3}), but nevertheless can be considered to be of the same order, in view of numerous uncertainties in the estimation of these parameters.
(e) The Mirid's mass is estimated to be of the order of 0.7–1.0 $\mathcal{M}$ (Baud, Habing, 1983), i.e. of just the same order as the suggested mass of the nuclei of PNs.
(f) The presence of a genetic relationship between OH/IR sources and PNs is also supported by one more fact of extreme importance: in both categories of objects there are dust particles. This should be considered serious confirmation of the primordial character of the dust particles in PNs. In fact, the maser OH sources at the same time are also powerful sources of infrared emission (Engels et al., 1983) caused exceptionally by radiation of dust particles (hence, the notation IR). In the present context this should be regarded as a factor supporting the presence of an evolutionary sequence between these objects maser OH sources and PNs. Mirids are the early stage, the next is OH/IR sources, and then follow PNs, though in some cases it is hard to distinguish OH/IR sources from Mirids.

The origin of the name "maser source" is as follows: from the four emission radio lines emitted by OH molecules at the frequencies 1 612, 1 665, 1 667 and 1 720 MHz, the first one, 1 612 GHz (18.6 cm), must be, according to calculations, the faintest line. In fact this line is the strongest one, which is possible only with the action of a maser effect, i.e. the pumping of the upper level by other photons from the far infrared region emitted by dust particles.

The Galactic distribution of maser sources is the same as for PNs. The dispersion of the radial velocities for maser sources equals 125 km s^{-1} – close to that of PNs. The age of the maser sources is estimated to be of the order of 10^4 years – an order smaller than the duration of the life of the Mirids.

The total number of discovered maser OH/IR sources in the Galaxy exceeds 500, and their total number in the Galaxy is estimated to be of the order of 50 000 – again of the same order as the total number of PNs in the Galaxy.

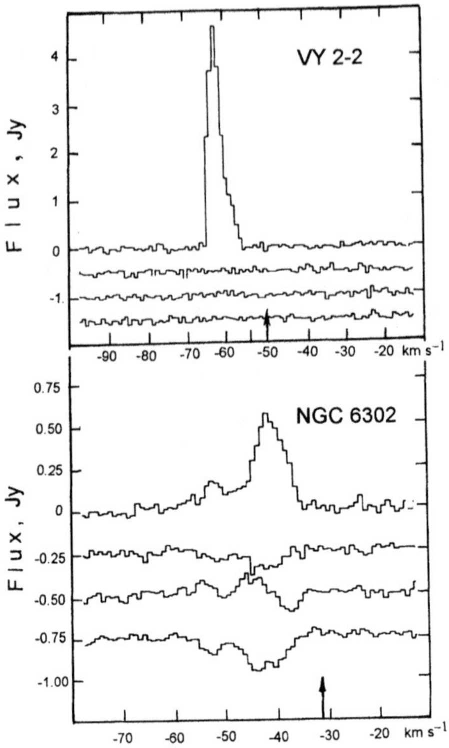

Fig. 18.2. Radio spectra for two planetary nebulae, NGC 6302 and Vy 2-2, at the frequencies (from top to bottom) 1612, 1665, 1667 and 1720 MHz of the OH molecule. In both cases the maser line 1612 MHz is in emission and shifted in the direction of negative velocities. The arrows indicate the position of radial velocities of optical nebulae

18.8 Planetary Nebulae as Maser Sources

If the OH/IR sources are the previous stage of PNs, then we cannot exclude the possibility of discovering signs of maser radiation as a rudimentary event in certain PNs.

The first maser OH emission was discovered in the late 1970s, with high flux (4.64 Jy) in the PN Vy 2-2 (Davis et al., 1979). Later large scale observations of over 100 PNs (Payne et al., 1988) revealed only two objects – the same Vy 2-2 and NGC 6302 – as sources of OH emission. Figure 18.2 shows recordings at four frequencies for these two nebulae; only in one case, at 1612 MHz, is the maser mechanism – the pumping of the upper sublevel of OH molecules under the action of infrared radiation resulting in this line, observed in emission. At three other frequencies, 1665, 1667 and 1720 MHz, no signs of emission can be detected.

In both nebulae, the emission line 1 612 MHz is shifted in the direction of negative velocities, relative to the radial velocity vector of the given nebula. This means that the line is excited only on the front side of the nebula relative to the observer, while its opposite side is absent, presumably, owing to complete absorption in the nebula itself. Can one conclude from this that both nebulae are optically thick at frequencies of the maser OH radiation?

These two nebulae differ from each other strongly. One of them, Vy 2–2, is compact, its angular radius is $0''.3$, and, hence, it is very young. As a rule, young PNs are objects of low excitation class; this is just the case for Vy 2–2, where $p = 2$. In such conditions the amount of OH molecules can still be preserved.

The situation with NGC 6 302 is quite different. This is a high excitation nebula, $p - 10$, with a hot nucleus of a temperature higher than 100 000 K, and with an anomalous form – bipolar – and of an enormous size – 3.7×10^{17} cm. Obviously, it cannot be a young object. Then the question arises, how can the amount of OH molecules be preserved for so long a time in such a high excitation nebula? A possible explanation is that the nucleus of this nebula is a double system with one of the companions being a cool giant and simultaneously the source of OH molecules and dust particles. Then both the maser OH emission and infrared radiation are generated under the action of the high temperature component.

18.9 Dispersion of the Mass of a Planetary Nebula

The mass of any PN cannot be determined without knowledge of its distance D: this dependence is strong, $\mathcal{M} \sim D^2$. However, the distances of individual nebulae are known badly, and therefore it is difficult not only to obtain the nebula's mass with confidence, but even to obtain a true idea of the dispersion of PN masses. The importance of knowing the dispersion of PN masses is obvious, first of all from a cosmogonical point of view. It is therefore necessary to find a way to solve this problem without using the distance as an independent parameter.

One way might be the use of the PN population in another galaxy, for example, in the Magellanic Clouds. Then, the distances of all PNs from us will be the same and, hence, the distance will no longer be an independent parameter.

In both Magellanic Clouds, according to data up to 1988, 103 PNs have been discovered, 28 in the Small Cloud (SMC) and 75 in the Large Cloud (LMC). For the majority of these nebulae the fluxes $F(\mathrm{H}_\beta)$ in the H_β emission line have been measured (Meatheringham et al., 1988). Then the mass $\mathcal{M}$ of the ionized part of the nebula can be determined with the help of the following expression:

$$\mathcal{M}_n = 4\pi D^2 \left(1 - 4\frac{\mathrm{He}}{\mathrm{H}}\right) \frac{F_{\mathrm{H}_\beta}}{h\nu\,\alpha_{\mathrm{eff}}} ,$$

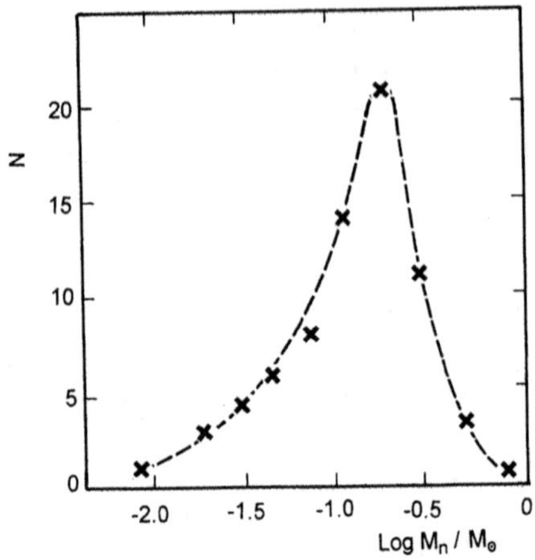

Fig. 18.3. Distribution of the number of planetary nebulae N versus their mass $\mathcal{M}$ in both Magellanic Clouds. The maximum for the number of nebulae corresponds to $0.17\mathcal{M}_\odot$

where $\alpha_{\rm eff}$ is the effective coefficient of recombination causing the H_β emission line and $h\nu$ is the energy of the H_β photon. Setting $h\nu\,\alpha_{\rm eff} = 1.24 \times 10^{-25}$ ergs cm^3 s^{-1} and He/H = 0.1, we obtain

$$\log \mathcal{M}_n = \log F_{\mathrm{H}_\beta} - \log n_{\rm e} + C\,, \tag{4}$$

where $C = 15.52$ for PNs in the LMC ($D = 53$ kpc) and $C = 15.71$ for the SMC ($D = 66$ kpc).

Equation (4) also includes $n_{\rm e}$ – the electron concentration in the nebula, which is known for 54 PNs in both clouds. From these data the masses of the ionized part were obtained for those 54 nebulae: they fell in the limits from $0.01\mathcal{M}_\odot$ to $0.35\mathcal{M}_\odot$ with a sharp maximum at $0.17\mathcal{M}_\odot$, as follows from Fig. 18.3. This result is in agreement with the mean value of the masses of 121 Galactic PNs (Maciel, Pottasch, 1980). If this coincidence is not accidental, then it can be interpreted as evidence that the distances of Galactic PNs have not been determined so poorly as might first have been thought.

As we can see, the scatter (dispersion) in the values of the ionized masses of PNs is too large – more than an order within the limiting values of the mass. The question is whether this dispersion is also related to the real mass of the nebula. In the concrete case there is reason to answer this question positively. The explanation is clear: the electron concentration in all 54 nebulae used in Fig. 18.3, are less than 10^4 cm^{-3}, and, hence, all these nebulae must with a high probability be optically thin at the frequencies of ionization of hydrogen atoms, i.e. of complete ionization of all the nebular mass.

The cosmogonic content of this result is clear: the observed dispersion in nebula masses should be related first of all to the dispersion in the initial masses of the stars embryos during their inflation. If this is the case, then the curve in Fig. 18.3 must enter as an independent parameter into the theory of the creation of gaseous envelopes from the PNs embryos.

18.10 Two Maxima on the Expansion Velocity Curve

The expansion velocities are known for 250 PNs (Weinberger, 1989). The distribution of the nebulae by their expansion velocities reveals an unexpected result, namely, the existence of two groups of PNs, I and II, with different values of the mean expansion velocities; see Fig. 18.4 which is based on the data in Table 18.2. Here the observed distribution of the number of PNs, N($2V_0$), is given as a function of the expansion velocity $2V_0$, based only on [OIII] data. This Figure shows two maxima of the distribution curve, corresponding to mean expansion velocities of 30 km s^{-1} (group I) and 80 km s^{-1} (group II). The curve can be represented as the sum of two Gaussians, I and II (Gurzadyan, Terzian, 1991):

$$N(2V_0) = N_0 \exp\left[-\frac{(2V_0 - 2\bar{V}_0)^2}{2\sigma^2}\right], \qquad (5)$$

with different dispersions $2\sigma^2$ and different maximum velocities $2V_0$; here N_0 is the number of nebulae at the maximum of the Gaussian.

Curves I and II for both groups correspond to the following numerical parameters:

	$2V_0$, km s^{-1}	$2\sigma^2$, km^2s^{-2}	N_0	N
Group I	30	625	56	218
Group II	80	160	15	34

Here N is the total number of PNs in each group.

Most nebulae belong to group I (80%), while group II contains one-seventh of the total. The Gaussians in Fig. 18.4 are shown separately for each group,

Table 18.2. Distribution of the number of planetary nebulae N according to the expansion velocity $2V_0$ (based only on [OIII] line measurements)

$2V_0$, km s^{-1}	N	$2V_0$, km s^{-1}	N
Group I		Group II	
5–15	29	65–75	12
15–25	46	75–85	15
25–35	54	85–95	6
35–45	36	95–105	1
45–55	34	> 105	2
55–65	19		

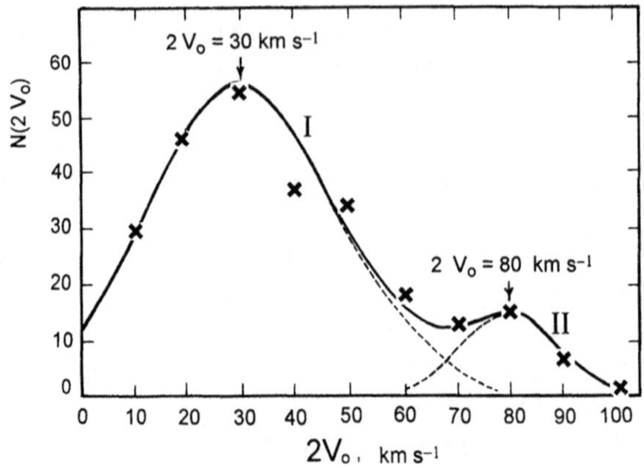

Fig. 18.4. Distribution of the number of planetary nebulae $N(2V_0)$ over the expansion velocity $2V_0$. Crosses are observations; the solid line represents the sum of two Gaussian distribution curves for two groups, I and II, of planetary nebulae with mean expansion velocities $2V_0 = 30$ km s^{-1} (group I) and $2V_0 = 80$ km s^{-1} (group II)

and the transition region, from I to II, corresponds to the velocity interval 60–75 km s^{-1}.

The existence of two groups of PNs with different expansion velocities can be a reflection of the mechanism of their origin and evolution. Therefore there may exist either two mechanisms for the origin of PNs from the same type of protogenitor or two categories of progenitors. As we shall see below, the second possibility seems more probable.

18.11 More About the Expansion Velocities

If the primordial objects from which PNs originate is homogeneous and if there is a common mechanism for their origin, the mean velocity of expansion for any sample of PNs at given distances from the Galactic plane should obviously be the same. Is this the case?

We have determined the average expansion velocities $2V_0$ for several PN samples containing n nebulae located at various distances from the Galactic plane based on the catalogue by Weinberger (1989), including PN distances mainly from Maciel (1984) and Kaler (1983). The results are given in Table 18.3 and Fig. 18.5 (Gurzadyan, Terzian, 1991).

As we can see, the average expansion velocity $\overline{2V_0}$ is too great to be the same for each sample of PNs at different heights above the Galactic plane. Numerically $2V_0$ is the smallest (~ 37 km s^{-1}) near the Galactic plane ($z = 0$), and increases almost monotonically with distance from the Galactic plane. At Galactic heights $|z| = 1.8$–2.0 kpc the value $2V_0$ is of the order

Table 18.3. Dependence of average expansion velocities $\overline{2V_0}$ (in km s^{-1}) of planetary nebulae on the absolute height above the Galactic plane z. $N(z)$ is the number of nebulae in the given interval of $|z|$

$\|z\|$, kpc	$\overline{2V_0}$	$N(z)$	$\|z\|$, kpc	$\overline{2V_0}$	$N(z)$
0–0.2	36.6	118	0.8–1.0	42.5	12
0.2–0.4	38.5	51	1.0–1.4	44.6	9
0.4–0.6	43.5	25	1.4–2.0	49.4	11
0.6–0.8	41.5	17			

of 50 km s^{-1} (in the region $z \sim 2$ kpc, 11 PNs have distances that are too uncertain and therefore are not included).

The existence of two groups of PNs with different expansion velocities can be related to the mechanism of their origin and evolution. Therefore we must consider that there exist two mechanisms for the origin of planetary nebulae from the same type of star or protoplanetary object; or that there exist two categories of stellar protoplanetary objects that produce the planetary nebulae. As we shall see below, the second possibility seems more probable.

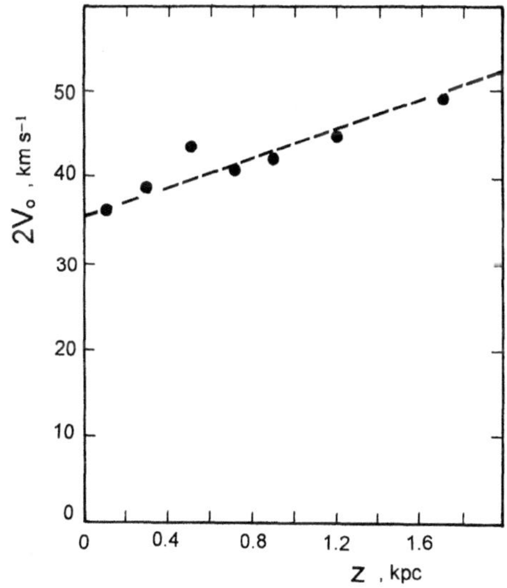

Fig. 18.5. Distribution of the average expansion velocities $\overline{2V_0}$ of planetary nebulae with absolute height $|z|$ from the Galactic plane. Crosses are observations; the broken line represents the mean relationship between $2V_0$ and $|z|$

18.12 Two-Component Systems of Progenitors

From the above discussion we see that the observed PNs can represent a mixture of at least two types of stellar progenitor with different initial escape velocities from their outer layers.

The known distribution of PNs above or below the Galactic plane is neither flat nor intermediate. The "dispersion" factor or the half-thickness, β, in the exponential law

$$N(z) = N(0)e^{-z/\beta}, \tag{6}$$

is assumed to be $\beta \sim 200$ pc (Table 18.1). The flat and intermediate subsystems have $\beta \sim 100$ pc and $\beta \sim 400$ pc, respectively.

Assuming two different types of PN progenitor A and B, we then have "dispersion" factors $\beta_{\rm A}$ and $\beta_{\rm B}$. We now proceed to estimate these parameters for both subsystems, i.e. $\beta_{\rm A}$, $N_{\rm A}(0)$ and $\beta_{\rm B}$, $N_{\rm B}(0)$. We possess the following data: (a) the distribution of the number of PN, $N(z)$, as a function of z (Table 18.3); (b) the distribution of the average expansion velocities $V(z)$ as a function of z (Fig. 18.5); (c) the values of the mean expansion velocities $2V_{\rm A} = 30$ km s^{-1} and $2V_{\rm B} = 80$ km s^{-1} for two PN subsystems (Fig. 18.4).

The total number of PNs, $N(z)$, at a given height z is (Gurzadyan, Terzian, 1991)

$$N(z) = N_{\rm A}(z) + N_{\rm B}(z). \tag{7}$$

Since the function $N(z)$ is represented by an exponential law (6), both functions $N_{\rm A}(z)$ and $N_{\rm B}(z)$ have also to be represented by an exponential law with parameters $\beta_{\rm A}$, $N_{\rm A}(0)$ and $\beta_{\rm B}$, $N_{\rm B}(0)$:

$$N_{\rm A}(z) = N_{\rm A}(0)\exp(-z/\beta_{\rm A}), \tag{8}$$

$$N_{\rm B}(z) = N_{\rm B}(0)\exp(-z/\beta_{\rm B}). \tag{9}$$

Then we can write for the mean velocity of expansion for a two-component mixture of nebulae

$$\overline{V_0(z)} = \frac{V_{\rm A}N_{\rm A}(z) + V_{\rm B}N_{\rm B}(z)}{N_{\rm A}(z) + N_{\rm B}(z)}, \tag{10}$$

or, using (7),

$$\overline{V_0(z)} = V_{\rm A}\frac{N_{\rm A}(z)}{N(z)} + V_{\rm B}\frac{N_{\rm B}(z)}{N(z)}. \tag{11}$$

Dividing (7) by $N(z)$ and inserting the ratio $N_{\rm B}(z)/N_{\rm A}(z)$ into (11), we find the relative components, $N_{\rm A}(z)/N(z)$ and $N_{\rm B}(z)/N(z)$, denoting the progenitors of types A and B, respectively:

$$\frac{N_{\rm A}(z)}{N(z)} = \frac{V_{\rm B} - \overline{V_0(z)}}{V_{\rm B} - V_{\rm A}}, \tag{12}$$

and

$$\frac{N_{\mathrm{B}}(z)}{N(z)} = 1 - \frac{N_{\mathrm{A}}(z)}{N(z)} \,. \tag{13}$$

In these expressions the numerical values of $\overline{V_0(z)}$ are taken from Fig. 18.5 (broken line) and are given in Table 18.4 (second column). Setting also $2V_{\mathrm{A}} = 30$ km s^{-1} and $2V_{\mathrm{B}} = 80$ km s^{-1}, we obtain $N_{\mathrm{A}}(z)/N_{\mathrm{B}}(z)$ and the functions $N_{\mathrm{A}}(z)$ and $N_{\mathrm{B}}(z)$; they are presented in last two columns of Table 18.4. For calculations, the values of $N(z)$ from Table 18.3 are used.

Table 18.4. The numbers of planetary nebulae $N_{\mathrm{A}}(z)$ and $N_{\mathrm{B}}(z)$ originating from type A and B progenitors

z [kpc]	$2V_0(z)$ [km s^{-1}]	$\frac{N_{\mathrm{A}}(z)}{N(z)}$	$\frac{N_{\mathrm{B}}(z)}{N(z)}$	$N_{\mathrm{A}}(z)$	$N_{\mathrm{B}}(z)$
0–0.2	36	0.88	0.12	105	14
0.2–0.4	38	0.84	0.16	43	8
0.4–0.6	39	0.82	0.18	20	5
0.6–0.8	41	0.78	0.22	13	4
0.8–1.0	43	0.74	0.26	9	3
1.0–1.4	46	0.68	0.32	6	3
1.4–2.0	50	0.61	0.39	7	4

We now obtain the numerical values of the parameters of the Gaussian distributions, β_{A}, $N_{\mathrm{A}}(0)$ and β_{B}, $N_{\mathrm{B}}(0)$, using the data of the last two columns of Table 18.4:

$$\beta_{\mathrm{A}} = 230\,\mathrm{pc}, \qquad N_{\mathrm{A}}(0) = 160\,,$$

$$\beta_{\mathrm{B}} - 370\,\mathrm{pc}, \qquad N_{\mathrm{B}}(0) = 18\,.$$

Therefore, we have for the distributions $N_{\mathrm{A}}(z)$ and $N_{\mathrm{B}}(z)$

$$N_{\mathrm{A}}(z) = 160 \exp(-z/230)\,, \tag{14}$$

$$N_{\mathrm{B}}(z) = 80 \exp(-z/370)\,. \tag{15}$$

Note that β_{B} is nearly one and half times larger than β_{A}, though in terms of orders of magnitude they are equal. These data still do not allow the identification of protoplanetary objects with any definite category of stellar populations (maser OH/IR sources, Mirids?). Obviously, more data are required. In any case, the contributions of both categories A and B are not equivalent and the progenitors of type B must yield about $N_{\mathrm{A}}(0)/N_{\mathrm{B}}(0) = 18/160$, or 8 times smaller than those of type A.

18.13 Comparison with OH/IR Sources

Recall that for the two populations of PNs defined by the two systems of expansion velocities the OH/IR sources are suggested as the possible population of PN progenitors. The OH 1 612 MHz observations of 350 infrared maser sources – OH/IR objects – by the Arecibo radio telescope (Eder et al., 1988), show that the distribution of expansion velocities of these objects also have two maximums. The first is at $2V_0 = 25$–$30\,\mathrm{km\ s^{-1}}$, in complete agreement with the above result for the population of PNs. The second maximum is very weak and corresponds to $2V_0 = 60\,\mathrm{km\ s^{-1}}$, noticeably smaller than was recorded above for PNs.

Although these distributions need further investigation, some preliminary conclusions can be already drawn. Indeed, the first maximum at $2V_0 = 25 - 30\,P\mathrm{km\ s^{-1}}$ is rather clear – it is based on data from over three hundred OH/IR sources. This fact not only confirms the above suggestion on the evolutionary relation of OH/IR objects and PNs. Presumably, it also shows that PNs not only "accepted" from OH/IR sources their main physical properties – the mass, chemical composition, aggregate condition, dust particles etc. – but even their main kinematic property, i.e. their initial expansion velocity. This implies the preservation of the expansion velocity at a constant level of $\sim 30\,\mathrm{km\ s^{-1}}$ during the lifetimes of both categories of objects – maser sources and PNs – i.e. over 50 000 or even 100 000 years.

18.14 "Helium" Nebulae

There are PNs that have an anomalously large abundance of helium. Typically, for the majority of PNs He/H ~ 0.10 (Table 2.6), while in "helium" nebulae the He/H ratio is of the order of 0.15. In certain cases the He/H ration approaches 0.20 or even 0.23 (NGC 6 445, NGC 6 302), i.e. the helium will form nearly half of the nebula's mass. It should be true, therefore, to speak of "helium" nebulae as special kind of PN (Peimbert, 1978).

The list of helium PNs, 20 objects with He/H > 0.125, is given in Table 18.5 (Aller, 1978; Kaler, 1981a; Peimbert, Torres-Peimbert, 1983; Pottasch, 1984).

With respect to a number of properties and parameters differing from normal PN, helium nebulae can be considered a separate category of PNs:

(a) Helium PNs are objects of the highest excitation classes, $p = 9$–10 or higher. The attribution of a little group of low excitation class nebulae, $p = 2$–6, to the category of helium nebulae – the last six objects in Table 18.5 – is doubtful. The observed values of He/H observed for these objects are probably related to their nuclei.

(b) The mean expansion velocity $\overline{2V_0(\mathrm{He})}$ for helium nebulae is too high – 60 km s^{-1}, two times larger than the value derived above, 30 km s^{-1},

Table 18.5. "Helium" planetary nebulae

Planetary nebula	He/H	$2V_0$ [km s^{-1}]	Excit. class	Planetary nebula	He/H	$2V_0$ [km s^{-1}]	Excit. class
NGC 650-1	0.131	77	10	NGC 7293	0.180	48:	7
NGC 2346	0.130	26	9	NGC 7354	0.132	54:	10
NGC 2371-2	0.134	85	11	IC 2165	0.133	–	10
NGC 2440	0.153	45	10	Hu 1–2	0.170	64:	12
NGC 2818	0.143	73	11	M 2–27	0.144	–	11
NGC 3132	0.127	42	9	PB 4	0.139	–	9
NGC 4361	0.135	64	12	PB 6	0.187	–	12
NGC 6302	0.225	–	10	NGC 5315	0.127	80	4
NGC 6445	0.228	76	10	NGC 6629	0.143	13	2
NGC 6537	0.140	36	11	NGC 6778	0.143	42	2
NGC 6741	0.136	44	9	IC 4406	0.133:	28	6
NGC 6853	0.133	63	10	Cn 2–1	0.137	–	3
NGC 7008	0.140	80	11	Me 2–2	0.161	31	2

for normal PNs, and closer to the second maximum, 80 km s^{-1}. Do the helium nebulae comprise an independent or intermediate (third) group?

(c) Of special importance is the absence of a correlation between the helium abundance, He/H, and the linear sizes of the nebula, R_n, which follows even from the following comparisons:

	He/H	R_n, pc
NGC 650-1	0.13	0.223
NGC 6741	0.14	0.027
Hu 1–2	0.17	0.017
NGC 7293	0.18	0.326

(d) In helium nebulae an anomalous abundance of nitrogen is also observed (Peimbert, 1978). This property strengthens the above conclusion about the primordial nature of the anomalous abundance of helium in these nebulae.

(e) The helium nebulae are distinguished also from the morphological point of view, i.e. by a strongly developed filamentary structure (Peimbert, Torres-Peimbert, 1983). Examples, in particular, are the helium nebulae NGC 6302, NGC 2818, NGC 2440, Hu 1-2. The filamentary structure may be interpreted as the result of specific behaviour of the intrinsic magnetic fields of nebulae (Sect. 17.15). If this is the case, then a question arises: can helium be an indicator promoting the formation of filamentary structure in nebulae?

Helium PNs with He/H > 0.125 have also been discovered in both Magellanic Clouds (Webster, 1976; Kaler, 1981a).

Thus, these facts seem to indicate that the progenitors of PNs themselves represent a mixture of stars of various populations and subsystems of the Galaxy.

A typical helium nebula, A 30, was examined and analysed in Sect. 15.4 as a purely physical phenomenon of a separate category of PNs. There we saw that under the general title of "helium" nebulae the *cosmogonical* essence of the PN problem can be hidden.

18.15 Distribution of Planetary Nebulae by Excitation Class

The system of spectral classification or, which is the same, the method of determining the excitation class of the nebula p, was described in Sect. 1.7. The main advantages of this system are, first, that it is of a purely quantitative character, and hence, is simpler to apply, and, second, that it is based on the use of the most widely available emission lines, $N_1 + N_2$ [OIII], 4 686 HeII and H_β. At the same time, this is the minimum information with which a gaseous nebula may be accepted to be a PN.

In the case of stars, the spectral class is not an independent parameter, since under the given spectral class we always have a definite combination of its effective temperature and luminosity. In the case of gaseous nebulae, the spectral class or excitation class p has a collective meaning and in the hidden form depends on practically all the parameters – the temperature and the luminosity of the central star, and the size, mass, shape, structure, electron temperature and concentration, inner kinematics, chemical composition, etc. of the nebula. In the case of so many physical and geometrical parameters, the excitation class acquires the status of an independent parameter.

Our problem is now to apply this system to the known population of PNs, to obtain their excitation classes and to carry out an analysis of the results with an aim to establish, in particular, some regularities of evolutionary character.

Two interpretations are possible. First, the excitation class is not an evolutionary parameter. In this case, the nebula remains in the same, i.e. the initial excitation class is practically invariant during its lifetime. Second, the excitation class is an evolutionary parameter. In this case, the nebula is born of low excitation class, and then, after subsequently passing through all classes, it finishes its life in one of the highest excitation classes. As we shall see below, the first case does not correspond to reality.

Using the data of the PN catalogue by Acker et al. (1992), the excitation classes p have been obtained for 750 PNs. The results are presented at the end of this book in the form of a *Catalogue of Excitation Classes for 750 Galactic Planetary Nebulae*. Graphically the distribution of PNs, $N(p)$, with respect to the excitation class p is presented in Fig. 18.6.

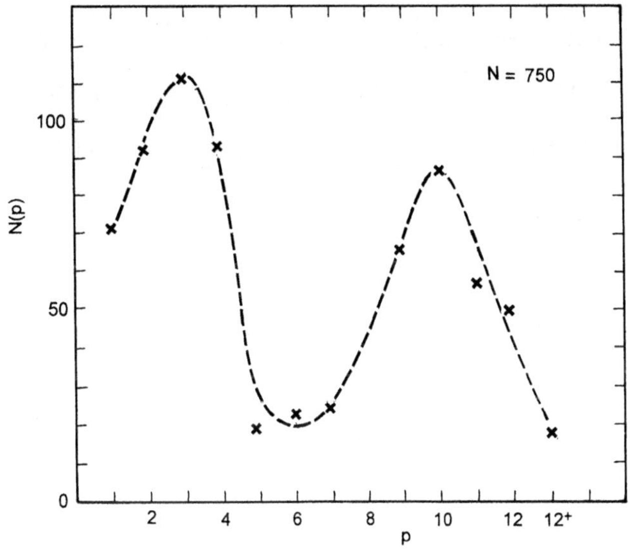

Fig. 18.6. Distribution of the number of planetary nebulae $N(p)$ with respect to the excitation class p

Figure 18.6 is rather remarkable since it offers a simple interpretation, i.e. a kind of a marathon run of the nebula from $p = 1$ up to $p = 12^+$ occurs that is too non-uniform. The nebula spends too long a time in the lowest classes of excitation, $p = 1$–3, then quickly slips through classes 4, 5 and 6 up to class 10, where the lifetime is again longer and similar to those of states 1, 2 or 3.

It is not difficult to obtain the lifetime $\Delta T(p)$ of the nebula in the given state of excitation class p from the following obvious relationship:

$$\Delta T(p) = T \frac{N(p)}{\sum N(p)}, \tag{16}$$

where T is the lifetime of the nebula and $\sum N(p) = 750$ is the total number of all PNs used in Fig. 18.6. For example, when $T = 30\,000$ years, we have, as follows from the data of Table 18.6, the longest duration of $\sim 4\,500$ years in the states $p = 1$, 2 and 3, and the shortest one, 800–900 years, in the states $p = 4$, 5 and 6, as well as in the highest excitation states, $p = 12^+$

Thus, the empirical curve in Fig. 18.6 can be modified literally into an evolution curve, in view of its time distribution with respect to various states of excitation p. This curve, of course, must "enter" into the theory of the evolution of PNs.

Two important conclusions follow.

(a) About 80% of PNs reach the state $p = 10$, i.e. the overwhelming majority, after which their rate of disappearance grows.

Table 18.6. The distribution of planetary nebulae $N(p)$ and the duration of their stay $\Delta T(p)$ in the excitation class p over the full lifetime $T = 30\,000$ year

Excit. class [p]	$N(p)$	$\Delta T(p)$ [year]	Excit. class [p]	$N(p)$	$\Delta T(p)$ [year]	Excit class [p]	$N(p)$	$\Delta T(p)$ [year]
1	72	2 800	5	19	800	9	66	2 600
2	91	3 600	6	22	900	10	87	3 500
3	112	4 500	7	23	900	11	55	2 200
4	94	3 800	8	43	1 700	12	49	2 200
						12^+	17	700

(b) The highest degree of excitation, $p = 12^+$, is reached by only 15% of PNs, the rest disappear long before reaching this state.

18.16 Dependence of the Excitation Class on Nebula Size

Does the excitation class depend on the linear size of the nebula? Either answer to this question can suggest a definite interpretation (Gurzadyan, Egikyan, 1991).

In Fig. 18.7, the distribution of the relative number N of PNs (vertical scale) with the nebula's radius, R_n, in pc (horizontal scale) is given for each excitation class separately, from $p = 1, 2 \ldots$ up to 12 and 12^+. The radii of the nebulae R_n were determined from known angular radii φ'' and distance D_n taken mainly from distance catalogues by Zhang (1995) and Cahn et al. (1992), in some cases also from those by Daub (1984) and Maciel (1982). In all, data for 404 PN have been used in Fig. 18.7.

This brief survey already shows the existence of an obvious relationship, i.e. of a tendency of the mean radius of the nebulae for a given excitation class p to increase from low excitation classes, $p = 1, 2$ to high excitation classes, $p = 12^+$. The smallest mean radius, $\overline{R_n} \sim 0.06$ pc, occurs for $p = 1$, 2 and the largest one, $\overline{R_n} \sim 0.35$ pc, for $p = 12^+$. In this case the question is to do with the *ascending* branch of the dependence of N on the nebula's radius R_n; it is indicated schematically indicated in Fig. 18.8 by a solid line.

Although there is a large scatter of points, one more relationship seems noticeable: the existence of a *descending* branch from $p = 12^+$ up to $p = 3, 2$, 1 with an increase of the mean radius from $\overline{R_n} \sim 0.35$ pc to $\overline{R_n} \sim 1$ pc; this is represented in Fig. 18.8 by a broken line. The mean values of $\overline{R_n}$ depending on p derived from these both curves are presented in Table 18.7 separately for three excitation groups, high ($p = 12^+$–10), middle (9–5), and low (4–1). Here N is the number of nebulae with a given p.

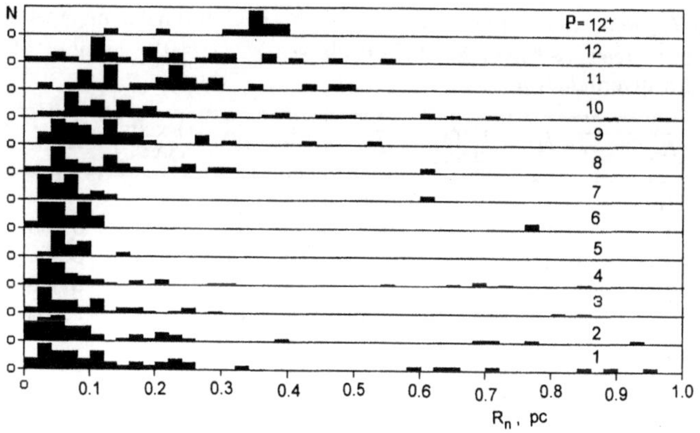

Fig. 18.7. Distribution of the relative number N of planetary nebulae with radius R_n (pc) for each excitation class, p, from 1 up to 12^+, separately

Large uncertainties and errors in the determination of the distances of PNs obviously limit the reliability of the curves in Fig. 18.8 and of the data in Table 18.7, and more careful examination is required.

Between these two moves, ascending, from $p = 1$ to $p = 12^+$, and descending, from $p = 12^+$ to $p = 1$, there is an extended area, *an avoidance zone,* in the lower part of Fig. 18.7.

The evolutionary content of the curves in Fig. 18.8 is obvious. Hence, the excitation class should be regarded as an evolutionary criterion. The existence of a void in Fig. 18.8 indicates that the evolution takes place not from the

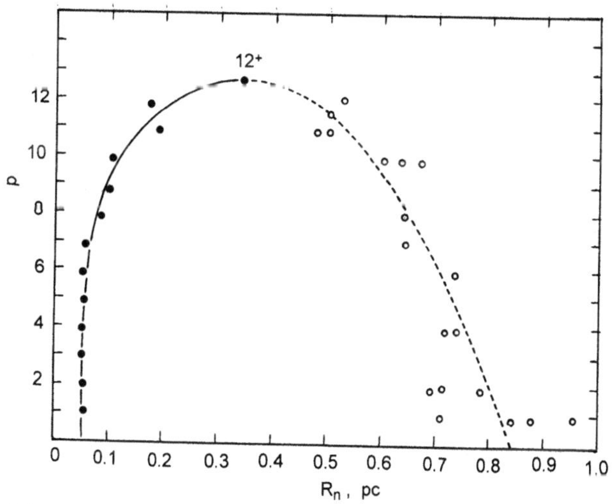

Fig. 18.8. Schematic representation of Fig. 18.7; both ascending (*solid curve*) and descending (*broken curve*) correlation branches of the type $R_n \sim p$ are shown. The arrows denote the possible direction of the nebula's evolution

Table 18.7. Mean radii of planetary nebulae $\overline{R_n}$ (pc) and mean expansion velocities $\overline{2V_0}$ (km s^{-1}) averaged over excitation classes p. N is the number of planetary nebulae of a given class p

Excit. class, p	$\overline{R_n}$	$\overline{2V_0}$	N	Excit. class, p	$\overline{R_n}$	$\overline{2V_0}$	N	Excit. class, p	$\overline{R_n}$	$\overline{2V_0}$	N
	High				Middle				Low		
12^+	0.34	73	15	9	0.10	45	61	4	0.053	29	85
12	0.18	61	41	8	0.086	42	39	3	0.068	40	102
11	0.18	55	44	7	0.061	33	22	2	0.068	31	102
10	0.11	49	84	6	0.060	31	19	1	0.064	28	110
				5	0.065	29	17				

left to the right, i.e. with increasing sizes, but with constant excitation class during the life of a nebula. The evolution occurs from bottom to top, from $p = 1, 2...$ up to $p =$ 8–9, when the mean size of the nebula increases only two times, from $R_n \sim 0.05$ pc up to $R_n \sim 0.10$ pc. Then the nebula moves in direction to higher excitation classes – $p =$ 8–9 or higher. The PN finishes its life after reaching the highest level – the apogee of excitation class 12^+. Judging from the small number of PNs in $p = 12^+$, not all PNs reach this limiting class.

The temperatures of the nuclei of nearly all PNs of the highest excitation class $p = 12^+$ in Table 18.7 are of the order of 300 000 K or higher (Sect. 5.7). Hence, the evolution of PNs takes place parallel with the increase in nuclear temperature.

A certain number of PNs seem to reach class 12^+ before disappearing and their nuclei eventually become high temperature stars. But we have not detected stars with temperatures of 300 000 K or higher in the common Galactic background. This may be a result of observational selection – we are not able to measure such high temperatures without using the emission lines of various atoms and ions. Note that such high temperature stars themselves typically will not be nuclei of nebulae nor possess a proper gaseous envelope, while the gradient or the declination of continuous spectra even in the ultraviolet (1 000– 2 000 Å) loses its sensitivity to temperature when $T > 100\,000$ K.

On the other hand, the possibility that a certain fraction of PNs can continue their evolution after reaching apogee at $p = 12^+$ in the direction of lower excitation classes, i.e. through the descending branch in Fig. 18.8 with further increase of size, is not excluded either. In this case the question is to with the group of *evolved* PNs, with nebulae of extremely large radii and low surface brightness (Gieseking et al., 1986; Hippelein, Weinberger, 1990).

The existence of a large dispersion in the relationship $p \sim R_n$ in Fig. 18.7 shows the possibility of the existence of various families of curves $p \sim R_n$ in Fig. 18.8, paths of PN evolution. These paths appear to come together near the point $p = 12^+$.

The fact that the diagram $p \sim R_n$ in Fig. 18.8 admits an evolutionary interpretation enables it to be regarded as an analogy of the Hertzsprung–Russell diagram for PNs. The interrelation of the excitation classes at the moment nebulae appear on the ascending branch of the diagram and those at the moment they are destroyed on the descending branch demands special examination. At present we can establish only that the PNs should be of excitation class from 1 up to 6 at their birth and of classes 12^+ up to 1, 2 or 3 at their deaths.

18.17 Dependence of Expansion Velocity on Excitation Class

In Table 18.7, the numerical values of the mean expansion velocity $\overline{2V_0}$ averaged over excitation classes p are presented (third column) according to the catalogue of Weinberger (1989); in total 120 PNs have been used. Again, despite expectations, the existence of some correlation between $\overline{2V_0}$ and p, i.e. the increase on average of the expansion velocity as one moves to higher excitation nebulae and vice versa, is clear from Fig. 18.9. Combined with Fig. 18.8, this implies an increase in expansion velocity with an increase in nebular size. It is likely that the signs of such behavior of PNs have been noticed before (Sabbadin et al., 1986). However, here the point is the establishment of a real dependence of $\overline{2V_0}$ on p, which can be fitted by empirical formulae (broken line in Fig. 18.9),

$$\overline{2V_0} = 31.5 - 1.93p + 0.37p^2 \,, \tag{17}$$

where $\overline{2V_0}$ is expressed in km s^{-1} and p in discrete numbers 1, 2, ..., 13 (= 12^+). Obviously, both the curve in Fig. 18.9 and (17) can be modified if new observational data appear.

Relationship (17) reflects a crucial fact, that the expansion of a nebula takes place, on average, at an accelerated rate during the PNs' lifetime.

According to the theoretical discussions developed in Sect. 16.2, the expansion of PNs must occur with an acceleration with the increase in their sizes. However, observational data do not directly indicate the existence of such a correlation between $2V_0$ and p, which follows from the structure of Fig. 16.2. However, in the same section it was mentioned that the desirable correlation may be revealed for PN samples with definite parameters. Now we see that the population of PNs selected by excitation class in fact offers such a possibility – the revealing of a positive correlation between $\overline{2V_0}$ and p, as follows from Fig. 18.9.

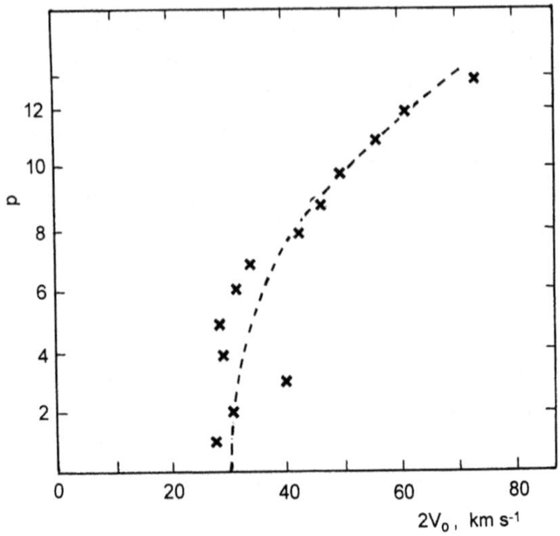

Fig. 18.9. Empirical dependence between the excitation class of a nebula p and the mean expansion velocity $\overline{2V_0}$ (Table 18.7). Crosses are observations; the broken line, empirical formula (17)

18.18 The Macrostructure of Expansion

The following three stages or periods can be distinguished in the the process of the birth and the evolution of PNs.

Period I. The inflation of the outer layers of the star. This period continues up to the moment when a noticeable part of the star's mass, of the order of 0.1–0.2 $\mathcal{M}_\odot$, departs from the star.

Period II. The expansion of the separated mass into an envelope. The expansion occurs up to sizes or concentrations when the excitation of the forbidden lines becomes possible. The duration of this period is not large – less than a thousand years.

Period III. The evolution of the rarefied gaseous envelope. The duration of this period is determined by the expansion and decay of the envelope up to its complete disappearance, i.e. of the order of 10^4–10^5 years.

Period I is completely attributed to the progenitors, in the present case, to Mirids, the same as OH/IR maser sources.

Period II is related to the evolution of the mass separated from the star-progenitor. This corresponds to the protoplanetary state, i.e. to a transition phase from progenitor (I) to PN (III).

Thus, the macrostructure of the evolution emerges in the following form:

$$Mirid \rightarrow Protoplanetary\ object \rightarrow Planetary\ nebula,$$

where Mirid is identified with OH/IR maser source.

The moment of separation of a mass of $\sim 0.1\mathcal{M}_\odot$ from the star's body is still not the birth of a PN, although the separation itself signifies a departure from the previous stage. The moment of birth of the physical PN should be defined as the appearance of the first forbidden lines, i.e. when the electron concentration reaches 10^7 cm^{-3} or even 10^8 cm^{-3}.

18.19 Spectral Variability of the Nebula IC 4 997

Spectral variations have been established in a number of PNs. Rich observational data have been accumulated for IC 4 997, a dense, young, compact PN $1''.6$ in diameter, with an electron concentration $> 10^6$ cm^3, and age ~ 200 years. These data have implications first of all for the intensity variations in the emission line 4 363 [OIII]. In the Harvard and Lick spectrograms of this nebula of 1895, this line was stronger than the hydrogen line H_γ (4 360 Å). In 1938–1939, the line 4 363 [OIII] was measured, for the first time, photometrically, and was 60% brighter than H_γ. In 1940 this line was still brighter than H_γ by 40%, and only in 1949 were both the lines, 4 363 Å and H_γ of equal intensity.

The decline in the intensity of the line 4 363 Å continued, and in 1956 it was only 76% of the intensity of H_γ. During a period nearly ten years or so, the intensity of the line 4 363 Å remained at a more or less constant level. The lowest level of the ratio $R = I(4\,363)/I(H_\gamma)$, 62% was reached in 1969. Afterwards, a slow but steady increase of R took place, and in 1980 this ratio became of the order of unity: $R \approx 1$ (the summary of early references see in Feibelman, 1979; Hyung et al., 1994).

The results of two decades (1970–1990) of systematic spectrophotometric observations of IC 4 997 carried out by Kostyakova (1989, 1991) are presented in Fig. 18.10 (as crosses) along with other observations since 1938. The vertical bars denote the results of Purgathofer and Stoll (1981). The largest magnitude $R = 1.51$ was reached in 1988. Then, a fall in the magnitude of R took place again, the last point in Fig. 18.10 corresponding to a magnitude of $R = 1.30$.

Because of the non-homogeneity of these data, it is hard to judge whether the scatter in the magnitudes of R is real and is not caused by small scale variations in the intensity of 4 363 [OIII]. Without excluding the possibility of such a bias, the fact of the large scale and especially the smooth variation, of R since 1938, should be regarded as certain.

Note the proximity of two values of R, 1.51 and 1.6, for 1988 and 1938, respectively. Does this mean that the maximum of R in 1938 was of the same type as was established in 1988? If yes, one has to consider the possibility of a 50 year periodicity of the maxima of the curve of R.

As mentioned above, the Harvard and Lick plates indicate that, at least from eye estimations, the line 4 363 [OIII] was stronger than H_γ, so that R was obviously larger than unity, probably not less than 1.2–1.3. This could mean

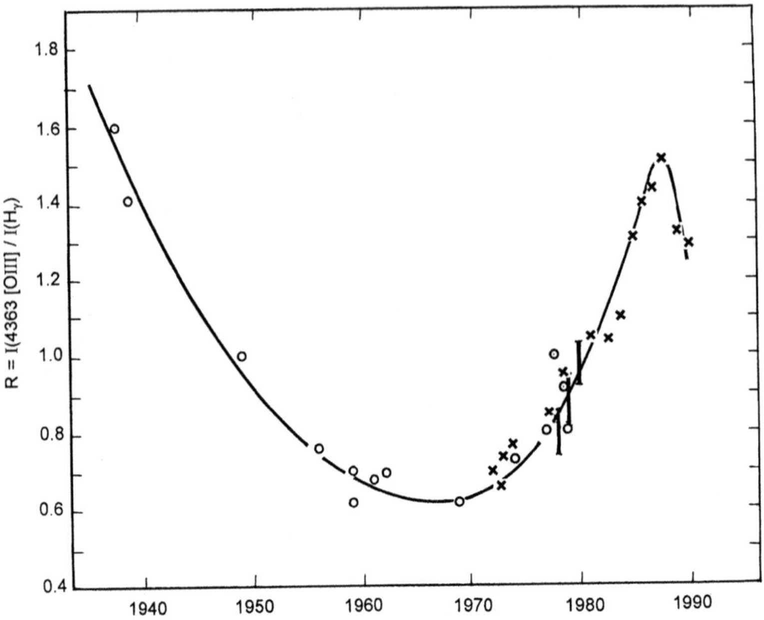

Fig. 18.10. Variations of the ratio R of the intensities of the emission lines 4 363 [OIII] and H_γ in the spectrum of the star-like planetary nebula IC 4 997 in the period 1938–1990. Circles denote the data from various observations (see Feibelman et al. 1979); crosses, those by Kostyakova (1990); and vertical bars, those by Purgathofer, Stoll (1981)

that in 1895 the curve was not far from its maximum at $R \sim 1.5$, and the previous period 1938–1895 = 43 year was not far from the above mentioned period (50 years) for the last cycle (1938–1988). Obviously, this is a proposal, though it is not groundless. The discovery of even a single measurement of R among the unknown archive materials of the previous cycle, 1895–1938, might be decisive.

It can be shown that the intensity variations of the line 4 363 [OIII] are caused by variations of electron temperature T_e of the nebula and, in the long run, by variations of the effective temperature T_* of the central star (Gurzadyan, 1958, 1961; Feibelman et al., 1979; Purgathofer, Stoll, 1981; Aller et al., 1994). Now we can see that these variations may also be of cyclic nature. However, it is first necessary to have reliable observational evidence of the reality of the cyclic nature of the variations of the line 4 363 [OIII] itself.

18.20 The Paradox of the Nebula IC 4 997

For over two decades, spectrophotometric observations of IC 4 997 have led to unexpected results: the sharp disagreement between the variations of the

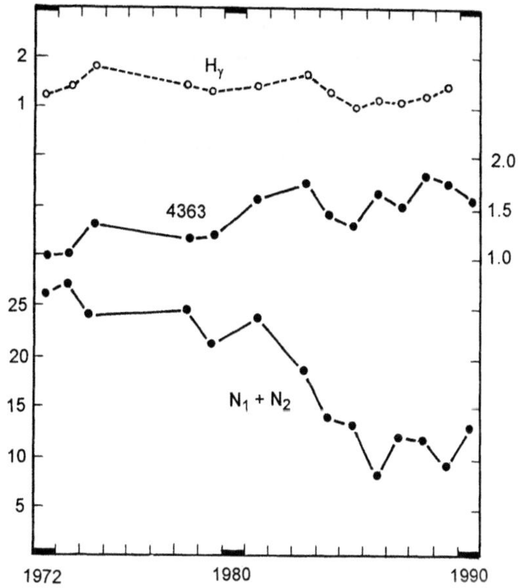

Fig. 18.11. Variations of intensities (in units of 10^{-11} ergs cm^{-2} s^{-1}) of the emission lines 4 363 [OIII] and $N_1 + N_2$ in the spectrum of the planetary nebula IC 4 997 for the observational period 1972–1990. For comparison, the curve for the emission line H_γ, which has a practically constant intensity, is also shown

line 4 363 [OIII], on the one hand, and of the doublet $N_1 + N_2$ [OIII], on the other, although theoretically one should expect a strong synchronicity between the variations of these emission lines. As follows from Fig. 18.11, the variations are definitely in opposite directions – the rise in the intensity of 4 363 [OIII] is accompanied by a clear decrease in the intensity of $N_1 + N_2$.

To solve this paradox in the framework of the usual theory of the luminescence of gaseous nebulae in forbidden lines seems impossible. As a detailed quantitative analysis shows, there is no relation between the electron temperature and electron concentration when the variations in the intensities of 4 363 [OIII] and $N_1 + N_2$ can occur in opposite directions.

The paradox of the nebula IC 4 997 can be removed if we assume the possibility of transition of ions O^{++} from the state ^1D by non-radiative means to levels higher than ^1S. Here the nearest levels are $^1D_2^0$ and $^1P_1^0$, both of configuration $2s\,2p^3$, with excitation potentials of 23.2 and 26.1 eV, respectively (Fig. 18.12). Hence, if we assume the possibility of the appearance of certain continuous radiation, or at least radiation in a more or less wide band, including the lines 599.6 Å and 525.8 Å, then, in principle, the transfer of some of the ions O^{++} from the level ^1D to $^1D_2^0$ or $^1P_1^0$ can occur under the action of this hypothetical radiation. As a result, we should observe a decrease in the number of ions O^{++} at the level ^1D and, hence, the weakening of the doublet $N_1 + N_2$ without a weakening of the line 43 663 [OIII].

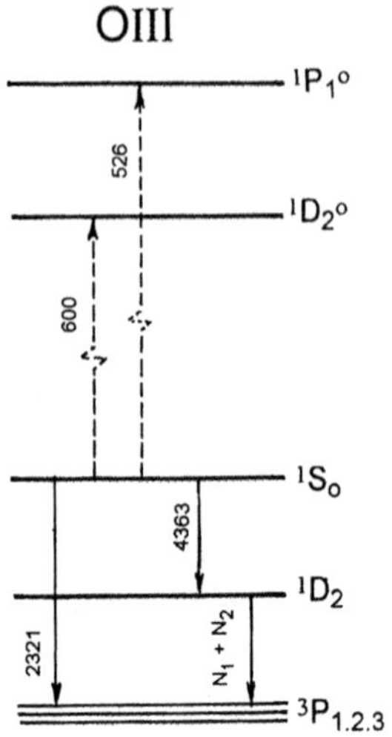

Fig. 18.12. Scheme of energy levels of OIII. The lower three levels are of the configuration $2p^2$; the upper two levels, of $2s\,2p^3$

In the scenario just depicted, the behaviour of the doublet $N_1 + N_2$ becomes completely independent of the behaviour of the line 4363 [OIII]. Our problem now is to find the source of such hypothetical radiation with a density at which the observed effect – a decrease in the intensity of the doublet $N_1 + N_2$ with the intensification of the line 4363 [OIII] – can be explained.

18.21 Transition Radiation

We suggest that the hypothetical radiation discussed in the previous section can be of nature of *transition radiation*. For the physics of PNs this is a new concept, the main point being is as follows.

When a charged particle, e.g. a relativistic electron, transits from one medium to another, or in terms of electrodynamics when it crosses the boundary between media of different dielectric capabilities, a reconstruction of the electromagnetic field of the particle takes place. As a result, part of this field is emitted in the form of radiation. Assuming a vacuum and, e.g., a the dust particle to be two different media, then one can expect the appearance of transition radiation twice: first, on the passage of the electron from the vacuum into the dust particle, and second, on its exit from the dust particle to the vacuum. The amount of energy escaping during each passage is very small. Nevertheless, the final effect can be non-negligible owing to the large number of transitions of electrons via the dust particles.

Transition radiation can be generated on the interaction of dust particles not only with relativistic electrons but also with non-relativistic electrons,

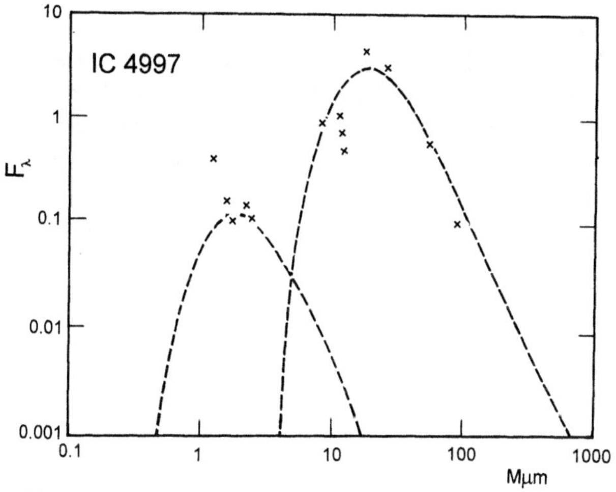

Fig. 18.13. Observed spectrum of IC 4997 in the infrared according to the *IRAS* catalogue. Observed fluxes F_λ are in units of 10^{-12} Wm^{-2}. Two maxima are visible at $\sim 2\,\mu$m and $\sim 15\,\mu$m. The long wavelength maximum corresponds to an effective temperature of 185 K; the short wavelength maximum, to 1 500 K

and this is one of the remarkable features of transition radiation. In this case, the lower limit of the electron energy is determined by the condition of its penetration into the dust particle up to a certain depth. In practice, this corresponds to an electron energy of the order of a few keV. Hence, in order to generate transition radiation it is enough even to have electrons which are not "suitable" for the generation, say, of synchrotron radiation.

Transition radiation was discovered theoretically in 1946 by Ginzburg and Frank (1946) and was subsequently confirmed experimentally. A number of attempts were carried out to apply this theory in relation to astrophysical objects including dust particles, and where relativistic electrons can also exist (see, for example, Janikova et al., 1969; Gurzadyan, 1985).

The spectrum of transition radiation presents special interest. It depends strongly on the dielectric capabilities of dust particles, as well as on the electron energy. For example, at a plasma frequency of dust particles $\omega_0 = 10^{16}$ s^{-1} and electron energy $E \sim 5$ MeV, the maximum of the transition radiation spectrum corresponds to wavelengths of ~ 500 Å. For real particles, different types of silicates, graphites, etc., the spectrum of transition radiation displays a number of cusps or even sharp maxima in the region of the far ultraviolet shorter than 1 000 Å.

IC 4997 is included in the *IRAS* catalogue (Gezari et al., 1987) as a powerful source of infrared radiation. In the observed infrared spectrum of this nebula, as given in Fig. 18.13, two maxima are observed – one at $\sim 2\,\mu$m, the other at $\sim 15\,\mu$m. Hence, the first assumption is that, this nebula includes at least two types of dust particle.

Assume also the possibility of generation or spontaneous appearance of relativistic electrons with an energy of a few MeV in the outer regions of the nucleus of this nebula. Such electrons can be captured by the magnetic field of the nebula and wander for a long time within the nebula: they therefore have to undergo number of collisions with the dust particles. Each such collision will be accompanied by the generation of transition radiation. As a result, we have a transition radiation field in the nebula with the definite density of radiation. We assume that this can be radiation required to realize the induced transitions of ions O^{++} from the level ^1D and to lead to the decrease in the intensity of the $N_1 + N_2$ doublet.

The problem is now to find the parameters of both, the relativistic electrons and dust particles, to explain the observed behaviour of the line 4 363 [OIII] and doublet $N_1 + N_2$ (Gurzadyan, 1992).

The long wavelength maximum ($\sim 15\mu$m) of the infrared spectrum of IC 4 997 is incomparably more powerful, and has been studied in more detail (Pottasch et al.,1984). As a result, an effective temperature of the dust particles $T_{\mathrm{D}} = 185$ K has been obtained. The observed infrared flux, integrated at all wavelengths is $F_{\mathrm{IR}} = 7.8 \times 10^{-12}$ W m^{-2}. At the distance of the nebula $d = 1\,000$ pc, we get for the infrared luminosity $L_{\mathrm{IR}}/L_{\odot} = 234$.

For the integral infrared luminosity of the nebula we have

$$L_{\mathrm{IR}} = 4\pi\, d^2 F_{\mathrm{IR}}\,. \tag{18}$$

If we assume the radiation of dust particles is of black body type, we can write for the same infrared integral luminosity

$$L_{\mathrm{IR}} = \pi\alpha^2\sigma T_{\mathrm{D}}^4 \frac{4\pi}{3} R_n^3 n_p\,, \tag{19}$$

where α and n_p are the diameter and concentration of dust particles and R_n is the radius of the nebula.

From (18) and (19) we find that

$$\alpha^2 n_p = \frac{3}{\pi}\frac{F_{\mathrm{IR}}}{\sigma T_{\mathrm{D}}^4}\,. \tag{20}$$

All parameters on the right hand side of this expression, F_{IR}, T_{D}, d and R_n, are known from observations. Upon closer examination of the problem, the so-called emission factor Q_ν can also be taken into account when deriving (20).

Denote by numbers 1, 2, 3 and 4 the energy levels ^3P, ^1D, ^1S, and $^1D_2^0$ (or $^1P_1^0$), respectively, for the ion O^{++}, as shown in Fig. 18.12. Usually the strength of the doublet $N_1 + N_2$ is determined by the number of spontaneous transitions 2–1 and is equal to $n_2 A_{21}$. In our case, however, with the presence of the field of transition radiation with density ρ_{24}, corresponding to the frequency of the transition 2–4, induced transitions 2–4 at a rate $n_2\, B_{24}\, \rho_{24}$

should also be realized. If q is the fraction of transitions having induced the radiation of the doublet $N_1 + N_2$, we can write

$$q = \frac{A_{21}}{A_{21} + B_{24}\rho_{24}}, \tag{21}$$

where

$$B_{24} = \frac{g_4}{g_2} \frac{c^3}{8\pi h\nu^3} A_{42}, \tag{22}$$

where ν is the frequency of the line arising at the transition 2–4, and A_{24} is the Einstein probability coefficient for the spontaneous transition 4–2 for O^{++}.

From (21) we obtain for the frequency interval $\Delta\nu$

$$\rho_{24}\Delta\nu = \frac{1-q}{q} \frac{A_{21}}{A_{42}} \frac{8\pi h\nu^3}{c^3} \Delta\nu. \tag{23}$$

The doublet $N_1 + N_2$ can be observed in its full strength with the absence of the induced transition 2–4; in this case $q = 1$ and $\rho_{24} = 0$. In the case of IC 4 997, more than a three-fold weakening of the doublet $N_1 + N_2$ is detected (Fig. 18.11), and, hence, $q \approx 0.3$.

For optically thin nebulae we have for the intensity I_ν of transition radiation

$$I_\nu = \frac{\varepsilon_\nu}{\alpha} (1 - e^{-t}), \tag{24}$$

where ε_ν is the volume emission coefficient of the radiation at the same frequency ν.

For the density of radiation we can write

$$I_\nu 2R_n\varepsilon_\nu, \tag{25}$$

where ε_ν is now the energy of transition radiation emitted per unit volume per unit interval of frequency and per unit of time. From the general theory of transition radiation in the case of astrophysical problems we have for ε_ν:

$$\varepsilon_\nu = J_\nu \alpha^2 n_p n_e c \text{ ergs cm}^{-3}\text{ s}^{-1}\text{ Hz}^{-1}, \tag{26}$$

where J_ν stands for the intensity of transition radiation at the frequency ν, and n_p and n_e are the concentration of dust particles and relativistic electrons, respectively.

We are interested, however, in the amount of transition radiation in the frequency interval $\Delta\nu$. Therefore, we can write from (26)

$$\varepsilon_\nu \Delta\nu = pJ\alpha^2 n_p n_e c, \tag{27}$$

where J = full (integral, at all frequencies) intensity of the transition and is given by the expression

$$J = \int_0^\infty J_\nu \mathrm{d}\nu = \frac{2e^2}{\pi c}\omega_0\gamma\,, \tag{28}$$

where ω_0 is a plasma frequency of the dust particles and γ is the dimensionless energy of the electron.

The factor p in (27) is the fraction of the full intensity of transition radiation j in the frequency interval $\delta\nu$.

Setting $\gamma = E/mc^2$ in (28), we have

$$J = \frac{2e^2}{\pi mc^2}\omega_0 E = 5.97 \times 10^{-24} E\omega_0\,, \tag{29}$$

where the electron energy E is expressed in ergs.

From (26), (27) and (29) we obtain, setting also $\alpha^3 n_p$ from (24), in ergs cm^{-3}

$$\rho_{24} = 1.14 \times 10^{-23} p\omega_0 \frac{F_{\mathrm{IR}}}{\sigma T_{\mathrm{D}}^4} n_{\mathrm{e}} E \left(\frac{d}{R_n}\right)^2 . \tag{30}$$

Here $n_{\mathrm{e}}E$ is, obviously, the total energy of relativistic electrons $\varepsilon_{\mathrm{rel}}$ per unit volume. Writing

$$\varepsilon_{\mathrm{rel}} = n_{\mathrm{e}}\,E\,, \tag{31}$$

we have, instead of (30),

$$\rho_{24} = 1.14 \times 10^{-23} p\omega_0\,\varepsilon_{\mathrm{rel}} \frac{F_{\mathrm{IR}}}{\sigma T_{\mathrm{D}}^4} \left(\frac{d}{R_n}\right)^2 . \tag{32}$$

Equating (32) and (24), we obtain for $\varepsilon_{\mathrm{rel}}$

$$\varepsilon_{\mathrm{rel}} = 3.08 \times 10^{-38}\, \frac{1-q}{q}\, \frac{g_2}{g_4}\, \frac{A_{21}}{A_{42}}\, \frac{T_{\mathrm{D}}^4}{F_{\mathrm{IR}}} \left(\frac{R_n}{d}\right)^2 \frac{\nu^3}{p\omega_0} . \tag{33}$$

The ratio R_n/d is the angular radius of the nebula. If we therefore use $R_n/d = \varphi''/206\,265$, we shall have instead of (33)

$$\varepsilon_{\mathrm{rel}} = 0.725 \times 10^{-48} \frac{1-q}{q}\, \frac{g_2}{g_4}\, \frac{A_{21}}{A_{42}}\, \frac{T_{\mathrm{D}}^4}{F_{\mathrm{IR}}}\, \frac{\nu^3}{p\omega_0} (\varphi'')^2 . \tag{34}$$

In such a form the magnitude $\varepsilon_{\mathrm{rel}}$ turns out to be independent of the distance d of the nebula as well as of its linear radius R_n; this has to be credited as the main advantage of (34).

For the ion O^{++} we have (Fig. 18.12) $g_2/g_4 = 1$, $A_{21} = 2 \times 10^{-2}\,\mathrm{s}^{-1}$, $A_{42} = 8.8 \times 10^9$ s^{-1} (in the Coulomb approximation) and $\nu = 5 \times 10^{15}$ s^{-1} (for $\lambda = 600$ Å). With these data we find from (34) that

$$\varepsilon_{\mathrm{rel}} = 2.17 \times 10^{-13} \frac{1-q}{q}\, \frac{(\varphi'')^2}{p\omega_0}\, \frac{T_{\mathrm{D}}^4}{F_{\mathrm{IR}}} \text{ ergs cm}^{-3} . \tag{35}$$

With the aid of this relationship we obtain the volume energy of relativistic electrons $\varepsilon_{\mathrm{rel}}$ in the nebula when the following quantities are known: the total infrared flux $F_{\rm IR}$ detected in the outer parts of the Earth's atmosphere, the effective infrared temperature $T_{\rm IR}$ derived from the infrared spectra, the angular radius of the nebula φ'', and the quantity q – the divergence between the curves of variations of intensities of the emission lines 4 363 [OIII] and $\rm N_1 + N_2$. All these parameters can be found from observations. In fact, in (35) we have only two unknown parameters: the plasma frequency of the dust particles ω_0 and p, the fraction of the complete transition radiation spent for the excitation of the level ^1D of ion $\rm O^{++}$.

As for the parameter p, it can be shown (Gurzadyan, 1992) that its possible magnitudes are likely to be within the limits $p = 10^{-4}$–10^{-5}.

We can now apply (35) to the nebula IC 4 997. We have $T_{\rm D} = 185$ K, $F_{\rm IR} = 7.8 \times 10^{-9}$ ergs cm^{-2}s^{-1}, $q = 0.3$, and $\varphi'' = 0''.8$. Then we obtain from (35), taking $\omega_0 = 10^{16}\,\rm s^{-1}$,

$$\varepsilon_{\rm rel} = 2.09 \times 10^{-8}\,\text{ergs cm}^{-3} \quad \text{at} \quad p = 10^{-4}\,,$$

$$\varepsilon_{\rm rel} = 2.09 \times 10^{-7}\,\text{ergs cm}^{-3} \quad \text{at} \quad p = 10^{-5}\,.$$

To determine the concentration of relativistic electrons $n_{\rm e}$ or their energy E, an additional condition is required. However, the determining factor in our problem is not the concentration of relativistic electrons and not even the energy of a single electron E, but only the product $\varepsilon_{\rm rel}$, the energy density.

It is necessary to mention that, in principle, induced transitions 3–4 (Fig. 18.12) under the action of transition radiation are also possible, which might seem to lead, by analogy with the doublet $\rm N_1 + N_2$, to the weakening of 4 363 [OIII]. However, as the final effect depends on the ratio A_{32}/A_{43}, in our case this ratio is of the order of $10^{-9}(A_{32} = 1.6\ \rm s^{-1}$, $A_{43} = 1.8 \times 10^9\ \rm s^{-1})$, which is nearly four hundred times larger than the ratio $A_{21}/A_{42} (= 2.3 \times 10^{-12})$. Hence, the effect of the weakening of 4 363 [OIII] must be four hundred times smaller than in the case of the doublet $\rm N_1 + N_2$, obviously at one and the same concentration of relativistic electrons.

18.22 Relativistic Electrons in Planetary Nebulae?

From (31), we have for the concentration of relativistic electrons

$$n_{\rm e} = \frac{\varepsilon_{\rm rel}}{E}\,, \tag{36}$$

where E is expressed in ergs, or

$$n_{\rm e} = 0.625 \times 10^6\,\frac{\varepsilon_{\rm rel}}{E}\,, \tag{37}$$

where E is expressed in MeV. From here we have for the concentration of relativistic electrons

$$n_{\rm e} \approx 10^{-3}–10^{-4}\ \rm cm^{-3}$$

for $p = 10^{-5}$ and $p = 10^{-4}$, respectively. This magnitude of n_e should be regarded as extraordinarily high.

In order to reveal the role of relativistic electrons in the energy balance of nebulae, we introduce a new parameter – the ratio of the specific energy of relativistic electrons $\varepsilon_{\rm rel}$ to the specific kinetic energy of the nebula, $\varepsilon_{\rm kin}$, conditioned only by thermal motions,

$$\varepsilon_{\rm kin} = 2kT_e N_e\,, \tag{38}$$

where N_e is the concentration of thermal electrons, T_e is the electron temperature of the nebula, and the factor 2 takes into account the proton part of the kinetic energy. From (35) and (38) we have

$$\frac{\varepsilon_{\rm rel}}{\varepsilon_{\rm kin}} = 0.79 \times 10^3 \,\frac{1-q}{q}\,\frac{(\varphi'')^2}{p w_0}\,\frac{1}{N_e T_e}\,\frac{T_D^4}{F_{\rm IR}}\,. \tag{39}$$

If we take $N_e \sim 10^4\ \mathrm{cm}^{-3}$, $T_e \sim 10^4$ K, we find, at $E = 1$ MeV and $w_0 = 10^{16}\mathrm{s}^{-1}$, that

$$\frac{\varepsilon_{\rm rel}}{\varepsilon_{\rm kin}} \approx 1 \qquad \text{at} \qquad p = 10^{-4}\,,$$

$$\frac{\varepsilon_{\rm rel}}{\varepsilon_{\rm kin}} \approx 10 \qquad \text{at} \qquad p = 10^{-5}\,,$$

i.e. in the condition of energy balance of the nebula, the relativistic electrons can play a dominating role.

At first glance this conclusion seems hardly probable, though there are reasons to think that it cannot be groundless. Recall the unusually high infrared luminosity of this nebula. According to one of the existing points of view (see Chap. 10), the heating of dust particles in nebulae is realized by the Lα radiation of the nebula. If this is the case, then the condition $L_{\rm IR}/L_\alpha < 1$ should be satisfied. However, in the overwhelming majority of cases the observed ratio is larger than unity, e.g. for IC 4997 it is $9.3 \sim 10$ (Cohen, Barlow, 1974). This means that L$_\alpha$ radiation cannot be the only source of heating of dust particles in the nebula.

Moreover, at least in the case of IC 4997 the observed infrared luminosity can be explained neither by L$_c$ radiation heating ($L_c/L_\alpha = 1.33$), nor even by the complete bolometric radiation of nuclei, $L_{\rm IR}/L_{\rm bol} = 4$ (at a temperature of the central star of $T_* = 50\,000$ K). A possible way out is to consider that the dust particles are heated by *corpuscular* streams, probably, by relativistic or fast electrons.

18.23 Protoplanetary Objects

A simple determination of the term "protoplanetary objects" is not really possible. Usually by this term one denotes a conditional star combining the characteristics of M type giants and hot stars. It is known also that such a combination is possible in the case of binary systems with one of the components a cool giant – the source of ejection of the gaseous nebular cloud – the other being a hot star – the source of high temperature radiation ensuring the luminescence of this nebular cloud. However, this is the case also for symbiotic stars.

All symbiotic stars, without exception, are binary or even triple systems, and only this fact is enough to reject the main possibility of identifying the ancestors of ordinary PNs with symbiotic stars. Symbiotic stars can give birth to PNs only sporadically, since the nebula formed as a result of the dynamical and physical interaction of the components of the symbiotic system will not be an ordinary nebula. With respect to a number of properties it should appear as an anomalous nebula (see the next section).

By the term protoplanetary objects we should understand objects intermediate between the Mirids or OH/IR maser sources and typical PNs, however, less like the maser source and more like the PNs. Just this condition turns out to be the determining factor when defining the group of protoplanetary objects, which include such anomalous stars as V 1016 Cyg, V 1329 Cyg, V 348 Sgr, HM Sge, CRL 618, as well as CW Leo, an extremely bright object in infrared rays (in the region of 5–60 μm).

The similarity of these objects with PNs is so clear that the first four objects, V 1016 Cyg, V 1329 Cyg, V 348 Sgr and HM Sge, are even included in the catalogue by Acker et al. (1992) as possible PNs. In *IUE* spectra, for example of V 1016 Cyg not only are the traditional emission lines, 1550 CIV, 1640 HeII or 1240 NV, observed with too high intensities, but also such exotic and extremely high excitation forbidden lines as 2784 [MgV] and 2929 [MgV] along with the lowest excitation doublet 2800 MgII are. Such rare combination have also been registered in the case of NGC 7027 (see Table 7.3).

In protoplanetary objects there has also been detected a powerful radio emission – a continuum at a in the frequency of 10.6 GHz – generated by electron gas (free–free transitions). At least one of these objects, V 1016 Cyg, is also a source of intense X-ray emission, according to *Einstein* observations (Allen, 1981). With the help of the high excitation emission lines NV, MgV, NeV the main physical parameters of V 1016 Cyg, i.e. on the stage of protoplanetary nebulae, have been obtained: $T_* = 160\,000$ K, $R_*/R_\odot = 0.06$, $L_*/L_\odot = 3.3$ for the central star, and $n_e = 3 \times 10^6$ cm^{-3} and $8\,000 < T_e < 27\,000$ K for the nebular envelope (Nussbaumer, Schild, 1981). The list of observed characteristics of these and other protoplanetary objects can be continued (for details, see Zuckerman et al., 1976; Westerbrook et al.,

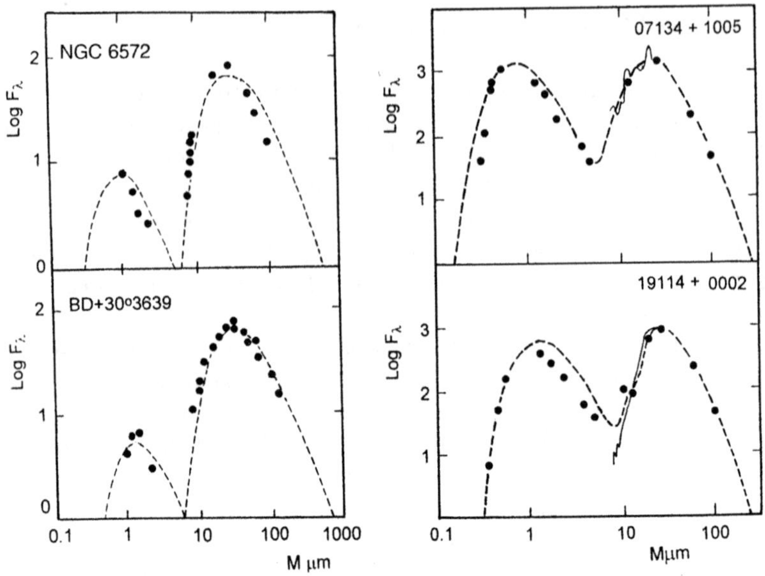

Fig. 18.14. Infrared spectra of two planetary nebulae – NGC 6572 and BD +30°3639 (*left*) – and of two protoplanetary objects – 07134 +1005 and 19114 +0002 (*right*) according to *IRAS* data. In both cases two maxima in the infrared spectrum are at one and the same wavelength, namely, at ~ 2 μm and ~ 30 μm. Dots are *IRAS* observations, broken lines are model computations, and F_λ is the infrared flux in arbitrary units

1975; Carsenty, Solf, 1982; Hrivank et al., 1994; Likkel et al., 1994; Ortiz, Maciel, 1993).

Thus, the whole set of observational data on the protoplanetary objects seems to indicate that, first, protoplanetary objects are single star and, second, being at an intermediate stage between the maser sources and PNs, protoplanetary objects are much nearer to the PNs and are much further from the maser OH/IR sources. More precisely this can be formulated in the following manner: the protoplanetary object represents the very earliest stage of a typical PN.

A more concrete criterion for protoplanetary objects to be the previous stage of PNs, can be formulated on the basis of the manner of the existence of *two maxima* in the infrared spectrum of the nebula IC 4997 (Fig. 18.13). This two-maxima structure of the infrared spectrum seems to be a common property both for PNs on the one hand, and for protoplanetary objects, on the other. This crucial fact is illustrated on Fig. 18.14, where the infrared spectra are given for two PNs – NGC 6 6972 and BD+30°3639 (on the left), and for two protoplanetary objects, IRAS 07 134 +1 005 and IRAS 19 114 +0002 (on the right). Up to 1992, at least 20 protoplanetary objects had been registered with two-maxima structure of their infrared spectra (Hrivank et al., 1989, 1992).

It is remarkable that in both PNs and protoplanetary objects both maxima in their infrared spectra correspond practically to one and the same wavelength – $\sim$ 1–2 μm and $\sim$ 30 μm.

So, the two-maxima structure in infrared spectra alone should be considered a decisive criterion, a kind of "blood test", which unifies PNs with protoplanetary objects under a common genetic origin.

18.24 The Nebula NGC 7 027 – An Evolved Symbiotic System?

Thus, it is hard to doubt the origin of PNs as the result of the evolutionary development of Mirids or OH/IR sources. Nevertheless, there is reason to think that this cannot be the only origin and means of forming PNs. We have in mind the puzzle of the well known PN NGC 7 027.

NGC 7 027 has constantly been the subject of unexplained mysterious phenomena. Everything here is extraordinary, contradictory and incomparable: the existence of highly excited lines, NV, CIV, SiIV, NeV, MgV, etc. along with low excitation lines, MgII, CII, SII, and even OI, SI etc., the combination of powerful ultraviolet emission with no less powerful emission in the infrared, quite a high level of radio emission both in the continuum and lines, and very small sizes, 0.009 pc (at a distance of 270 pc, Cahn et al., 1992) or 0.021 pc (at 640 pc, Zhang, 1995). For a long time, there was an absence of observational information on its central star – it was detected only in 1988, and turned out to be an object of visual magnitude $V = 16.3$ (Jacoby, 1988). This latter corresponds to an absolute magnitude of its nucleus of $M_{\mathrm{V}} = +\,7.3$ (at a distance of 640 pc), which indicates a K0 V type star of the main sequence or a G0 type subdwarf – both these stars cannot be the source of the nebula's luminescence. If we attribute its luminosity to a white dwarf ($M_v \sim +14$), then we will have for its apparent magnitude $V \sim 24$ (!). The results of determining the temperature of its nucleus have been rather contradictory: 150 000 K using the [OIII]/[OII] method, 184 000 K with HeII/H_β (Gurzadyan, 1988), $T_z(\mathrm{H}) = 500\,000$ K, $T_z = 470\,000$ K with Zanstra's method and $T_s = 295\,000$ K with Stoy's method (Pottasch, 1984). This nebula is of a high excitation class, $p = 10$, but very small linear size, ~ 0.01 pc, and therefore, is in sharp contrast with the general relationship in Fig. 18.8. The list of peculiarities of NGC 7 027 can be continued.

Of special importance is the fact of the nebula's being of a high excitation class while also being small. The small size indicates the obvious youth of the nebula: at an expansion velocity of 21.5 km s^{-1} (Weinberger, 1989), its age cannot be more than 300 or 1 500 years (!). This means that the temperature of its nucleus was extremely high – of the order of 150 000–200 000 K – at the moment of the nebula's birth. This cannot happen in the case of normal PNs if they are really born as a result of the evolution of Mirid or maser sources.

The extremely high temperature of nuclei in this phase is possible only if the progenitors of the nebulae form a symbiotic system, with an extremely high temperature star as a companion acting as a high energy source of excitation. In this case, the main component of the system must be a more or less cool giant, ejecting low temperature matter – a mixture of gas, molecules and dust particles. In this case the mass of the nebula itself will not be so large.

The binary nature of the nucleus of NGC 7027 cannot be established by the usual observational means because of the extreme faintness of the high temperature component. As for the cool component, after losing the outer part of its mass, it cannot be anything other than a typical giant of $M_V \sim 0$, with essentially low luminosity, which again does not contradict the observations.

Finally, if we identify the nucleus of NGC 7027 with an evolved symbiotic star, the possibility of having a white dwarf with an accretion disk as a hot component of the system is not excluded. Then we can determine the effective temperature of the outer parts of the accretion disk by a method applied in analogous cases, based on the maintenance of the balance between the summed intensity $E_e = \sum E_i$ of all ultraviolet emission lines emitted by the nebula, on the one hand, and the definite part of the black-body radiation of the accretion disk (in the region of $\lambda < \lambda_0$) at the temperature T_*, on the other hand (Gurzadyan, 1989).

The essence of the method is as follows. First, we determine the numerical magnitude of the function $\Phi(T_*)$ with the help of the relationship

$$\Phi(T_*) = 1.06 \times 10^{14} \frac{1}{D} \left(\frac{\gamma}{E_e}\right)^{1/2} \left(\frac{\mathcal{M}_*}{\mathcal{M}_\odot}\right)^{1/2} \left(\frac{\dot{\mathcal{M}}}{10^{-8}\mathcal{M}_\odot}\right)^{1/3}, \tag{40}$$

and then, with the known value of $\Phi(T_*)$, we obtain T_* from the following relationship:

$$\Phi(T_*) = T_*^{4/3} \left[\int_0^{\lambda_0} \lambda^{-5} \exp\left(-\frac{hc}{\lambda k T_*}\right) d\lambda\right]^{-1/2}. \tag{41}$$

Here D is the distance of the nebula from the Sun, $\mathcal{M}_*$ is the mass of white dwarf, and $\dot{\mathcal{M}}$ is the accretion rate. Taking $\mathcal{M}/\mathcal{M}_\odot = 1$ and $\dot{\mathcal{M}} = 10^{-8}\mathcal{M}_\odot$ per year, $\lambda_0 = 180$ Å and setting also $E_e = 4.7 \times 10^{-9}$ ergs cm^{-2} s^{-1} from the results of *IUE* measurements of 23 emission lines in the spectrum of this nebula (Table 7.3), we obtain from (40) $\Phi(T_*) = 1.7 \times 10^{-4}$. In this case $\gamma = 0.1$ is assumed, in view of the weak dependence of the temperature T_* on this parameter, where γ is the fraction of ionizing radiation of the accretion disk in the region shorter than λ_0 used for the excitation of ultraviolet emission lines.

Then from known values of $\Phi(T_*)$ we obtain from (19) $T_* = 180\,000$ K which coincides with the above value of T_*, obtained by the HeII/H_β method.

At such temperatures, the accretion disk can turn out, at least in principle, to be a source of X-ray radiation. However, X-rays have not been detected from NGC 7027.

So, NGC 7027 is not a PN in the ordinary sense. It is an evolved symbiotic system with the origin of the nebulosity being one of the results of its evolution. This conclusion is supported also by the fact that this object is one of 13 PNs, among 76 observed at the radio frequency of 1612 MHz, in which the maser OH emission has not been detected even at 3σ limits (Payne et al., 1988).

The origin of nebulae from OH/IR maser sources must be considered the first and the main mechanism for their origin. Now a second mechanism has appeared, namely, that PNs develop from symbiotic systems. However, this second way needs additional arguments to confirm it. Such an example may be the object AS 201, previously classified as a symbiotic system with an invisible hot component and a G2 III giant. The application of CCD techniques, revealed a faint nebulosity around this system (Schwarz, 1991). Even some parameters of the invisible hot component were determined: $T_* \sim 80\,000$ K, $L_*/L_\odot \sim 1\,000$, $\mathcal{M}/\mathcal{M}_\odot \approx 0.55$. At the estimated distance of this object, ~ 2 kpc, the radius of the nebula is $R_n \sim 0.05$ pc

Five other symbiotic systems – the symbiotic nebulae: He 2-106, He 2-127, He 2-171, H 1-36, and SS 122 – have been considered as being an early stage in the formation of PNs (Costa, Pacheco, 1994). Two of them, He 2-171 and H 1-36, are listed in the catalogue by Acker et al.(1992) as genuine PNs, and one, He 2-106, as a possible PN.

An interesting fact is that those PNs known to have a symbiotic origin – a symbiotic nebula – are all of the highest excitation class. They are as follows:

Symbiot. Neb.	H 1-36	He 2-106	He 2-127	He 2-171	He 2-104	NGC 7027
Excit. class, p	12	12	12	12	11	11

That is not all. In the catalogue by Acker et al., (1992), there is a list including 40 more symbiotic systems that are rejected as PNs, some of which appear under a question mark. This is remarkable: in certain cases, it appears that astrophysicists cannot simply distinguish the symbiotic system and the PN – any given object may be neither a symbiotic system nor a PN. Presumably, here we have some *transition* type symbiotic stars, the transition being from a typical symbiotic system to a typical PN. The necessity of studying this problem further is obvious.

Thus, the population of known PNs should be divided into at least two categories (Gurzadyan, 1993). To the first category should be attributed those PNs which have been born as a result of the evolution of Mirids or OH/IR sources. To this category belong the overwhelming majority of PNs.

PNs of the second category were born as a result of the evolution of symbiotic systems, or more precisely, as a result of a dynamical interaction between both components of the system, one of which is a white dwarf with an accretion disk. At the same time we cannot insist that the evolution of any symbiotic system should be completed by the birth of a PN.

18.25 Condensations in Planetary Nebulae. Stellar Wind

Condensations in PNs are globule-like, very bright formations of very small size, 1–3″ in diameter. Many of them are visible on the large scale image of NGC 7 293, perhaps, because this nebula is one of the nearest (~ 150 pc) to us. These condensations, over 100 in number, are scattered far from the central regions of the nebula within a relatively narrow zone near the outer half of the nebula. The absolute linear sizes of the condensations are of the order of 10^{15} cm.

The nature of these condensations is obscure. There is a suggestion according to which they are of low temperature and, hence, of high concentration regions of matter determined by the pressure balance at the boundary of their separation from the surrounding nebula. According to another version the origin of these condensations should be connected with a thermal instability or an instability of Rayleigh–Taylor type (Capriotti, 1973). Finally, there is also the so-called dynamical interpretation (Gurzadyan, 1968, 1994) according to which the luminescence of these condensations should be connected with the particle streams – stellar winds flowing out from the central star of the nebula; this follows from the radially directed filaments as well as streams flowing from the condensations and directed away from the nucleus of the nebula.

The high resolution *IUE* spectral recordings for the nuclei of a number of PN give persuasive information about the nature of stellar winds in these objects. We have in mind P Cyg type profiles which were recorded for a number of ultraviolet lines of high potential ions, such as CIV, SiIV, NIV, NV, OV etc. As an example, in Fig. 18.15 two profiles of the line 1 371 OV, both of clearly expressed P Cyg type, for NGC 1 535 and NGC 6 210 are shown. The computed model profiles are also drawn (solid lines), and their agreement with the observations is obvious. However, note that the computed model profiles depend on a large number of parameters – over ten (!) – both physical and dynamical, and therefore it is not difficult in principle to reach the proper accordance with the observed profiles by successful selection of these parameters (Lamers, 1991; Castor, 1991; Patriarchi, Perinotto, 1991; Werner, Koesterke, 1992; Kudritzki et al., 1992).

However, in the given case, of special importance is the fact that these computations give us the possibility of obtaining two crucial parameters of the stellar wind, i.e. the edge velocities V_{ed}, as well as the observed mass loss rates dM/dt for a number of nebulae. The observed values of V_{ed} range from about 600 to 3 600 km s^{-1}. The following values of V_{ed} were considered: 2 700,

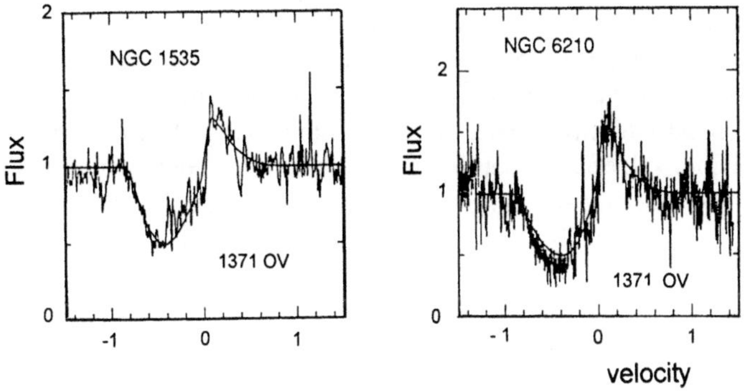

Fig. 18.15. *IUE* high resolution recordings for the ultraviolet line 1 371.29 OV with a P Cyg type profile in the spectrum of the planetary nebulae NGC 1 535 and NGC 6 210

1 700, 1 550, 1 350 km s^{-1} in NGC 6 210, 6 826, 6 891 and IC 4 593, respectively (Bianchi, De Francesco, 1993), 3 600 km s^{-1} in NGC 6 905 (Heap, 1986). Moreover, the value of about 500 km s^{-1} in NGC 6 302 has been derived from the optical data (Meaburn, Walsh, 1980).

The observed mass loss rates range between 2×10^{-6} and 5×10^{-11} $\mathcal{M}_\odot$ yr^{-1}, but values by different authors for the same object scatter by 1–2 orders of magnitude, owing to the different methods and different values of the fundamental stellar parameters adopted. A more realistic range for the mass loss rate is 10^{-8}–10^{-9} $\mathcal{M}_\odot$ yr^{-1} for the majority of PN nuclei. For a number of PNs the most accurate data for V_{ed} and dM/dt (in $\mathcal{M}_\odot$ yr^{-1}) are as follows (Perinotto, 1993):

	V_{ed}, km s^{-1}	dM/dt		V_{ed}, km s^{-1}	dM/dt
NGC 1 535	1 900	1.4×10^{-9}	NGC 7 009	12 770	2.8×10^{-9}
NGC 6 210	12 180	2.2×10^{-9}	IC 418	1 940	0.3×10^{-9}
NGC 6 543	11 900	4.0×10^{-8}	IC 4 593	11 000	4.2×10^{-8}
NGC 6 826	11 750	6.4×10^{-8}	A 78	13 700	2.5×10^{-8}

The highest value of V_{ed} is detected for the central K 1-16 type star equal to 3 800 km s^{-1}, with an extremely low mass loss rate, lower than $2 \times 10^{-11} \mathcal{M}_\odot$ per year (Patriarchi, Perinotto, 1995).

At the same time there are PNs (IC 418, NGC 7 662 etc.) without any sign of the presence of condensations in their structure. Two possibilities exist. First, the condensations are formed and evolve only at definite periods of the nebula's life; hence, the condensations are a transient phenomenon. Second, as a physical formation, the condensations always exist; however, they can be "illuminated" only at periods of enhanced activity of the nucleus in the form of intense stellar wind.

Many years ago, Minkowski had obtained on the 5 m telescope of the Palomar Observatory excellent images of the nebula NGC 7 293. The condensations alone were the first and most intriguing impression on these pictures, condensations with bright heads and less bright tails. All such streams without exception, are directed radially and outward from the centre. An important and characteristic feature is that the width of the stream with strongly linear edges increases slightly while moving off from the head of the condensation. Obviously, condensations undergo a powerful bombardment by some agent acting strongly radially from the nebula's nucleus. This agent most likely is the stellar wind or particle stream. As a result of the bombardment, something sweeps out the condensation forming a stream moving away via a kind of thermal diffusion. The stellar wind or particle stream can cause luminescence both of the condensation and of the stream, it being either purely gaseous or a mixture with dust particles. Thus, although we know nothing about the composition of the condensations nor about the mechanism of their illumination, it is hard to doubt the decisive role of particle streams. It seems that the problem of condensations cannot be solved separately from the problem of the streams.

The observational data on NGC 7 293 seem to support the above mentioned role of the stellar wind from the PN nucleus. Indeed, in the atlas of *IUE* spectra of PNs and of their nuclei (Feibelman et al., 1988), the line 1 550 CIV with a P Cyg type profile in the spectrum of the nucleus of this nebula is visible. Strongly expressed radial trains are visible behind the second (outer) envelope on the impressive *Hubble Space Telescope* image of NGC 6 543 (Harrington, Borkowski, 1995); here the very thin and long trains are nothing other than the matter swept away by stellar wind from the outer envelope of this remarkable nebula.

18.26 Former Nuclei of Planetary Nebulae

Ten percent of the stars in the Galaxy pass through the stage of PN rather quickly – over a few tens of thousands of years. During this time, the ancestors of PNs – Mirids and OH/IR sources – are transformed into hot stars now with the status of nuclei of PNs, either with emission lines (WR), or absorption lines (O), or with a continuous spectrum without lines (Cont.). With the expansion and the development of the nebulae, their nuclei evolve as well. According to Fig. 18.1, the evolution proceeds in the direction of classical white dwarfs, i.e. with a marked, 10 or 100 times, decrease in luminosity with the preservation, however, of the high effective temperature.

At the time when PNs dissipate within the interstellar medium and they completely disappear, their nuclei, however, preserve their high luminosity – within limits of $L_*/L_\odot \approx 10$–100 – and high temperature – of the order of 100 000 K. At this time, the nuclei of PNs are mixed within the common stellar field of the Galaxy or more strictly, with its flat and intermediate subsystems. At that epoch the absolute magnitudes of the nuclei M_v are not very large – of the order of – 2^{m} or $+2^{\mathrm{m}}$ or even smaller, and in all cases they are much fainter than the classical WR and O stars with an absolute luminosity of the order of $-5^{\rm m}$ or $-6^{\rm m}$ and higher, though with respect to the structure of their spectra the nuclei of PNs and WR or O stars have something common.

The question, therefore is what kind of Galactic object, as a transient state, should be identified with the former nuclei of a PN after the complete loss of the nebular envelope and long before the white dwarf state is reached. At least four possibilities can be imagined.

I. Galactic O and WR Stars. The classical WR and O stars, as a rule, are present in the Galaxy in groups, in O associations. It is clear that the former nuclei of PNs cannot have anything common with these WR or O stars. At the same time, the individually scattered WR and O type stars in various regions of the Galaxy might well be former nuclei of PNs. Thus, these WR and O stars should be of essentially lower luminosities than those of PN nuclei.

However, a convincing answer can be reached if we examine the statistics of spatial or even the visible distribution of these two categories of objects, PNs and galactic WR or O stars. The results are as follows: 80% of PNs with O type nuclei are located above $b = 10°$, while only 10% of known O type stars are above $b = 10°$. Also, only 1% of known WR stars, including those which are in O associations, have $|b| > 10°$. Meanwhile 35% of PNs with WR type nuclei are scattered above $b = 10°$.

The absence of any genetic connection between the former nuclei of PN and the Galactic O type and WR stars is obvious.

II. Hot Stars at High Galactic Latitudes. There are stars of early classes, mainly of O–Op types, which with respect to their spatial distribution, kinematic characteristics and low luminosities differ from the usual representatives of the given class of objects. In particular, all these objects are scattered higher than $z = 1\,000$ pc from the plane of the Galaxy. Some of these objects may be the former nuclei of PNs.

III. Humason–Zwicky Objects. There is a special category of low luminosity stars of early spectral classes, so-called "faint blue stars", scattered at high Galactic latitudes – those are known as Humason–Zwicky objects, and are related to the type II stellar population. These are the blue stars mainly of spectral type B, sometimes of Op, with quite low absolute luminosity, $M_v \sim +1^{\rm m}.5$. More than several dozen such objects are known, and all belong to the stellar population of type II. Therefore the possibility that some of these Humason–Zwicky objects may be the former nuclei of PNs is not be excluded.

If this conjecture is true, then the nuclei of PNs must be "hot dwarfs of early classes", located at 4–$8^{\rm m}$ lower than the hot giants, and at 10–$12^{\rm m}$ lower than the usual white dwarfs.

IV. Galactic X-Ray Sources. The possibility of identifying a certain fraction of Galactic X-ray sources with the former nuclei of PNs seems very likely. Apart from their high temperature, of the order of 100 000 K, one more parameter, the low luminosity, should be involved in the identification of the given X-ray source with the former nuclei of PNs.

Here four possibilities of identifying of the former nuclei of PNs have been discussed. There are other possibilities as well. However, the most important problem remains the determination of the effective temperature for an individual and extremely high temperature star, of the order of 100 000 K or higher, supplied with all spectral lines; in this case the maximum of its radiation will be at 300 Å. This problem cannot be solved even with the help of space observatories like the *IUE*. This problem can be solved only through space observations in the extreme ultraviolet, at wavelengths shorter than 300 Å.

The problem of the former nuclei of PNs still awaits a solution.

18.27 The Last Sigh of the Dying Star...

At the present stage of our knowledge about PNs and their nuclei, summarized in the eighteen chapters of the present monograph, the the cosmogonical mission of PN, seems to be revealed.

The PN is nothing other than a kind of evidence of the illness of a star before its death; an illness coming towards the end of its life, an illness occurring in an extremely sharp form in a very short time and without any hope of a cure – its fate is predefined.

For all its long life, of the order of five billion years or somewhat longer, the star has secured its existence by a solid resource of nuclear fuel: hydrogen. However, the period of exhaustion and then of the complete ceasing of generation of nuclear energy comes, and the star is left without hydrogen and the hydrogen engine is terminated. The epoch of global cooling of the star starts: its gravitational contraction begins, which later will transform to a collapse. The generation of energy in the central parts of the star's interior is renewed, however, now owing to gravitation.

The collapse is accompanied by a slow increase of temperature in the central parts of the star, and at a definite moment the helium flash starts, since there is helium in the interior of the star along with other heavy elements – the helium engine starts. The peculiarity of the situation is that for a certain category of stars, namely, for cool giants, or more precisely, for Mirids or OH/IR sources with a mass slightly smaller the solar mass, this process can have a non-explosive character contrary to novae or supernovae, and can proceed quietly. Then, instead of the explosion of the star we will observe only slow, non-catastrophic expansion, or more precisely, its inflation with rather small velocities – two–three tens of km s^{-1}.

The star's inflation proceeds for fairly long time, while comparatively quickly, over a few hundred years, the outer layers of the star, reaching the radii where the parabolic velocity of the escape turns out to be smaller than their expansion velocity, find themselves isolated from the star, isolated now forever. The departed mass possesses the form of a dense envelope, continues its motion away from the star by inertia, and after some time forms the so-called *planetary nebula.*

The helium engine does not function for too long, for a few million years, i.e. just an instant – a thousandth part of the star's lifetime. Hence, the appearance of the PN should be considered as a signal that the star is entering the final stage of its life, and is proceeding toward its death.

Thus, the main conclusion is that PNs are the most impressive product – only a product (!) – and at the same time, are an inevitable stage on the path of evolution of a certain category of the stellar giants, the stage which occurs at the end of the star's life. In other words, PNs are the external expression of deep processes proceeding in the interior of the dying star during its natural evolution in the direction of white dwarfs.

And this, a sort of farewell to life, takes place very stormily and with great noise, as if the star is in agony with the forced removal of an essential fraction, 15–20%, of the initial mass of the star, which then will form the future PN. Then, the star passes to a state of highest temperature, of the order of a hundred thousand degrees, unprecedented for the whole of the previous period of its life, which turns out to a powerful source of ultraviolet radiation, modifying the nebula itself to be a no less powerful source of infrared radiation. It becomes a kind of a lighthouse, quite bright on the background of galactic radio emission. Finally, in the star of the nebula the puzzling generator of X-ray radiation becomes silent for a long time – thus all bands of electromagnetic frequencies are covered, accompanied by no less powerful particle emission in the form of a stellar wind.

Such is the cosmogonical mission of the planetary nebula – the last sigh of the dying star...

Catalogue of Excitation Classes *p* for 750 Galactic Planetary Nebulae

Name	*p*	Name	*p*	Name	*p*	Name	*p*
NGC 40	1	NGC 6072	9	NGC 6881	10	IC 4663	11
NGC 246	12+	NGC 6153	3	NGC 6884	7	IC 4673	10
NGC 650-1	10	NGC 6210	4	NGC 6886	9	IC 4699	9
NGC 1360	12	NGC 6302	10	NGC 6891	4	IC 4732	5
NGC 1501	10	NGC 6309	10	NGC 6894	10	IC 4776	2
NGC 1514	8	NGC 6326	9	NGC 6905	11	IC 4846	3
NGC 1535	8	NGC 6337	11	NGC 7008	11	IC 4997	8
NGC 2022	12	NGC 6369	4	NGC 7009	7	IC 5117	6
NGC 2242	12+	NGC 6439	8	NGC 7026	9	IC 5148-50	6
NGC 2346	9	NGC 6445	10	NGC 7027	11	IC 5217	6
NGC 2371-2	12	NGC 6537	11	NGC 7048	11	A 1	1
NGC 2392	10	NGC 6543	5	NGC 7094	12	A 2	10
NGC 2438	10	NGC 6563	8	NGC 7139	9	A 4	10
NGC 2440	10	NGC 6565	7	NGC 7293	7	A 12	4
NGC 2452	10	NGC 6567	4	NGC 7354	10	A 15	12+
NGC 2610	12	NGC 6572	7	NGC 7662	10	A 20	12+
NGC 2792	11	NGC 6578	2	IC 289	12	A 21	1
NGC 2818	11	NGC 6620	8	IC 351	10	A 23	4
NGC 2867	9	NGC 6629	5	IC 418	1	A 24	1
NGC 2899	10	NGC 6644	7	IC 972	10	A 30	12+
NGC 3132	9	NGC 6720	10	IC 1295	10	A 33	11
NGC 3195	9	NGC 6741	9	IC 1297	9	A 35	1
NGC 3211	10	NGC 6751	9	IC 1454	10	A 36	12+
NGC 3242	9	NGC 6765	10	IC 1747	9	A 40	2
NGC 3587	8	NGC 6772	9	IC 2003	10	A 41	1
NGC 3699	9	NGC 6778	9	IC 2149	2	A 43	2
NGC 3918	9	NGC 6781	8	IC 2165	10	A 46	2
NGC 4071	11	NGC 6790	4	IC 2448	9	A 49	4
NGC 4361	12+	NGC 6803	5	IC 2501	3	A 50	10
NGC 5189	10	NGC 6804	12	IC 2553	8	A 51	12
NGC 5307	9	NGC 6807	4	IC 2621	9	A 54	12
NGC 5315	2	NGC 6818	10	IC 3568	3	A 55	4
NGC 5873	10	NGC 6826	11	IC 4191	6	A 57	3
NGC 5882	6	NGC 6833	2	IC 4406	4	A 60	2
NGC 5879	12	NGC 6842	2	IC 4593	6	A 61	3
NGC 5979	12	NGC 6852	12	IC 4634	3	A 63	1
NGC 6026	11	NGC 6853	10	IC 4637	3	A 65	1
NGC 6058	11	NGC 6879	7	IC 4642	12	A 67	3

Name	p	Name	p	Name	p	Name	p
A 69	3	H 1-41	9	H 4-1	8	He 2-83	1
A 70	3	H 1-42	4	Hb 4	7	He 2-84	4
A 72	12^+	H 1-45	3	Hb 5	9	He 2-85	9
A 75	2	H 1-46	2	Hb 6	9	He 2-86	3
A 77	10	H 1-50	6	Hb 7	2	He 2-88	2
A 78	12	H 1-51	2	Hb 8	4	He 2-90	1
A 79	2	H 1-52	9	Hb 12	2	He 2-96	3
A 83	2	H 1-53	2	He 1-1	9	He 2-97	2
BD+30°	1	H 1-54	2	He 1-3	4	He 2-102	9
Cn 1-1	7	H 1-56	5	He 1-4	8	He 2-103	8
Cn 1-3	2	H 1-57	2	He 1-5	1	He 2-104	11
Cn 1-4	4	H 1-58	1	He 1-6	3	He 2-105	1
Cn 1-5	1	H 1-59	11	He 2-5	2	He 2-108	10
Cn 2-1	5	H 1-60	3	He 2-7	4	He 2-109	7
Cn 3-1	1	H 1-61	2	He 2-9	3	He 2-111	10
H 1-1	7	H 1-63	2	He 2-11	10	He 2-112	8
H 1-2	8	H 1-64	1	He 2-15	10	He 2-114	10
H 1-3	3	H 1-66	10	He 2-18	8	He 2-115	2
H 1-4	3	H 1-67	10	He 2-21	9	He 2-116	2
H 1-5	4	H 2-1	1	He 2-25	2	He 2-117	3
H 1-6	3	H 2-7	3	He 2-26	5	He 2-118	4
H 1-7	3	H 2-10	4	He 2-28	7	He 2-119	10
H 1-8	3	H 2-11	2	He 2-29	10	He 2-120	3
H 1-9	1	H 2-12	1	He 2-32	4	He 2-123	1
H 1-11	3	H 2-13	5	He 2-34	1	He 2-128	2
H 1-12	4	H 2-14	12	He 2-35	3	He 2-129	4
H 1-13	4	H 2-15	3	He 2-36	11	He 2-132	10
H 1-14	8	H 2-16	11	He 2-37	9	He 2-133	4
H 1-15	2	H 2-18	4	He 2-39	10	He 2-136	12^+
H 1-16	6	H 2-23	4	He 2-41	3	He 2-141	10
H 1-18	3	H 2-24	2	He 2-47	1	He 2-143	8
H 1-19	2	H 2-26	2	He 2-48	3	He 2-145	4
H 1-21	4	H 2-27	3	He 2-50	9	He 2-146	1
H 1-22	2	H 2-29	1	He 2-51	10	He 2-147	3
H 1-23	4	H 2-30	12	He 2-55	12	He 2-149	3
H 1-24	1	H 2-32	12	He 2-57	2	He 2-152	11
H 1-26	9	H 2-33	2	He 2-62	3	He 2-153	8
H 1-27	3	H 2-35	8	He 2-63	10	He 2-155	6
H 1-28	11	H 2-36	12	He 2-64	1	He 2-157	1
H 1-29	4	H 2-37	3	He 2-67	7	He 2-158	2
H 1-30	4	H 2-39	9	He 2-68	1	He 2-159	9
H 1-31	4	H 2-40	2	He 2-70	10	He 2-161	2
H 1-32	4	H 2-41	11	He 2-71	1	He 2-163	4
H 1-33	3	H 2-42	11	He 2-72	3	He 2-164	12
H 1-34	1	H 2-43	12	He 2-73	7	He 2-165	11
H 1-35	2	H 2-44	11	He 2-76	10	He 2-169	3
H 1-36	10	H 2-45	3	He 2-77	2	He 2-170	3
H 1-37	11	H 2-46	3	He 2-78	1	He 2-171	12
H 1-38	2	H 2-48	12	He 2-81	3	He 2-175	6
H 1-40	3	H 3-75	8	He 2-82	3	He 2-182	1

Name	p	Name	p	Name	p	Name	p
He 2-185	3	K 3-35	3	KFL 3	2	M 1-44	1
He 2-186	9	K 3-36	11	KFL 4	9	M 1-47	5
He 2-187	2	K 3-37	4	KFL 7	12	M 1-48	8
He 2-207	10	K 3-40	2	KFL 8	12	M 1-50	7
He 2-248	3	K 3-41	4	KFL 10	4	M 1-51	2
He 2-250	9	K 3-43	10	KFL 11	3	M 1-52	9
He 2-262	3	K 3-46	2	KFL 12	3	M 1-53	5
He 2-406	8	K 3-51	12	KFL 13	11	M 1-54	9
He 2-418	12	K 3-52	2	KFL 14	11	M 1-56	6
He 2-428	2	K 3-53	4	KFL 15	3	M 1-57	8
He 2-429	3	K 3-56	12	KFL 16	12	M 1-58	10
He 2-430	3	K 3-57	10	KFL 19	3	M 1-59	8
He 2-434	3	K 3-58	3	Lo 4	12^+	M 1-60	4
He 2-436	3	K 3-60	9	Lo 5	2	M 1-61	4
He 2-442	11	K 3-61	3	Lo 6	3	M 1-62	6
Hu 1-1	8	K 3-62	4	Lo 10	1	M 1-63	6
Hu 1-2	12	K 3-63	12	Lo 16	9	M 1-64	3
Hu 2-1	1	K 3-64	10	Lo 17	2	M 1-66	6
J 320	4	K 3-65	4	LoTr 7	11	M 1-69	9
J 900	10	K 3-66	1	LoTr 8	4	M 1-71	2
Jn 1	11	K 3-67	4	M 1-1	12	M 1-72	1
Jn Er 1	4	K 3-68	10	M 1-2	10	M 1-73	6
K 1-1	1	K 3-69	10	M 1-4	4	M 1-74	3
K 1-2	12	K 3-70	10	M 1-6	1	M 1-75	3
K 1-4	1	K 3-71	12	M 1-7	8	M 1-78	1
K 1-7	3	K 3-73	10	M 1-8	10	M 1-79	10
K 1-12	12	K 3-76	4	M 1-9	2	M 1-80	10
K 1-14	12^+	K 3-78	2	M 1-13	8	M 2-2	3
K 1-16	12^+	K 3-79	4	M 1-16	9	M 2-4	3
K 1-21	3	K 3-80	3	M 1-17	6	M 2-6	2
K 1-23	4	K 3-81	11	M 1-18	4	M 2-7	1
K 1-27	12^+	K 3-82	12	M 1-19	2	M 2-8	9
K 2-5	3	K 3-84	7	M 1-20	3	M 2-9	1
K 2-7	4	K 3-87	12	M 1-22	10	M 2-10	1
K 2-15	1	K 3-88	4	M 1-23	10	M 2-11	8
K 2-17	3	K 3-90	12^+	M 1-24	2	M 2-13	3
K 3-1	11	K 3-91	2	M 1-25	2	M 2-14	1
K 3-2	1	K 3-92	4	M 1-28	3	M 2-15	9
K 3-4	4	K 3-93	8	M 1-29	8	M 2-16	7
K 3-13	4	K 3-94	11	M 1-30	1	M 2-17	9
K 3-16	3	K 4-5	10	M 1-31	3	M 2-18	2
K 3-19	4	K 4-8	4	M 1-32	2	M 2-19	1
K 3-21	3	K 4-10	4	M 1-33	5	M 2-20	2
K 3-24	4	K 4-16	10	M 1-34	9	M 2-21	8
K 3-26	11	K 4-41	4	M 1-35	5	M 2-22	10
K 3-27	12^+	K 4-48	4	M 1-39	1	M 2-24	2
K 3-29	4	K 4-53	3	M 1-40	8	M 2-25	8
K 3-30	3	K 4-55	3	M 1-41	4	M 2-26	10
K 3-31	4	KFL 1	11	M 1-42	9	M 2-27	3
K 3-34	10	KFL 2	11	M 1-43	2	M 2-28	9

Name	p	Name	p	Name	p	Name	p
M 2-29	5	M 3-30	11	PC 22	6	Th 3-26	10
M 2-30	9	M 3-31	2	PC 23	5	Th 3-27	3
M 2-31	2	M 3-32	8	PC 24	9	Th 3-35	2
M 2-32	4	M 3-33	8	Pe 1-1	4	Th 3-55	3
M 2-33	2	M 3-34	9	Pe 1-2	7	Th 4-1	1
M 2-34	9	M 3-36	3	Pe 1-3	4	Th 4-2	11
M 2-35	8	M 3-37	4	Pe 1-6	3	Th 4-5	10
M 2-36	6	M 3-38	7	Pe 1-8	10	Th 4-6	3
M 2-37	1	M 3-39	4	Pe 1-9	1	Th 4-7	10
M 2-40	1	M 3-42	10	Pe 1-11	10	Th 4-10	1
M 2-41	11	M 3-43	11	Pe 1-12	12	Th 4-11	1
M 2-42	3	M 3-45	8	Pe 1-13	12	Vd 1-1	2
M 2-43	2	M 3-46	2	Pe 1-14	3	Vd 1-3	1
M 2-44	9	M 3-47	3	Pe 1-15	3	Vd 1-4	4
M 2-45	3	M 3-48	2	Pe 1-16	10	Vd 1-5	8
M 2-47	5	M 3-49	2	Pe 1-17	10	Vd 1-6	10
M 2-48	4	M 3-50	12	Pe 1-18	4	Vd 1-8	4
M 2-49	4	M 3-51	2	Pe 1-19	3	Vd 1-9	3
M 2-50	4	M 3-53	3	Pe 1-20	11	Vo 2	3
M 2-51	3	M 3-54	9	Pe 1-21	12	Vo 3	3
M 2-52	11	M 3-55	2	Pe 2-4	4	Vo 4	4
M 2-53	3	M 4-2	11	Pe 2-5	1	Vy 1-1	5
M 2-55	10	M 4-3	3	Pe 2-7	10	Vy 1-2	9
M 3-1	3	M 4-4	3	Pe 2-8	3	Vy 1-4	7
M 3-2	12	M 4-6	4	Pe 2-10	1	Vy 2-1	2
M 3-3	9	M 4-9	2	Pe 2-11	1	Vy 2-2	12
M 3-4	11	M 4-10	4	Pe 2-13	11	Vy 2-3	5
M 3-5	9	M 4-11	11	Pe 2-14	9	Wray 16-20	1
M 3-6	4	M 4-14	10	Pe 2-15	4	Wray 16-22	11
M 3-7	2	M 4-17	9	Sa 1-5	4	Wray 16-93	12^+
M 3-8	2	Mz 1	3	Sa 1-6	1	Wray 16-120	11
M 3-9	9	Mz 2	10	Sa 1-8	3	Wray 16-121	10
M 3-10	6	Mz 3	1	Sa 2-21	2	Wray 16-122	11
M 3-11	1	PB 1	4	Sa 2-23 0	12	Wray 16-128	12
M 3-12	9	PB 2	1	Sa 2-23 7	9	Wray 16-189	4
M 3-13	1	PB 3	10	Sa 3-92	3	Wray 16-266	12
M 3-14	7	PB 4	4	Sa 4-1	4	Wray 16-278	2
M 3-15	6	PB 5	3	Sn 1	4	Wray 16-286	4
M 3-16	3	PB 6	5	Sp 1	12	Wray 16-363	2
M 3-19	2	PB 8	3	Sp 3	1	Wray 16-385	4
M 3-20	4	PB 9	12	Sp 4-1	5	Wray 16-411	12^+
M 3-21	7	PB 10	11	Th 2-3	2	Wray 16-423	7
M 3-22	11	PC 11	8	Th 3-1	8	Wray 17-1	11
M 3-23	4	PC 12	6	Th 3-4	3	Wray 17-18	10
M 3-23	11	PC 13	9	Th 3-6	1	Wray 17-31	2
M 3-24	8	PC 14	6	Th 3-10	3	Wray 17-40	1
M 3-26	10	PC 17	8	Th 3-12	1	Wray 17-59	3
M 3-27	2	PC 19	9	Th 3-13	3	Wray 17-75	11
M 3-28	4	PC 20	8	Th 3-24	2		
M 3-29	2	PC 21	12	Th 3-25	2		

Plates

The images of planetary nebulae shown in Plates **I–XIV** were obtained in the following conditions:

NGC 7293 (**I**) – by T. Gull in red light with the Kitt Peak National Observatory 4 m telescope

NGC 7293 (**II**) – by K.B. Kwitter in the H_α line with the Kitt Peak National Observatory Schmidt telescope (Courtesy of K.B. Kwitter)

NGC 6826 (**II**) – by G. Jacoby with the Kitt Peak National Observatory 4 m telescope (Courtesy of Y.H. Chu and G. Jacoby)

Jn Er 1 (**III**) – by R. Minkowski and L.H. Aller with the Palomar Observatory Schmidt telescope

NGC 2346 (**III**) – by L. Kohoutek and S. Laustsen with the Palomar Observatory Schmidt telescope

NGC 7094 (**IV**), NGC 6804 (**IV,IX**) – by Y.H. Chu and G. Jacoby with the Kitt Peak National Observatory 2.1 m telescope (Courtesy of Y.H. Chu and G. Jacoby)

NGC 6720 (**V**), NGC 6543 (**V**) – by B. Balick with the Kitt Peak National Observatory 2.1 m telescope (Courtesy of B. Balick)

IC 1295 (**VI**), NGC 3587 (**VI**) – by Y.H. Chu and G. Jacoby with the Kitt Peak National Observatory 2.1 and 0.9 m telescopes (Courtesy of Y.H. Chu and G. Jacoby)

NGC 1501 (**XI**), NGC 1535 (**VIII**), NGC 2022 (**VIII**), NGC 2392 (**VIII**), NGC 3242 (**IX**), NGC 4361 (**XIII**), NGC 6210 (**XIV**), NGC 6309 (**XIII**), NGC 6337 (**X**), NGC 6369 (**X,XIV**), NGC 6563 (**VII,IX**), NGC 6853 (**XIII**), NGC 6905 (**XIV**), BD +30°3639 (**X**) – by Y.H. Chu and G. Jacoby with the Kitt Peak National Observatory 2.1 m telescope

IC 418 (**X**), M 2-9 (**XII**), NGC 6302 (**XII**) – by R. Minkowski and L. Kohoutek with the Palomar Observatory Schmidt telescope

Lo 5 (**X**), Wray 17-1 (**XIV**) – by A.J. Longmore with the UK Schmidt telescope, Australia

IC 289 (**VIII**), NGC 2452 (**XI**), NGC 6826 (**IX**), NGC 6543 (**XII**), NGC 7354 (**VIII**), NGC 7662 (**VIII**) – by B. Balick with the Kitt Peak National Observatory 2.1 m telescope

NGC 7009 (**IX,XIV**) – with the Kitt Peak National Observatory 2.1 m telescope

A 24 (**VII**), A 43 (**XI**), A 72 (**XI**), A 79 (**XIII**), A 80 (**VII**), Hb 5 (**XIII**), Hf 48 (**XII**), IC 4406 (**XII**), K 1-10 (**VII**), Mz 2 (**VII**), NGC 246 (**XI**), NGC 6058 (**VII**), NGC 7026 (**XIII**) – with the Palomar Observatory Schmidt telescope

Plate I

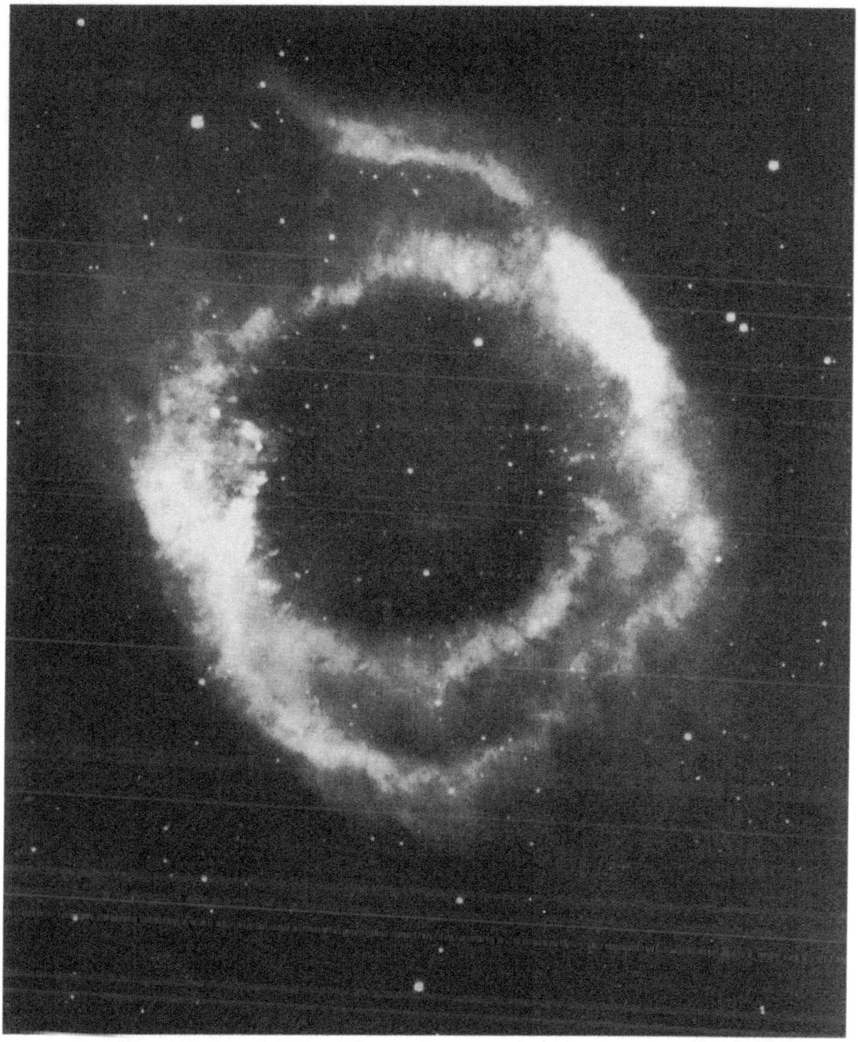

NGC 7293 Giant bipolar planetary nebula, Helix, in Aquarius. Note the tiny interior blobs from which short jets stream outward

Plate II

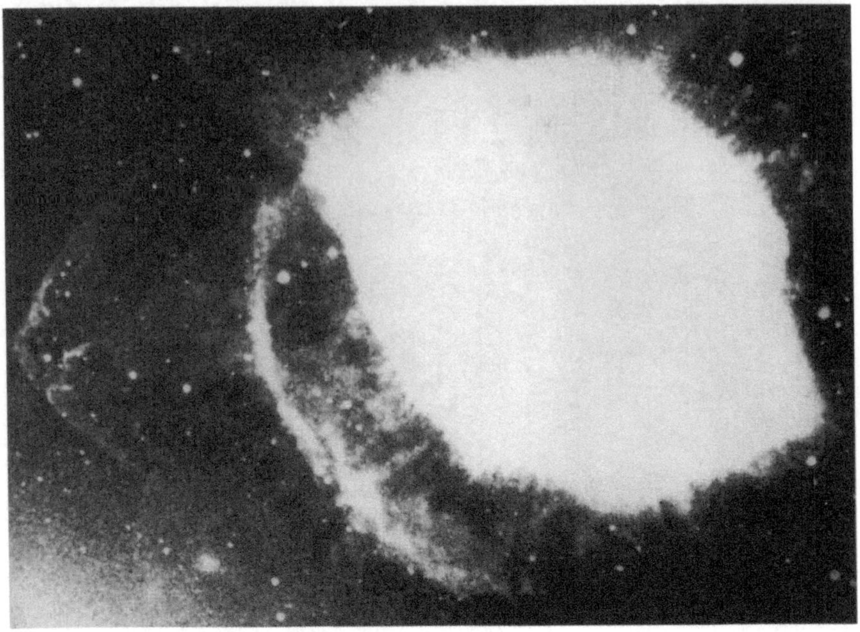

NGC 7293   Planetary nebula with three envelopes

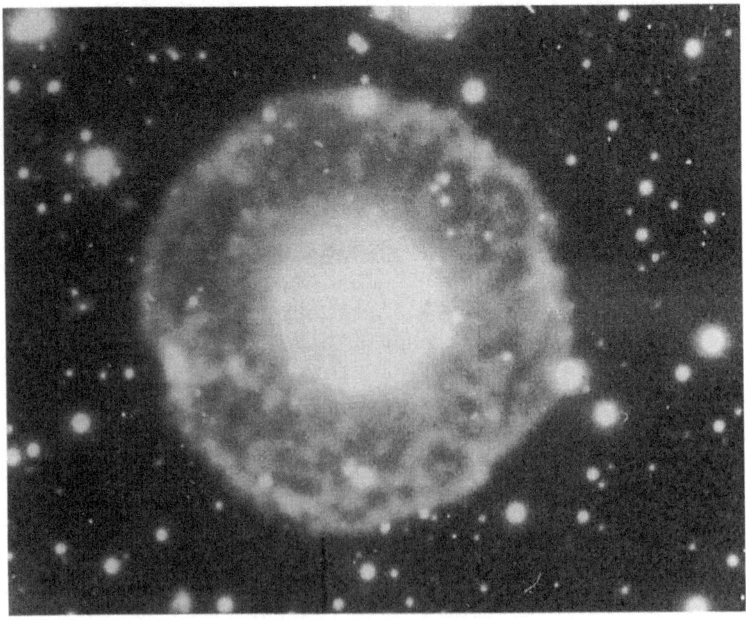

NGC 6826   Two-envelope planetary nebula with extended halo

Plate III

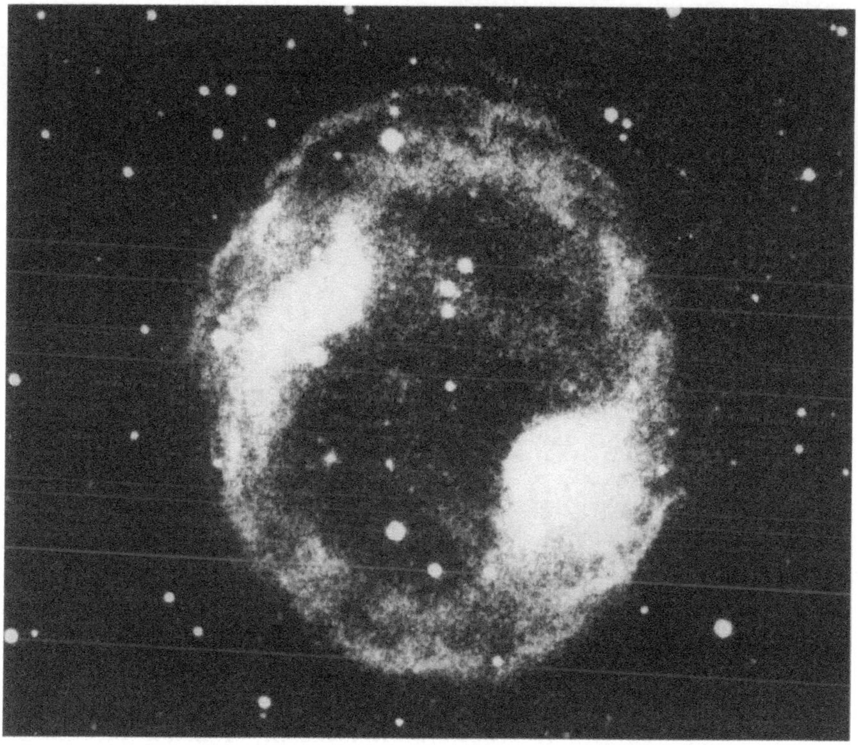

Jn Er 1   Giant bipolar planetary nebula in Lynx

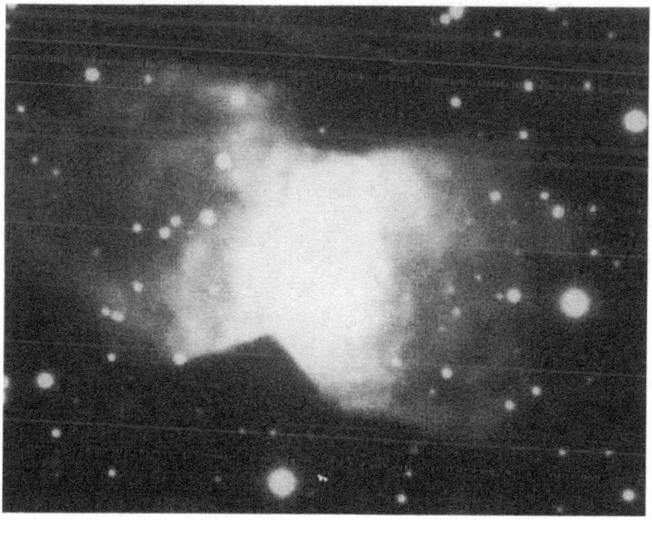

NGC 2346   Anomalous bipolar planetary nebula

Plate IV

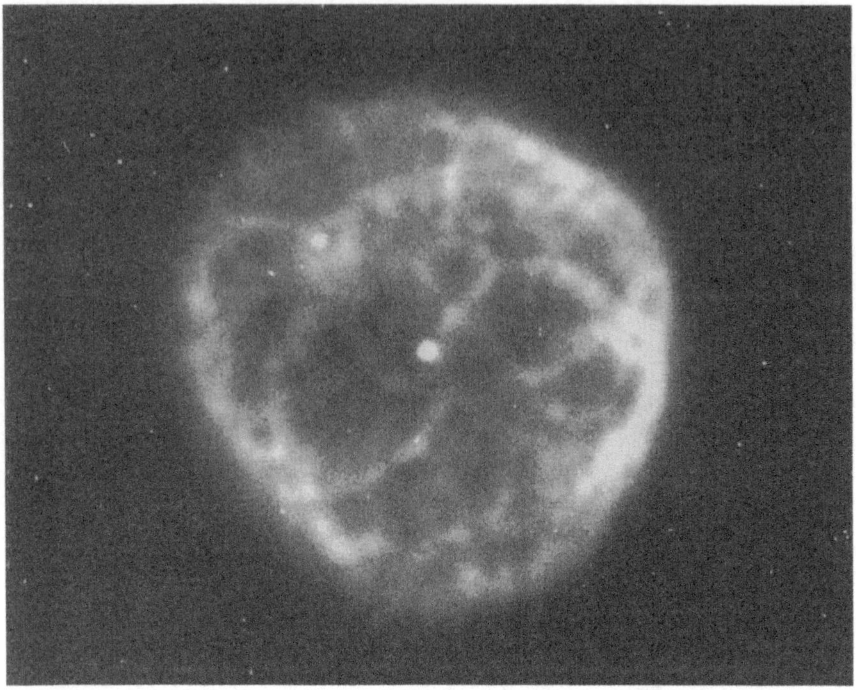

NGC 7094 Planetary nebula with filamentary structure

NGC 6804 Two-envelope bipolar planetary nebula

Plate V

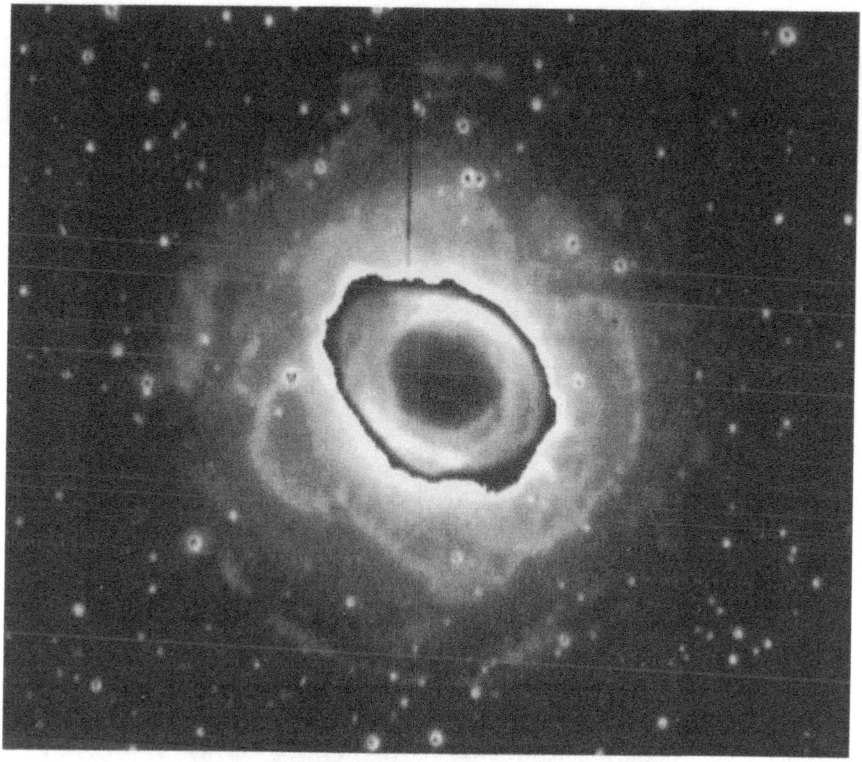

NGC 6720 Two-envelope bipolar planetary nebula with extended halo

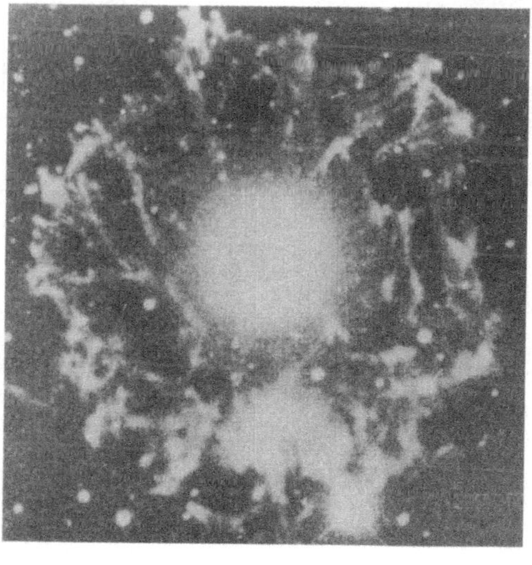

NGC 6543 Two-envelope anomalous planetary nebula with extremely extended halo

Plate VI

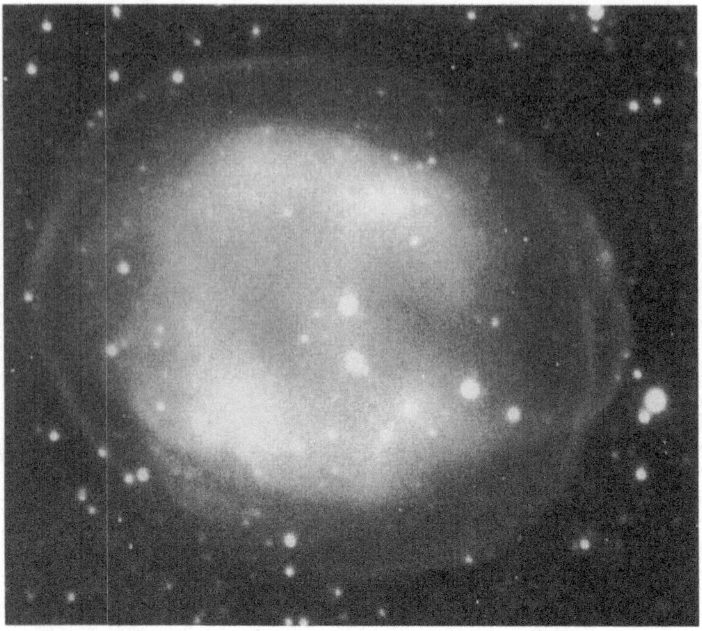

IC 1295 Bipolar two-envelope planetary nebula with very thin outer envelope

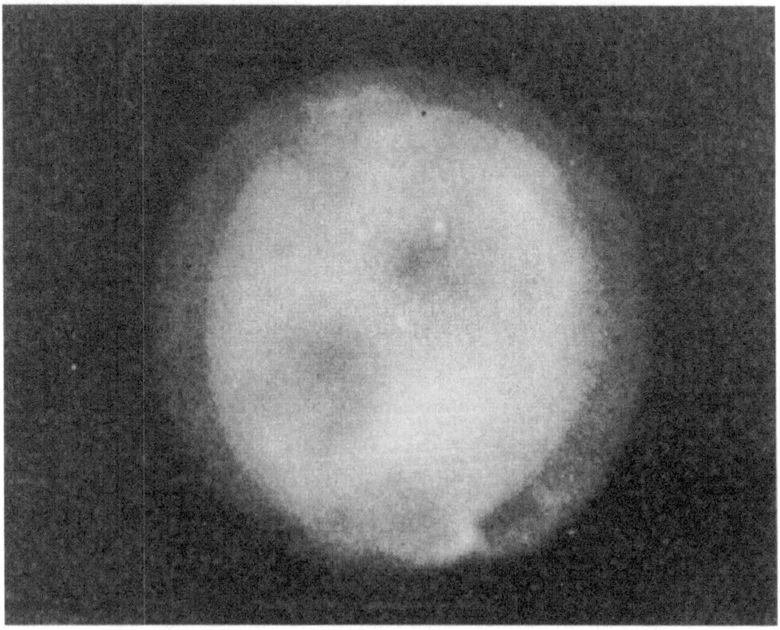

NGC 3587 Bipolar two-envelope planetary nebula with thin outer envelope

Plate VII

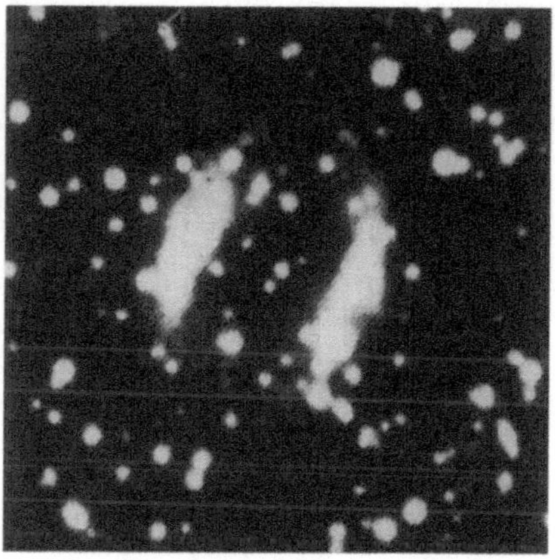

A 80

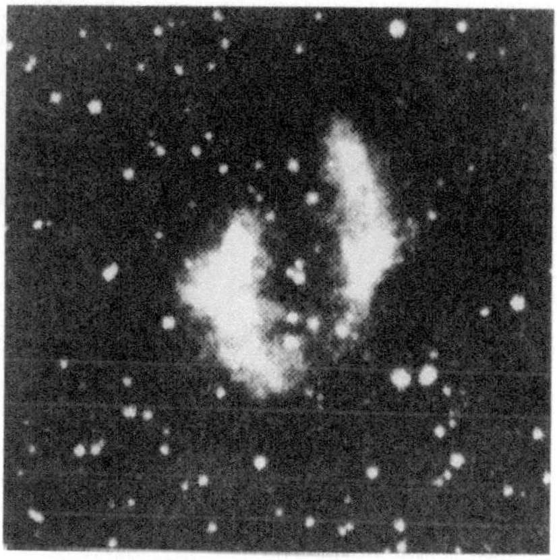

A 24

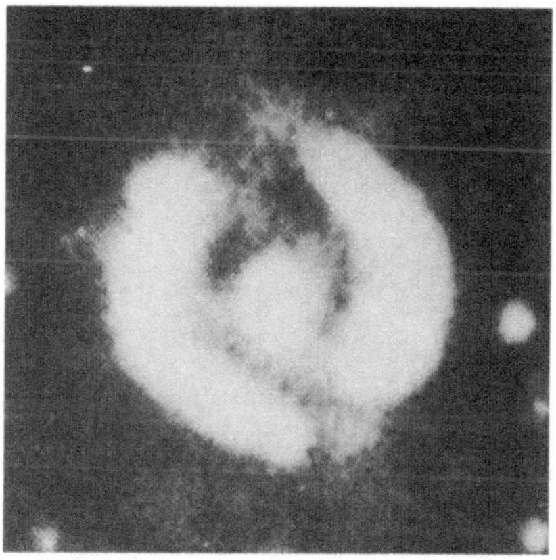

Mz 2

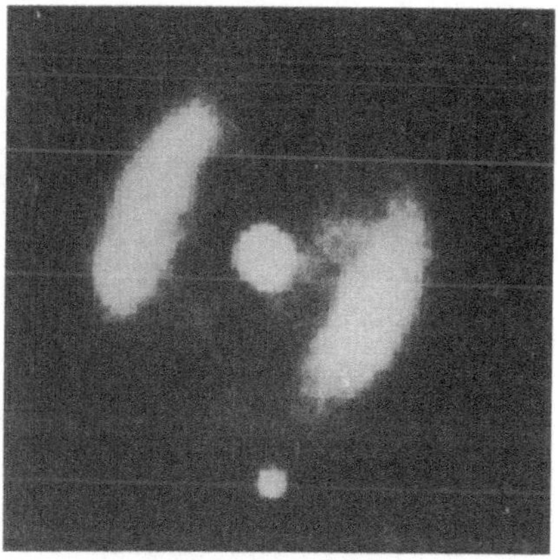

NGC 6058

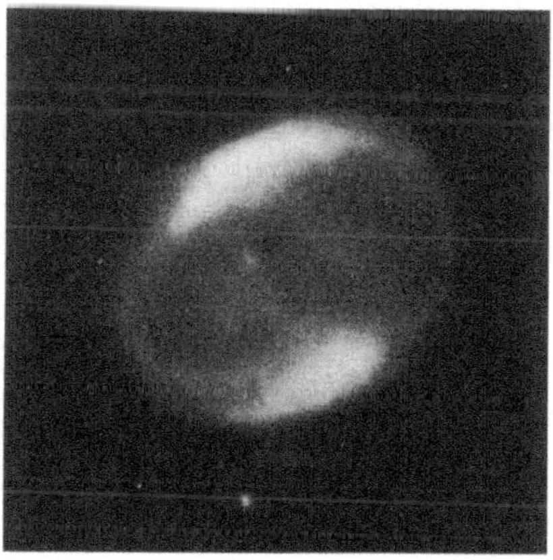

NGC 6563

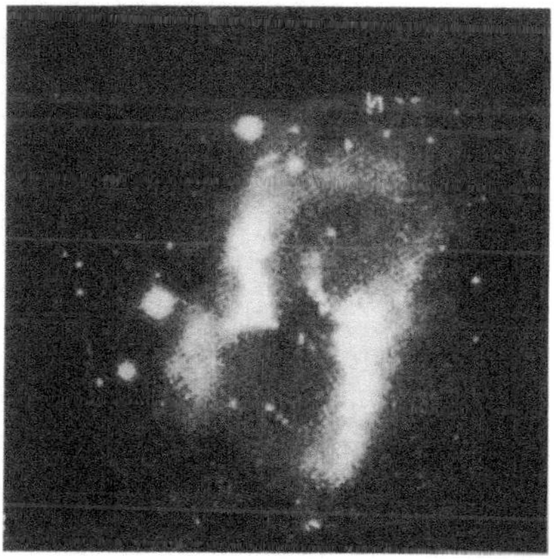

K 1-10

Bipolar planetary nebulae

Plate VIII

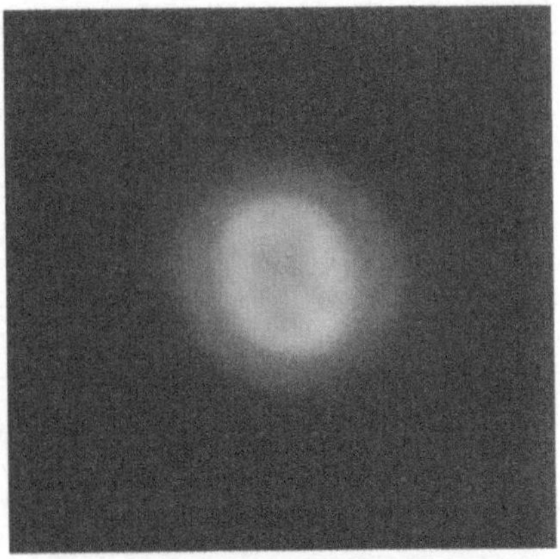

NGC 2022

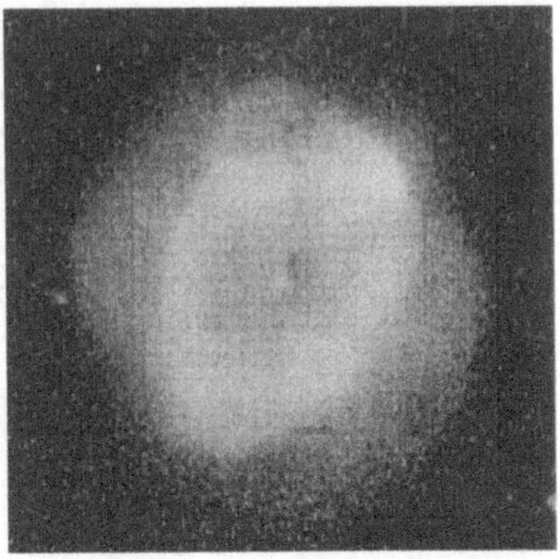

IC 289

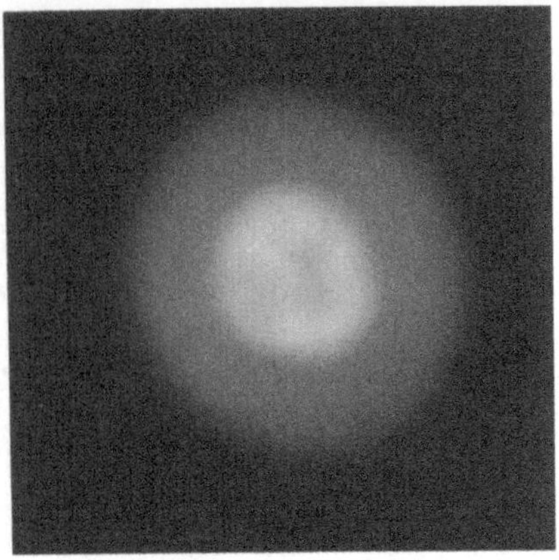

NGC 1535

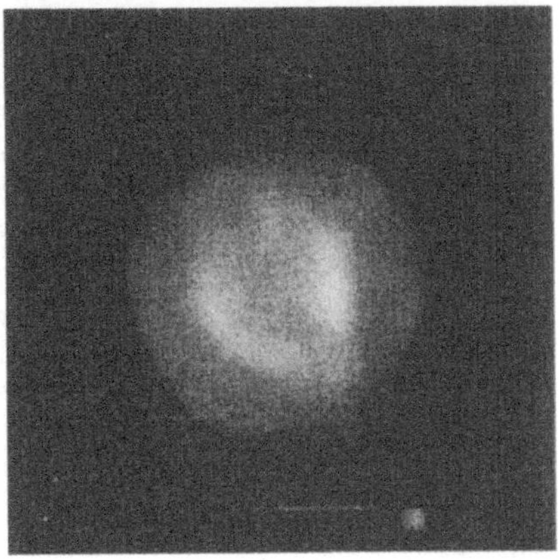

NGC 7354

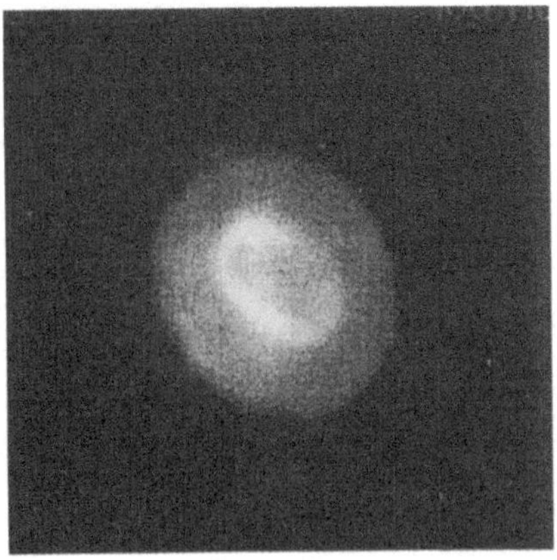

NGC 7662

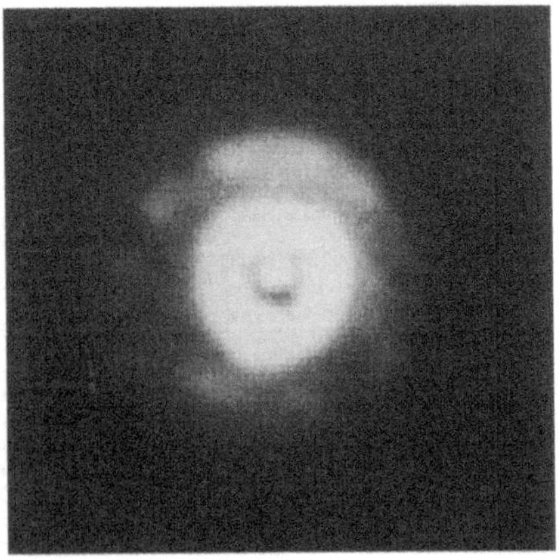

NGC 2392

Two-envelope planetary nebulae

Plate IX

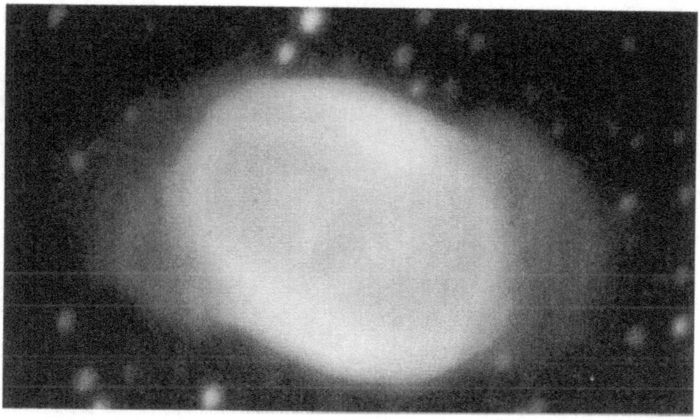

NGC 6563

NGC 6826

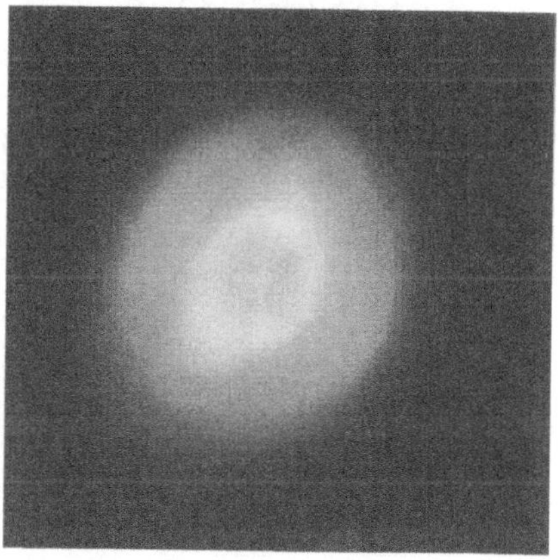

NGC 3242

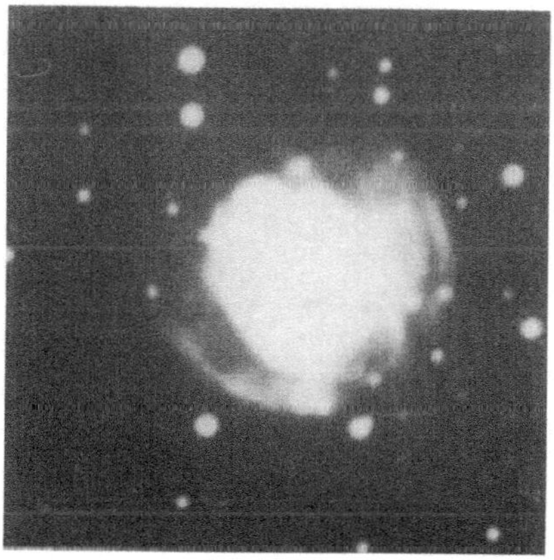

NGC 6804

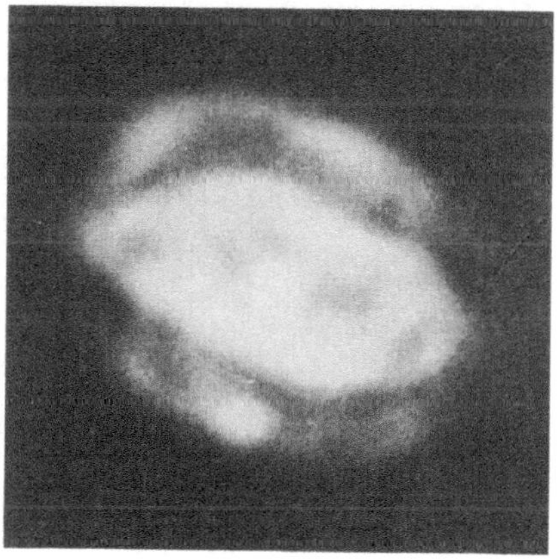

NGC 7009

Bipolar two-envelope planetary nebulae

Plate X

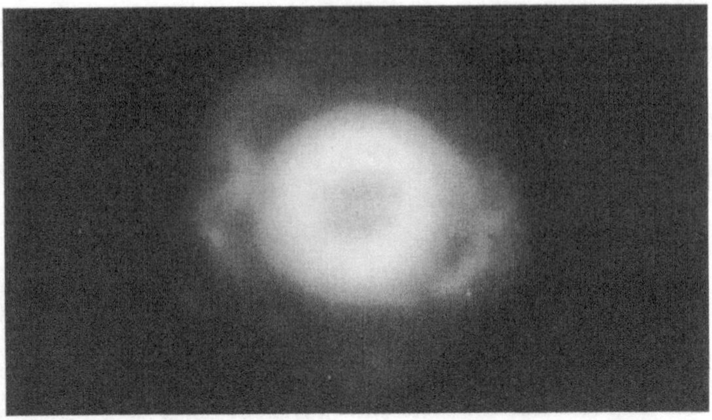

NGC 6369

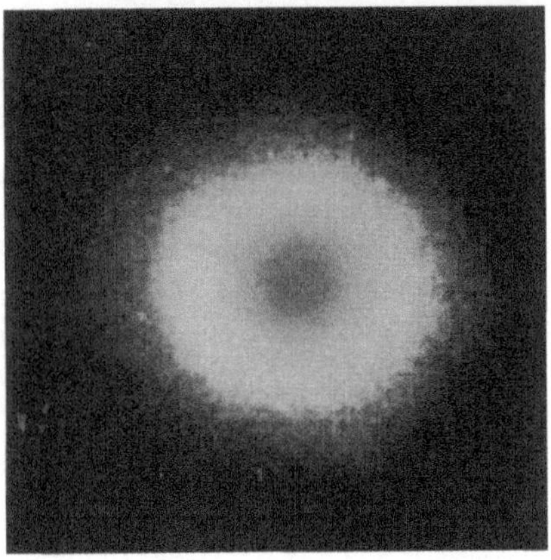

BD +30°3639

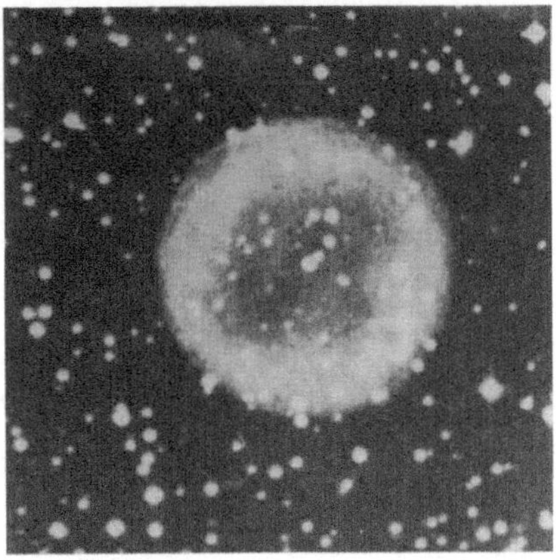

Lo 5

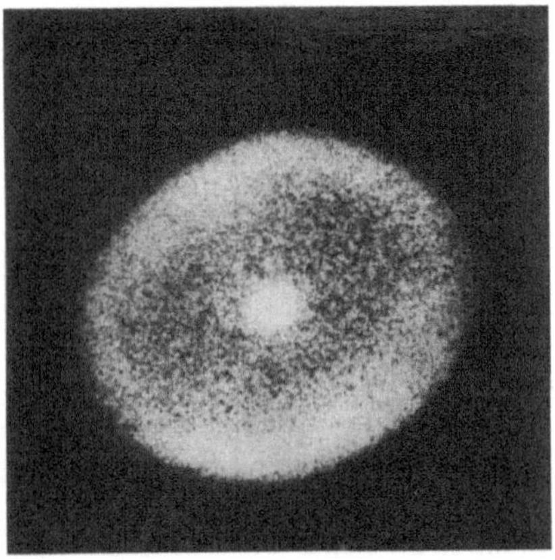

IC 418

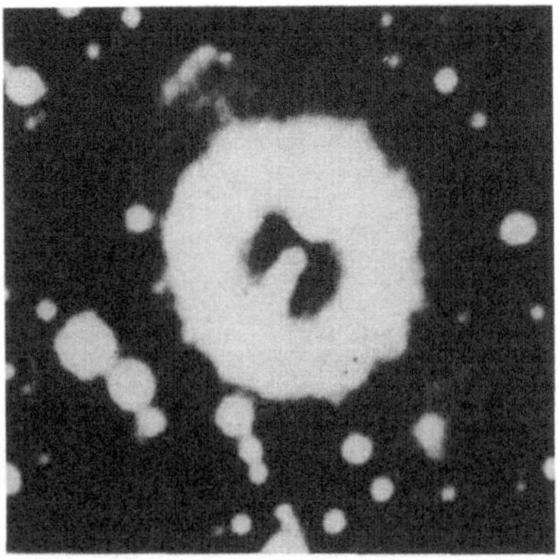

NGC 6337

Spherically symmetrical planetary nebulae

Plate XI

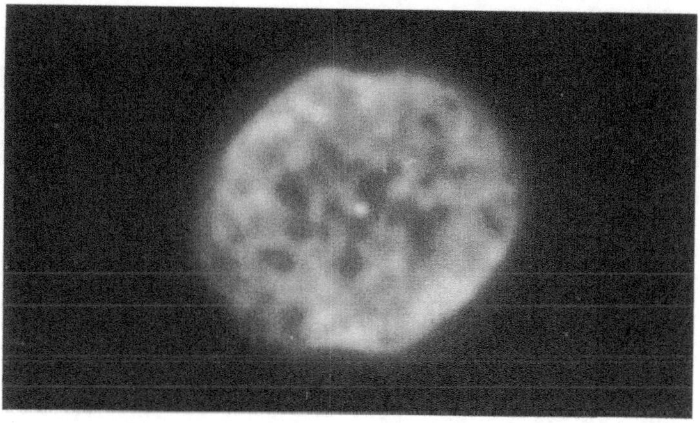

NGC 1501

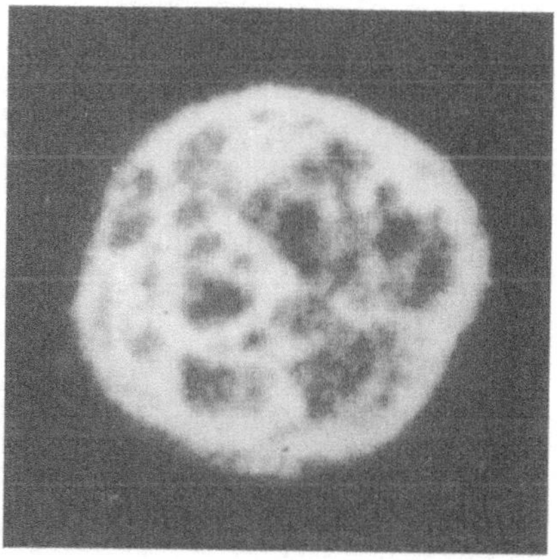

A 43

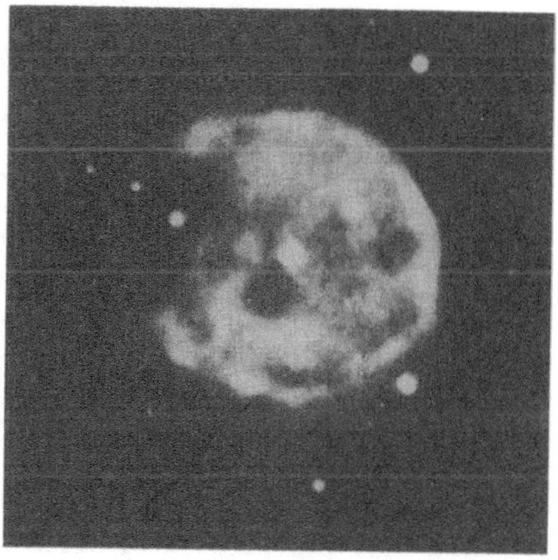

NGC 246

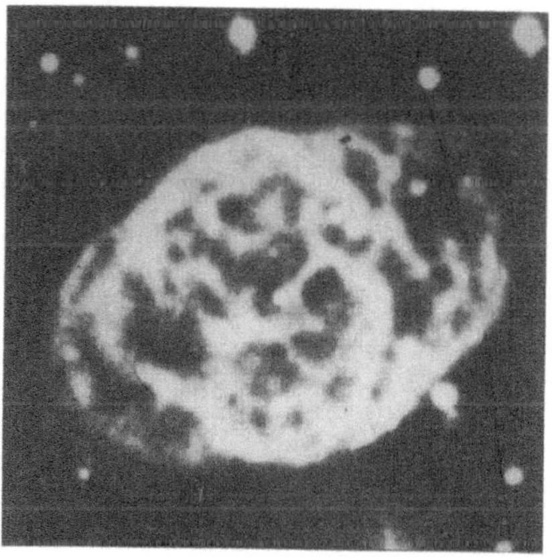

A 72

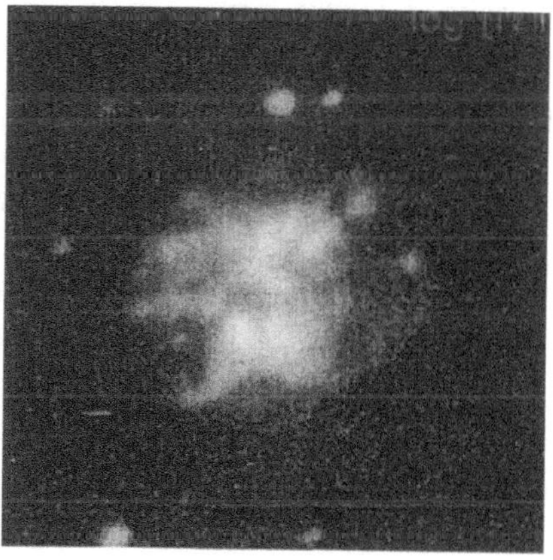

NGC 2452

Filamentary planetary nebulae

Plate XII

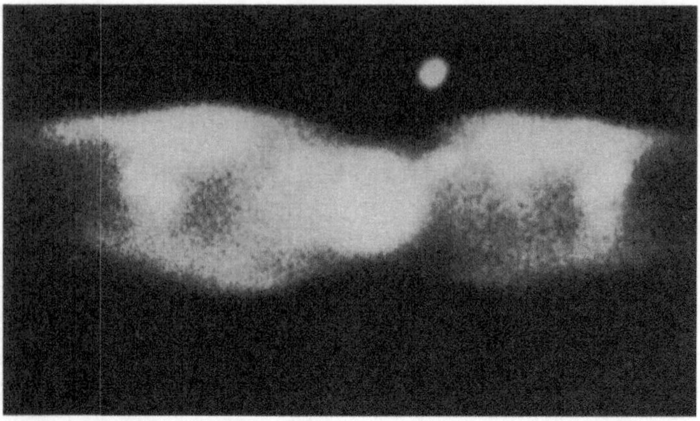

M 2-9

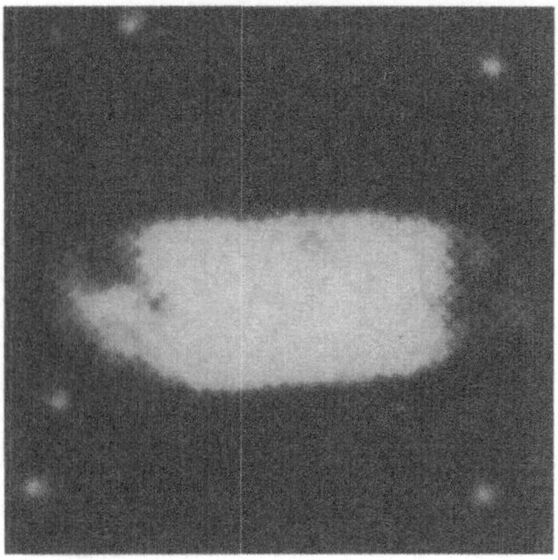

IC 4406

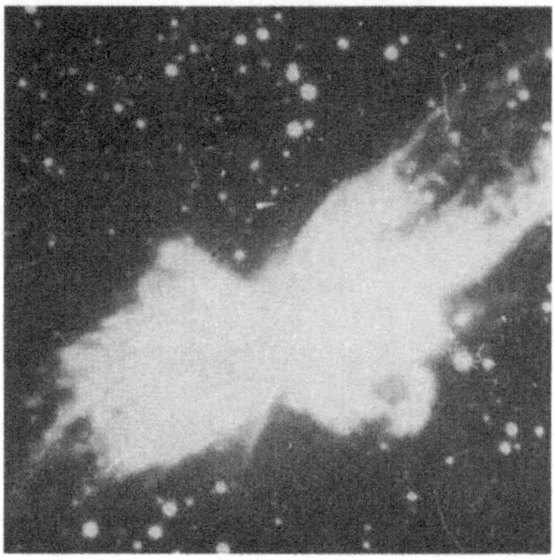

NGC 6302

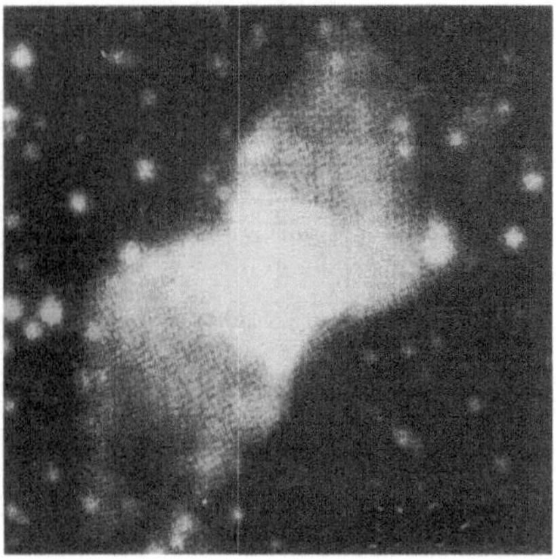

Hf 48

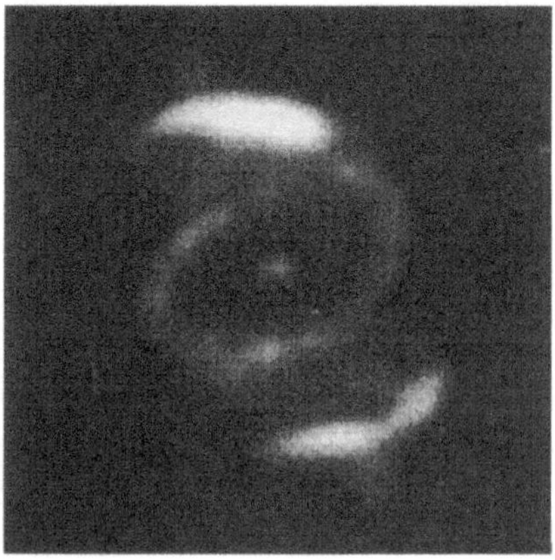

NGC 6543

Anomalous planetary nebulae

Plate XIII

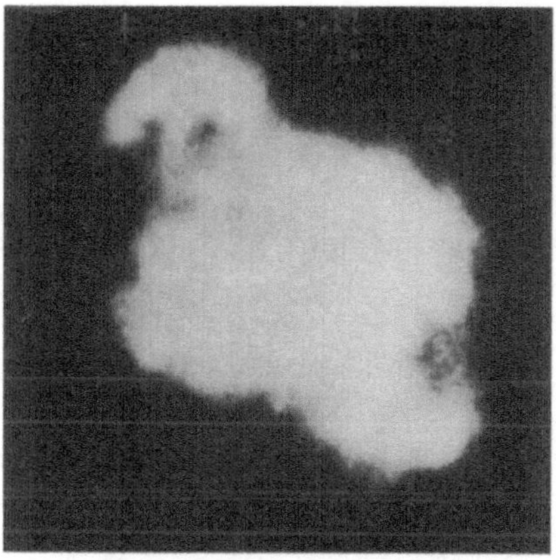

NGC 4361

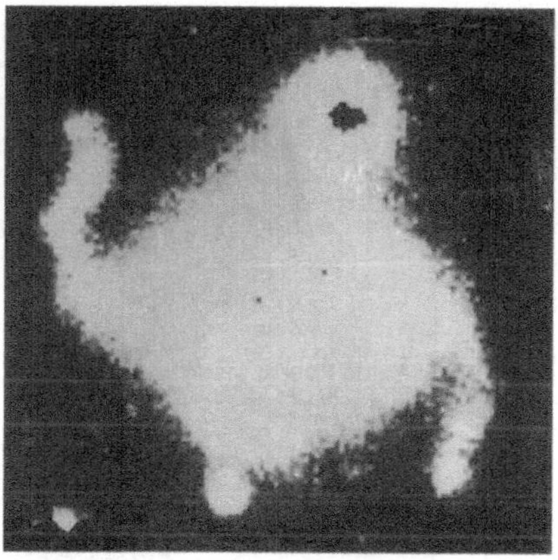

NGC 6309

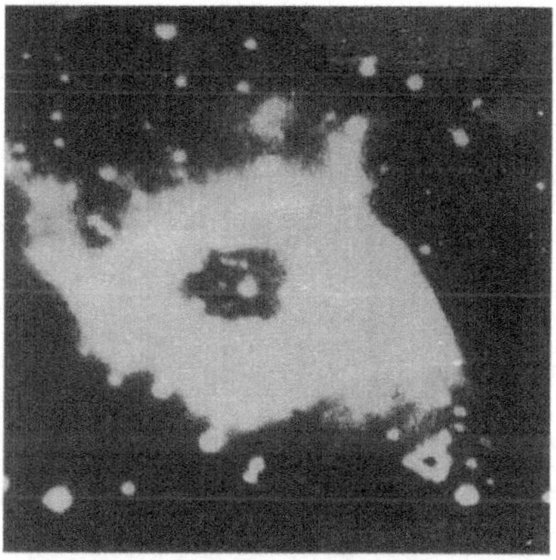

A 79

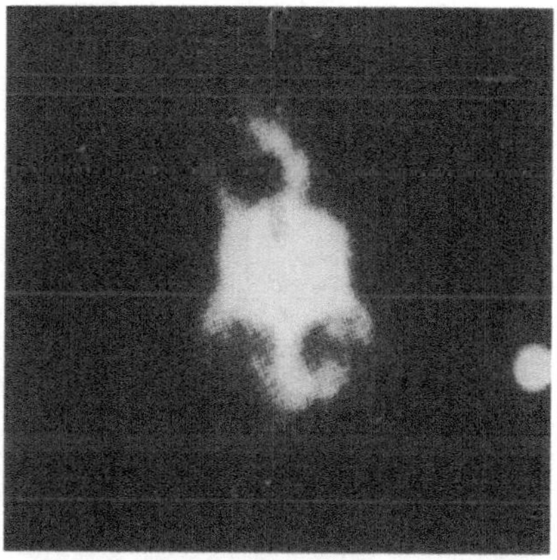

NGC 7026

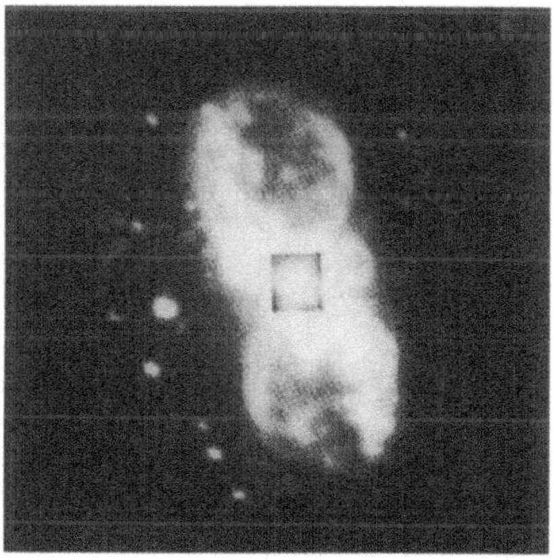

Hb 5

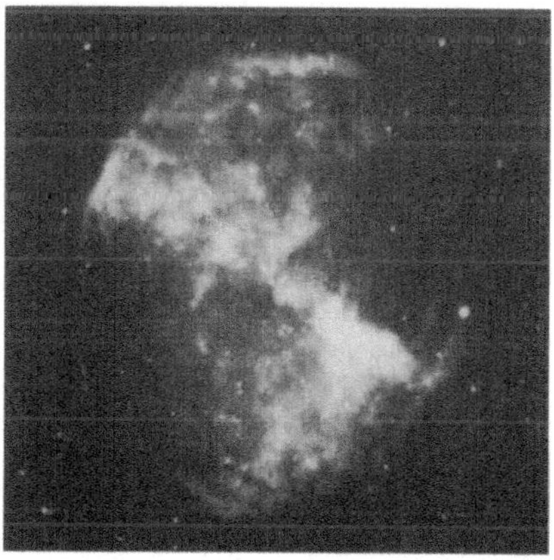

NGC 6853

Spiral type planetary nebulae

Plate XIV

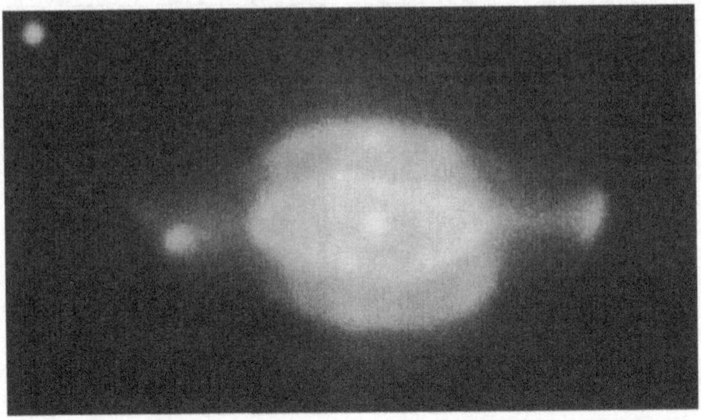

NGC 7009

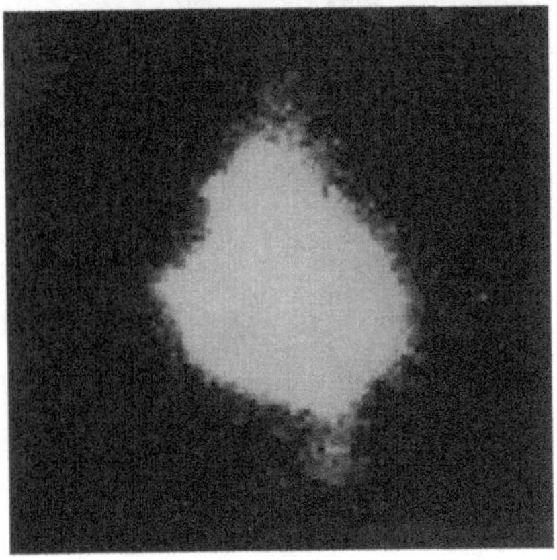

NGC 6210

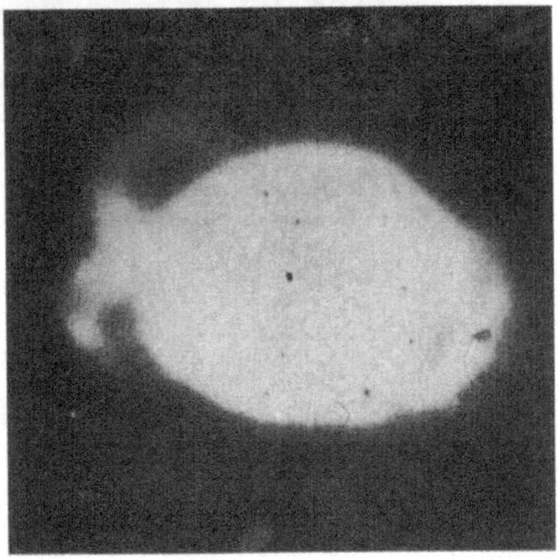

NGC 6369

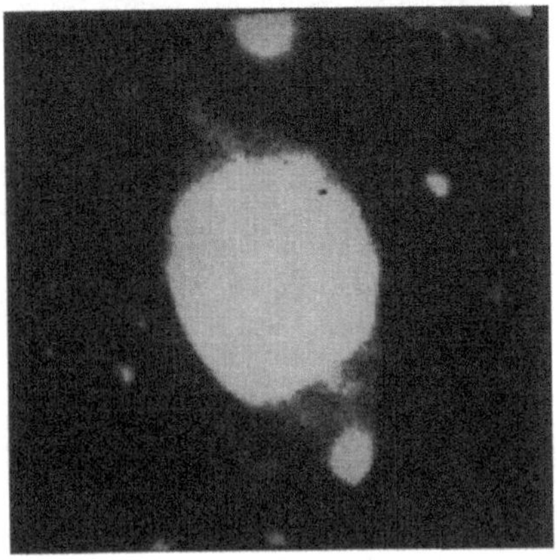

NGC 6905

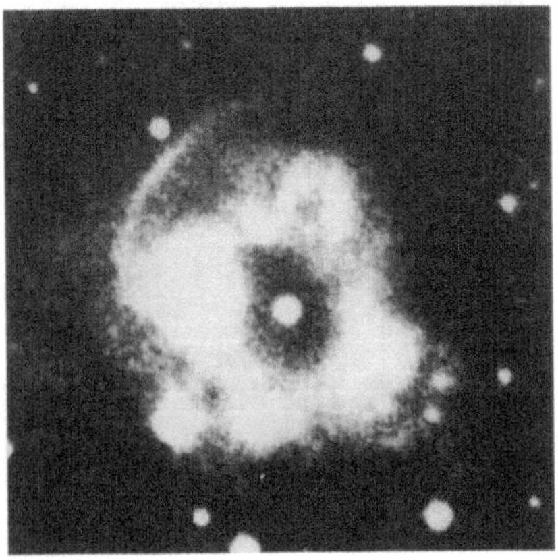

Wray 17-1

Planetary nebulae with polar outflow

References

Books, Catalogues, Proceedings

Acker A., Ochsenbein F., Stenholm B., Tylenda R., Marcout J.: 1992, The Strasburg ESO Catalogue of Galactic Planetary Nebulae. ESO

Aller L.H.: 1956, Gaseous Nebulae. Chapman, London

Aller L.H.: 1961, The Abundance of the Elements. Interscience, N.Y.

Aller L.H., Liller W.: 1968, Stars and Stellar Systems. Vol. 7. Planetary Nebulae. Chicago University Press

Busbridge I.W.: 1960, The Mathematics of Radiative Transfer, Cambridge

Cahn J.H., Kaler J.B., Stanghelini L.: 1992, Catalogue of Distances of Planetary nebulae. A&A Suppl.94, 399

Chandrasekhar S.: 1961, Hydrodynamics and Hydromagnetic Stability. Clarendons, Oxford

Daub C.T.: 1982, A Statistical Survey of Local Planetary Nebulae. ApJ **260**, 612

Feibelman W.A., Oliverson N.A., Nichols J., Garhart M.P.: 1988, IUE Spectral Atlas of Planetary Nebulae, Central Stars, and Related Objects. NASA, Ref. Publ. No. 1203

Gezari D.Y., Schmitz M., Mead J.M.: 1987, Catalog of Infrared Observations, NASA

Gurzadyan G.A.: 1954, Problems of the Dynamics of Planetary Nebulae (in Russian). Armenian Acad. Sci., Yerevan

Gurzadyan G.A.: 1969, Planetary Nebulae. Gordon and Breach, N.Y.

Gurzadyan G.A.: 1980, Flare Stars. Pergamon, London

Gurzadyan G.A.: 1984, Stellar Chromospheres (in Russian), Nauka, Moscow

Gurzadyan G.A.: 1985, Stellar Flares. Physics, Cosmogony (in Russian). Nauka, Moscow

Gurzadyan G.A.: 1993, Planetary Nebulae. Physics, Dynamics (in Russian). Nauka, Moscow

Gurzadyan G.A.: 1996 Catalogue of Excitation Classes for 750 Planetary Nebulae. Present book

IAU Symposium No 34. Planetary Nebulae. Eds. D.E.Osterbrock, C.R. O'Dell. Reidel, Dordrecht = 1968, PN Symp. 34

IAU Symposium No. 76. Planetary Nebulae. Ed. Y.Terzian. Reidel, Dordrecht = 1978, PN Symp. 76

IAU Symposium No. 103. Planetary Nebulae. Ed. D.R. Fowler. Reidel, Dordrecht = 1983, PN Symp. 103

IAU Symposium No. 131. Planetary Nebulae. Ed. S. Toress-Peimbert. Kluwer, Dordrecht = 1989, PN Symp. 131

IAU Symposium No. 155. Planetary Nebulae. Eds. R. Weinberger, A. Acker. Kluwer, Dordrecht = 1993, PN Symp.155

Kaler J.B.: 1976, A Catalog of Relative Emission Line Intensities Observed in Planetary and Diffuse Nebulae. ApJ Suppl. **31**, 517
Lang K.R.: 1988, Astrophysical Formulae, Springer, Berlin, Heidelberg
Maciel W.J.: 1984, A&A Suppl. **55**, 253
Osterbrock D.E.: 1974, Astrophysics of Gaseous Nebulae. Freeman, San Francisco
Perek L., Kohoutek L.: 1967, Catalogue of Galactic Planetary Nebulae. Academia, Prague
Pottasch S.R.: 1984, Planetary Nebulae. Reidel, Dordrecht
Sabbadin F.: 1984, A Catalogue of Expansion Velocities of Planetary Nebulae. A&A Suppl. **58**, 273
Sobolev V.V.: 1960, Moving Envelopes of Stars. Harvard University. Press
Sobolev V.V.: 1986, Theoretical Astrophysics. (in Russian). Nauka, Moscow
Spitzer L.: 1978, Physical Processes in the Interstellar Medium. Wiley, N.Y.
Van de Steene G.C., Zijlstra A.A.: 1994, Catalogue with Statistical Distances to Planetary Nebulae. A&A Suppl. **108**, 485
Weinberger R.: 1989, A Catalogue of Expansion Velocities of Galactic Planetary Nebulae. A&A Suppl. **78**, 301
Zhang C.Y.: 1995, A Statistical Distance Scale for Galactic Planetary Nebulae. ApJ Suppl. (in press)

Early Period, up to 1969

Abell G.O.: 1966, ApJ **144**, 259
Abell G.O., Goldreich R.: 1966, PASP **78**, 232
Aller L.H., Bowen I.S., Minkowski R.: 1955, ApJ **122**, 62
Aller L.H., Bowen I.S., Wilson O.C.: 1963, ApJ **138**, 1013
Aller L.H., Czyzak S.J.: 1968, PN Symp. 34, p. 209
Aller L.H., Kaler J.B., Bowen I.S.: 1966, ApJ **144**, 291
Ambartsumian V.A.: 1932, Poulkovo Obs. Circ. **4**, 8
Ambartsumian V.A.: 1933a, Poulkovo Obs. Circ. **6**, 10
Ambartsumian V.A.: 1933b, Poulkove Obs. Circ. **13**, 3
Bowen I.S.: 1927, PASP **39**, 295
Bowen I.S.: 1928, ApJ **67**, 1
Bowen I.S.: 1935, ApJ **81**, 1
Breit G., Teller E.: 1940, ApJ **91**, 215
Burgess A.: 1958, MNRAS **118**, 477
Burgess A.: 1965, ApJ **141**, 1588
Burke P.G., Ormonde S., Whiataker W.: 1967, Proc. Phys. Soc. **92**, 319
Capriotti E.R., Kovach W.S.: 1968, ApJ 106, 307
Chamberlain J.W.: 1953, ApJ 117, 387
Chandrasekhar S.: 1935, Z. f.Ap. 9, 266
Chandrasekhar S., Fermi E.: 1953, ApJ **118**, 116
Code A.D.: 1960, AJ **65**, 278
Cox D.P., Mathews W.G.: 1969, ApJ **155**, 859
Delmar T.N., Gould R.J., Ramsey W.: 1967, ApJ **149**, 495
Dravskich Z.V., Dravskich A.F.: 1964, Astron. Circ. **282**, 2
Eddington A.S.: 1927, MNRAS **88**, 134
Eissner W., Martins P., Nussbaumer H., Saraph H.E., Seaton M.J.: 1969, MNRAS **146**, 63
Evans D.: 1950, MNRAS **110**, 37
Evans D., Thackeray A.: 1950, MNRAS **110**, 429

Feinberg G.: 1958, Phys.Rev. **112**, 1637
Flower D.R.: 1969, MNRAS **146**, 171
Garstang R.H.: 1968, PN Symp. 34, p. 143
Geltmen S.S.: 1956, Phys. Rev. **102**, 171
Gillett F.C., Low F.J., Stein W.A.: 1967, ApJ **149**, L97
Ginzburg V.L., Frank I.M.: 1946, JETP **16**, 15
Göppert-Mayer M.G.: 1929, Naturwiss., **19**, 932
Greenstein J.: 1952, PASP **64**, 256
Gurzadyan G.A.: 1953, Commun. Armenian Acad. Sci., Ser.Phys.Math. **6**, 19; **6**, 53
Gurzadyan G.A.: 1955, Commun. Byurakan Obs., **16**, 3
Gurzadyan G.A.: 1956, Commun. Byurakan Obs., **18**, 3
Gurzadyan G.A.: 1957, Dokl. AN SSSR, **113**, 1231
Gurzadyan G.A.: 1957, Astron.Zh. **34**, 820
Gurzadyan G.A.: 1958, Commun. Byurakan Obs., **24**, 33
Gurzadyan G.A.: 1958, Commun. Byurakan Obs., **25**, 101
Gurzadyan G.A.: 1958, Dokl. AN SSSR, **120**, 734
Gurzadyan G.A.: 1958, Astron.Zh. **35**, 520
Gurzadyan G.A.: 1960, Commun. Byurakan Obs. **27**, 73
Gurzadyan G.A.: 1960, Dokl. AN SSSR **130**, 47
Gurzadyan G.A.: 1960, Dokl. AN SSSR **130**, 287
Gurzadyan G.A.: 1960, Dokl. AN SSSR **133**, 1053
Gurzadyan G.A.: 1961, Dokl. AN SSSR **141**, 1061
Gurzadyan G.A.: 1962, Dokl. AN SSSR **144**, 1011
Gurzadyan G.A.: 1962, Vistas in Astronomy **5**, 40
Gurzadyan G.A.: 1963, Commun. Byurakan Obs. **34**, 61
Gurzadyan G.A.: 1964, Commun. Byurakan Obs. **35**, 59
Gurzadyan G.A.: 1965, Astrofiz. **1**, 91
Gurzadyan G.A.: 1965, Astrofiz. **1**, 319
Gurzadyan G.A.: 1967, Dokl. AN SSSR **172**, 1046
Gurzadyan G.A.: 1968, PN Symp. 34, p. 450
Gurzadyan G.A., Razmadze N.A.: 1959, Commun. Byurakan Obs. **35**, 43
Hall R.G.: 1953, PASP **65**, 154
Harman R.J., Seaton M.J.: 1966, MNRAS **132**, 15
Harrington J.P.: 1968, ApJ **152**, 943
Harrington J.P.: 1969, ApJ **156**, 903
Henize K.G.: 1964, PASP **73**, 159
Hickok F.R., Morton D.C.: 1968, ApJ **152**, 203
Hummer D.G.: 1962, MNRAS **125**, 21
Hummer D.G., Seaton M.J.: 1963, MNRAS **125**, 437
Janikova E., Janout Z., Lehar F., Pavlovic P., Zrelov V.: 1969, Nucl. Instrum. and Meth. **74**, 61
Johnson H.M., Hogg D.E.: 1965, ApJ **142**, 1033
Kardashev N.S.: 1959, Astron.Zh. **36**, 838
Kipper A.Ya.: 1952, Publ. Tartu Obs. **32**, 63
Kipper A.Ya., Tiyth V.M.: 1958, Voprosy Cosmogonii **6**, 98
Klemola R.C.: 1962, PASP **67**, 740
Kohoutek L.: 1964, PASP **76**, 435
Kohoutek L.: 1966, BAC **17**, 318
Kohoutek L.: 1967, BAC **18**, 103
Lewis D.: 1950, Proc. Roy. Soc. A, **202**, 81
Liller W., Aller L.H.: 1957, Sky & Tel. **5**, 222
Liller W., Walther B., Liller M.: 1966, ApJ **144**, 280

Lutz G.R.: 1961, Publ. NRAO **1**, 85
Mathews W.C.: 1966, ApJ **143**, 173
Menzel D.H.: 1968, PN Symp. 34, p. 279
Mihalas D.: 1965, ApJ Suppl. **9**, 321
Milford S.M.: 1960, ApJ **131**, 407
Minin I.N.: 1958, Voprosy Cosmogonii, **6**, 211
Minkowski R.: 1964, PASP 76, 197
Minkowski R., Aller L.H.: 1954, ApJ **120**, 261
Minkowski R., Aller L.H.: 1956, ApJ **124**, 93
Minkowski R., Osterbrock D.E.: 1960, ApJ **131**, 537
Munch G.: 1958, Rev. Mod. Phys. **30**, 1035
O'Dell C.R.: 1965, ApJ **142**, 1093
O'Dell C.R.: 1966a, ApJ **143**, 168
O'Dell C.R.: 1966b, ApJ **145**, 487
Oort J.: 1946, MNRAS **106**, 159
Oster L.F.: 1961, ApJ **134**, 1010
Osterbrock D.E.: 1960, ApJ **131**, 541
Pengelly R.M.: 1964, MNRAS **127**, 145, 165
Purcell E.: 1952, ApJ **116**, 457
Robbins R.R.: 1968, ApJ **151**, 497
Rubin R.H.: 1968, ApJ **153**, 761
Salpeter E.E.: 1958, Phys. Rev. **112**, 1642
Schwarz H.E., Corradi R.L.M., Melnick J.: 1992, A Catalogue of Narrow Band Images of Planetary Nebulae. A&AS **96**, 23
Seaton M.J.: 1955, MNRAS **115**, 279
Seaton M.J.: 1959, MNRAS **119**, 90
Seaton M.J.: 1960, Rep. Progr. Phys. **23**, 313
Seaton M.J.: 1966, MNRAS **132**, 113
Shklovskii I.S.: 1956, Astron. Zh. **33**, 315
Slee O.B., Orchiston D.W.: 1965, Australian J. Phys. **18**, 187
Sobolev V.V.: 1957, Astron.Zh. 34, 694
Sobolev V.V.: 1958, Voprosy Cosmogonii **6**, 112
Spitzer L.: 1954, ApJ **120**, 1
Spitzer L., Greenstein J.: 1951, ApJ **114**, 407
Stein T., Carrico J., Lipworth E., Weisskopf M.: 1967, Phys. Rev. **19**, L741
Stoy P.: 1933, MNRAS **93**, 588
Taylor G.: 1950, Proc. Roy. Soc. A, **201**, 192
Terzian Y.: 1966, ApJ **144**, 657
Terzian Y.: 1968, PN Symp. No. 34, p. 87
Terzian Y.: 1989, PN Symp. 131, p. 17
Thomas R.N.: 1957, ApJ **125**, 260
Thompson A.R.: 1968, PN Symp. 34, p. 112
Thompson A.R., Colvin R.S.: 1967, ApJ **150**, 345
Treffers R.R., Fink U., Larson H.P., Gautier T.N.: 1967, ApJ **209**, 793
Ueno S.: 1956, Contr. Inst. Astrophys. Kyoto, No **64**
Unno W.: 1952, PASJ **4**, 100
Vitjazev V.V: 1969, Astrofiz. **5**, 83
Weymann R., Williams R.E.: 1969, ApJ **157**, 1201
Wilson O.C.: 1950, ApJ **111**, 279
Wilson O.C.: 1958, Rev. Mod. Phys. **30**, 1025
Wilson O.C., Aller L.H.: 1951, ApJ **114**, 421
Wilson O.C., Aller L.H.: 1954, ApJ **119**, 243
Yada J.: 1954, PASJ **6**, 76

Zanstra H.: 1927, ApJ **65**, 50
Zanstra H.: 1934, MNRAS **95**, 81
Zanstra H.: 1936, MNRAS **97**, 37
Zanstra H.: 1951, BAN **11** N 429
Zanstra H.: 1956, BAN **12** N 465

Later Period, 1970–1989

Acker A.: 1978, A&A **33**, 367
Acker A.: 1978, A&A Suppl. **33**, 367
Acker A., Marcout J., Ochsenbein F.: 1981, A&A Suppl. **43**, 265
Ahern R.J.: 1978, ApJ **223**, 901
Aldrovandi S.M.V., Pequignot D.: 1973, A&A **25**, 137
Aller L.H.: 1978, PN Symp. 76, p. 225
Aller L.H.: 1985, ApJ **296**, 492
Aller L.H., Keyes C.D.,: 1980, Proc. Nat. Acad. Sci. USA **77**, 1231
Aller L.H., Keyes C.D., Czyzak S.J.: 1981, ApJ **250**, 596
Aller L.H., Keyes C.D., Feibelman W.A.: 1986, ApJ **311**, 930
Aller L.H., Keyes C.D., Maran S.P. et al.: 1987, ApJ **320**, 159
Aller L.H., Ross J.E., O'Mara B.J.: 1981, MNRAS **197**, 95N
Altschuler D.R., Scheider S.E., Giovanardi C., Silverglate P.R.: 1986, ApJ **305**, L88
Appararo K.M.V., Tarafder S.P.: 1989, ApJ **344**, 826
Ashley M.C.B., Hyland A.R.: 1988, ApJ **331**, 532
Balick B.: 1989 PN Symp. 131, p. 83
Band B., Habing H.J.: 1983, A&A **127**, 73
Barfield W.D., Koontz G.D., Hueber W.F.: 1972, J. Quant. Spectr. & Rad. Transf. **12**, 1409
Barker T.: 1978, ApJ **219**, 914
Barker T.: 1980, ApJ **249**, 99
Barker T.: 1982, ApJ **253**, 167
Barker T.: 1983, ApJ **267**, 630
Barker T.: 1984, ApJ **284**, 589
Barker T.: 1985, ApJ **294**, 193
Barker T.: 1986, ApJ **308**, 314
Barker T., Tuchmann Y.: 1980, ApJ **237**, 105
Barlow M.J.: 1989, PN Symp. 131, p. 319
Bates G.N., Altick P.L.: 1973, J. Phys. B **136**, 653
Becker S.A., Iben I.Jh. : 1980, ApJ **237**, 111
Bentley A.F.: 1982, AJ **87**, 1810
Bentley A.F., Hackwell J.A., Grasdalen G.L., Gehrz R.D.: 1984, ApJ **278**, 665
Bianchi L., Cerrato S., Grewing M.: 1986, A&A **169**, 277
Blair W.P., Stencel R.E., Feibelman W.A., Michalitsianos A.G.: 1983, ApJ Suppl. **53**, 573
Blanco V.M., McCarthy M.F.: 1981, Physical Processes in Red Giants. Eds. I.Iben, A. Renzini. Reidel, Dordrecht, p. 147
Bogges, A., Carr, F.A., Evans, D.C., Fischel, D., Freeman, H.R., et. al., 1978, Nature, **275**, 2.
Bohlin R.C., Harrington J.P., Stecher T.P.: 1978, ApJ **219**, 575
Bohlin R.C., Harrington J.P., Stecher T.P.: 1982, ApJ **252**, 635
Bohlin R.C., Marionni P.A., Stecher T.P.: 1975, ApJ **202**, 415
Boroson T.A., Liebert J.: 1989, ApJ **339**, 844

Brocklehurst M.: 1971, MNRAS **153**, 471
Brocklehurst M.: 1972, MNRAS **157**, 211
Brown R.T.: 1971, ApJ **164**, 387
Brown R.T., Mathews W.G.: 1970, ApJ **160**, 939
Cahn J.H.: 1976, AJ **81**, 407
Cahn J.H., Kaler J.B.: 1971, ApJ Suppl. **22**, 319
Cahn J.H., Rubin R.H.: 1974, AJ **79**, 128
Canto J., Alliot K.H., Meaburn J., Theokas A.C.: 1980, MNRAS **193**, 911
Carsently U., Solf J.: 1982, A&A **106**, 307
Chevalier R.A., Klein R.I.: 1978, ApJ **219**, 994
Chu Y.-H.: 1989, PN Symp. 131, p. 105
Chu Y.-H., Jacoby G.H., Arendt R.: 1987, ApJ Suppl. **64**, 529
Chu Y.-H., Jacoby G.H., Arendt R.: 1989, ApJ Suppl. **64**, 529
Chu Y.-H., Treffers R.R.: 1981, ApJ **249**, 586
Clegg R.E.S., Harrington J.P., Barlow M.J., Walsh J.R.: 1987, ApJ **314**, 551
Cohen M., Barlow M.J.: 1974, ApJ **193**, 401
Davis L.E., Lewis B.M., Parton C.R.: 1979, ApJ **230**, 434
Dinerstein H.L.: 1984, ApJ **284**, 702
Dinerstein H.L., Lester D.F., Werner M.W.: 1985, ApJ **291**, 561
Dinerstein H.L., Sneden C.: 1988, ApJ **353**, L23
Dufour R.J.: 1984, ApJ **287**, 341
Eder J., Lewis B.M., Terzian Y.: 1988, ApJ Suppl. **66**, 183
Egikyan A.G.: 1989, Astrofiz. **30**, 270
Engels D., Kreysa E., Schultz G.V., Sherwood W.A.: 1983, A&A **124**, 123
Ericson E.F., Swift C.D., Witteborn F.C. et al.: 1973, ApJ **183**, 535
Feibelman W.A.: 1982a, ApJ **258**, 548
Feibelman W.A.: 1982b, ApJ **258**, 562
Feibelman W.A.: 1982c, ApJ **263**, L69
Feibelman W.A.: 1984, ApJ **287**, 353
Feibelman W.A., Aller L.H., Keyes C.D., Czyzak S.J.: 1985, Proc. Nat. Acad. Sci. USA, **82**, 2202
Feibelman W.A., Boggess A., Hobbs R.W., McCracken C.W.: 1980, ApJ **241**, 725
Feibelman W.A., Boggess A., McCracken C.W., Hobbs R.W.: 1981, ApJ **246**, 807
Feibelman W.A., Hobbs R.W., McCracken C.W., Brown L.W.: 1979, ApJ **231**, 111
Field G.: 1982, Symp. Orion Neb. N.Y., p. 290.
Flower D.R., Nussbaumer H., Schild H.: 1979, A&A **72**, L1
Flower D.R., Penn C.J.: 1981, MNRAS **194**, 13p
Ford H.C.: 1983, PN Symp. 103, p. 443
Ford H.C., Ciardullo R., Jacoby G.H., Hui X. 1989, PN Symp. 131, p. 335
Garay G., Moran J.M., Reid M.J.: 1987, ApJ **314**, 535
Gathier R.: 1987, A&A Suppl. **71**, 245
Geballe T.R., Rank D.M.: 1973, ApJ **182**, L113
Gieseking F., Becker I., Solf J.: 1985, ApJ **295**, L17
Gieseking F., Hippelein H., Weinberger R.: 1986, A&A **156**, 101
Gieseking F., Solf J.,: 1986, A&A **163**, 174
Gillett F.C., Forrest W.J.: 1973a, ApJ **179**, 483
Gillett F.C., Forrset W.J., Merrill K.M.: 1973b, ApJ **183**, 87
Greenhouse M., Hayward T., Thronson H.: 1988, Sky & Tel. **75**, No. 2
Grewing M., Boksenberg A., Seaton M.J. et al., 1978a, Nature **275**, 575
Grewing M., Boksenberg A., Seaton M.J. et al., 1978b, Nature **275**, 394
Gurzadyan G.A.: 1975, MNRAS **172**, 249
Gurzadyan G.A.: 1976, PASP **88**, 891
Gurzadyan G.A.: 1977a, PN Symp. 76, p. 79

Gurzadyan G.A.: 1977b, ApSS **46**, 471
Gurzadyan G.A.: 1979, Astrofiz. **15**, 461
Gurzadyan G.A.: 1979, Vistas in Astronomy **23**, 45
Gurzadyan G.A.: 1981, ApSS **80**, 189
Gurzadyan G.A.: 1987, ApSS **133**, 307
Gurzadyan G.A.: 1988, ApSS **149**, 343
Gurzadyan G.A.: 1989, ApSS **158**, 123
Gurzadyan G.A.: 1987, Dokl. AN SSSR, **292**, 304
Gurzadyan G.A., Egikyan A.,G.: 1988, Dokl. AN SSSR, **300**, 316
Hamilton N., Liller W.: 1973, Mem. Soc. Roy. Liege **6**, 213
Harrington J.P., Feibelman W.A.: 1983, ApJ **265**, 258
Harrington J.P., Feibelman W.A.: 1984, ApJ **277**, 716
Harrington J.P., Lutz J.H., Seaton M.J., Stickland D.J.: 1980 MNRAS **191**, 13
Harrington J.P., Seaton M.J., Adams S., Lutz J.H.: 1982, MNRAS **199**, 517
Hazard C., Terlevich R., Morton D.C., Sargent W.L., Ferland G.: 1980 Nature **285**, 463
Heap S.R.: 1979, IAU Symp. No. 83: Mass Loss and Evolution of O-Type Stars. Reidel, Dordrecht, p. 99
Heap S.R.: 1983, PN Symp. 103, p. 375
Heap S.R.: 1986 in New Insights in Astrophys., ESA SP-263, p. 291
Heiligman G.M.: 1980, MNRAS **191**, 761
Henry R.J.W.: 1970, ApJ **161**, 1153
Higgs L.A.: 1971, MNRAS **153**, 315
Hippelein H.H., Baessgen M., Grewing M.: 1985, A&A **152**, 213
Holz J.Z., Geballe T.R., Rank D.M.: 1971, ApJ **164**, L29
Hrivank B.J., Kwok S., Volk K.M.: 1989, ApJ **346**, 265
Hummer D.G., Mihalas D.: 1970, MNRAS **147**, 339
Iben I., Kaler J.B., Truran J.W., Renzini A.: 1983, ApJ **264**, 605
Jacoby G.H.: 1979, PASP **91**, 754
Jacoby G.H.: 1983, PN Symp. 103, p. 427
Jacoby G.H.: 1988, ApJ **333**, 193
Jacoby G.H., Ford H.C.: 1983, ApJ **266**, 298
Jacoby G.H., Lesser M.P.: 1981, AJ **86**, 185
Jacoby G., Ford H., Ciardullo R.: 1989, PN Symp. 131, p. 357
Jewell P.P., Webber J.C., Snijder L.E.: 1980, ApJ **242**, L29
Jewitt D.C., Danielson G.E., Kupferman P.N.: 1986, ApJ **302**, 727
Johnson H.M.: 1981, ApJ **250**, 590
Johnson H.R.: 1978, ApJ Suppl. **36**, 217
Kafatos M., Lynch J.P.: 1980, ApJ Suppl. **42**, 611
Kahn F.D.: 1983, PN Symp. 103, p. 305
Kahn F.D.: 1989, PN Symp. 131, p. 411
Kaler J.B.: 1970, ApJ **160**, 887
Kaler J.B.: 1976b, ApJ **210**, 843
Kaler J.B.: 1981a, ApJ **250**, L31
Kaler J.B.: 1981b, ApJ **249**, 201
Kaler J.B.: 1983a, ApJ **264**, 594
Kaler J.B.: 1983b, ApJ **271**, 188
Kaler J.B.: 1986, ApJ **308**, 322
Kaler J.B., Feibelman W.A.: 1985, ApJ **297**, 724
Kaler J.B., Hartkopf W.I.: 1981, ApJ **249**, 602
Kindl C., Marxer N., Nussbaumer H.: 1982, A&A **116**, 265
Kirpatrik R.C.: 1972, ApJ **176**, 381

Kostyakova E.B., Archipova V.P., Saveleva M.V.: 1973, Mem. Soc. Roy. Liege. **5**, 473
Kostyakova E.B.: 1989, Astron. Circ., No 1537
Krueger T.K., Aller L.H., Czyzak S.J.: 1970, ApJ **160**, 921
Kwok S.: 1982, ApJ **258**, 280
Kwok S.: 1983, PN Symp. 103, p. 293
Kwok S.: 1985, AJ **90**, 49
Kwok S.: 1985, ApJ: **290**, 568
Kwok S., Purton C.R., Fitzgerald P.M.: 1978, ApJ **219**, L125
Kwok S., Purton C.R., Keenan D.W.: 1981, ApJ **250**, 230
Köppen J., Tarafder S.P.: 1978, A&A **69**, 363
Lamers H.J., Cerruti-Sola M., Perinotto M.: 1987, ApJ **314**, 726
Lenzuni P., Natta A., Panagia N.: 1989, ApJ **345**, 306
Leroy J.L., Le Borgen J.F., Arrand J.: 1986, A&A **160**, 171
Lester D.F., Dinerstein H.L.: 1984, ApJ **281**, L67
LeVan P.D., Rudy R.J.: 1983, ApJ **272**, 137
Louergue M., Nussbaumer H.: 1974, A&A **34**, 225
Lutz J.H., Carnochan D.J.: 1979, MNRAS **189**, 701
Maciel W.J.: 1989, PN Symp. 131, p. 73
Maciel W., Pottasch S.R.:1980, A&A **88**, 1
Maciel W.J., Pottasch S.R.: 1980, A&A **88**, 1,
Masson C.R.: 1986, ApJ **302**, L27
Meabarn J., Walsh J.R.: 1980, MNRAS **191**, 5p
Meatheringham S.J., Wood P.R., Faulkner D.J.: 1988, ApJ **334**, 862
Mendez R.H., Kudrizki R.P., Herrero A., Husfeld D., Groth H.G.: 1988, A&A **190**, 113
Mendoza C.: 1978, PN Symp. 76, p. 279
Mendoza C.: 1983, PN Symp. 103, p. 143
Merill K.M., Soifer B.T., Russell R.W.: 1975, ApJ **200**, L37
Mihalas D.: 1972, ApJ **177**, 115
Miller J.S., Mathews W.G.: 1972, ApJ **172**, 593
Milne D.K., Aller L.H.: 1975, A&A **38**, 183
Milne D.K., Aller L.H.: 1982, A&A Suppl. **50**, 209
Melnick G., Harwit M. : 1975, MNRAS **171**, 441
Moseley H.: 1980, ApJ **238**, 892
Nandy K., Thompson G.J., Jamar G., Monfils A., Wilson R.: 1975, A&A **44**, 195
Netzer H., Penston M.: 1976, MNRAS **174**, 319
Neugebauer G., Habing H.J., van Duinen R. et al.: 1984, ApJ **278**, L1
Nussbaumer H., Schild H.: 1981, A&A **101**, 118
O'Dell C.R.: 1989 PN Symp. 131, p. 458
Paczynski B.: 1971, Acta Astron. **21**, 417
Parsson S.E.: 1970, ApJ **161**, L51
Pascoli G.: 1985, A&A **147**, 257
Payne H.E., Phillips J.A., Terzian Y.,: 1988, ApJ **326**, 368
Peimbert M.: 1978, PN Symp. 76, p. 215
Peimbert M., Torres-Peimbert S.: 1983, PN Symp. 103, p. 233
Pena M., Torres-Peimbert S.: 1981, Rev. Mex. A&A **6**, 309
Pequignot D., Aldrovandi S.M.V., Stasinska G.: 1977, A&A **58**, 411
Pequignot D., Baluteau J.P.: 1988, A&A **206**, 298
Pequignot D., Stasinska G.: 1980, A&A **81**, 121
Perinotto M.: 1983, PN Symp. 103, p. 323
Perinotto M., Benvenuti P.: 1981a, A&A **100**, 241
Perinotto M., Benvenuti P.: 1981b, A&A **101**, 88

Peterson A.W.: 1976, A&A **53**, 442
Petrosian V.: 1970, ApJ **159**, 833
Phillips J.P.: 1984, A&A **137**, 92
Phillips J.P.: 1989, PN Symp. 131, p. 425
Pottasch S.R.: 1978, PN Symp. 76, p. 93
Pottasch S.R.: 1980, A&A **89**, 336
Pottasch S.R., Band B., Beintema D. et al.: 1984a, A&A **138**, 10
Pottasch S.R., Beintema D.A., Raimond E. et al.: 1984b, ApJ **278**, L33
Pottasch S.R., Gathier R., Girla D.P., Wesseilus P.R.: 1981, A&A **102**, 237
Pottasch S.R., Preite-Martinze A., Olnon F.M., Jing-Er Mo, Kongma S.: 1986, A&A **161**, 363
Preite-Martinez A., Pottasch R.S.: 1983, A&A **126**, 31.
Purgathofer A., Stoll M.: 1981, A&A **99**, 218
Pwa T.H., Pottasch S.R., Mo S.E.: 1986, A&A **164**, 184
Rank D.M.: 1978, PN Symp. 76, p. 103
Reay N.K., Atherton P.D., Taylor K.: 1983, MNRAS **203**, 1087
Recillas-Crus, E., Pişmiş P.: 1984, MNRAS **210**, 57
Renzini A.: 1981, in Physical Processes in Red Giants. Reidel, Dordrecht. p. 267
Renzini A.: 1983, PN Symp. 103, p. 267
Robbins R.R.: 1970, ApJ **160**, 519
Rodriguez L.F., Moran J.M.: 1982, Nature **299**, 323
Roxburg I.W.: 1978, PN Symp. 76, p.295
Russell R.W., Soifer B.T., Willner S.P.: 1977, ApJ **217**, 217, L143
Sabbadin F., Bianchini A., Hamzaoglu E.: 1984a, A&A **136**, 193
Sabbadin F., Falomo R., Ortolani S.: 1984b, A&A Suppl. **65**, 259
Saraph H.E., Seaton M.J.: 1974, MNRAS **148**, 367
Savage B.D., Mathis J.S.: 1979, ARA&A **17**, 73
Schneider S.E., Silverglate P.R., Altschuler D.R., Giovanardi C.: 1987, ApJ **314**, 572
Schneider S.E., Terzian Y., Perinotto M.: 1983, ApJ Suppl. **52**, 399
Schwarz R.D.: 1974, ApJ **191**, 419
Schönberner D.: 1979, A&A **79**, 108
Schönberner D.: 1981, A&A **103**, 119
Schönberner D.: 1983, ApJ **272**, 708
Scrimger J.N.: 1984, ApJ **280**, 170
Seaton M.J.: 1978, MNRAS **185**, 5p
Seaton M.J.: 1979, MNRAS **187**, 785
Shakura N.I., Sunyaev R.A.: 1973, A&A **24**, 337
Shevelko V.P.: 1973, Opt. & Spectr. **36**, 7
Shields G.A.: 1978, ApJ **219**, 565
Shields G.A., Aller L.H., Keyes C.D., Czyzak S.J.: 1981, ApJ **248**, 568
Shull J.M., Van Steenberg M.: 1982, ApJ Suppl. **48**, 95
Simpson J.P.: 1975, A&A **39**, 43
Simpson J.P., Rubin R.H., Erikson E.F., Haas M.R.: 1986, ApJ **311**, 895
Slik J., Brown R.L.: 1971, ApJ **163**, 495
Smith L.I., Willis A.J., Wilson R.: 1979, The First Year of IUE, p. 18
Soker N., Livio N.: 1989, ApJ **339**, 268
Storey P.J.: 1981, MNRAS **195**, 27p
Tarafter S.P., Appararo K.M.: 1989, ApJ **327**, 342
Taylor A.R., Gussie G.T., Goss W.M.: 1989, ApJ **340**, 932
Terzian Y.: 1974, Vistas in Astron. **16**, 279
Terzian Y.: 1977, Sky & Tel. **54**, 459
Terzian Y.: 1978, PN Symp. 76, p. 111

Terzian Y., Balick B., Bignell C.: 1974, ApJ **188**, 257
Terzian Y., Dickey J.: 1973, AJ **78**, 875
Thompson A.R., Colvin R.S.: 1970, ApJ **160**, 363
Thompson A.R., Davies F.G.: 1970, MNRAS **150**, 359
Torres-Peimbert S., Peimbert M.: 1977, Rev. Mex. A&A **2**, 181
Torres-Peimbert S., Pena M.: 1981, Rev. Mex. A&A **6**, 301
Triffers R.R., Fink U., Larson H.P., Gautier T.N.,: 1977, A&A **58**, 411
Tylenda R.: 1986, A&A **156**, 217
Tylenda R., Stasinska G.: 1989, A&A **217**, 209
Watson D.M., Storey J.W.V., Townes C.H.: 1981, ApJ **250**, 605
Webster B.L.: 1976, MNRAS **174**, 513
Weinberger G.: 1989 PN Symp. 131, p. 93
Wesemael F., Auer L.H., Van Horn H.M., Savedoff M.P.: 1980, ApJ Suppl. **43**, 159
Westerbrook W.E., Becklin E.E., Merill K.M. et al.: 1975, ApJ **202**, 407
Wiedmann V.: 1977, A&A **61**, L27
Wood P.R., Faulkner D.J.: 1986, ApJ **307**, 659
Word D.B.: 1975, ApJ **200**, L41
Zipoy D.M.: 1975, ApJ **201**, 397
Zuckerman B., Gilra D.P., Turner B.E., Morris M., Palmer P.: 1976, ApJ **205**, L15

The Most-Recent Period, 1990–1996

Aaqust O.B. : 1993, A&A **267**, 160
Aller L.H.: 1991, in: From Asteroids to Quasars. Ed. P.M. Lugger. Cambridge University Press p. 85
Balick B., Perinotto M., Maccioni A., Terzian Y., Hajian A.: 1994, ApJ **424**, 800
Bassgen M., Diesch C., Grewing M.: 1995, A&A **297**, 828
Bedding T.R., Zijlstra A.A.: 1994, A&A **283**, 955
Bianchi L., de Francesco G.: 1993, IAU Symp. 155, p. 84
Blades J.C., Barlow M.J., Osmer S.: 1992, ApJ **398**, L41
Boffi F.R., Stanghellini L.: 1994, A&A **284**, 248
Borkowski K.J., Sarazin C.E., Sokker, N.: 1992, ApJ **360**, 173
Borkowski K.J., Tsvetanov Z., Harrington J.P.: 1993a, ApJ **402**, L57
Borkowski K.J., Harrington J.P., Tsvetanov Z., Clegg R.E.S.: 1993b, ApJ **415**, L47
Bottinelli L., Gouguenheim L., Paturel G., Teerikopi P.: 1991, A&A **252**, 550
Bowyer S.: 1994, Sci. American, **271**, 32.
Bowyer S., Lieu R., Lampton M., Lewis J., Wu X., Drake J.J., Malina. R.F.: 1994, ApJ.Suppl. **93**, 569.
Bowyer S., Malina R.F.: 1991, The EUVE Mission in Extreme Ultraviolet Astronomy, ed. R.F.Malina, S.Bowyer, N.Y. Pergamon, p. 391
Breidschwerdt D., Kahn F.D.: 1990, MNRAS **244**, 521
Brown T., Cordova F., Ciardullo R., Thompson R., Bond H.: 1994, ApJ **422**, 118
Bruce M., Mellema G., Clayton C.A., Meaburn J., Balick B., Lopez J.A.: 1996, A&A **307**, 253.
Chevalier R.A., Luo D.: 1994, ApJ **421**, 225
Ciardullo R., Jacoby G.H., Tonry J.L.: 1993, ApJ **419**, 479
Conlon E.S., Dufton P.L., McCausland R.J.H., Keenan F.P.: 1993, ApJ **408**, 593
Corny S.K., Tylenda R., Szczerba R.: 1994, A&A **284**, 949
Corradi A., Schwarz H.E.: 1993, IAU Symp. 103, p. 216
Dopita M.A., Meatheringham S.J., Wood P.R., Ford H.C., Bohlin R.C., Stecher T.P., Maran S., Harrington J.P.: 1993, PN Symp. 155, p. 212.

Dopita M.A., Vassiliadis E., Meatheringham S.J., Ford H.C., Bohlin R. Wood P.R., Stecher T.P., Maran S.P., Harrington J.P. 1994, ApJ **426**, 150
Frank A., Balick B., Icke V., Mellema G.: 1993, ApJ **404**, L25
Frank A., Mellema G.: 1994, A&A **289**, 937
Frank A., Mellema G.: 1994, ApJ **430**, 800
Frank A., van der Veen W.E.C.J., Balick B.: 1994, A&A **282**, 554
Frank A.: 1993, PN Symp. 155, p. 311
Frank A.: 1994, AJ **107**, 261
Freitas Pacheco J.A. de, Costa, R.D.D.: 1996, A&A **309**, 629.
Fruscione A., Drake J.J., McDonald K., Malina, R.: 1996 ApJ (in press)
Gomez Y., Rodriguez L.F., Moran J.M.: 1993, ApJ **416**, 620
Graham T.M., Herbst T.M., Matthews K., Neugebauer G. Soifer B.T., Serabyn E., Beckwith S.: 1993, ApJ **408**, L105
Groenewegen M.A.T., de Jong T.: 1994, A&A **282**, 127
Gurzadyan G.A.: 1991, ApSS **179**, 293
Gurzadyan G.A.: 1992, ApSS **190**, 219
Gurzadyan G.A.: 1993, ApSS **208**, 69
Gurzadyan G.A.: 1994, ApSS **211**, 179
Gurzadyan G.A., Egikyan A.G., Terzian Y.: 1991a, ApSS **176**, 9
Gurzadyan G.A., Egikyan A.G., Terzian Y.: 1991b, ApSS **175**, 191
Gurzadyan G.A., Egikyan A.G.: 1991a, ApSS **175**, 15
Gurzadyan G.A., Egikyan A.G.: 1991b, ApSS **181**, 73
Gurzadyan G.A., Terzian Y.: 1991, AJ **101**, 1952
Gurzadyan G.A.: 1996, A&A **312** (in press)
Gussie G.T.: 1993, PN Symp. 155, p. 148
Hajian A.P., Terzian Y., Bignell, C.: 1995a, AJ **106**, 1965.
Hajian A.P., Terzian Y., Bignell, C.: 1995b, AJ **109**, 2600.
Han Z., Podsaidlowski P., Egglteton P.P.: 1995, MNRAS **272**, 800
Harrington J.P., Borkowski K.J.: 1995, News, Photo Release No. STScI- PRC 95-01. Hubble Space Telescope, NASA
Harrington J.P., Paltoglou G.: 1993, ApJ **411**, L103
Hippelein H.H., Weinberger R.: 1990, A&A **232**, 129
Hrivank B.J., Kwok S., Geballe T.R.: 1993, PN Symp. 155, p. 335
Hrivank B.J., Kwok S., Geballe T.R.: 1994, ApJ **420**, 783
Huggins P.J.: 1993, PN Symp. 155, p. 147
Huggins P.J., Bachiller R., Cox P., Foreile T.:1993, PN Symp. 155, p. 148
Hui X., Ford H.C., Ciardullo R., Jacoby G.H.: 1993, ApJ **414**, 463
Hyung S., Aller L.H., Feibelman W.A.: 1994, ApJ Suppl. **93**, 465
Jacoby G., Ciardullo B : 1993, PN Symp. 155, p. 503
Jacoby G.H., Walker A.R., Ciardullo R.: 1990, ApJ **365**, 471
Jasniewicz G., Therenin F., Monier R., Skiff, B.A.: 1996, A&A **307**, 200.
Kahn F.D., Breitschwerdt D.: 1990, MNRAS **242**, 505
Kastner J.H., Gatly I, Merill K.M., Probst R., Weintranb A : 1994, ApJ **421**, 600
Kostyakova E.B.: 1991, Trans. Sternberg Astr. Inst. **62**, 58
Köppen R.T.: 1993, PN Symp. 155, p. 557
Köppen R.T., Werner K.: 1994, A&A **286**, 543
Köppen R.T., Arimoto N.: 1990 A&A **240**, 22
Kreysing H.C., Diesch C., Zweigle J., Staubert R., Grewing M.: 1992, A&A **240**, 1000
Kwitter K.B., Chu Y.-H., Downes R.A.: 1993, PN Symp. 155, p. 209
Kwok S.: 1994, PASP **106**, 344
Lamers H.J.G.L.M.: 1991, in: Stellar Atmosphers: Beyond Classical Models. NATO ASI Ser.Confer. Kluwer, Dordrecht, p. 311

Leahy D.A., Zhang C.Y., Kwok S.: 1994, ApJ **422**, 205
Maciel W.J., Köppen J.: 1994, A&A **282**, 436
Marten H.: 1993, A&A **277**, L9
Marten H., Gesicki K., Szczerba R.: 1993, PN Symp. 155, p. 315
Marten H., Schönberner D.: 1991, A&A **248**, 590
Martin W.: 1994, A&A **281**, 49
McMillan R., Ciardullo R., Jacoby G.: 1993, ApJ **416**, 62
Meaburn J., Lopez J.A., Bruce M., Mellema G.: 1996, A&A **307**, 579.
Mellema G.: 1994, A&A **290**, 915
Mendez R.H., Kudritzki R.P., Herrero A.: 1992, A&A **260**, 329
Meyssonnier M., Azzopardi M., Lequeux J.: 1993, PN Symp. 155, p. 351
Modigliani A., Patriarchi P., Perinotto M.: 1993, ApJ **415**, 258
Morgan D.H.: 1994, A&A **282**, 349
Ortiz R., Maciel W.J.: 1994, A&A **287**, 552
Pascoli G.: 1990, A&A 232, 604
Patriarchi P., Perinotto M.: 1991, A&A **91**, 325
Patriarchi P., Perinotto M.: 1996, A&A **309**, 895.
Peimbert M.: 1993, PN Symp. 155, p. 523
Perinotto M.: 1993, PN Symp. 155, p. 57
Perinotto M., Purgathofer A., Pasquali A., Patriarchi P.: 1994, A&A **292**, 355
Phillips J.P., Cuesta L.: 1994, A&A **284**, 701
Pişmiş P., Manteiga M., Mampaso A, Gruz-Gonzalez G.: 1993, PN Symp. 155, p. 202
Pottasch S.R.: 1996, A&A **307**, 567.
Pottasch S.R., Zijlstra A.A.: 1994, A&A **284**, 261
Pottasch S.R., Zijlstra A.A.: 1994, A&A **289**, 261
Rouch, T., Köppen, J., Werner, K.: 1996, A&A **310**, 613.
Saurer W. : 1995, A&A **297**, 261
Schild H.: 1995, A&A **297**, 246
Schönberner D.: 1993, PN Symp. 155, p. 415
Schwarz H.E., Corradi R.L.M., Melnick J.: 1992, A&A **96**, 23
Schwarz R.D.: 1991, A&A **243**, 496
Soifer B.T., Serabyn E., Beckwith S.: 1993, ApJ **408**, L105
Soffner T., Mendez R.H., Jacoby G.H., Ciardullo R., Roth M.M., Kudritzki, R.P.: 1996, A&A **306**, 9.
Stasinska G., Tylenda R., Acker A., Stenholm B.: 1991 A&A 247, 173
Stasinska G., Tylenda R., Acker A., Stenholm B.: 1993 PN Symp. 155, p. 177
Storey P.J.: 1994, A&A **282**, 199
Tammann G.A.: 1993, PN Symp. No. 155, p. 515
Taylor A.R., Gussie G.T., Pottasch S.R.: 1990, ApJ **351**, 515
Terzian Y.: 1993, PN Symp. 155, p. 109
Tweedy R.W., Kwitter K.B.: 1994, ApJ **433**, L93
Xilouris K.M., Papamastorakis J., Paleologou E., Terzian Y.: 1996, A&A **610**, 603.
Young K., Phillips T.G., Knapp G.R.: 1993, ApJ **409**, 725
Zhang C.Y.: 1993, ApJ **410**, 239
Zhang C.: 1995, ApJ Suppl.(in press))
Zhang C.Y., Kwok S.: 1993, ApJ Suppl. **88**, 137
Zhekov S.A., Perinotto M.: 1996, A&A **309**, 648.
Zucker D.B., Soker N.: 1993, ApJ **408**, 579
Zuckerman B., Kastner T.H., Balick B., Gatley I.: 1990, ApJ **356**, L59

Object Index

Planetary Nebulae

Diffuse Nebulae

Galaxies

Globular Clusters

Large Magellanic Cloud

Small Magellanic Cloud

Protoplanetary Objects

Stars

Supernova

Subject Index

-&-

LIBRARY

Theory of Orbits (2 volumes)
Volume 1: Integrable Systems and Non-perturbative Methods
Volume 2: Perturbative and Geometrical Methods
By D. Boccaletti and G. Pucacco

Galaxies and Cosmology
By F. Combes, P. Boissé, A. Mazure and A. Blanchard

The Solar System 2nd Edition By T. Encrenaz and J.-P. Bibring

Nuclear and Particle Physics of Compact Stars
By N. K. Glendenning

The Physics and Dynamics of Planetary Nebulae By G. A. Gurzadyan

Astrophysical Concepts 2nd Edition By M. Harwit

Stellar Structure and Evolution By R. Kippenhahn and A. Weigert

Modern Astrometry By J. Kovalevsky

Supernovae Editor: A. Petschek

General Relativity, Astrophysics, and Cosmology
By A. K. Raychaudhuri, S. Banerji and A. Banerjee

Tools of Radio Astronomy 2nd Edition
By K. Rohlfs and T. L. Wilson

Atoms in Strong Magnetic Fields
Quantum Mechanical Treatment and Applications
in Astrophysics and Quantum Chaos
By H. Ruder, G. Wunner, H. Herold and F. Geyer

The Stars By E. L. Schatzman and F. Praderie

Gravitational Lenses By P. Schneider, J. Ehlers and E. E. Falco

Relativity in Astrometry, Celestial Mechanics and Geodesy By M. H. Soffel

The Sun An Introduction By M. Stix

Galactic and Extragalactic Radio Astronomy 2nd Edition
Editors: G. L. Verschuur and K. I. Kellermann

Reflecting Telescope Optics (2 volumes)
Volume I: Basic Design Theory and its Historical Development
Volume II: Manufacture, Testing, Alignment, Modern Techniques
By R. N. Wilson

Printing: Mercedesdruck, Berlin
Binding: Buchbinderei Lüderitz & Bauer, Berlin

GPSR Compliance

The European Union's (EU) General Product Safety Regulation (GPSR) is a set of rules that requires consumer products to be safe and our obligations to ensure this.

If you have any concerns about our products, you can contact us on ProductSafety@springernature.com

In case Publisher is established outside the EU, the EU authorized representative is:

Springer Nature Customer Service Center GmbH
Europaplatz 3
69115 Heidelberg, Germany

Batch number: 10682187

Printed by Printforce, the Netherlands